MW01628380

Welcome!

Thank you for joining us! Since its first publication in 2014, this textbook has become a bestseller in the field of computational biology, having been adopted by dozens of universities in twenty countries (page xxxiv). Yet we are equally proud that the textbook also powers our popular online courses in bioinformatics, which have been completed by thousands of learners around the world. We hope that you will sign up for our courses and explore our material online as well.

You can also find lecture videos, PowerPoint slides, and answers to FAQs at the textbook website, http://bioinformaticsalgorithms.org.

The subtitle of this book is *An Active Learning Approach*. This is because as you read, you will find a number of active learning components that will help you learn the material interactively and at your own pace.

1. **Code Challenges** ask you to implement the algorithms that you will encounter (in any programming language you like). These code challenges are hosted in the "Bioinformatics Textbook Track" location on Rosalind (http://rosalind.info), a website that will automatically test your implementations.

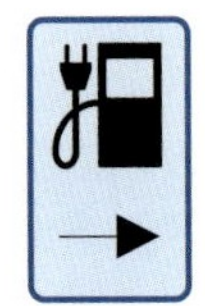

2. **Charging Stations** provide additional insights on implementing the algorithms you encounter. However, we suggest trying to solve a Code Challenge before you visit a Charging Station.

3. **Exercise Breaks** offer "just in time" assessments testing your understanding of a topic before moving to the next one.

4. **STOP and Think** questions invite you to slow down and contemplate the current material before continuing to the next topic.

5. **Detours** provide extra content that didn't quite fit in the main text.

6. **Challenge Problems** ask you to apply what you have learned to real experimental datasets.

7. **FAQs** are questions that we encountered many times from the learners who have taken our online courses and the students we have taught using this book in our own offline courses. When you see the question mark symbol in the margin, it indicates that there is an FAQ on the current subject; answers to FAQs are hosted at the textbook website.

Finally, we have tried to make the book as fun to read as possible. For example, since bioinformatics is still an ever-changing frontier, we decided to implement a Wild West theme for the book. Accordingly, you will notice that each chapter begins with a cartoon image showing the authors entangled in some adventure. What do these cartoons have to do with bioinformatics? Continue reading to find out!

Bioinformatics Algorithms: An Active Learning Approach

3rd Edition

Phillip Compeau & Pavel Pevzner

http://bioinformaticsalgorithms.org

Printed in the United States of America

Third Printing, 2020

ISBN: 978-0-9903746-3-3

Active Learning Publishers, LLC
3520 Lebon Dr. #5208
San Diego, CA 92122

To my family. — P. C.

To my parents. — P. P.

Overview

Where in the Genome Does DNA Replication Begin?
Algorithmic Warmup
DNAA

WHICH DNA PATTERNS PLAY THE ROLE OF MOLECULAR CLOCKS?
Randomized Algorithms
DOESN'T RISE
DosR 2 to 1
CONVERGE
36
16

CHAPTER 3 — p. 115

CHAPTER 4 — p. 184

CHAPTER 5 — p. 224

CHAPTER 6 — p. 296

CHAPTER 7 — p. 352

CHAPTER 8 — p. 416

CHAPTER 9 — p. 468

CHAPTER 10 — p. 530

CHAPTER 11 — p. 586

Contents

List of Code Challenges

Meet the Authors

PHILLIP COMPEAU is the Assistant Department Head for Education in the Computational Biology Department at Carnegie Mellon University's School of Computer Science. He is excited about the future of online education and how it can improve our teaching offline. He received a Ph. D. in mathematics from the University of California, San Diego in 2014 and later served as a postdoctoral researcher in the Computer Science & Engineering Department. He co-founded Rosalind with Nikolay Vyahhi in 2012. A retired tennis player, he dreams of one day going pro in golf.

PAVEL PEVZNER is Ronald R. Taylor Professor of Computer Science at the University of California, San Diego. He holds a Ph. D. from Moscow Institute of Physics and Technology, Russia and an Honorary Degree from Simon Fraser University. He is a Howard Hughes Medical Institute Professor (2006), an Association for Computing Machinery Fellow (2010), and an International Society for Computational Biology Fellow (2012). He has authored the textbooks *Computational Molecular Biology: An Algorithmic Approach* (2000) and *An Introduction to Bioinformatics Algorithms* (2004) (jointly with Neil Jones).

Meet the Development Team

Olga Botvinnik works as a Bioinformatics Data Scientist at the Chan Zuckerberg Biohub. She received her Ph. D. in Bioinformatics and Systems Biology from the University of California, San Diego. She holds an M.S. in Bioinformatics from the University of California, Santa Cruz and B.S. degrees in Mathematics and Biological Engineering from the Massachusetts Institute of Technology. She enjoys blogging about open-source software, dance, and playing cello.

Niema Moshiri is a Ph. D. student in the Bioinformatics and Systems Biology program at the University of California, San Diego. His research focuses on phylogenetics. Niema is also involved with teaching, and in 2016 he co-authored the online interactive textbook *Data Structures* (with Liz Izhikevich). In 2017, he received a Distinguished Teacher Award from UCSD.

Son Pham is the founder and Chief Scientific Officer at BioTuring (http://bioturing.com). He previously completed postdoctoral research at Salk Institute. He received his Ph. D. in Computer Science and Engineering from the University of California, San Diego. His research interests include graph theory, genome assembly, comparative genomics, and neuroscience. Besides research, he enjoys walking meditation and gardening.

Acknowledgments

This textbook was greatly improved by the efforts of a large number of individuals, to whom we owe a debt of gratitude.

The development team, as well as Laurence Bernstein, Ksenia Krasheninnikova, Max Shen, and Jeffrey Yuan, implemented coding challenges and exercises, rendered figures, helped typeset the text, and offered insightful feedback on the manuscript.

Glenn Tesler provided thorough chapter reviews and even implemented some software to catch errors in the early version of the manuscript!

Robin Betz, Petar Ivanov, James Jensen, and Yu Lin provided insightful comments on the manuscript in its early stages. David Robinson was kind enough to copy edit a few chapters and palliate our punctuation maladies.

Randall Christopher brought to life our crazy ideas for illustrations throughout the book, including the chapter header cartoons and the textbook cover.

Andrey Grigoriev and Max Alekseyev gave advice on the content of Chapter 1 and Chapter 6, respectively. Martin Tompa helped us develop the narrative in Chapter 2 by suggesting that we analyze latent tuberculosis infection.

Nikolay Vyahhi led a team composed of Andrey Balandin, Artem Suschev, Aleksey Kladov, and Kirill Shikhanov, who worked hard to support an online, interactive version of this textbook used in our online courses on Coursera.

Our students on Coursera, especially Mark Mammel, Isabel Lupiani, Erika Ramírez and Dmitry Kuzminov, found hundreds of typos in our preliminary manuscript.

Mikhail Gelfand, Uri Keich, Hosein Mohimani, Son Pham, and Glenn Tesler advised us on some of the book's "Open Problems" and led Massive Open Online Research projects (MOORs) in the first session of our online course.

The Big Data to Knowledge Program at the National Institutes of Health, as well as Howard Hughes Medical Institute and the Russian Ministry of Education and Science generously gave their support for the development of the online courses accompanying this textbook. The Bioinformatics and Systems Biology Program and the Computer Science & Engineering Department at the University of California, San Diego provided additional support.

Finally, our families gracefully endured the many long days and nights that we spent poring over manuscripts over the last several years; they helped us preserve our sanity along the way.

P. C. and P. P.
May 2018

Where in the Genome Does DNA Replication Begin?

Algorithmic Warmup

A Journey of a Thousand Miles...

Genome replication is one of the most important tasks carried out in the cell. Before a cell can divide, it must first replicate its genome so that each of the two daughter cells inherits its own copy. In 1953, James Watson and Francis Crick completed their landmark paper on the DNA double helix with a now-famous phrase:

> *It has not escaped our notice that the specific pairing we have postulated immediately suggests a possible copying mechanism for the genetic material.*

They conjectured that the two strands of the parent DNA molecule unwind during replication, and then each parent strand acts as a template for the synthesis of a new strand. As a result, the replication process begins with a pair of complementary strands of DNA and ends with two pairs of complementary strands, as shown in Figure 1.1.

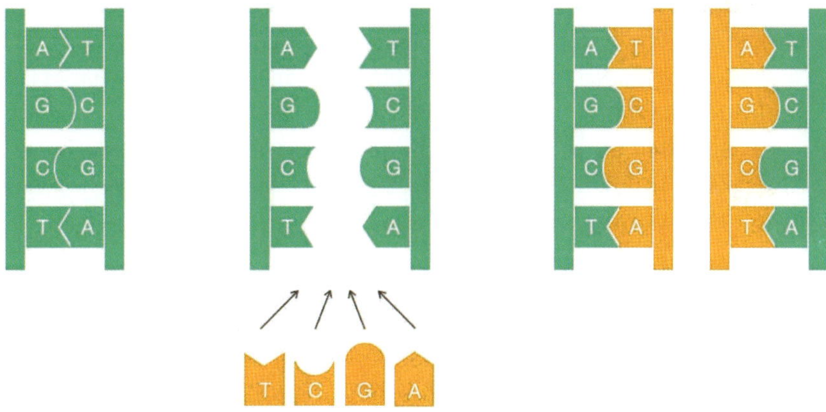

FIGURE 1.1 A naive view of DNA replication. Nucleotides adenine (A) and thymine (T) are complements of each other, as are cytosine (C) and guanine (G). Complementary nucleotides bind to each other in DNA.

Although Figure 1.1 models DNA replication on a simple level, the details of replication turned out to be much more intricate than Watson and Crick imagined; as we will see, an astounding amount of molecular logistics is required to ensure DNA replication.

At first glance, a computer scientist might not imagine that these details have any computational relevance. To mimic the process in Figure 1.1 algorithmically, we only need to take a string representing the genome and return a copy of it! Yet if we take the time to review the underlying biological process, we will be rewarded with new algorithmic insights into analyzing replication.

Replication begins in a genomic region called the **replication origin** (denoted *ori*) and is performed by molecular copy machines called **DNA polymerases**. Locating *ori* presents an important task not only for understanding how cells replicate but also for various biomedical problems. For example, some gene therapy methods use genetically engineered mini-genomes, which are called **viral vectors** because they are able to penetrate cell walls (just like real viruses). Viral vectors carrying artificial genes have been used in agriculture to engineer frost-resistant tomatoes and pesticide-resistant corn. In 1990, gene therapy was first successfully performed on humans when it saved the life of a four-year-old girl suffering from Severe Combined Immunodeficiency Disorder; the girl had been so vulnerable to infections that she was forced to live in a sterile environment.

The idea of gene therapy is to intentionally infect a patient who lacks a crucial gene with a viral vector containing an artificial gene that encodes a therapeutic protein. Once inside the cell, the vector replicates and eventually produces many copies of the therapeutic protein, which in turn treats the patient's disease. To ensure that the vector actually replicates inside the cell, biologists must know where *ori* is in the vector's genome and ensure that the genetic manipulations that they perform do not affect it.

In the following problem, we assume that a genome has a single *ori* and is represented as a **DNA string**, or a string of nucleotides from the four-letter alphabet {A, C, G, T}.

Finding Origin of Replication Problem:

Input: A DNA string *Genome*.
Output: The location of *ori* in *Genome*.

STOP and Think: Does this biological problem represent a clearly stated computational problem?

Although the Finding Origin of Replication Problem asks a legitimate biological question, it does not present a well-defined computational problem. Indeed, biologists would immediately plan an experiment to locate *ori*: for example, they might delete various short segments from the genome in an effort to find a segment whose deletion stops replication. Computer scientists, on the other hand, would shake their heads and demand more information before they can even start thinking about the problem.

Why should biologists care what computer scientists think? Computational methods are now the only realistic way to answer many questions in modern biology. First, these methods are much faster than experimental approaches; second, the results of many experiments cannot be interpreted without computational analysis. In particular, existing experimental approaches to *ori* prediction are rather time consuming. As a result, *ori* has only been experimentally located in a handful of species. Thus, we would like to design a computational approach to find *ori* so that biologists are free to spend their time and money on other tasks.

Hidden Messages in the Replication Origin

DnaA boxes

In the rest of this chapter, we will focus on the relatively easy case of finding *ori* in bacterial genomes, most of which consist of a single circular chromosome. Research has shown that the region of the bacterial genome encoding *ori* is typically a few hundred nucleotides long. Our plan is to begin with a bacterium in which *ori* is known, and then determine what makes this genomic region special in order to design a computational approach for finding *ori* in other bacteria. Our example is *Vibrio cholerae*, the bacterium that causes cholera; here is the nucleotide sequence appearing in its *ori*:

```
atcaatgatcaacgtaagcttctaagcatgatcaaggtgctcacacagtttatccacaac
ctgagtggatgacatcaagataggtcgttgtatctccttcctctcgtactctcatgacca
cggaaagatgatcaagagaggatgatttcttggccatatcgcaatgaatacttgtgactt
gtgcttccaattgacatcttcagcgccatattgcgctggccaaggtgacggagcgggatt
acgaaagcatgatcatggctgttgttctgtttatcttgttttgactgagacttgttagga
tagacggtttttcatcactgactagccaaagccttactctgcctgacatcgaccgtaaat
tgataatgaatttacatgcttccgcgacgatttacctcttgatcatcgatccgattgaag
atcttcaattgttaattctcttgcctcgactcatagccatgatgagctcttgatcatgtt
tccttaaccctctattttttacggaagaatgatcaagctgctgctcttgatcatcgtttc
```

How does the bacterial cell know to begin replication exactly in this short region within the much larger *Vibrio cholerae* chromosome, which consists of 1,108,250 nucleotides? There must be some "hidden message" in the *ori* region ordering the cell to begin replication here. Indeed, we know that the initiation of replication is mediated by ***DnaA***, a protein that binds to a short segment within the *ori* known as a ***DnaA* box**. You can think of the *DnaA* box as a message within the DNA sequence telling the *DnaA* protein: "bind here!" The question is how to find this hidden message without knowing what it looks like in advance — can you find it? In other words, can you find something that stands out in *ori*? This discussion motivates the following problem.

Hidden Message Problem:
Find a "hidden message" in the replication origin.

Input: A string *Text* (representing the replication origin of a genome).
Output: A hidden message in *Text*.

STOP and Think: Does this problem represent a clearly stated computational problem?

Hidden messages in "The Gold-Bug"

Although the Hidden Message Problem poses a legitimate intuitive question, it again makes absolutely no sense to a computer scientist because the notion of a "hidden message" is not precisely defined. The *ori* region of *Vibrio cholerae* is currently just as puzzling as the parchment discovered by William Legrand in Edgar Allan Poe's story "The Gold-Bug". Written on the parchment was the following:

```
53++!305))6*;4826)4+.)4+);806*;48!8‘60))85;1+(;:+*8
!83(88)5*!;46(;88*96*?;8)*+(;485);5*!2:*+(;4956*2(5
*-4)8‘8*;4069285);)6!8)4++;1(+9;48081;8:8+1;48!85:4
)485!528806*81(+9;48;(88;4(+?34;48)4+;161;:188;+?;
```

Upon seeing the parchment, the narrator remarks, "Were all the jewels of Golconda awaiting me upon my solution of this enigma, I am quite sure that I should be unable to earn them". Legrand retorts, "It may well be doubted whether human ingenuity can construct an enigma of the kind which human ingenuity may not, by proper application, resolve". He reasons that the three consecutive symbols "**;48**" appear with surprising frequency on the parchment:

```
53++!305))6*;4826)4+.)4+);806*;48!8‘60))85;1+(;:+*8
!83(88)5*!;46(;88*96*?;8)*+(;485);5*!2:*+(;4956*2(5
*-4)8‘8*;4069285);)6!8)4++;1(+9;48081;8:8+1;48!85;4
)485!528806*81(+9;48;(88;4(+?34;48)4+;161;:188;+?;
```

Legrand had already deduced that the pirates spoke English; he therefore assumed that the high frequency of ";48" implied that it encodes the most frequent English word, "THE". Substituting each symbol, Legrand had a slightly easier text to decipher, which would eventually lead him to the buried treasure. Can you decode this message too?

```
53++!305))6*THE26)H+.)H+)TE06*THE!E‘60))E5T1+(T:+*E
!E3(EE)5*!TH6(TEE*96*?TE)*+(THE5)T5*!2:*+(TH956*2(5
*-H)E‘E*TH0692E5)T)6!E)H++T1(+9THE0E1TE:E+1THE!E5TH
)HE5!52EE06*E1(+9THET(EETH(+?3HTHE)H+T161T:1EET+?T
```

Counting words

Operating under the assumption that DNA is a language of its own, let's borrow Legrand's method and see if we can find any surprisingly frequent "words" within the *ori* of *Vibrio cholerae*. We have added reason to look for frequent words in the *ori* because for various biological processes, certain nucleotide strings appear surprisingly often in small regions of the genome. This is because certain proteins can only bind to DNA if a specific string of nucleotides is present, and if there are more occurrences of the string, then it is more likely that binding will successfully occur. (It is also less likely that a mutation will disrupt the binding process.)

For example, **ACTAT** is a surprisingly frequent substring of

ACA**ACTAT**GCAT**ACTAT**CGGGA**ACTAT**CCT.

We use the term ***k*-mer** to refer to a string of length *k* and define COUNT(*Text*, *Pattern*) as the number of times that a *k*-mer *Pattern* appears as a substring of *Text*. Following the above example,

COUNT(ACA**ACTAT**GCAT**ACTAT**CGGGA**ACTAT**CCT, **ACTAT**) = 3.

Note that COUNT(CG**ATATA**TCC**ATA**G, **ATA**) is equal to 3 (not 2) since we should account for overlapping occurrences of *Pattern* in *Text*.

To compute COUNT(*Text*, *Pattern*), our plan is to "slide a window" down *Text*, checking whether each *k*-mer substring of *Text* matches *Pattern*. We will therefore refer to the *k*-mer starting at position *i* of *Text* as *Text*(i, k). Throughout this book, we will often use **0-based indexing**, meaning that we count starting at 0 instead of 1. In this case, *Text* begins at position 0 and ends at position $|Text| - 1$ ($|Text|$ denotes the number of symbols in *Text*). For example, if *Text* = GACCATACTG, then *Text*$(4, 3)$ = ATA. Note that the last k-mer of *Text* begins at position $|Text| - k$, e.g., the last 3-mer of GACCATACTG starts at position $10 - 3 = 7$. This discussion results in the following **pseudocode** for computing COUNT(*Text*, *Pattern*).

```
PATTERNCOUNT(Text, Pattern)
    count ← 0
    for i ← 0 to |Text| − |Pattern|
        if Text(i, |Pattern|) = Pattern
            count ← count + 1
    return count
```

In this text, we use pseudocode to describe the algorithms that we encounter for solving problems in modern biology. Pseudocode is a universal method of describing algorithms that is more precise than human language but does not require us to get bogged down in the syntax of a specific programming language. If you have not encountered pseudocode before, please consult the Appendix.

The Frequent Words Problem

We say that *Pattern* is a **most frequent *k*-mer** in *Text* if it maximizes COUNT(*Text*, *Pattern*) among all k-mers. You can see that **ACTAT** is a most frequent 5-mer for *Text* = ACA**ACTAT**GCAT**ACTAT**CGGGA**ACTAT**CCT, and **ATA** is a most frequent 3-mer for *Text* = CG**ATATA**TCC**ATA**G.

STOP and Think: Can a string have multiple most frequent k-mers?

We now have a rigorously defined computational problem.

Frequent Words Problem:
Find the most frequent k-mers in a string.

Input: A string *Text* and an integer k.
Output: All most frequent k-mers in *Text*.

A straightforward algorithm for finding the most frequent k-mers in a string *Text* checks all k-mers appearing in this string (there are $|Text| - k + 1$ such k-mers) and then computes how many times each k-mer appears in *Text*. To implement this algorithm, called

FrequentWords, we will need to generate an array Count, where Count(i) stores Count$(Text, Pattern)$ for $Pattern = Text(i, k)$ (see Figure 1.2).

Text	**A**	**C**	**T**	G	**A**	**C**	**T**	C	C	C	A	C	C	C	C
Count	**2**	1	1	1	**2**	1	1	3	1	1	1	3	3		

Figure 1.2 The array Count for *Text* = ACTGACTCCCACCCC and $k = 3$. For example, Count(0) = Count(4) = 2 because ACT (shown in boldface) appears twice in *Text* at positions 0 and 4.

```
FrequentWords(Text, k)
    FrequentPatterns ← an empty set
    for i ← 0 to |Text| − k
        Pattern ← the k-mer Text(i, k)
        Count(i) ← PatternCount(Text, Pattern)
    maxCount ← maximum value in array Count
    for i ← 0 to |Text| − k
        if Count(i) = maxCount
            add Text(i, k) to FrequentPatterns
    remove duplicates from FrequentPatterns
    return FrequentPatterns
```

STOP and Think: How fast is **FrequentWords**?

Although **FrequentWords** finds most frequent k-mers, it is not very efficient. Each call to **PatternCount**$(Text, Pattern)$ checks whether the k-mer *Pattern* appears in position 0 of *Text*, position 1 of *Text*, and so on. Since each k-mer requires $|Text| - k + 1$ such checks, each one requiring as many as k comparisons, the overall number of steps of **PatternCount**$(Text, Pattern)$ is $(|Text| - k + 1) \cdot k$. Furthermore, **FrequentWords** must call **PatternCount** $|Text| - k + 1$ times (once for each k-mer of *Text*), so that its overall number of steps is $(|Text| - k + 1) \cdot (|Text| - k + 1) \cdot k$. To simplify the matter, computer scientists often say that the runtime of **FrequentWords** has an upper bound of $|Text|^2 \cdot k$ steps and refer to the **complexity** of this algorithm as $\mathcal{O}(|Text|^2 \cdot k)$ (see DETOUR: Big-O Notation).

Charging Station (The Frequency Array): If $|Text|$ and k are small, as is the case when looking for *DnaA* boxes in the typical bacterial *ori*, then an algorithm with running time of $\mathcal{O}(|Text|^2 \cdot k)$ is perfectly acceptable. But once we find some new biological application requiring us to solve the Frequent Words Problem for a very long *Text*, we will quickly run into trouble. Check out this Charging Station to learn about solving the Frequent Words Problem using a frequency array, a data structure that will also help us solve new coding challenges later in the chapter.

Frequent words in Vibrio cholerae

Figure 1.3 reveals the most frequent k-mers in the *ori* region from *Vibrio cholerae*.

k	3	4	5	6	7	8	9
count	25	12	8	8	5	4	3
***k*-mers**	tga	atga	gatca	tgatca	atgatca	atgatcaa	atgatcaag
			tgatc				cttgatcat
							tcttgatca
							ctcttgatc

FIGURE 1.3 The most frequent k-mers in the *ori* region of *Vibrio cholerae* for k from 3 to 9, along with the number of times that each k-mer occurs.

STOP and Think: Do any of the counts in Figure 1.3 seem surprisingly large?

For example, the 9-mer **ATGATCAAG** appears three times in the *ori* region of *Vibrio cholerae* — is it surprising?

```
atcaatgatcaacgtaagcttctaagcATGATCAAGgtgctcacacagtttatccacaac
ctgagtggatgacatcaagataggtcgttgtatctccttcctctcgtactctcatgacca
cggaaagATGATCAAGagaggatgatttcttggccatatcgcaatgaatacttgtgactt
gtgcttccaattgacatcttcagcgccatattgcgctggccaaggtgacggagcgggatt
acgaaagcatgatcatggctgttgttctgtttatcttgttttgactgagacttgttagga
tagacggtttttcatcactgactagccaaagccttactctgcctgacatcgaccgtaaat
tgataatgaatttacatgcttccgcgacgatttacctcttgatcatcgatccgattgaag
atcttcaattgttaattctcttgcctcgactcatagccatgatgagctcttgatcatgtt
tccttaaccctctattttttacggaagaATGATCAAGctgctgctcttgatcatcgtttc
```

We highlight a most frequent 9-mer instead of using some other value of k because experiments have revealed that bacterial *DnaA* boxes are usually nine nucleotides long.

The probability that there exists a 9-mer appearing three or more times in a randomly generated DNA string of length 500 is approximately 1/1300 (see **DETOUR: Probabilities of Patterns in a String**). In fact, there are four different 9-mers repeated three or more times in this region: **ATGATCAAG**, CTTGATCAT, TCTTGATCA, and CTCTTGATC.

PAGE 53

The low likelihood of witnessing even one repeated 9-mer in the *ori* region of *Vibrio cholerae* leads us to the working hypothesis that one of these four 9-mers may represent a potential *DnaA* box that, when appearing multiple times in a short region, jump-starts replication. But which one?

STOP and Think: Is any one of the four most frequent 9-mers in the *ori* of *Vibrio cholerae* "more surprising" than the others?

Some Hidden Messages are More Surprising than Others

Recall that nucleotides **A** and **T** are complements of each other, as are **C** and **G**. Having one strand of DNA and a supply of "free floating" nucleotides as shown in Figure 1.1, one can imagine the synthesis of a **complementary strand** on a **template strand**. This model of replication was confirmed by Meselson and Stahl in 1958 (see **DETOUR: The Most Beautiful Experiment in Biology**). Figure 1.4 shows a template strand **AGTCGCATAGT** and its complementary strand **ACTATGCGACT**.

PAGE 58

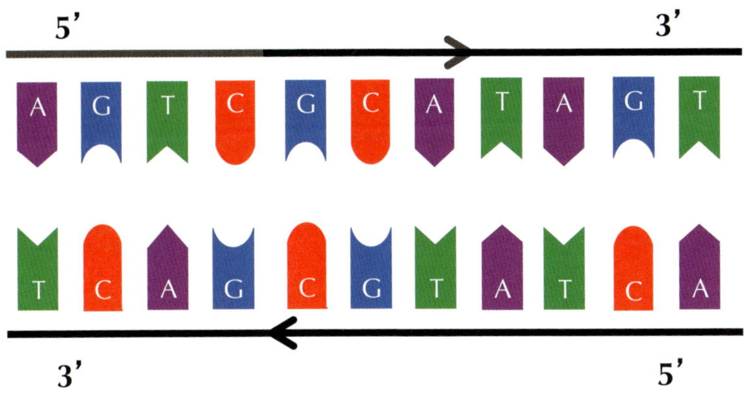

FIGURE 1.4 Complementary strands run in opposite directions.

At this point, you may think that we have made a mistake, since the complementary strand in Figure 1.4 reads out **TCAGCGTATCA** from left to right rather than **ACTATGCGACT**. We have not: each DNA strand has a direction, and the complementary strand runs in the opposite direction to the template strand, as shown by the arrows in Figure 1.4. Each strand is read in the 5′ → 3′ direction (see DETOUR: Directionality of DNA Strands to learn why biologists refer to the beginning and end of a strand of DNA using the terms 5′ and 3′).

PAGE 60

Given a nucleotide p, we denote its complementary nucleotide as $\overline{p}$. The **reverse complement** of a string $Pattern = p_1 \cdots p_n$ is the string $\overline{Pattern} = \overline{p_n} \cdots \overline{p_1}$ formed by taking the complement of each nucleotide in *Pattern*, then reversing the resulting string. We will need the solution to the following problem throughout this chapter.

1C

Reverse Complement Problem:
Find the reverse complement of a DNA string.

Input: A DNA string *Pattern*.
Output: $\overline{Pattern}$, the reverse complement of *Pattern*.

STOP

STOP and Think: Look again at the four most frequent 9-mers in the *ori* region of *Vibrio cholerae* from Figure 1.3. Now do you notice anything surprising?

Interestingly, among the four most frequent 9-mers in the *ori* region of *Vibrio cholerae*, **ATGATCAAG** and **CTTGATCAT** are reverse complements of each other, resulting in the following six occurrences of these strings.

```
atcaatgatcaacgtaagcttctaagcATGATCAAGgtgctcacacagtttatccacaac
ctgagtggatgacatcaagataggtcgttgtatctccttcctctcgtactctcatgacca
cggaaagATGATCAAGagaggatgatttcttggccatatcgcaatgaatacttgtgactt
gtgcttccaattgacatcttcagcgccatattgcgctggccaaggtgacggagcgggatt
acgaaagcatgatcatggctgttgttctgtttatcttgttttgactgagacttgttagga
tagacggtttttcatcactgactagccaaagccttactctgcctgacatcgaccgtaaat
tgataatgaatttacatgcttccgcgacgatttacctCTTGATCATcgatccgattgaag
atcttcaattgttaattctcttgcctcgactcatagccatgatgagctCTTGATCATgtt
tccttaaccctctattttttacggaagaATGATCAAGctgctgctCTTGATCATcgtttc
```

Finding a 9-mer that appears six times (either as itself or as its reverse complement) in a DNA string of length 500 is far more surprising than finding a 9-mer that appears three times (as itself). This observation leads us to the working hypothesis that **ATGATCAAG** and its reverse complement **CTTGATCAT** indeed represent *DnaA* boxes in *Vibrio cholerae*.

This computational conclusion makes sense biologically because the *DnaA* protein that binds to *DnaA* boxes and initiates replication does not care which of the two strands it binds to. Thus, for our purposes, both **ATGATCAAG** and **CTTGATCAT** represent *DnaA* boxes.

However, before concluding that we have found the *DnaA* box of *Vibrio cholerae*, the careful bioinformatician should check if there are other short regions in the *Vibrio cholerae* genome exhibiting multiple occurrences of **ATGATCAAG** (or **CTTGATCAT**). After all, maybe these strings occur as repeats throughout the entire *Vibrio cholerae* genome, rather than just in the *ori* region. To this end, we need to solve the following problem.

Pattern Matching Problem:
Find all occurrences of a pattern in a string.

Input: Strings *Pattern* and *Genome*.
Output: All starting positions in *Genome* where *Pattern* appears as a substring.

After solving the Pattern Matching Problem, we discover that **ATGATCAAG** appears 17 times in the following positions of the *Vibrio cholerae* genome:

116556, 149355, **151913**, **152013**, **152394**, 186189, 194276, 200076, 224527, 307692, 479770, 610980, 653338, 679985, 768828, 878903, 985368

With the exception of the three occurrences of **ATGATCAAG** in *ori* at starting positions **151913**, **152013**, and **152394**, no other instances of **ATGATCAAG** form *clumps*, i.e., appear close to each other in a small region of the genome. You may check that the same conclusion is reached when searching for **CTTGATCAT**. We now have strong statistical evidence that **ATGATCAAG**/**CTTGATCAT** may represent the hidden message to *DnaA* to start replication.

STOP and Think: Can we conclude that **ATGATCAAG**/**CTTGATCAT** also represents a *DnaA* box in other bacterial genomes?

An Explosion of Hidden Messages

Looking for hidden messages in multiple genomes

We should not jump to the conclusion that **ATGATCAAG**/**CTTGATCAT** is a hidden message for all bacterial genomes without first checking whether it even appears in known *ori* regions from other bacteria. After all, maybe the clumping effect of **ATGATCAAG**/**CTTGATCAT** in the *ori* region of *Vibrio cholerae* is simply a statistical fluke that has nothing to do with replication. Or maybe different bacteria have different *DnaA* boxes . . .

Let's check the proposed *ori* region of *Thermotoga petrophila*, a bacterium that thrives in extremely hot environments; its name derives from its discovery in the water beneath oil reservoirs, where temperatures can exceed 80° Celsius.

```
aactctatacctcctttttgtcgaatttgtgtgatttatagagaaaatcttattaactga
aactaaaatggtaggtttggtggtaggttttgtgtacattttgtagtatctgattttaa
ttacataccgtatattgtattaaattgacgaacaattgcatggaattgaatatatgcaaa
acaaacctaccaccaaactctgtattgaccattttaggacaacttcagggtggtaggttt
ctgaagctctcatcaatagactattttagtctttacaaacaatattaccgttcagattca
agattctacaacgctgttttaatgggcgttgcagaaaacttaccacctaaaatccagtat
ccaagccgatttcagagaaacctaccacttacctaccacttacctacccgggtggta
agttgcagacattattaaaaacctcatcagaagcttgttcaaaaatttcaatactcgaaa
cctaccacctgcgtcccctattatttactactactaataatagcagtataattgatctga
```

This region does not contain a single occurrence of **ATGATCAAG** or **CTTGATCAT**! Thus, different bacteria may use different *DnaA* boxes as "hidden messages" to the *DnaA* protein.

Application of the Frequent Words Problem to the *ori* region above reveals that the following six 9-mers appear in this region three or more times:

```
AACCTACCA    AAACCTACC    ACCTACCAC
CCTACCACC    GGTAGGTTT    TGGTAGGTT
```

Something peculiar must be happening because it is extremely unlikely that six different 9-mers will occur so frequently within a short region in a random string. We will cheat a little and consult with Ori-Finder, a software tool for finding replication origins in DNA sequences. This software chooses **CCTACCACC** (along with its reverse complement **GGTGGTAGG**) as a working hypothesis for the *DnaA* box in *Thermotoga petrophila*. Together, these two complementary 9-mers appear five times in the replication origin:

```
aactctatacctcctttttgtcgaatttgtgtgatttatagagaaaatcttattaactga
aactaaaatggtaggtttGGTGGTAGGttttgtgtacattttgtagtatctgatttttaa
ttacataccgtatattgtattaaattgacgaacaattgcatggaattgaatatatgcaaa
acaaaCCTACCACCaaactctgtattgaccattttaggacaacttcagGGTGGTAGGttt
ctgaagctctcatcaatagactattttagtctttacaaacaatattaccgttcagattca
agattctacaacgctgttttaatgggcgttgcagaaaacttaccacctaaaatccagtat
ccaagccgatttcagagaaacctaccacttacctaccacttaCCTACCACCcgggtggta
agttgcagacattattaaaaacctcatcagaagcttgttcaaaaatttcaatactcgaaa
CCTACCACCtgcgtcccctattatttactactactaataatagcagtataattgatctga
```

The Clump Finding Problem

Now imagine that you are trying to find *ori* in a newly sequenced bacterial genome. Searching for "clumps" of **ATGATCAAG**/**CTTGATCAT** or **CCTACCACC**/**GGTGGTAGG** is unlikely to help, since this new genome may use a completely different hidden message! Before we lose all hope, let's change our computational focus: instead of finding clumps of a specific k-mer, let's try to find *every* k-mer that forms a clump in the genome. Hopefully, the locations of these clumps will shed light on the location of *ori*.

Our plan is to slide a window of fixed length L along the genome, looking for a region where a k-mer appears several times in short succession. The parameter value $L = 500$ reflects the typical length of *ori* in bacterial genomes.

We defined a k-mer as a "clump" if it appears many times within a short interval of the genome. More formally, given integers L and t, a k-mer *Pattern* forms an **(L, t)-clump** inside a (longer) string *Genome* if there is an interval of *Genome* of length L in which this k-mer appears at least t times. (This definition assumes that the k-mer *completely* fits within the interval.) For example, **TGCA** forms a $(25, 3)$-clump in the following *Genome*:

```
gatcagcataagggtcccTGCAATGCATGACAAGCCTGCAGTtgttttac
```

From our previous examples of *ori* regions, **ATGATCAAG** forms a $(500, 3)$-clump in the *Vibrio cholerae* genome, and **CCTACCACC** forms a $(500, 3)$-clump in the *Thermotoga petrophila* genome. We are now ready to formulate the following problem.

Clump Finding Problem:
Find patterns forming clumps in a string.

Input: A string *Genome*, and integers k, L, and t.
Output: All distinct k-mers forming (L, t)-clumps in *Genome*.

Charging Station (Solving the Clump Finding Problem): You can solve the Clump Finding Problem by simply applying your algorithm for the Frequent Words Problem to each window of length L in *Genome*. However, if your algorithm for the Frequent Words Problem is not very efficient, then such an approach may be impractical. For example, recall that **FrequentWords** has $\mathcal{O}(L^2 \cdot k)$ running time. Applying this algorithm to each window of length L in *Genome* will result in an algorithm with $\mathcal{O}(L^2 \cdot k \cdot |Genome|)$ running time. Moreover, even if we use a faster algorithm for the Frequent Words Problem (like the one described when we introduce a frequency array on page 40), the running time remains high when we try to analyze a bacterial — let alone human — genome. Check out this Charging Station to learn about a more efficient approach for solving the Clump Finding Problem.

Let's look for clumps in the *Escherichia coli* (*E. coli*) genome, the workhorse of bacterial genomics. We find hundreds of different 9-mers forming $(500, 3)$-clumps in the *E. coli* genome, and it is absolutely unclear which of these 9-mers might represent a *DnaA* box in the bacterium's *ori* region.

STOP and Think: Should we give up? If not, what would you do now?

At this point, an unseasoned researcher might give up, since it appears that we do not have enough information to locate *ori* in *E. coli*. But a fearless veteran bioinformatician would try to learn more about the details of replication in the hope that they provide new algorithmic insights into finding *ori*.

The Simplest Way to Replicate DNA

We are now ready to discuss the replication process in more detail. As illustrated in Figure 1.5 (top), the two complementary DNA strands running in opposite directions around a circular chromosome unravel, starting at *ori*. As the strands unwind, they create two **replication forks**, which expand in both directions around the chromosome until the strands completely separate at the **replication terminus** (denoted *ter*). The replication terminus is located roughly opposite to *ori* in the chromosome.

An important thing to know about replication is that a DNA polymerase does not wait for the two parent strands to completely separate before initiating replication; instead, it starts copying *while* the strands are unraveling. Thus, just four DNA

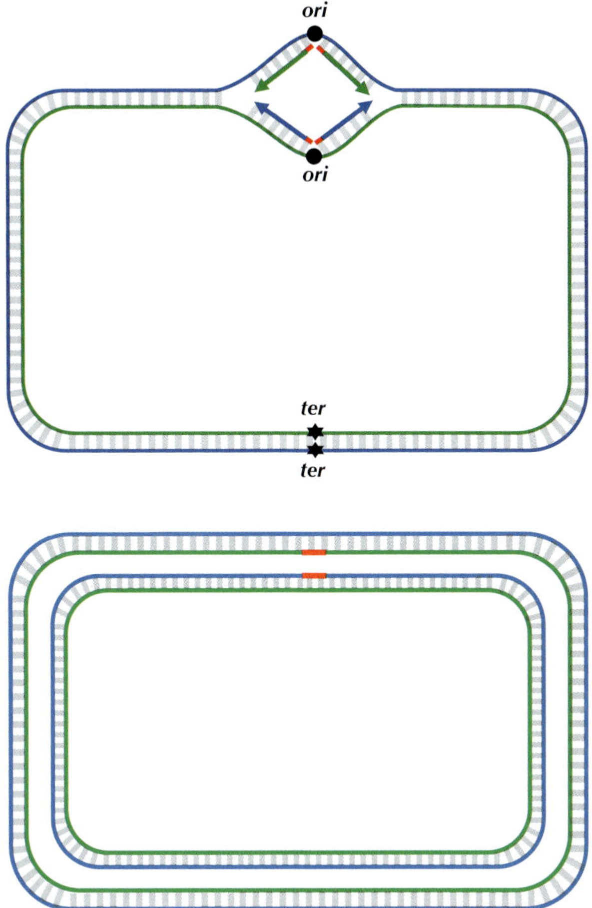

FIGURE 1.5 (Top) Four imaginary DNA polymerases at work replicating a chromosome as the replication forks extend from *ori* to *ter*. The blue strand is directed clockwise, and the green strand is directed counterclockwise. (Bottom) Replication is complete.

polymerases, each responsible for one half-strand, can all start at *ori* and replicate the entire chromosome. To start replication, a DNA polymerase needs a **primer**, a short complementary segment (shown in red in Figure 1.5) that binds to the parent strand and jump starts the DNA polymerase. After the strands start separating, each of the four DNA polymerases starts replication by adding nucleotides, beginning with the primer and proceeding around the chromosome from *ori* to *ter* in either the clockwise or counterclockwise direction. When all four DNA polymerases have reached *ter*, the chromosome's DNA will have been completely replicated, resulting in two pairs of complementary strands (Figure 1.5 (bottom)), and the cell is ready to divide.

Yet while you were reading the description above, biology professors were writing a petition to have us fired and sent back to Biology 101. And they would be right, because our exposition suffers from a major flaw; we only described the replication process in this way so that you can better appreciate what we are about to reveal.

The problem with our current description is that it assumes that DNA polymerases can copy DNA in *either* direction along a strand of DNA (i.e., both $5' \rightarrow 3'$ and $3' \rightarrow 5'$). However, nature has not yet equipped DNA polymerases with this ability, as they are **unidirectional**, meaning that they can only traverse a template strand of DNA in the $3' \rightarrow 5'$ direction. Notice that this is opposite from the $5' \rightarrow 3'$ direction of DNA.

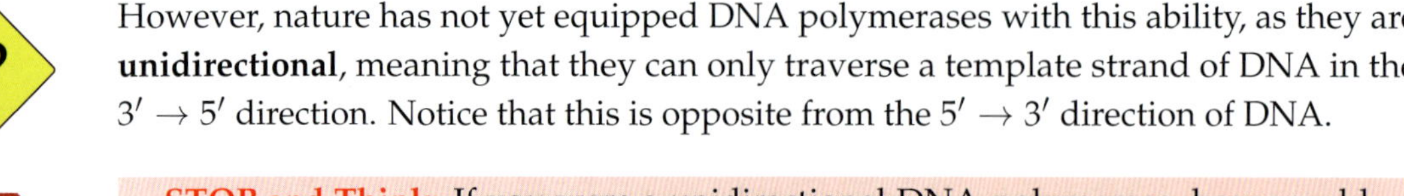

STOP and Think: If you were a unidirectional DNA polymerase, how would you replicate DNA? How many DNA polymerases would be needed to complete this task?

The unidirectionality of DNA polymerase requires a major revision to our naive model of replication. Imagine that you decided to walk along DNA from *ori* to *ter*. There are four different half-strands of parent DNA connecting *ori* to *ter*, as highlighted in Figure 1.6. Two of these half-strands are traversed from *ori* to *ter* in the $5' \rightarrow 3'$ direction

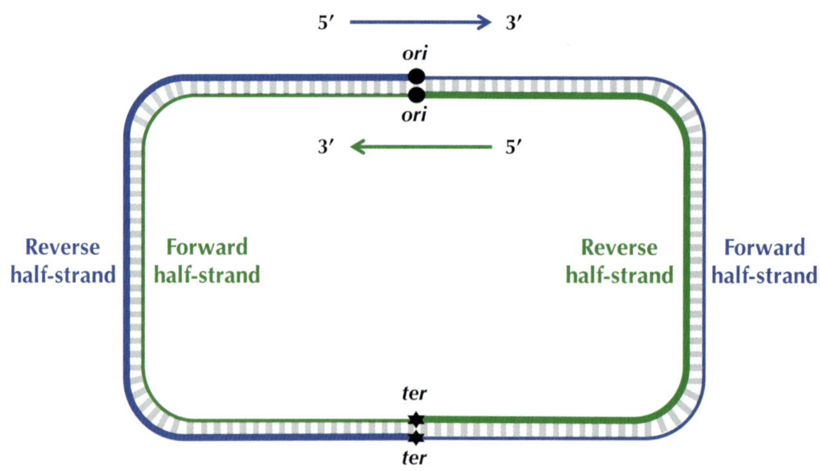

FIGURE 1.6 Complementary DNA strands with forward and reverse half-strands shown as thin and thick lines, respectively.

and are thus called **forward half-strands** (represented by thin blue and green lines in Figure 1.6). The other two half-strands are traversed from *ori* to *ter* in the $3' \rightarrow 5'$ direction and are thus called **reverse half-strands** (represented by thick blue and green lines in Figure 1.6).

Asymmetry of Replication

While biologists will feel at home with the following description of DNA replication, computer scientists may find it overloaded with new terms. If it seems too biologically complex, then feel free to skim this section, as long as you believe us that the replication process is **asymmetric**, i.e., that forward and reverse half-strands have very different fates with respect to replication.

Since a DNA polymerase can only move in the reverse $(3' \rightarrow 5')$ direction, it can copy nucleotides non-stop from *ori* to *ter* along reverse half-strands. However, replication on forward half-strands is very different because a DNA polymerase cannot move in the forward $(5' \rightarrow 3')$ direction; on these half-strands, a DNA polymerase must replicate *backwards* toward *ori*. Take a look at Figure 1.7 to see why this must be the case.

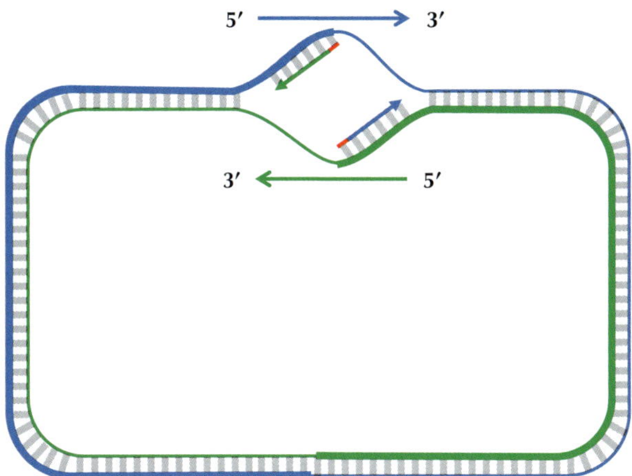

FIGURE 1.7 Replication begins at *ori* (primers shown in red) with the synthesis of fragments on the reverse half-strands (shown by thick lines). A DNA polymerase must wait until the replication fork has opened some (small) distance before it starts copying the forward half-strands (shown by thin lines) back toward *ori*.

On a forward half-strand, in order to replicate DNA, a DNA polymerase must wait for the replication fork to open a little (approximately 2,000 nucleotides) until a new primer is formed at the *end* of the replication fork; afterwards, the DNA polymerase starts replicating a small chunk of DNA starting from this primer and moving *backward* in the direction of *ori*. When the two DNA polymerases on forward half-strands reach *ori*, we have the situation shown in Figure 1.8. Note the contrast between this figure and Figure 1.5.

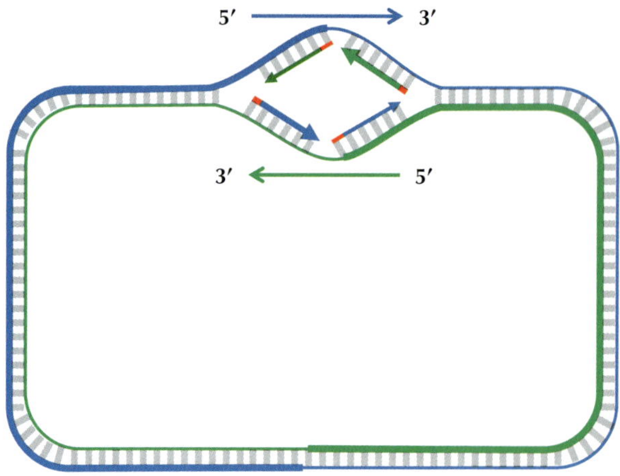

FIGURE 1.8 The daughter fragments are now synthesized (with some delay) on the forward half-strands (shown by thin lines).

After this point, replication on each reverse half-strand progresses continuously; however, a DNA polymerase on a forward half-strand has no choice but to wait again until the replication fork has opened another 2,000 nucleotides or so. It then requires a new primer to begin synthesizing another fragment back toward *ori*. On the whole, replication on a forward half-strand requires occasional stopping and restarting, which results in the synthesis of short **Okazaki fragments** that are complementary to intervals on the forward half-strand. You can see these fragments forming in Figure 1.9 (top).

When the replication fork reaches *ter*, the replication process is almost complete, but gaps still remain between the disconnected Okazaki fragments (Figure 1.9 (middle)).

Finally, consecutive Okazaki fragments are sewn together by an enzyme called **DNA ligase**, resulting in two intact daughter chromosomes, each consisting of one parent strand and one newly synthesized daughter strand, as shown in Figure 1.9 (bottom).

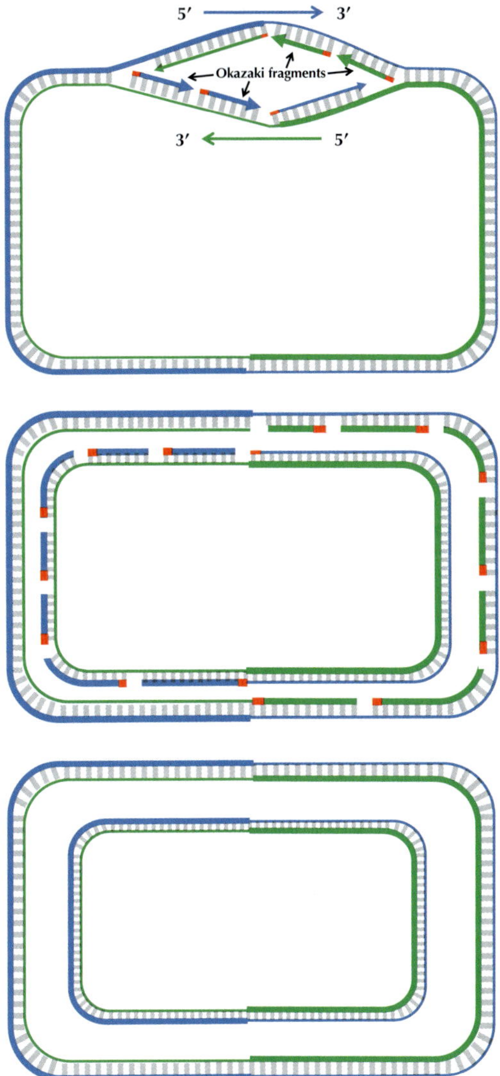

FIGURE 1.9 (Top) The replication fork continues growing. Only one primer is needed for each of the reverse half-strands (shown by thick lines), while the forward half-strands (shown by thin lines) require multiple primers in order to synthesize Okazaki fragments. Two of these primers are shown in red on each forward half-strand. (Middle) Replication is nearly complete, as all daughter DNA is synthesized. However, half of each daughter chromosome contains disconnected Okazaki fragments. (Bottom) Okazaki fragments have been sewn together, resulting in two intact daughter chromosomes.

In reality, DNA ligase does not wait until after all the Okazaki fragments have been replicated to start sewing them together.

Biologists call a reverse half-strand a **leading half-strand** since a single DNA polymerase traverses this half-strand non-stop, and they call a forward half-strand a **lagging half-strand** since it is used as a template by many DNA polymerases stopping and starting replication. If you are confused about the differences between the leading and lagging half-strands, you are not alone — we and legions of biology students are also confused. The confusion is exacerbated by the fact that different textbooks use different terminology depending on whether the authors intend to refer to a leading *template* half-strand or a leading half-strand *that is being synthesized* from a (lagging) template half-strand. You hopefully see why we have chosen the terms "reverse" and "forward" half-strands in an attempt to mitigate some of this confusion.

Peculiar Statistics of the Forward and Reverse Half-Strands

Lurking biological phenomenon or statistical fluke?

Figure 1.10 (upper left) reveals a surprising pattern. We have partitioned the *E. coli* genome into 46 equally sized fragments of approximately 100,000 nucleotides, starting at the experimentally verified terminus of replication, and then computed the frequency of cytosine in each window. The first 23 fragments (starting from *ter*) represent the reverse half-strand, and the last 23 fragments (starting at *ori*) represent the forward half-strand (consult Figure 1.6). Most fragments on the reverse half-strand have a high cytosine frequency (above 25%), whereas most fragments on the forward half-strand have a low cytosine frequency (below 25%). In contrast, as Figure 1.10 (upper right) illustrates, most fragments on the reverse half-strand have a low guanine frequency (below 25%), whereas most fragments on the forward half-strand have a high guanine frequency (above 25%).

Figure 1.10 (bottom left) shows the difference in frequencies of G and C in each genome fragment and presents an even more striking visualization of the peculiar statistics of nucleotide frequencies on the reverse and forward half-strands. Even if we assume that we do not know the location of *ori* in advance, the pattern still presents itself when starting at an arbitrary position of the *E. coli* genome, as shown in Figure 1.10 (bottom right).

If the pattern that we have found in Figure 1.10 isn't a statistical fluke, then we have uncovered a hint about how to find *ori* — we can simply walk along the genome and check where the difference between the frequency of guanine and cytosine switches from negative to positive! But why in the world would such a simple test allow us to find the replication origin of a bacterium?

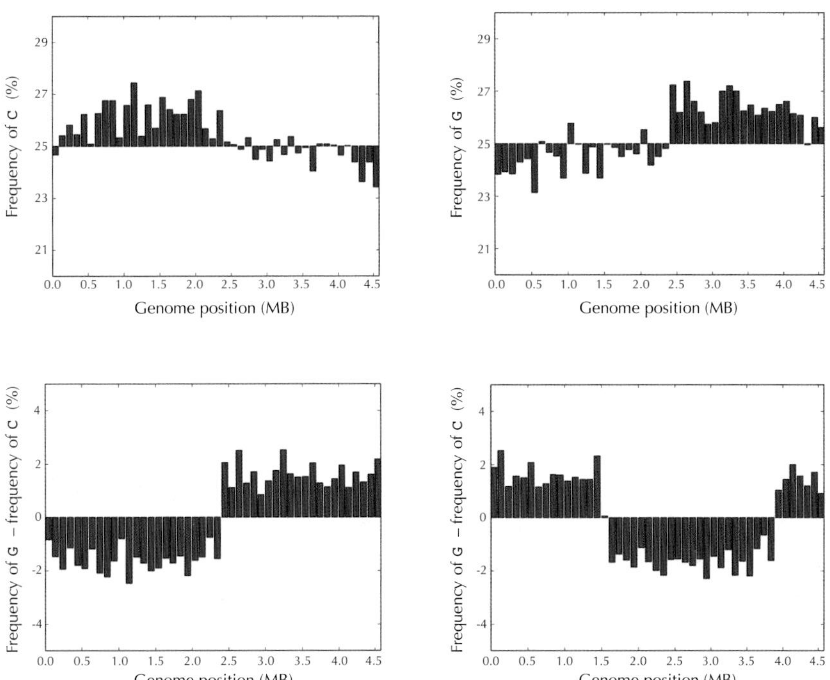

FIGURE 1.10 (Upper left) The frequency of cytosine in each of 46 equal-length disjoint fragments (of approximately 100,000 nucleotides each) covering the *E. coli* genome. The *ter* site is located at position 0, whereas the *ori* site is located roughly opposite to *ter* on the circular *E. coli* chromosome approximately 2.3 million nucleotides away from *ter*. The reverse half-strand spans the first half of the histogram (from 0 to *ori*), whereas the forward half-strand spans the second half of the histogram (starting from *ori*). (Upper right) The frequency of guanine in the same 46 fragments in the *E. coli* genome. (Bottom left) The difference between the frequencies of guanine and cytosine across the 46 fragments of the *E. coli* genome, assuming that the genome begins at the experimentally verified *ter* of *E. coli*. (Bottom right) The difference between the frequencies of guanine and cytosine across the 46 fragments of the *E. coli* genome, assuming that the genome begins at an arbitrary selected location.

Deamination

In the last section, we saw that as the replication fork expands, DNA polymerase synthesizes DNA quickly on the reverse half-strand but suffers delays on the forward half-strand. Notice that since the replication of a reverse half-strand proceeds quickly, it lives double-stranded for most of its life. Conversely, a forward half-strand spends a much larger amount of its life single-stranded, waiting to be used as a template for replication. This discrepancy between the forward and reverse half-strands is important because single-stranded DNA has a much higher mutation rate than double-stranded DNA. In particular, if one of the four nucleotides in single-stranded DNA has a greater tendency than other nucleotides to mutate in single-stranded DNA, then we should observe a shortage of this nucleotide on the forward half-strand.

Following up on this thought, let's examine the nucleotide counts of the reverse and forward half-strands. The nucleotide counts for *Thermotoga petrophila* are shown in Figure 1.11. Although the frequencies of A and T are practically identical on the two half-strands, C is more frequent on the reverse half-strand than on the forward half-strand, resulting in a difference of $219518 - 207901 = +11617$. Its complementary nucleotide G is less frequent on the reverse half-strand than on the forward half-strand, resulting in a difference of $201634 - 211607 = -9973$.

	#C	#G	#A	#T
Entire strand	427419	413241	491488	491363
Reverse half-strand	219518	201634	243963	246641
Forward half-strand	207901	211607	247525	244722
Difference	+11617	-9973	-3562	+1919

FIGURE 1.11 Counting nucleotides in the *Thermotoga petrophila* genome on the forward and reverse half-strands.

It turns out that we observe these discrepancies because cytosine (C) has a tendency to mutate into thymine (T) through a process called **deamination**. Deamination rates rise 100-fold when DNA is single-stranded, which leads to a decrease in cytosine on the forward half-strand, thus forming mismatched base pairs T−G. These mismatched pairs can further mutate into T−A pairs when the bond is repaired in the next round of replication, which accounts for the observed decrease in guanine (G) on the reverse half-strand (recall that a forward parent half-strand synthesizes a reverse daughter half-strand, and vice-versa).

STOP and Think: If deamination changes cytosine to thymine, why do you think that the forward half-strand still has some cytosine?

The skew diagram

Let's see if we can take advantage of these peculiar statistics caused by deamination to locate *ori* even more accurately than the approach that we illustrated in Figure 1.10. As Figure 1.11 illustrates, the difference between the total amount of guanine and the total amount of cytosine is negative on the reverse half-strand (201634 − 219518 = −17884) and positive on the forward half-strand (211607 − 207901 = 3706). Thus, our idea is to traverse the genome, keeping a running total of the difference between the counts of G and C. If this difference starts *increasing*, then we guess that we are on the forward half-strand; on the other hand, if this difference starts *decreasing*, then we guess that we are on the reverse half-strand. See Figure 1.12.

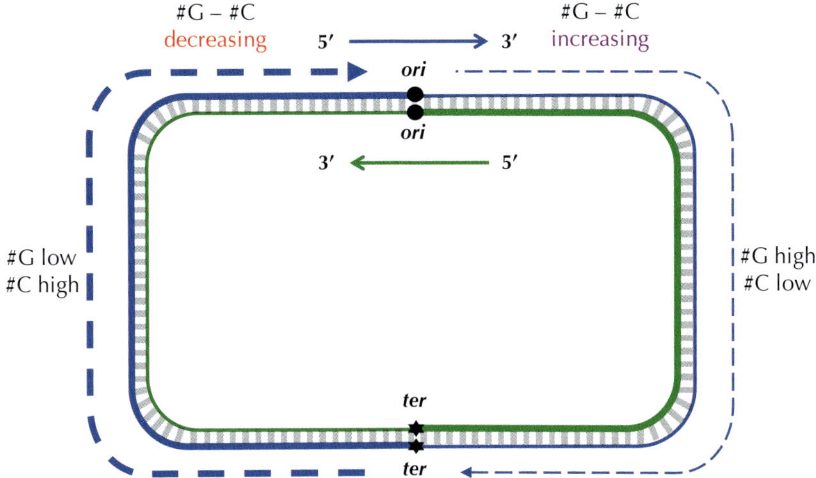

FIGURE 1.12 Because of deamination, each forward half-strand has a shortage of cytosine compared to guanine, and each reverse half-strand has a shortage of guanine compared to cytosine. The dashed blue line illustrates an imaginary walk along the outer strand of the genome counting the difference between the counts of G and C. We assume that the difference between these counts is positive on the forward half-strand and negative on the reverse half-strand.

STOP and Think: Imagine that you are reading through the genome (in the $5' \to 3'$ direction) and notice that the difference between the guanine and cytosine counts just switched its behavior from decreasing to increasing. Where in the genome are you?

Since we don't know the location of *ori* in a circular genome, let's linearize it (i.e., select an arbitrary position and pretend that the genome begins here), resulting in a linear string *Genome*. We define $\text{SKEW}_i(\textit{Genome})$ as the difference between the total number of occurrences of G and the total number of occurrences of C in the first i nucleotides of *Genome*. The **skew diagram** is defined by plotting $\text{SKEW}_i(\textit{Genome})$ as i ranges from 0 to $|\textit{Genome}|$, where $\text{SKEW}_0(\textit{Genome})$ is set equal to zero. Figure 1.13 shows a skew diagram for a short DNA string.

Note that we can compute $\text{SKEW}_{i+1}(\textit{Genome})$ from $\text{SKEW}_i(\textit{Genome})$ according to the nucleotide in position i of *Genome*. If this nucleotide is **G**, then $\text{SKEW}_{i+1}(\textit{Genome}) = \text{SKEW}_i(\textit{Genome}) + 1$; if this nucleotide is **C**, then $\text{SKEW}_{i+1}(\textit{Genome}) = \text{SKEW}_i(\textit{Genome}) - 1$; otherwise, $\text{SKEW}_{i+1}(\textit{Genome}) = \text{SKEW}_i(\textit{Genome})$.

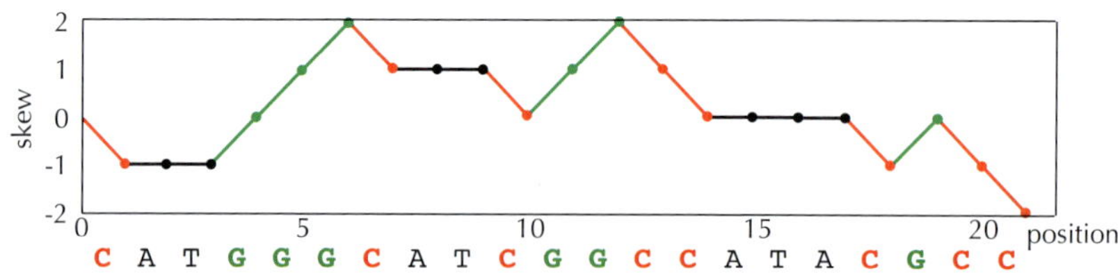

FIGURE 1.13 The skew diagram for *Genome* = CATGGGCATCGGCCATACGCC.

Figure 1.14 depicts the skew diagram for a linearized *E. coli* genome. Notice the very clear pattern! It turns out that the skew diagram for many bacterial genomes has a similar characteristic shape.

STOP and Think: After looking at the skew diagram in Figure 1.14, where do you think that *ori* is located in *E. coli*?

Let's follow the $5' \to 3'$ direction of DNA and walk along the chromosome from *ter* to *ori* (along a reverse half-strand), then continue on from *ori* to *ter* (along a forward half-strand). In Figure 1.12, we saw that the skew is decreasing along the reverse half-strand and increasing along the forward half-strand. Thus, the skew should achieve a

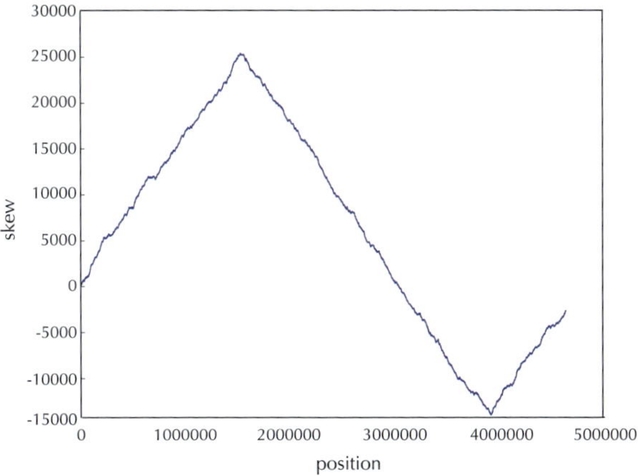

FIGURE 1.14 The skew diagram for *E. coli* achieves a maximum and minimum at positions 1550413 and 3923620, respectively.

minimum at the position where the reverse half-strand ends and the forward half-strand begins, which is exactly the location of *ori*! We have just developed an algorithm for locating *ori*: it should be found where the skew attains a minimum.

Minimum Skew Problem:
Find a position in a genome where the skew diagram attains a minimum.

Input: A DNA string *Genome*.
Output: All integer(s) i minimizing $\text{SKEW}_i(Genome)$ among all values of i (from 0 to $|Genome|$).

STOP and Think: Note that the skew diagram changes depending on where we start our walk along the circular chromosome. Do you think that the minimum of the skew diagram points to the same position *in the genome* regardless of where we begin walking to generate the skew diagram?

Some Hidden Messages are More Elusive than Others

Solving the Minimum Skew Problem now provides us with an approximate location of *ori* at position 3923620 in *E. coli*. In an attempt to confirm this hypothesis, let's look for a hidden message representing a potential *DnaA* box near this location. Solving the Frequent Words Problem in a window of length 500 starting at position 3923620 (shown below) reveals no 9-mers (along with their reverse complements) that appear three or more times! Even if we have located *ori* in *E. coli*, it appears that we still have not found the *DnaA* boxes that jump-start replication in this bacterium ...

```
aatgatgatgacgtcaaaaggatccggataaaacatggtgattgcctcgcataacgcggt
atgaaaatggattgaagcccgggccgtggattctactcaactttgtcggcttgagaaaga
cctgggatcctgggtattaaaaagaagatctatttatttagagatctgttctattgtgat
ctcttattaggatcgcactgccctgtggataacaaggatccggcttttaagatcaacaac
ctggaaaggatcattaactgtgaatgatcggtgatcctggaccgtataagctgggatcag
aatgaggggttatacacaactcaaaaactgaacaacagttgttctttggataactaccgg
ttgatccaagcttcctgacagagttatccacagtagatcgcacgatctgtatacttattt
gagtaaattaacccacgatcccagccattcttctgccggatcttccggaatgtcgtgatc
aagaatgttgatcttcagtg
```

STOP and Think: What would you do next?

Before we give up, let's examine the *ori* of *Vibrio cholerae* one more time to see if it provides us with any insights on how to alter our algorithm to find *DnaA* boxes in *E. coli*. You may have noticed that in addition to the three occurrences of **ATGATCAAG** and three occurrences of its reverse complement **CTTGATCAT**, the *Vibrio cholerae ori* contains additional occurrences of **ATGATCAAC** and **CATGATCAT**, which differ from **ATGATCAAG** and **CTTGATCAT** in only a single nucleotide:

```
atcaATGATCAACgtaagcttctaagcATGATCAAGgtgctcacacagtttatccacaac
ctgagtggatgacatcaagataggtcgttgtatctccttcctctcgtactctcatgacca
cggaaagATGATCAAGagaggatgatttcttggccatatcgcaatgaatacttgtgactt
gtgcttccaattgacatcttcagcgccatattgcgctggccaaggtgacggagcgggatt
acgaaagCATGATCATggctgttgttctgtttatcttgttttgactgagacttgttagga
tagacggtttttcatcactgactagccaaagccttactctgcctgacatcgaccgtaaat
tgataatgaatttacatgcttccgcgacgatttacctCTTGATCATcgatccgattgaag
atcttcaattgttaattctcttgcctcgactcatagccatgatgagctCTTGATCATgtt
tccttaaccctctatttttacggaagaATGATCAAGctgctgctCTTGATCATcgtttc
```

Finding eight *approximate* occurrences of our target 9-mer and its reverse complement in a short region is even more statistically surprising than finding the six *exact* occurrences of **ATGATCAAG** and its reverse complement **CTTGATCAT** that we stumbled upon in the beginning of our investigation. Furthermore, the discovery of these approximate 9-mers makes sense biologically, since *DnaA* can bind not only to "perfect" *DnaA* boxes but to their slight variations as well.

We say that position i in k-mers $p_1 \cdots p_k$ and $q_1 \cdots q_k$ is a **mismatch** if $p_i \neq q_i$. The number of mismatches between strings p and q is called the **Hamming distance** between these strings and is denoted HAMMINGDISTANCE(p, q).

Hamming Distance Problem:
Compute the Hamming distance between two strings.

Input: Two strings of equal length.
Output: The Hamming distance between these strings.

We say that a k-mer *Pattern* appears as a substring of *Text* with at most d mismatches if there is some k-mer substring *Pattern'* of *Text* having d or fewer mismatches with *Pattern*, i.e., HAMMINGDISTANCE$(Pattern, Pattern') \leq d$. Our observation that a *DnaA* box may appear with slight variations leads to the following generalization of the Pattern Matching Problem.

Approximate Pattern Matching Problem:
Find all approximate occurrences of a pattern in a string.

Input: Strings *Pattern* and *Text* along with an integer d.
Output: All starting positions where *Pattern* appears as a substring of *Text* with at most d mismatches.

Our goal now is to modify our previous algorithm for the Frequent Words Problem in order to find *DnaA* boxes by identifying frequent k-mers, possibly with mismatches. Given strings *Text* and *Pattern* as well as an integer d, we define COUNT$_d$(*Text*, *Pattern*) as the number of occurrences of *Pattern* in *Text* with at most d mismatches. For example,

$$\text{COUNT}_1(\texttt{AACAAGCATAAACATTAAAGAG}, \texttt{AAAAA}) = 4$$

because **AAAAA** appears four times in this string with at most one mismatch: **AACAA**, **ATAAA**, **AAACA**, and **AAAGA**. Notice that two of these occurrences overlap.

Exercise Break: Compute COUNT$_2$(AACAAGCATAAACATTAAAGAG, AAAAA).

Computing COUNT$_d$(*Text*, *Pattern*) simply requires us to compute the Hamming distance between *Pattern* and every k-mer substring of *Text*, as follows.

```
APPROXIMATEPATTERNCOUNT(Text, Pattern, d)
    count ← 0
    for i ← 0 to |Text| − |Pattern|
        Pattern′ ← Text(i, |Pattern|)
        if HAMMINGDISTANCE(Pattern, Pattern′) ≤ d
            count ← count + 1
    return count
```

Exercise Break: Implement **APPROXIMATEPATTERNCOUNT**. What is its running time?

A **most frequent *k*-mer with up to *d* mismatches** in *Text* is simply a string *Pattern* maximizing $\text{COUNT}_d(\textit{Text}, \textit{Pattern})$ among all *k*-mers. Note that *Pattern* does not need to actually appear as a substring of *Text*; for example, as we saw above, **AAAAA** is the most frequent 5-mer with 1 mismatch in AACAAGCATAAACATTAAAGAG, even though it does not appear exactly in this string. Keep this in mind while solving the following problem.

Frequent Words with Mismatches Problem:
Find the most frequent k-mers with mismatches in a string.

Input: A string *Text* as well as integers *k* and *d*.
Output: All most frequent *k*-mers with up to *d* mismatches in *Text*.

Charging Station (Solving the Frequent Words with Mismatches Problem): One way to solve the find frequent words with mismatches is to generate all 4^k *k*-mers *Pattern*, compute $\text{COUNT}_d(\textit{Text}, \textit{Pattern}, d)$ for each *k*-mer, and then output *k*-mers with the maximum number of approximate occurrences. This is an inefficient approach in practice, since many of the 4^k *k*-mers should not be considered because neither they nor their mutated versions (with up to *d* mismatches) appear in *Text*. Check out this Charging Station to learn about a better approach that generalizes our algorithm for finding frequent words without mismatches.

We now redefine the Frequent Words Problem to account for both mismatches and reverse complements. Recall that $\overline{\textit{Pattern}}$ refers to the reverse complement of *Pattern*.

Frequent Words with Mismatches and Reverse Complements Problem:
Find the most frequent k-mers (with mismatches and reverse complements) in a string.

> **Input**: A DNA string *Text* as well as integers k and d.
> **Output**: All k-mers *Pattern* that maximize the sum $\text{COUNT}_d(Text, Pattern) + \text{COUNT}_d(Text, \overline{Pattern})$ over all possible k-mers.

A Final Attempt at Finding *DnaA* Boxes in *E. coli*

We now make a final attempt to find *DnaA* boxes in *E. coli* by finding the most frequent 9-mers with mismatches and reverse complements in the region suggested by the minimum skew as *ori*. Although the minimum of the skew diagram for *E. coli* is found at position 3923620, we should not assume that its *ori* is found exactly at this position due to random fluctuations in the skew. To remedy this issue, we could choose a larger window size (e.g., $L = 1000$), but expanding the window introduces the risk that we may bring in other clumped 9-mers that do not represent *DnaA* boxes but appear in this window more often than the true *DnaA* box. It makes more sense to try a small window either starting, ending, or centered at the position of minimum skew.

Let's cross our fingers and identify the most frequent 9-mers (with 1 mismatch and reverse complements) within a window of length 500 starting at position 3923620 of the *E. coli* genome. Bingo! The experimentally confirmed *DnaA* box in *E. coli* (**TTATCCACA**) is a most frequent 9-mer with 1 mismatch, along with its reverse complement **TGTGGATAA**:

```
aatgatgatgacgtcaaaaggatccggataaaacatggtgattgcctcgcataacgcggt
atgaaaatggattgaagcccgggccgtggattctactcaactttgtcggcttgagaaaga
cctgggatcctgggtattaaaaagaagatctatttatttagagatctgttctattgtgat
ctcttattaggatcgcactgcccTGTGGATAAcaaggatccggcttttaagatcaacaac
ctggaaaggatcattaactgtgaatgatcggtgatcctggaccgtataagctgggatcag
aatgaggggTTATACACAactcaaaaactgaacaacagttgttcTTTGGATAActaccgg
ttgatccaagcttcctgacagagTTATCCACAgtagatcgcacgatctgtatacttattt
gagtaaattaacccacgatcccagccattcttctgccggatcttccggaatgtcgtgatc
aagaatgttgatcttcagtg
```

You will notice that we highlighted an interior interval of this sequence with darker text. This region is the experimentally verified *ori* of *E. coli*, which starts 37 nucleotides after position 3923620, where the skew reaches its minimum value.

We were fortunate that the *DnaA* boxes of *E. coli* are captured in the window that we chose. Moreover, while **TTATCCACA** represents a most frequent 9-mer with 1 mismatch and reverse complements in this 500-nucleotide window, it is not the only one:

GGATCCTGG, GATCCCAGC, GTTATCCAC, AGCTGGGAT, and CTGGGATCA also appear four times with 1 mismatch and reverse complements.

STOP and Think: In this chapter, every time we find *ori*, we seem to find some other surprisingly frequent 9-mers. Why do you think this is?

We do not know what purpose — if any — these other 9-mers serve in the *E. coli* genome, but we do know that there are *many different types* of hidden messages in genomes; these hidden messages have a tendency to cluster within a genome, and most of them have nothing to do with replication. One example is the regulatory DNA motifs responsible for gene expression that we will study in Chapter 2. The important lesson is that existing approaches to *ori* prediction remain imperfect and sometimes inconclusive. However, even providing biologists with a small collection of 9-mers as candidate *DnaA* boxes is a great aid as long as one of these 9-mers is correct.

Thus, the moral of this chapter is that even though computational predictions can be powerful, bioinformaticians should collaborate with biologists to verify their computational predictions. Or improve these predictions: the next question hints at how *ori* predictions can be carried out using **comparative genomics**, a bioinformatics approach that uses evolutionary similarities to answer difficult questions about genomes.

STOP and Think: *Salmonella enterica* is a close relative of *E. coli* that causes typhoid fever and foodborne illness. After having learned what *DnaA* boxes look like in *E. coli*, how would you search for *DnaA* boxes in *Salmonella enterica*?

You will have an opportunity to look for *DnaA* boxes in *Salmonella enterica* in the epilogue, which will feature a "Challenge Problem" asking you to apply what you have learned to a real dataset. Some chapters also have an "Open Problems" section outlining unanswered research questions.

Epilogue: Complications in *ori* Predictions

In this chapter, we have considered three genomes and found three different hypothesized 9-mers encoding *DnaA* boxes: **ATGATCAAG** in *Vibrio cholerae*, **CCTACCACC** in *Thermotoga petrophila*, and **TTATCCACA** in *E. coli*. We must warn you that finding *ori* is often more complex than in the three examples we considered. Some bacteria have even fewer *DnaA* boxes than *E. coli*, making it difficult to identify them. The *ter* region is often located not directly opposite to *ori* but may be significantly shifted, resulting in reverse and forward half-strands having substantially different lengths. The position of the skew minimum is often only a rough indicator of *ori* position, which forces researchers to expand their windows when searching for *DnaA* boxes, bringing in extraneous repeated substrings. Finally, skew diagrams do not always look as nice as that of *E. coli*; for example, the skew diagram for *Thermotoga petrophila* is shown in Figure 1.15.

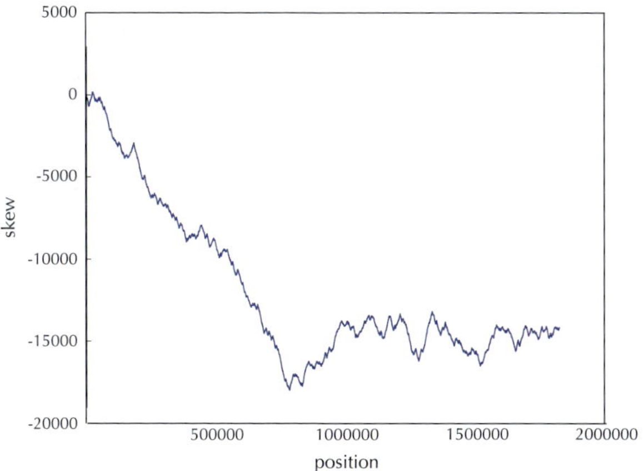

FIGURE 1.15 The skew diagram for *Thermotoga petrophila* achieves a minimum at position 787199 but does not have the same nice shape as the skew diagram for *E. coli*.

STOP and Think: What evolutionary process could possibly explain the shape of the skew diagram for *Thermotoga petrophila*?

Since the skew diagram for *Thermotoga petrophila* is complex and the *ori* for this genome has not even been experimentally verified, there is a chance that the region predicted by Ori-Finder as the *ori* region for *Thermotoga petrophila* (or even for *Vibrio cholerae*) is actually incorrect!

You now should have a good sense of how to locate *ori* and *DnaA* boxes computationally. We will take the training wheels off and ask you to solve a challenge problem.

Challenge Problem: Find *DnaA* boxes in *Salmonella enterica*.

Open Problems

Multiple replication origins in a bacterial genome

Biologists long believed that each bacterial chromosome has only one *ori*. Wang et al., 2011 genetically modified *E. coli* by inserting a synthetic *ori* a million nucleotides away from the bacterium's known *ori*. To their surprise, *E. coli* continued business as usual, starting replication at both locations!

Following the publication of this paper, the search for naturally occurring bacteria with multiple *ori*s immediately started. In 2012, Xia raised doubts about the "single *ori*" postulate and gave examples of bacteria with highly unusual skews. In fact, having more than one *ori* makes sense in the light of evolution: if the genome is long and replication is slow, then multiple replication origins would decrease the amount of time that the bacterium must spend replicating its DNA.

For example, *Wigglesworthia glossinidia*, a symbiotic bacterium living in the intestines of tsetse flies, has the atypical skew diagram shown in Figure 1.16. Since this diagram has at least two pronounced local minima, Xia argued that this bacterium may have two or more *ori* regions.

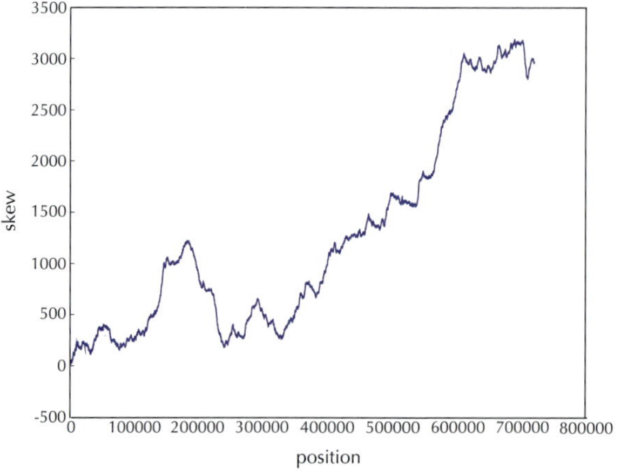

FIGURE 1.16 The skew diagram for *Wigglesworthia glossinidia*.

We should be careful with Xia's hypothesis that this bacterium has two *ori*s, as there may be alternative explanations for multiple local minima in the skew. For example, **genome rearrangements** (which we will study in Chapter 6) move genes within a genome and often reposition them from the forward to the reverse half-strand and vice-versa, thus resulting in irregularities in the skew diagram. One example of a genome rearrangement is a **reversal**, which flips around a segment of chromosome and switches it to the opposite strand; Figure 1.17 shows what happens to the skew diagram after a reversal.

Another difficulty is presented by the fact that different species of bacteria may exchange genetic material in **horizontal gene transfer**. If a gene from the forward half-strand of one bacterium is transferred to the reverse half-strand of another (or vice-versa), then we will observe an irregularity in the skew diagram. As a result, the question about the number of *oris* of *Wigglesworthia glossinidia* remains unresolved.

STOP and Think: After learning about genome assembly in Chapter 3, can you suggest another explanation for the unusual shape of the *Wigglesworthia glossinidia* skew diagram in Figure 1.16? (Hint: bacterial genomes are sometimes assembled incorrectly.)

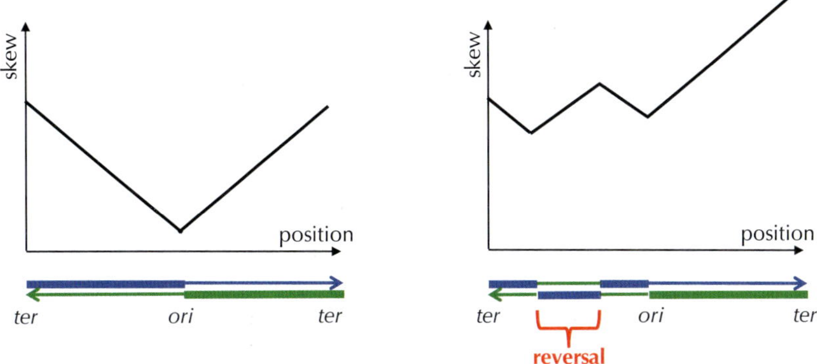

FIGURE 1.17 (Left) An "ideal" V-shaped skew diagram that achieves minimum skew at *ori*. The skew diagram decreases along the reverse half-strand (shown by a thick line) and increases along the forward half-strand (shown by a thin line). We assume that a circular chromosome was cut at *ter*, resulting in a linear chromosome that starts and ends at *ter*. (Right) A skew diagram after a reversal that switches segments between the reverse and forward strands and alters the skew diagram. As before, the skew diagram still decreases along the segments of the genome shown by thick lines and increases along the segments shown by thin lines.

However, if you could demonstrate that there exist two sets of identical *DnaA* boxes in the vicinity of two local minima in the skew diagram of *Wigglesworthia glossinidia*, then you would have the first solid evidence in favor of multiple bacterial *oris*. Maybe simply applying your solution for the Frequent Words with Mismatches and Reverse Complements Problem will reveal these *DnaA* boxes. Can you find other bacterial genomes where a single *ori* is in doubt and check whether they indeed have multiple *oris*?

Finding replication origins in archaea

Archaea are unicellular organisms so distinct from other life forms that biologists have placed them into their own **domain of life** separate from bacteria and eukaryotes. Although archaea are visually similar to bacteria, they have some genomic features that are more closely related to eukaryotes. In particular, the replication machinery of archaea is more similar to eukaryotes than bacteria. Yet archaea use a much greater variety of energy sources than eukaryotes, feeding on ammonia, metals, or even hydrogen gas.

Figure 1.18 shows the skew diagram of *Sulfolobus solfataricus*, a species of archaea growing in acidic volcanic springs in temperatures over 80° C. In its skew diagram, you can see at least three local minima, represented by deep valleys, in addition to many more shallow valleys.

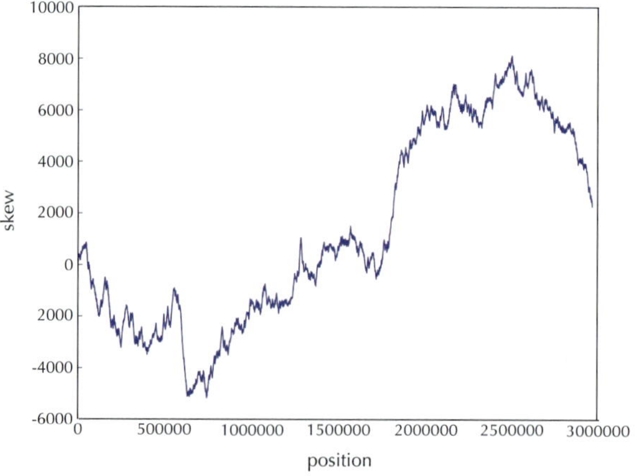

FIGURE 1.18 The skew diagram of *Sulfolobus solfataricus*.

Lundgren et al., 2004 demonstrated experimentally that *Sulfolobus solfataricus* indeed has three *oris*. Since then, multiple *oris* have been identified in many other archaea. However, no accurate computational approach has been developed to identify multiple *oris* in a newly sequenced archaea genome. For example, the methane-producing archaea *Methanococcus jannaschii* is considered the workhorse of archaea genomics, but its *ori*(s) still remain unidentified! Its skew diagram (shown in Figure 1.19) suggests that it may have multiple replication origins: can you find them?

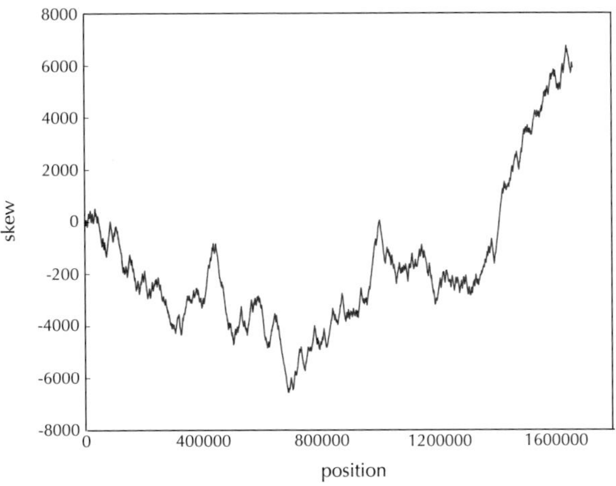

FIGURE 1.19 The skew diagram for *Methanococcus jannaschii.*

Finding replication origins in yeast

If you think that finding replication origins in bacteria is a complex problem, wait until you analyze replication origins in more complex organisms like yeast or humans, which have hundreds of replication origins. Among various yeast species, the yeast *Saccharomyces cerevisiae* has the best characterized replication origins. It has approximately 400 different *oris*, many of which may be used during the replication of any single yeast cell.

Having a large number of *oris* results in dozens of replication forks hurtling towards each other from different locations in the genome in ways that are not yet completely understood. However, researchers have discovered that the replication origins of *S. cerevisiae* share a (somewhat variable) pattern called the **ARS Consensus Sequence (ACS)**. The ACS is the binding site for the so-called **Origin Recognition Complex**, which initiates the loading of additional proteins required for origin firing. Many ACSs correspond to the following canonical thymine-rich pattern of length 11.

TTTAT(G/A)TTT(T/A)(G/T)

Here, the notation (X/Y) indicates that either nucleotide X or nucleotide Y may appear in that position.

However, various ACSs may differ from this canonical pattern, with lengths varying from 11 to 17 nucleotides. For example, the 11-nucleotide long pattern shown above is often part of a 17-nucleotide pattern:

(T/A)(T/A)(T/A)(T/A)**TTTAT(G/A)TTT(T/A)(G/T)**(T/G)(T/C)

Recently, some progress has been made in characterizing the ACS in a few other yeast species. In some species like *S. bayanus*, the ACS is almost identical to that of *S. cerevisiae*, while in others such as *K. lactis*, it is very different. More alarmingly, at least for bioinformaticians, in some yeast species such as *S. pombe*, the Origin Recognition Complex binds to loosely defined AT-rich regions, which makes it next to impossible to find replication origins based on sequence analysis alone.

Despite recent efforts, finding *oris* in yeast remains an open problem, and no accurate software exists for predicting origins of replication from the sequence of yeast genomes. Can you explore this problem and devise an algorithm to predict replication origins in yeast?

Computing probabilities of patterns in a string

PAGE 53

In the main text, we told you that the probability that a random DNA string of length 500 contains a 9-mer appearing three or more times is *approximately* 1/1300. In **DETOUR: Probabilities of Patterns in a String**, we describe a method to estimate this probability, but it is rather inaccurate. This open problem is aimed at finding better approximations or even deriving exact formulas for probabilities of patterns in strings.

We start by asking a question: what is the probability that a specific *k*-mer *Pattern* will appear (at least once) as a substring of a random string of length *N*? This question proved to be not so simple and was first addressed by Solov'ev, 1966 (see also Sedgewick and Flajolet, 2013).

The first surprise is that different *k*-mers may have different probabilities of appearing in a random string. For example, the probability that *Pattern* = "01" appears in a random binary string of length 4 is 11/16, while the probability that *Pattern* = "11" appears in a random binary string of length 4 is 8/16. This phenomenon is called the **overlapping words paradox** because different occurrences of *Pattern* can overlap each other for some patterns (e.g., "11") but not others (e.g., "01"). See **DETOUR: The Overlapping Words Paradox**.

PAGE 63

We are interested in computing the following probabilities for a random N-letter string in an A-letter alphabet:

- $\Pr(N, A, Pattern, t)$, the probability that a string *Pattern* appears at least t times in a random string;
- $\Pr^*(N, A, Pattern, t)$, the probability that a string *Pattern* and its reverse complement $\overline{Pattern}$ appear at least t total times in a random string.

Note that the above two probabilities are relatively straightforward to compute. Several variants of these questions are open:

- $\Pr_d(N, A, Pattern, t)$, the probability that a string *Pattern* approximately appears at least t times in a random string (with at most d mismatches);
- $\Pr(N, A, k, t)$, the probability that there exists *any* k-mer appearing at least t times in a random string;
- $\Pr_d(N, A, k, t)$, the probability that there exists *any* k-mer with at least t approximate occurrences in a random string (with at most d mismatches).

Charging Stations

The frequency array

To make **FREQUENTWORDS** faster, we will think about why this algorithm is slow in the first place. It slides a window of length k down *Text*, identifying a k-mer *Pattern* of *Text* at each step. For each such k-mer, it must slide a window down the entire length of *Text* in order to compute PATTERNCOUNT(*Text*, *Pattern*). Instead of doing all this sliding, we aspire to slide a window down *Text* only once. As we slide this window, we will keep track of the number of times that each k-mer *Pattern* has already appeared in *Text*, updating these numbers as we proceed.

To achieve this goal, we will first order all 4^k k-mers **lexicographically** (i.e., according to how they would appear in the dictionary) and then convert them into the 4^k different integers between 0 and $4^k - 1$. Given an integer k, we define the **frequency array** of a string *Text* as an array of length 4^k, where the i-th element of the array holds the number of times that the i-th k-mer (in lexicographic order) appears in *Text* (Figure 1.20).

k-mer	AA	AC	AG	AT	CA	CC	CG	CT	GA	GC	GG	GT	TA	TC	TG	TT
index	0	1	2	3	4	5	6	7	8	9	10	11	12	13	14	15
frequency	3	0	2	0	1	0	0	0	0	1	3	1	0	0	1	0

FIGURE 1.20 The lexicographic order of DNA 2-mers (top), along with the index of each 2-mer in this order (middle), and the frequency array for AAGCAAAGGTGGG (bottom). For example, the frequency array at index 10 is equal to 3 because GG, the tenth DNA 2-mer according to lexicographic order, occurs three times in AAGCAAAGGTGGG.

To compute the frequency array, we need to determine how to transform a k-mer *Pattern* into an integer using a function PATTERNTONUMBER(*Pattern*). We also should know how to reverse this process, transforming an integer between 0 and $4^k - 1$ into a k-mer using a function NUMBERTOPATTERN(*index*, k). Figure 1.20 illustrates that PATTERNTONUMBER(GT) = 11 and NUMBERTOPATTERN(11, 2) = GT.

Exercise Break: Compute the following:

1. PATTERNTONUMBER(ATGCAA)
2. NUMBERTOPATTERN(5437, 7)
3. NUMBERTOPATTERN(5437, 8)

Charging Station (Converting Patterns Into Numbers and Vice-Versa): Check out this Charging Station to see how to implement PATTERNTONUMBER and NUMBERTOPATTERN.

The pseudocode below generates a frequency array by first initializing every element in the frequency array to zero (4^k operations) and then making a single pass down *Text* (approximately $|Text| \cdot k$ operations). For each k-mer *Pattern* that we encounter, we add 1 to the value of the frequency array corresponding to *Pattern*. As before, we refer to the k-mer beginning at position i of *Text* as $Text(i, k)$.

```
COMPUTINGFREQUENCIES(Text, k)
    for i ← 0 to 4^k − 1
        FREQUENCYARRAY(i) ← 0
    for i ← 0 to |Text| − k
        Pattern ← Text(i, k)
        j ← PATTERNTONUMBER(Pattern)
        FREQUENCYARRAY(j) ← FREQUENCYARRAY(j) + 1
    return FREQUENCYARRAY
```

We now have a faster algorithm for the Frequent Words Problem. After generating the frequency array, we can find all most frequent k-mers by simply finding all k-mers corresponding to the maximum element(s) in the frequency array.

```
FASTERFREQUENTWORDS(Text , k)
    FrequentPatterns ← an empty set
    FREQUENCYARRAY ← COMPUTINGFREQUENCIES(Text, k)
    maxCount ← maximal value in FREQUENCYARRAY
    for i ← 0 to 4^k − 1
        if FREQUENCYARRAY(i) = maxCount
            Pattern ← NUMBERTOPATTERN(i, k)
            add Pattern to the set FrequentPatterns
    return FrequentPatterns
```

Charging Station (Finding Frequent Words by Sorting): Although **FASTERFREQUENTWORDS** is fast for small k (i.e., you can use it to find *DnaA* boxes in an *ori* region), it becomes impractical when k is large. If you are familiar with sorting algorithms and are interested in seeing a faster algorithm, check out this Charging Station.

Exercise Break: Our claim that **FASTERFREQUENTWORDS** is faster than **FREQUENTWORDS** is only correct for certain values of $|Text|$ and k. Estimate the running time of **FASTERFREQUENTWORDS** and characterize the values of $|Text|$ and k when **FASTERFREQUENTWORDS** is indeed faster than **FREQUENTWORDS**.

Converting patterns to numbers and vice-versa

Our approach to computing PATTERNTONUMBER(*Pattern*) is based on a simple observation. If we remove the final symbol from all lexicographically ordered k-mers, the resulting list is still ordered lexicographically (think about removing the final letter from every word in a dictionary). In the case of DNA strings, every $(k-1)$-mer in the resulting list is repeated four times (Figure 1.21).

AAA	**AA**C	**AA**G	**AA**T	**AC**A	**AC**C	**AC**G	**AC**T
AGA	**AG**C	**AG**G	**AG**T	**AT**A	**AT**C	**AT**G	**AT**T
CAA	**CA**C	**CA**G	**CA**T	**CC**A	**CC**C	**CC**G	**CC**T
CGA	**CG**C	**CG**G	**CG**T	**CT**A	**CT**C	**CT**G	**CT**T
GAA	**GA**C	**GA**G	**GA**T	**GC**A	**GC**C	**GC**G	**GC**T
GGA	**GG**C	**GG**G	**GG**T	**GT**A	**GT**C	**GT**G	**GT**T
TAA	**TA**C	**TA**G	**TA**T	**TC**A	**TC**C	**TC**G	**TC**T
TGA	**TG**C	**TG**G	**TG**T	**TT**A	**TT**C	**TT**G	**TT**T

FIGURE 1.21 If we remove the final symbol from all lexicographically ordered DNA 3-mers, we obtain a lexicographic order of (red) 2-mers, where each 2-mer is repeated four times.

Thus, the number of 3-mers occurring before **AG**T is equal to four times the number of 2-mers occurring before **AG** plus the number of 1-mers occurring before T. Therefore,

$$\begin{aligned}\text{PATTERNTONUMBER}(\texttt{AGT}) &= 4 \cdot \text{PATTERNTONUMBER}(\texttt{AG}) + \text{SYMBOLTONUMBER}(\texttt{T}) \\ &= 8 + 3 = 11\,,\end{aligned}$$

where SymbolToNumber(*symbol*) is the function transforming symbols A, C, G, and T into the respective integers 0, 1, 2, and 3.

If we remove the final symbol of *Pattern*, denoted LastSymbol(*Pattern*), then we will obtain a $(k-1)$-mer that we denote as Prefix(*Pattern*). The preceding observation therefore generalizes to the formula

$$\text{PatternToNumber}(\textit{Pattern}) = 4 \cdot \text{PatternToNumber}(\text{Prefix}(\textit{Pattern})) + \text{SymbolToNumber}(\text{LastSymbol}(\textit{Pattern})) . \quad (*)$$

This equation leads to the following **recursive algorithm**, i.e., a program that calls itself. If you want to learn more about recursive algorithms, see **DETOUR: The Towers of Hanoi**. PAGE 61

```
PatternToNumber(Pattern)
    if Pattern contains no symbols
        return 0
    symbol ← LastSymbol(Pattern)
    Prefix ← Prefix(Pattern)
    return 4 · PatternToNumber(Prefix) + SymbolToNumber(symbol)
```

In order to compute the inverse function NumberToPattern(*index*, *k*), we return to (*) above, which implies that when we divide *index* = PatternToNumber(*Pattern*) by 4, the remainder will be equal to SymbolToNumber(*symbol*), and the quotient will be equal to PatternToNumber(Prefix(*Pattern*)). Thus, we can use this fact to peel away symbols at the end of *Pattern* one at a time, as shown in Figure 1.22.

STOP and Think: Once we have computed NumberToPattern(9904, 7) in Figure 1.22, how would you compute NumberToPattern(9904, 8)?

In the pseudocode below, we denote the quotient and the remainder when dividing integer n by integer m as Quotient(n, m) and Remainder(n, m), respectively. For example, Quotient(11, 4) = 2 and Remainder(11, 4) = 3. This pseudocode uses the function NumberToSymbol(*index*), which is the inverse of SymbolToNumber and transforms the integers 0, 1, 2, and 3 into the respective symbols A, C, G, and T.

n	QUOTIENT$(n, 4)$	REMAINDER$(n, 4)$	NUMBERTOSYMBOL
9904	2476	0	A
2476	619	0	A
619	154	3	T
154	38	2	G
38	9	2	G
9	2	1	C
2	0	2	G

FIGURE 1.22 When computing *Pattern* = NUMBERTOPATTERN(9904, 7), we divide 9904 by 4 to obtain a quotient of 2476 and a remainder of 0. This remainder represents the final nucleotide of *Pattern*, or NUMBERTOSYMBOL(0) = A. We then iterate this process, dividing each subsequent quotient by 4, until we obtain a quotient of 0. The symbols in the nucleotide column, read upward from the bottom, yield *Pattern* = GCGGTAA.

1M

```
NUMBERTOPATTERN(index , k)
    if k = 1
        return NUMBERTOSYMBOL(index)
    prefixIndex ← QUOTIENT(index, 4)
    r ← REMAINDER(index, 4)
    symbol ← NUMBERTOSYMBOL(r)
    PrefixPattern ← NUMBERTOPATTERN(prefixIndex, k − 1)
    return concatenation of PrefixPattern with symbol
```

Exercise Break: Apply the algorithm illustrated in Figure 1.22 to compute NUMBERTOPATTERN(11, 3). Is the answer what you expect? Explain what is wrong with the stopping rule described in the Figure 1.22 caption, and then correct the rule.

Finding frequent words by sorting

To see how sorting can help us find frequent k-mers, we will consider a motivating example when $k = 2$. Given a string *Text* = AAGCAAAGGTGGG, list all its 2-mers in the order they appear in *Text*, and convert each 2-mer into an integer using PATTERNTONUMBER to produce an array INDEX, as shown below.

2-mer	AA	AG	GC	CA	AA	AA	AG	GG	GT	TG	GG	GG
INDEX	0	2	9	4	0	0	2	10	11	14	10	10

We will now sort INDEX to generate an array SORTEDINDEX, as shown in Figure 1.23.

2-mer	AA	AA	AA	AG	AG	CA	GC	GG	GG	GG	GT	TG
SORTEDINDEX	0	0	0	2	2	4	9	10	10	10	11	14
COUNT	1	2	3	1	2	1	1	1	2	3	1	1

FIGURE 1.23 Lexicographically sorted 2-mers in AAGCAAAGGTGG (top), along with arrays SORTEDINDEX (middle) and COUNT (bottom).

STOP and Think: How can the sorted array in Figure 1.23 help us find frequent words?

Since identical k-mers clump together in the sorted array (e.g., $(0, 0, 0)$ for AA or $(10, 10, 10)$ for GG in Figure 1.23), frequent k-mers are the longest runs of identical integers in SORTEDINDEX. This insight leads to **FINDINGFREQUENTWORDSBYSORTING**, whose pseudocode is shown below. This algorithm uses an array COUNT for which COUNT(i) computes the number of times that the integer at position i in the array SORTEDINDEX appears in the first i elements of this array (Figure 1.23 (bottom)). In the pseudocode for **FINDINGFREQUENTWORDSBYSORTING**, we assume that you already know how to sort an array using an algorithm **SORT**.

```
FINDINGFREQUENTWORDSBYSORTING(Text, k)
    FrequentPatterns ← an empty set
    for i ← 0 to |Text| − k
        Pattern ← Text(i, k)
        INDEX(i) ← PATTERNTONUMBER(Pattern)
        COUNT(i) ← 1
    SORTEDINDEX ← SORT(INDEX)
    for i ← 1 to |Text| − k
        if SORTEDINDEX(i) = SORTEDINDEX(i − 1)
            COUNT(i) = COUNT(i − 1) + 1
    maxCount ← maximum value in the array COUNT
    for i ← 0 to |Text| − k
        if COUNT(i) = maxCount
            Pattern ← NUMBERTOPATTERN(SORTEDINDEX(i), k)
            add Pattern to the set FrequentPatterns
    return FrequentPatterns
```

Solving the Clump Finding Problem

Note: This Charging Station assumes that you have read **Charging Station: The Frequency Array**.

The pseudocode below slides a window of length L down *Genome*. After computing the frequency array for the current window, it identifies (L, t)-clumps simply by finding which k-mers occur at least t times within the window. To keep track of these clumps, our algorithm uses an array CLUMP of length 4^k whose values are all initialized to zero. For each value of i between 0 and $4^k - 1$, we will set CLUMP(i) equal to 1 if NUMBERTOPATTERN(i, k) forms an (L, t)-clump in *Genome*.

```
ClumpFinding(Genome, k, L, t)
    FrequentPatterns ← an empty set
    for i ← 0 to 4^k − 1
        Clump(i) ← 0
    for i ← 0 to |Genome| − L
        Text ← the string of length L starting at position i in Genome
        FrequencyArray ← ComputingFrequencies(Text, k)
        for index ← 0 to 4^k − 1
            if FrequencyArray(index) ≥ t
                Clump(index) ← 1
    for i ← 0 to 4^k − 1
        if Clump(i) = 1
            Pattern ← NumberToPattern(i, k)
            add Pattern to the set FrequentPatterns
    return FrequentPatterns
```

Exercise Break: Estimate the running time of **CLUMPFINDING**.

CLUMPFINDING makes $|Genome| - L + 1$ iterations, generating a frequency array for a string of length L at each iteration. Since this task takes roughly $4^k + L \cdot k$ time, the overall running time of **CLUMPFINDING** is $\mathcal{O}\Big(|Genome| \cdot (4^k + L \cdot k)\Big)$. As a result, when searching for *DnaA* boxes ($k = 9$) in a typical bacterial genome ($|Genome| > 1000000$), **CLUMPFINDING** becomes too slow.

STOP and Think: Can you speed up **CLUMPFINDING** by eliminating the need to generate a new frequency array at every iteration?

To improve **CLUMPFINDING**, we observe that when we slide our window of length L one symbol to the right, the frequency array does not change much, and so regenerating the frequency array from scratch is inefficient (Figure 1.24). This observation helps us modify **CLUMPFINDING** as shown below. Note that we now only call **COMPUTINGFREQUENCIES** once, updating the frequency array as we go along.

```
BETTERCLUMPFINDING(Genome, k, t, L)
    FrequentPatterns ← an empty set
    for i ← 0 to 4^k − 1
        CLUMP(i) ← 0
    Text ← Genome(0, L)
    FREQUENCYARRAY ← COMPUTINGFREQUENCIES(Text, k)
    for i ← 0 to 4^k − 1
        if FREQUENCYARRAY(i) ≥ t
            CLUMP(i) ← 1
    for i ← 1 to |Genome| − L
        FirstPattern ← Genome(i − 1, k)
        index ← PATTERNTONUMBER(FirstPattern)
        FREQUENCYARRAY(index) ← FREQUENCYARRAY(index) − 1
        LastPattern ← Genome(i + L − k, k)
        index ← PATTERNTONUMBER(LastPattern)
        FREQUENCYARRAY(index) ← FREQUENCYARRAY(index) + 1
        if FREQUENCYARRAY(index) ≥ t
            CLUMP(index) ← 1
    for i ← 0 to 4^k − 1
        if CLUMP(i) = 1
            Pattern ← NUMBERTOPATTERN(i, k)
            add Pattern to the set FrequentPatterns
    return FrequentPatterns
```

k-mer	**AA**	AC	AG	AT	CA	CC	CG	CT	GA	**GC**	GG	GT	TA	TC	TG	TT
INDEX	0	1	2	3	4	5	6	7	8	9	10	11	12	13	14	15
frequency	**3**	0	2	0	1	0	0	0	0	**1**	3	1	0	0	1	0
frequency'	**2**	0	2	0	1	0	0	0	0	**2**	3	1	0	0	1	0

FIGURE 1.24 The frequency arrays for two consecutive substrings of length 13 starting at positions 0 and 1 of *Text* = **AA**GCAAAGGTGG**GC** differ in only two elements corresponding to the first *k*-mer in *Text* (**AA**) and the last *k*-mer in *Text* (**GC**).

Solving the Frequent Words with Mismatches Problem

Note: This Charging Station uses some notation from **Charging Station: The Frequency Array**.

To prevent having to generate all 4^k *k*-mers in order to solve the Frequent Words with Mismatches Problem, our goal is to consider only those *k*-mers that are **close** to a *k*-mer in *Text*, i.e., those with Hamming distance at most *d* from this *k*-mer. Given a *k*-mer *Pattern*, we therefore define its ***d*-neighborhood** NEIGHBORS(*Pattern*, *d*) as the set of all *k*-mers that are close to *Pattern*. For example, NEIGHBORS(ACG, 1) consists of ten 3-mers:

ACG **C**CG **G**CG **T**CG A**A**G A**G**G A**T**G AC**A** AC**C** AC**T**

Exercise Break: Estimate the size of NEIGHBORS(*Pattern*, *d*).

Given integers *k* and *d*, we generalize the concept of the frequency array to account for approximate matches of *k*-mers with up to *d* mismatches. The **frequency array with up to *d* mismatches** of a string *Text* is an array of length 4^k, where the *i*-th element of the array holds the number of times that the *i*-th *k*-mer (in lexicographic order) appears in *Text* with up to *d* mismatches (Figure 1.25). For example, although CT does not occur in AAGCAAAGGTGGG, it appears twice (as C**A** and G**T**) if we allow for one mismatch.

k-mer	AA	AC	AG	AT	CA	CC	CG	CT	GA	GC	GG	GT	TA	TC	TG	TT
index	0	1	2	3	4	5	6	7	8	9	10	11	12	13	14	15
frequency	6	6	9	6	4	2	7	2	9	5	8	5	5	2	6	2

FIGURE 1.25 The lexicographic order of DNA 2-mers (top), along with the index of each 2-mer in this order (middle), and the frequency array with 1 mismatch for the 14 nucleotide-long "genome" AAGCAAAGGTGGG (bottom).

The pseudocode below computes the frequency array with up to d mismatches and differs from **COMPUTINGFREQUENCIES** only by the three lines shown in blue.

```
COMPUTINGFREQUENCIESWITHMISMATCHES(Text, k, d)
    for i ← 0 to 4^k − 1
        FREQUENCYARRAY(i) ← 0
    for i ← 0 to |Text| − k
        Pattern ← Text(i, k)
        Neighborhood ← NEIGHBORS(Pattern, d)
        for each string ApproximatePattern in Neighborhood
            j ← PATTERNTONUMBER(ApproximatePattern)
            FREQUENCYARRAY(j) ← FREQUENCYARRAY(j) + 1
    return FREQUENCYARRAY
```

We can now generalize **FASTERFREQUENTWORDS** into an algorithm for finding frequent words with d mismatches (which we call **FREQUENTWORDSWITHMISMATCHES**) by substituting the call to **COMPUTINGFREQUENCIES** by a call to our new algorithm **COMPUTINGFREQUENCIESWITHMISMATCHES**.

Charging Station (Generating the Neighborhood of a String): **COMPUTINGFREQUENCIESWITHMISMATCHES** calls **NEIGHBORS**(*Pattern*, *d*), a function that generates the *d*-neighborhood of a *k*-mer *Pattern*. Check out this Charging Station to learn how to implement this function.

STOP and Think: Although **FREQUENTWORDSWITHMISMATCHES** is faster than the naive algorithm described in the main text for the typical parameters used in *ori* searches, it is not necessarily faster for all parameter values. For which parameter values is **FREQUENTWORDSWITHMISMATCHES** slower than the naive algorithm?

Charging Station (Finding Frequent Words with Mismatches by Sorting): If you are familiar with sorting and are interested in seeing an even faster algorithm for the Frequent Words with Mismatches Problem, check out this Charging Station.

Generating the neighborhood of a string

Our goal is to generate the d-neighborhood NEIGHBORS$(Pattern, d)$, the set of all k-mers whose Hamming distance from *Pattern* does not exceed d. We will first generate the 1-neigborhood of *Pattern* using the following pseudocode.

```
IMMEDIATENEIGHBORS(Pattern)
    Neighborhood ← the set consisting of the single string Pattern
    for i = 1 to |Pattern|
        symbol ← i-th nucleotide of Pattern
        for each nucleotide x different from symbol
            Neighbor ← Pattern with the i-th nucleotide substituted by x
            add Neighbor to Neighborhood
    return Neighborhood
```

Our idea for generating NEIGHBORS$(Pattern, d)$ is as follows. If we remove the first symbol of *Pattern* (denoted FIRSTSYMBOL(*Pattern*)), then we will obtain a $(k-1)$-mer that we denote by SUFFIX$(Pattern)$.

STOP and Think: If we know NEIGHBORS(SUFFIX$(Pattern), d)$, how does it help us construct NEIGHBORS$(Pattern, d)$?

Now, consider a $(k-1)$-mer *Pattern′* belonging to NEIGHBORS(SUFFIX$(Pattern), d)$. By the definition of the d-neighborhood NEIGHBORS(SUFFIX$(Pattern), d)$, we know that HAMMINGDISTANCE$(Pattern'$, SUFFIX$(Pattern))$ is either equal to d or less than d. In the first case, we can add FIRSTSYMBOL$(Pattern)$ to the beginning of *Pattern′* to obtain a k-mer belonging to NEIGHBORS$(Pattern, d)$. In the second case, we can add any symbol to the beginning of *Pattern′* and obtain a k-mer belonging to NEIGHBORS$(Pattern, d)$.

For example, to generate NEIGHBORS(CAA, 1), first form NEIGHBORS(AA, 1) = {AA, CA, GA, TA, AC, AG, AT}. The Hamming distance between AA and each of six of these neighbors is 1. Firstly, concatenating C with each of these patterns results in six patterns (CAA, CCA, CGA, CTA, CAC, CAG, CAT) that belong to NEIGHBORS(CAA, 1). Secondly, concatenating any nucleotide with AA results in four patterns (AAA, CAA, GAA, and TAA) that belong to NEIGHBORS(CAA, 1). Thus, NEIGHBORS(CAA, 1) comprises ten patterns.

In the following pseudocode, we use the notation *symbol* • *Text* to denote the concatenation of a character *symbol* and a string *Text*, e.g., A • GCATG = AGCATG.

```
NEIGHBORS(Pattern, d)
    if d = 0
        return {Pattern}
    if |Pattern| = 1
        return {A, C, G, T}
    Neighborhood ← an empty set
    SuffixNeighbors ← NEIGHBORS(SUFFIX(Pattern), d)
    for each string Text from SuffixNeighbors
        if HAMMINGDISTANCE(SUFFIX(Pattern), Text) < d
            for each nucleotide x
                add x • Text to Neighborhood
        else
            add FIRSTSYMBOL(Pattern) • Text to Neighborhood
    return Neighborhood
```

STOP and Think: Consider the following questions.

1. What is the running time of **NEIGHBORS**?
2. **NEIGHBORS** generates all *k*-mers of Hamming distance at most *d* from *Pattern*. Modify **NEIGHBORS** to generate all *k*-mers of Hamming distance exactly *d* from *Pattern*.

If you are still learning how recursive algorithms (like **NEIGHBORS**) work, you may want to implement an iterative version of **NEIGHBORS** instead, shown below.

```
ITERATIVENEIGHBORS(Pattern, d)
    Neighborhood ← set consisting of single string Pattern
    for j = 1 to d
        for each string Pattern' in Neighborhood
            add IMMEDIATENEIGHBORS(Pattern') to Neighborhood
            remove duplicates from Neighborhood
    return Neighborhood
```

Finding frequent words with mismatches by sorting

Note: This Charging Station uses some notation from **Charging Station: Finding Frequent Words by Sorting**.

The following pseudocode reduces the Frequent Words with Mismatches Problem to sorting. It first generates all neighbors (with up to d mismatches) for all k-mers in *Text* and combines them all into an array NEIGHBORHOODARRAY. Note that a k-mer *Pattern* appears $\text{COUNT}_d(Text, Pattern)$ times in this array. The only thing left is to sort this array, count how many times each k-mer appears in the sorted array, and then return the k-mers that occur the maximum number of times.

```
FINDINGFREQUENTWORDSWITHMISMATCHESBYSORTING(Text, k, d)
    FrequentPatterns ← an empty set
    Neighborhoods ← an empty list
    for i ← 0 to |Text| − k
        add NEIGHBORS(Text(i, k), d) to Neighborhoods
    form an array NEIGHBORHOODARRAY holding all strings in Neighborhoods
    for i ← 0 to |Neighborhoods| − 1
        Pattern ← NEIGHBORHOODARRAY(i)
        INDEX(i) ← PATTERNTONUMBER(Pattern)
        COUNT(i) ← 1
    SORTEDINDEX ← SORT(INDEX)
    for i ← 0 to |Neighborhoods| − 2
        if SORTEDINDEX(i) = SORTEDINDEX(i + 1)
            COUNT(i + 1) ← COUNT(i) + 1
    maxCount ← maximum value in array COUNT
    for i ← 0 to |Neighborhoods| − 1
        if COUNT(i) = maxCount
            Pattern ← NUMBERTOPATTERN(SORTEDINDEX(i), k)
            add Pattern to FrequentPatterns
    return FrequentPatterns
```

Detours

Big-O notation

Computer scientists typically measure an algorithm's efficiency in terms of its **worst-case running time**, which is the largest amount of time an algorithm can take for the most difficult input of a given size. The advantage to considering the worst-case running time is that we are guaranteed that our algorithm will never behave worse than our worst-case estimate.

Big-O notation compactly describes the running time of an algorithm. For example, if your algorithm for sorting an array of n numbers takes roughly n^2 operations for the most difficult dataset, then we say that the running time of your algorithm is $\mathcal{O}(n^2)$. In reality, depending on your implementation, it may use any number of operations, such as $1.5n^2$, $n^2 + n + 2$, or $0.5n^2 + 1$; all these algorithms are $\mathcal{O}(n^2)$ because big-O notation only cares about the term that grows the fastest with respect to the size of the input. This is because as n grows very large, the difference in behavior between two $\mathcal{O}(n^2)$ functions, like $999 \cdot n^2$ and $n^2 + 3n + 9999999$, is negligible when compared to the behavior of functions from different classes, say $\mathcal{O}(n^2)$ and $\mathcal{O}(n^6)$. Of course, we would prefer an algorithm requiring $1/2 \cdot n^2$ steps to an algorithm requiring $1000 \cdot n^2$ steps.

When we write that the running time of an algorithm is $\mathcal{O}(n^2)$, we technically mean that it does not grow faster than a function with a leading term of $c \cdot n^2$, for some constant c. Formally, a function $f(n)$ is Big-O of function $g(n)$, or $\mathcal{O}(g(n))$, when $f(n) \leq c \cdot g(n)$ for some constant c and sufficiently large n.

Probabilities of patterns in a string

We mentioned that the probability that some 9-mer appears 3 or more times in a random DNA string of length 500 is approximately 1/1300. We assure you that this calculation does not appear out of thin air. Specifically, we can generate a **random string** modeling a DNA strand by choosing each nucleotide for any position with probability 1/4. The construction of random strings can be generalized to an arbitrary alphabet with A symbols, where each symbol is chosen with probability $1/A$.

Exercise Break: What is the probability that two randomly generated strings of length n in an A-letter alphabet are identical?

We now ask a simple question: what is the probability that a specific k-mer *Pattern* will appear (at least once) as a substring of a random string of length N? For example, say that we want to find the probability that "01" appears in a random **binary string** ($A = 2$) of length 4. Here are all possible such strings.

0000 0001 0010 0011 0100 0101 0110 0111
1000 1001 1010 1011 1100 1101 1110 1111

Because "01" is a substring of 11 of these 4-mers, and because each 4-mer could be generated with probability 1/16, the probability that "01" appears in a random binary 4-mer is 11/16.

STOP and Think: What is the probability that *Pattern* = "11" appears as a substring of a random binary 4-mer?

Surprisingly, changing *Pattern* from "01" to "11" changes the probability that it appears as a substring of a random binary string. Indeed, "**11**" appears in only 8 binary 4-mers:

0000 0001 0010 0011 0100 0101 0110 0111
1000 1001 1010 1011 1100 1101 1110 1111

As a result, the probability of "11" appearing in a random binary string of length 4 is $8/16 = 1/2$.

STOP and Think: Why do you think that "11" is less likely than "01" to appear as a substring of a random binary 4-mer?

Let $\text{Pr}(N, A, Pattern, t)$ denote the probability that a string *Pattern* appears t or more times in a random string of length N formed from an alphabet of A letters. We saw that $\text{Pr}(4, 2, \text{"01"}, 1) = 11/16$ while $\text{Pr}(4, 2, \text{"11"}, 1) = 1/2$. Interestingly, when we make t greater than 1, we see that "01" is *less* likely to appear multiple times than "11". For example, the probability of finding "01" twice or more in a random binary 4-mer is given by $\text{Pr}(4, 2, \text{"01"}, 2) = 1/16$ because "0101" is the only binary 4-mer containing "01" twice, and yet $\text{Pr}(4, 2, \text{"11"}, 2) = 3/16$ because binary 4-mers "0111", "1110" and "1111" all have at least two occurrences of "11".

Exercise Break: Compute $\text{Pr}(100, 2, \text{"01"}, 1)$.

We have seen that different k-mers have different probabilities of occurring multiple times as a substring of a random string. In general, this phenomenon is called the overlapping words paradox because different substring occurrences of *Pattern* can overlap each other for some choices of *Pattern* but not others (see **DETOUR: The Overlapping Words Paradox**). PAGE 63

For example, there are two overlapping occurrences of "11" in "1110", and three overlapping occurrences of "11" in "1111"; yet occurrences of "01" can never overlap with each other, and so "01" can never occur more than twice in a binary 4-mer. The overlapping words paradox makes computing $\Pr(N, A, Pattern, t)$ a rather complex problem because this probability depends heavily on the particular choice of *Pattern*. In light of the complications presented by the overlapping words paradox, we will try to *approximate* $\Pr(A, N, Pattern, t)$ rather than compute it exactly.

To approximate $\Pr(N, A, Pattern, t)$, we will assume that the k-mer *Pattern* is not overlapping. As a toy example, say we we wish to count the number of **ternary strings** ($A = 3$) of length 7 that contain "01" at least twice. Apart from the two occurrences of "01", we have three remaining symbols in the string. Let's assume that these symbols are all "2". The two occurrences of "01" can be inserted into "222" in ten different ways to form a 7-mer, as shown below.

0101222 0120122 0122012 0122201 2010122
2012012 2012201 2201012 2201201 2220101

We inserted these two occurrences of "01" into "222", but we could have inserted them into any other ternary 3-mer. Because there are $3^3 = 27$ ternary 3-mers, we obtain an approximation of $10 \cdot 27 = 270$ for the number of ternary 7-mers that contain two or more instances of "01". Because there are $3^7 = 2187$ ternary 7-mers, we estimate the probability $\Pr(7, 3, \text{"01"}, 2)$ as $270/2187$.

STOP and Think: Is $270/2187$ a good approximation for $\Pr(7, 3, \text{"01"}, 2)$? Is the true probability $\Pr(7, 3, \text{"01"}, 2)$ larger or smaller than $270/2187$?

To generalize the above method to approximate $\Pr(N, A, Pattern, t)$ for arbitrary parameter values, consider a string *Text* of length N having at least t occurrences of a k-mer *Pattern*. If we select exactly t of these occurrences, then we can think about *Text* as a sequence of $n = N - t \cdot k$ symbols interrupted by t insertions of the k-mer *Pattern*. If we fix these n symbols, then we wish to count the number of different strings *Text* that can be formed by inserting t occurrences of *Pattern* into a string formed by these n symbols.

For example, consider again the problem of embedding two occurrences of "01" into "222" ($n = 3$), and note that we have added five copies of a capital "X" below each 7-mer.

0101222	0120122	0122012	0122201	2010122
X X XXX	X XX XX	X XXX X	X XXXX	XX X XX
2012012	2012201	2201012	2201201	2220101
XX XX X	XX XXX	XXX X X	XXX XX	XXXX X

What do the "X" mean? Instead of counting the number of ways to insert two occurrences of "01" into "222", we can count the number of ways to select two of the five "X" to color blue.

XXXXX XXXXX XXXXX XXXXX XXXXX

XXXXX XXXXX XXXXX XXXXX XXXXX

In other words, we are counting the number of ways to choose 2 out of 5 objects, which can be counted by the **binomial coefficient** $\binom{5}{2} = 10$. More generally, the binomial coefficient $\binom{m}{k}$ represents the number of ways to choose k out of m objects and is equal to $m!/k!(m-k)!$

STOP and Think: How many ways are there to implant t instances of a (nonoverlapping) k-mer into a string of length n to produce a string of length $n + t \cdot k$?

To approximate $\Pr(N, A, Pattern, t)$, we want to count the number of ways to insert t instances of a k-mer *Pattern* into a fixed string of length $n = N - t \cdot k$. We will therefore have $n + t$ occurrences of "X", from which we must select t for the placements of *Pattern*, giving a total of $\binom{n+t}{t}$. We then need to multiply $\binom{n+t}{t}$ by the number of strings of length n into which we can insert t instances of *Pattern* to have an approximate total of $\binom{n+t}{t} \cdot A^n$ (the actual number will be smaller because of over-counting). Dividing by the number of strings of length N, we have our desired approximation,

$$\Pr(N, A, Pattern, t) \approx \frac{\binom{n+t}{t} \cdot A^n}{A^N} = \frac{\binom{N-t\cdot k+t}{t} \cdot A^{N-t\cdot k}}{A^N} = \frac{\binom{N-t\cdot(k-1)}{t}}{A^{t\cdot k}}.$$

We will now compute the probability that the specific 5-mer ACTAT occurs at least $t = 3$ times in a random DNA string ($A = 4$) of length $N = 30$. Since $n = N - t \cdot k = 15$, our estimated probability is

$$\Pr(30, 4, \texttt{ACTAT}, 3) \approx \frac{\binom{30-3\cdot 4}{3}}{4^{15}} = \frac{816}{1073741824} \approx 7.599 \cdot 10^{-7}.$$

The exact probability is closer to $7.572 \cdot 10^{-7}$, illustrating that our approximation is relatively accurate for non-overlapping patterns. However, it becomes inaccurate for overlapping patterns, e.g., $\Pr(30, 4, \texttt{AAAAA}, 3) \approx 1.148 \cdot 10^{-3}$.

We should not be surprised that the probability of finding ACTAT in a random DNA string of length 30 is so low. However, remember that our original goal was to approximate the probability that there exists *some* 5-mer appearing three or more times. In general, the probability that some k-mer appears t or more times in a random string of length N formed over an A-letter alphabet is written $\Pr(N, A, k, t)$.

We approximated $\Pr(N, A, \mathit{Pattern}, t)$ as

$$p = \frac{\binom{N - t\cdot(k-1)}{t}}{A^{t\cdot k}}.$$

Notice that the approximate probability that *Pattern* does *not* appear t or more times is therefore $1 - p$. Thus, the probability that *all* A^k patterns appear fewer than t times in a random string of length N can be approximated as

$$(1-p)^{A^k}.$$

Moreover, the probability that there exists a k-mer appearing t or more times should be 1 minus this value, which gives us the following approximation:

$$\Pr(N, A, k, t) \approx 1 - (1-p)^{A^k}.$$

Your calculator may have difficulty with this formula, which requires raising a number close to 1 to a very large power and can cause round-off errors. To avoid this, if we assume that p is about the same for any *Pattern*, then we can approximate $\Pr(N, A, k, t)$ by multiplying p by the total number of k-mers A^k,

$$\Pr(N, A, k, t) \approx p \cdot A^k = \frac{\binom{N - t\cdot(k-1)}{t}}{A^{t\cdot k}} \cdot A^k = \frac{\binom{N - t\cdot(k-1)}{t}}{A^{(t-1)\cdot k}}.$$

We acknowledge again that this approximation is a gross over-simplification, since the probability $\Pr(N, A, \mathit{Pattern}, t)$ varies across different choices of k-mers and because it assumes that occurrences of different k-mers are independent events. For example, in the main text, we wish to approximate $\Pr(500, 4, 9, 3)$, and the above formula results in the approximation

$$\Pr(500, 4, 9, 3) \approx \frac{\binom{500-3\cdot 8}{3}}{4^{(3-1)\cdot 9}} = \frac{17861900}{68719476736} \approx \frac{1}{3847}.$$

Because of overlapping strings, this approximation deviates from the true value of $\Pr(500, 4, 9, 3)$, which is closer to $1/1300$.

The most beautiful experiment in biology

The Meselson-Stahl experiment, conducted in 1958 by Matthew Meselson and Franklin Stahl, is sometimes called "the most beautiful experiment in biology". In the late 1950s, biologists debated three conflicting models of DNA replication, illustrated in Figure 1.26. The **semiconservative hypothesis** (recall Figure 1.1 from page 3), suggested that each parent strand acts as a template for the synthesis of a daughter strand. As a result, each of the two daughter molecules contains one parent strand and one newly synthesized strand. The **conservative hypothesis** proposed that the entire double-stranded parent DNA molecule serves as a template for the synthesis of a new daughter molecule, resulting in one molecule with two parent strands and another with two newly synthesized strands. The **dispersive hypothesis** proposed that some mechanism breaks the DNA backbone into pieces and splices intervals of synthesized DNA, so that each of the daughter molecules is a patchwork of old and new double-stranded DNA.

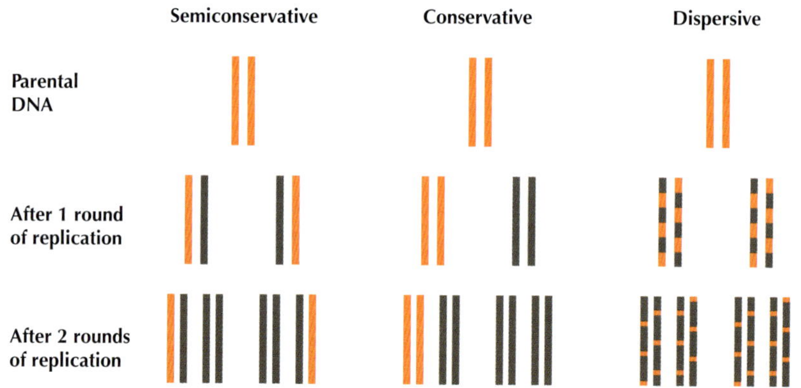

FIGURE 1.26 Semiconservative, conservative, and dispersive models of DNA replication make different predictions about the distribution of DNA strands after replication. Yellow strands indicate ^{15}N (heavy) segments of DNA, and black strands indicate ^{14}N (light) segments. The Meselson-Stahl experiment began with DNA consisting of 100% ^{15}N.

Meselson and Stahl's insight was that one isotope of nitrogen, **Nitrogen-14** (^{14}N), is lighter and more abundant than **Nitrogen-15** (^{15}N). Knowing that DNA naturally contains ^{14}N , Meselson and Stahl grew *E. coli* for many rounds of replication in a ^{15}N medium, which caused the bacteria to gain weight as they absorbed the heavier isotope into their DNA. When they were confident that the bacterial DNA was saturated with ^{15}N, they transferred the heavy *E. coli* cells to a less dense ^{14}N medium.

STOP and Think: What do you think happened when the "heavy" *E. coli* replicated in the "light" ^{14}N medium?

The brilliance of the Meselson-Stahl experiment is that all newly synthesized DNA would contain exclusively ^{14}N, and the three existing hypotheses for DNA replication predicted different outcomes for how this ^{14}N isotope would be incorporated into DNA. Specifically, after one round of replication, the conservative model predicted that half the *E. coli* DNA would still have only ^{15}N and therefore be heavier whereas the other half would have only ^{14}N and be lighter. Yet when they attempted to separate the *E. coli* DNA according to weight by using a centrifuge after one round of replication, all of the DNA had the same density! Just like that, they had refuted the conservative hypothesis once and for all.

Unfortunately, this experiment was not able to eliminate either of the other two models, as both the dispersive and semiconservative hypotheses predicted that all of the DNA after one round of replication would have the same density.

STOP and Think: What would the dispersive and semiconservative models predict about the density of *E. coli* DNA after two rounds of replication?

Let's first consider the dispersive model, which says that each daughter strand of DNA is formed by half mashed up pieces of the parent strand, and half new DNA. If this hypothesis were true, then after two replication cycles, any daughter strand of DNA should contain about 25% ^{15}N and about 75% ^{14}N. In other words, all the DNA should still have the same density. And yet when Meselson and Stahl spun the centrifuge after two rounds of *E. coli* replication, this is not what they observed!

Instead, they found that the DNA divided into two different densities. This is exactly what the semiconservative model predicted: after one cycle, every cell should possess one ^{14}N strand and one ^{15}N strand; after two cycles, half of the DNA molecules should have one ^{14}N strand and one ^{15}N strand, while the other half should have two ^{14}N strands, producing the two different densities they noticed.

STOP and Think: What does the semi-conservative model predict about the density of *E. coli* DNA after three rounds of replication?

Meselson and Stahl had rejected the conservative and dispersive hypotheses of replication, and yet they wanted to make sure that the semiconservative hypothesis was confirmed by further *E. coli* replication. This model predicted that after three rounds of replication, one-quarter of the DNA molecules should still have a ^{15}N strand, causing 25% of the DNA to have an intermediate density, whereas the remaining 75% should be lighter, having only ^{14}N. This is indeed what Meselson and Stahl witnessed in the lab, and the semiconservative hypothesis has stood strong to this day.

Directionality of DNA strands

The sugar component of a nucleotide has a ring of five carbon atoms, which are labeled as 1′, 2′, 3′, 4′, and 5′ in Figure 1.27 (left). The 5′ atom is joined onto the phosphate group in the nucleotide and eventually to the 3′ end of the neighboring nucleotide. The 3′ atom is joined onto another neighboring nucleotide in the nucleic acid chain. As a result, we call the two ends of the nucleotide the **5′-end** and the **3′-end** (pronounced "five prime end" and "three prime end", respectively).

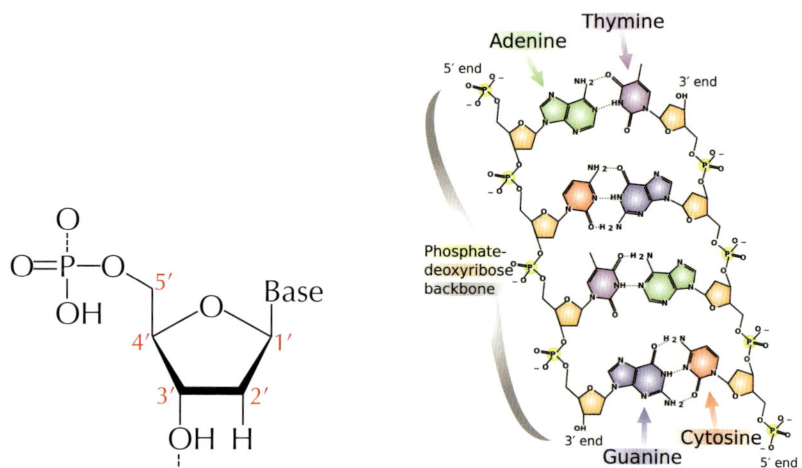

FIGURE 1.27 A nucleotide with sugar ring carbon atoms labeled 1′, 2′, 3′, 4′, and 5′.

When we zoom out to the level of the double helix, we can see in Figure 1.27 (right) that any DNA fragment is oriented with a 3′ atom on one end and a 5′ atom on the other end. As a standard, a DNA strand is always read in the $5' \to 3'$ direction. Note that the orientations run opposite to each other in complementary strands.

The Towers of Hanoi

The **Towers of Hanoi** puzzle consists of three vertical pegs and a number of disks of different sizes, each with a hole in its center so that it fits on the pegs. The disks are initially stacked on the left peg (peg 1) so that disks increase in size from the top down (Figure 1.28). The puzzle is played by moving one disk at a time between pegs, with the goal of moving all disks from the left peg (peg 1) to the right peg (peg 3). However, you are not allowed to place a disk on top of a smaller disk.

FIGURE 1.28 The Towers of Hanoi puzzle.

Towers of Hanoi Problem:
Solve the Towers of Hanoi puzzle.

> **Input**: An integer n.
> **Output**: A sequence of moves that will solve the Towers of Hanoi puzzle with n disks.

STOP and Think: What is the minimum number of steps needed to solve the Towers of Hanoi Problem for three disks?

STOP

Let's see how many steps are required to solve the Towers of Hanoi Problem for four disks. The first important observation is that sooner or later you will have to move the largest disk to the right peg. However, in order to move the largest disk, we first have to move all three smallest disks off the first peg. Furthermore, these three smallest disks must all be on the same peg because the largest disk cannot be placed on top of another disk. Thus, we first have to move the top three disks to the middle peg (7 moves), then move the largest disk to the right peg (1 move), then again move the three smallest disks from the middle peg to the top of the largest disk on the right peg (another 7 moves), for a total of 15 moves.

More generally, let $T(n)$ denote the minimum number of steps required to solve the Towers of Hanoi puzzle with n disks. To move n disks from the left peg to the right peg, you first need to move the $n-1$ smallest disks from the left peg to the middle peg ($T(n-1)$ steps), then move the largest disk to the right peg (1 step), and finally move the $n-1$ smallest disks from the middle peg to the right peg ($T(n-1)$ steps). This yields the recurrence relation

$$T(n) = 2T(n-1) + 1\,.$$

STOP and Think: Using the above recurrence relation, can you find a formula for $T(n)$ that does not require recursion?

We now have a recursive algorithm to move n disks from the left peg to the right peg. We will use three variables (each taking a different value from 1, 2, and 3) to denote the three pegs: *startPeg*, *destinationPeg*, and *transitPeg*. These three variables always represent different pegs, and so *startPeg* + *destinationPeg* + *transitPeg* is always equal to $1+2+3=6$. **HANOITOWERS**(*n*, *startPeg*, *destinationPeg*) moves *n* disks from *startPeg* to *destinationPeg* (using *transitPeg* as a temporary destination).

```
HANOITOWERS(n, startPeg, destinationPeg)
    if n = 1
        Move top disk from startPeg to destinationPeg
        return
    transitPeg = 6 – startPeg – destinationPeg
    HANOITOWERS(n – 1, startPeg, transitPeg)
    Move top disk from startPeg to destinationPeg
    HANOITOWERS(n – 1, transitPeg, destinationPeg)
    return
```

Even though this algorithm may seem straightforward, moving a 100-disk tower would require more steps than the number of atoms in the universe! The fast growth of the number of moves required by **HANOITOWERS** is explained by the fact that every time

HanoiTowers is called for n disks, it calls itself twice for $n-1$, which in turn triggers four calls for $n-2$, and so on. For example, a call to **HanoiTowers**$(4,1,3)$ results in calls **HanoiTowers**$(3,1,2)$ and **HanoiTowers**$(3,2,3)$; these calls, in turn, call **HanoiTowers**$(2,1,3)$, **HanoiTowers**$(2,3,2)$, **HanoiTowers**$(2,2,1)$, and **HanoiTowers**$(2,1,3)$.

The overlapping words paradox

We illustrate the overlapping words paradox with a two-player game called "Best Bet for Simpletons". Player 1 selects a binary k-mer A, and Player 2, knowing what A is, selects a different binary k-mer B. The two players then flip a coin multiple times, with coin flips represented by strings of "1" ("heads") and "0" ("tails"); the game ends when A or B appears as a block of k consecutive coin flips.

STOP and Think: Do the two players always have the same chance of winning?

STOP

At first glance, you might guess that every k-mer has an equal chance of winning. Yet suppose that Player 1 chooses "00" and Player 2 chooses "10". After two flips, either Player 1 wins ("00"), Player 2 wins ("10"), or the game continues ("01" or "11"). If the game continues, then Player 1 should surrender, since Player 2 will win as soon as "tails" ("0") is next flipped. Player 2 is therefore three times more likely to win!

It may seem that Player 1 should have the advantage by simply selecting the "strongest" k-mer. However, an intriguing feature of Best Bet for Simpletons is that if $k > 2$, then Player 2 can always choose a k-mer B that beats A, *regardless* of Player 1's choice of A. Another surprise is that Best Bet for Simpletons is a **non-transitive game**: if A defeats B, and B defeats C, then we cannot automatically conclude that A defeats C (c.f. rock-paper-scissors).

To analyze Best Bet for Simpletons, we say that B ***i*-overlaps** with A if the last i digits of A coincide with the first i digits of B. For example, "110110" 1-overlaps, 2-overlaps, and 5-overlaps with "011011", as shown in Figure 1.29.

Given two k-mers A and B, the **correlation** of A and B, denoted $\text{Corr}(A,B) = (c_0, \ldots, c_{k-1})$, is a k-letter binary word such that $c_i = 1$ if B $(k-i)$-overlaps with A, and 0 otherwise. The **correlation frequency** of A and B is defined as

$$K_{A,B} = c_0 + c_1 \cdot \frac{1}{2} + c_2 \cdot \left(\frac{1}{2}\right)^2 + \cdots + c_{k-1} \cdot \left(\frac{1}{2}\right)^{k-1}.$$

For the strings A and B in Figure 1.29, their correlation is "010011" and their correlation frequency is $K_{A,B} = (1/2) + (1/2)^4 + (1/2)^5 = 19/32$.

	$\text{CORR}(A, B)$
B = 110110	**0**
B = **11011**0	**1**
B = 110110	**0**
B = 110110	**0**
B = **110**110	**1**
B = **1**10110	**1**
A = **011011**	

FIGURE 1.29 The correlation of k-mers A = "011011" and B = "110110" is the string "**010011**".

John Conway suggested the following deceivingly simple formula to compute the odds that B will defeat A:

$$\frac{K_{A,A} - K_{A,B}}{K_{B,B} - K_{B,A}}$$

Conway never published a proof of this formula, and Martin Gardner, a leading popular mathematics writer, said the following about the formula:

> *I have no idea why it works. It just cranks out the answer as if by magic, like so many of Conway's other algorithms.*

Bibliography Notes

Using the skew to find replication origins was first proposed by Lobry, 1996 and also described in Grigoriev, 1998. Grigoriev, 2011 provides an excellent introduction to the skew approach, and Sernova and Gelfand, 2008 gave a review of algorithms and software tools for finding replication origins in bacteria. Lundgren et al., 2004 demonstrated that archaea may have multiple *oris*. Wang et al., 2011 inserted an artificial *ori* into the *E. coli* genome and showed that it triggers replication. Xia, 2012 was the first to conjecture that bacteria may have multiple replication origins. Gao and Zhang, 2008 developed the Ori-Finder software program for finding bacterial replication origins.

Liachko et al., 2013 provided the most comprehensive description of the replication origins of yeast. Solov'ev, 1966 was the first to derive accurate formulas for approximating the probabilities of patterns in a string. Gardner, 1974 wrote an excellent introductory article about the Best Bet for Simpletons paradox. Guibas and Odlyzko, 1981 provided an excellent coverage of the overlapping words paradox that illustrates the complexity of computing the probabilities of patterns in a random text. They also derived a rather complicated proof of Conway's formula for Best Bet for Simpletons. Sedgewick and Flajolet, 2013 gave an overview of various approaches for computing the probabilities of patterns in a string.

WHICH DNA PATTERNS PLAY THE ROLE OF MOLECULAR CLOCKS?

Randomized Algorithms

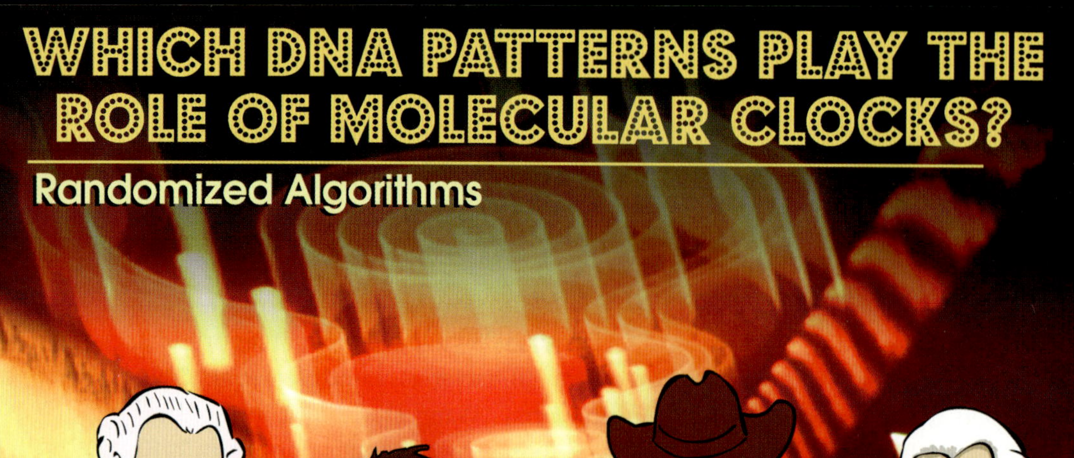

Do We Have a "Clock" Gene?

The daily schedules of animals, plants, and even bacteria are controlled by an internal timekeeper called the **circadian clock**. Anyone who has experienced the misery of jet lag knows that this clock never stops ticking. Rats and research volunteers alike, when placed in a bunker, naturally maintain a roughly 24-hour cycle of activity and rest in total darkness. And, like any timepiece, the circadian clock can malfunction, resulting in a genetic disease known as **delayed sleep-phase syndrome (DSPS)**.

The circadian clock must have some basis on the molecular level, which presents many questions. How do *individual cells* in animals and plants (let alone bacteria) know when they should slow down or increase the production of certain proteins? Is there a "clock gene"? Can we explain why heart attacks occur more often in the morning, while asthma attacks are more common at night? And can we identify genes that are responsible for "breaking" the circadian clock to cause DSPS?

In the early 1970s, Ron Konopka and Seymour Benzer identified mutant flies with abnormal circadian patterns and traced the flies' mutations to a single gene. Biologists needed two more decades to discover a similar clock gene in mammals, which was just the first piece of the puzzle. Today, many more circadian genes have been discovered; these genes, having names like *timeless*, *clock*, and *cycle*, orchestrate the behavior of hundreds of other genes and display a high degree of evolutionary conservation across species.

We will first focus on plants, since maintaining the circadian clock in plants is a matter of life and death. Consider how many plant genes should pay attention to the time when the sun rises and sets; indeed, biologists estimate that over a thousand plant genes are circadian, including the genes related to photosynthesis, photo reception, and flowering. These genes must somehow know what time it is in order to change their gene transcript production, or **gene expression**, throughout the day (see **DETOUR: Gene Expression**). PAGE 109

It turns out that every plant cell keeps track of day and night independently of other cells, and that just three plant genes, called LHY, CCA1, and TOC1, are the clock's master timekeepers. Such regulatory genes, and the **regulatory proteins** that they encode, are often controlled by external factors (e.g., nutrient availability or sunlight) in order to allow organisms to adjust their gene expression.

For example, regulatory proteins controlling the circadian clock in plants coordinate circadian activity as follows. TOC1 promotes the expression of LHY and CCA1, whereas LHY and CCA1 repress the expression of TOC1, resulting in a **negative feedback loop**. In the morning, sunlight activates the transcription of LHY and CCA1, triggering the

repression of TOC1 transcription. As light diminishes, so does the production of LHY and CCA1, which in turn do not repress TOC1 any more. Transcription of TOC1 peaks at night and starts promoting the transcription of LHY and CCA1, which in turn repress the transcription of TOC1, and the cycle begins again.

LHY, CCA1, and TOC1 are able to control the transcription of other genes because the regulatory proteins that they encode are **transcription factors**, or master regulatory proteins that turn other genes on and off. A transcription factor regulates a gene by binding to a specific short DNA interval called a **regulatory motif**, or **transcription factor binding site**, in the gene's **upstream region**, a 600-1000 nucleotide-long region preceding the start of the gene. For example, CCA1 binds to AAAAAATCT in the upstream region of many genes regulated by CCA1.

The life of a bioinformatician would be easy if regulatory motifs were completely conserved, but the reality is more complex, as regulatory motifs may vary at some positions, e.g., CCA1 may instead bind to AA**G**AA**C**TCT. But how can we locate these regulatory motifs without knowing what they look like in advance? We need to develop algorithms for **motif finding**, the problem of discovering a "hidden message" shared by a collection of strings.

Motif Finding Is More Difficult Than You Think

Identifying the evening element

In 2000, Steve Kay used **DNA arrays** (see **DETOUR: DNA Arrays**) to determine which genes in the plant *Arabidopsis thaliana* are activated at different times of the day. He then extracted the upstream regions of nearly 500 genes that exhibited circadian behavior and looked for frequently appearing patterns in their upstream regions. If you concatenated these upstream regions into a single string, you would find that AAAATATCT is a surprisingly frequent word, appearing 46 times.

STOP and Think: What is the possible downside of concatenating all the upstream regions into a single string and looking for frequent words in order to identify a motif?

Exercise Break: What is the expected number of occurrences of a 9-mer in 500 random DNA strings, each of length 1000?

Kay named `AAAATATCT` the **evening element** and performed a simple experiment to prove that it is indeed the regulatory motif responsible for circadian gene expression in *Arabidopsis thaliana*. After he mutated the evening element in the upstream region of one gene, the gene no longer exhibited circadian behavior.

Whereas the evening element in plants is very conserved, and thus easy to find, motifs having many mutations are more elusive. For example, if you infect a fly with a bacterium, the fly will switch on its **immunity genes** to fight the infection. Thus, some of the genes with elevated expression levels after the infection are likely to be immunity genes. Indeed, some of these genes have 12-mers similar to **TCGGGGATTTCC** in their upstream regions, the binding site of a transcription factor called **NF-ϰB** that activates various immunity genes in flies. However, NF-ϰB binding sites are nowhere near as conserved as the evening element. Figure 2.1 shows ten NF-ϰB binding sites from the *Drosophila melanogaster* genome; the most popular nucleotides in every column are shown by upper case colored letters.

```
 1   T C G G G G g T T T t t
 2   c C G G t G A c T T a C
 3   a C G G G G A T T T t C
 4   T t G G G G A c T T t t
 5   a a G G G G A c T T C C
 6   T t G G G G A c T T C C
 7   T C G G G G A T T c a t
 8   T C G G G G A T T c C t
 9   T a G G G G A a c T a C
10   T C G G G t A T a a C C
```

FIGURE 2.1 The ten candidate NF-ϰB binding sites appearing in the *Drosophila melanogaster* genome. The upper case colored letters indicate the most frequent nucleotide in each column.

Hide and seek with motifs

Our aim is to turn the biological challenge of finding regulatory motifs into a computational problem. Below, we have implanted a 15-mer hidden message at a randomly selected position in each of ten randomly generated DNA strings. This example mimics a transcription factor binding site hiding in the upstream regions of ten genes.

```
1   atgaccgggatactgataaaaaaaagggggggggcgtacacattagataaacgtatgaagtacgttagactcggcgccgccg
2   acccctatttttgagcagatttagtgacctggaaaaaaaatttgagtacaaaacttttccgaataaaaaaaaaggggggga
3   tgagtatccctgggatgacttaaaaaaaagggggggtgctctcccgattttgaatatgtaggatcattcgccagggtccga
4   gctgagaattggatgaaaaaaaagggggggtccacgcaatcgcgaaccaacgcggacccaaaggcaagaccgataaaggaga
5   tcccttttgcggtaatgtgccgggaggctggttacgtagggaagccctaacggacttaataaaaaaaagggggggcttatag
6   gtcaatcatgttcttgtgaatggatttaaaaaaaaggggggggaccgcttggcgcacccaaattcagtgtgggcgagcgcaa
7   cggttttggcccttgttagaggcccccgtaaaaaaaagggggggcaattatgagagagctaatctatcgcgtgcgtgttcat
8   aacttgagttaaaaaaaagggggggctggggcacatacaagaggagtcttccttatcagttaatgctgtatgacactatgta
9   ttggcccattggctaaaagcccaacttgacaatggaagatagaatccttgcataaaaaaaaggggggggaccgaaagggaag
10  ctggtgagcaacgacagattcttacgtgcattagctcgcttccggggatctaatagcacgaagcttaaaaaaaaggggggga
```

STOP and Think: Can you find the implanted hidden message?

This is a simple problem: applying an algorithm for the Frequent Words Problem to the concatenation of these strings will immediately reveal the most frequent 15-mer shown below as the implanted pattern. Since these short strings were randomly generated, it is unlikely that they contain other frequent 15-mers.

```
1   atgaccgggatactgatAAAAAAAAGGGGGGGggcgtacacattagataaacgtatgaagtacgttagactcggcgccgccg
2   acccctatttttgagcagatttagtgacctggaaaaaaaatttgagtacaaaacttttccgaataAAAAAAAAGGGGGGGa
3   tgagtatccctgggatgacttAAAAAAAAGGGGGGGtgctctcccgattttgaatatgtaggatcattcgccagggtccga
4   gctgagaattggatgAAAAAAAAGGGGGGGtccacgcaatcgcgaaccaacgcggacccaaaggcaagaccgataaaggaga
5   tcccttttgcggtaatgtgccgggaggctggttacgtagggaagccctaacggacttaatAAAAAAAAGGGGGGGcttatag
6   gtcaatcatgttcttgtgaatggatttAAAAAAAAGGGGGGGaccgcttggcgcacccaaattcagtgtgggcgagcgcaa
7   cggttttggcccttgttagaggcccccgtAAAAAAAAGGGGGGGcaattatgagagagctaatctatcgcgtgcgtgttcat
8   aacttgagttAAAAAAAAGGGGGGGctggggcacatacaagaggagtcttccttatcagttaatgctgtatgacactatgta
9   ttggcccattggctaaaagcccaacttgacaatggaagatagaatccttgcatAAAAAAAAGGGGGGGaccgaaagggaag
10  ctggtgagcaacgacagattcttacgtgcattagctcgcttccggggatctaatagcacgaagcttAAAAAAAAGGGGGGGa
```

Now imagine that instead of implanting exactly the same pattern into all sequences, we mutate the pattern before inserting it into each sequence by randomly changing the nucleotides at four randomly selected positions within each implanted 15-mer, as shown below.

```
1   atgaccgggatactgatAgAAgAAAGGttGGGggcgtacacattagataaacgtatgaagtacgttagactcggcgccgccg
2   acccctatttttgagcagatttagtgacctggaaaaaaaatttgagtacaaaacttttccgaatacAAtAAAAcGGcGGGa
3   tgagtatccctgggatgacttAAAAtAAtGGaGtGGtgctctcccgattttgaatatgtaggatcattcgccagggtccga
4   gctgagaattggatgcAAAAAAAGGGattGtccacgcaatcgcgaaccaacgcggacccaaaggcaagaccgataaaggaga
5   tcccttttgcggtaatgtgccgggaggctggttacgtagggaagccctaacggacttaatAtAAtAAAGGaaGGGcttatag
6   gtcaatcatgttcttgtgaatggatttAAcAAtAAGGGctGGaccgcttggcgcacccaaattcagtgtgggcgagcgcaa
7   cggttttggcccttgttagaggcccccgtAtAAAcAAGGaGGGccaattatgagagagctaatctatcgcgtgcgtgttcat
8   aacttgagttAAAAAAtAGGGaGccctggggcacatacaagaggagtcttccttatcagttaatgctgtatgacactatgta
9   ttggcccattggctaaaagcccaacttgacaatggaagatagaatccttgcatActAAAAAGGaGcGGaccgaaagggaag
10  ctggtgagcaacgacagattcttacgtgcattagctcgcttccggggatctaatagcacgaagcttActAAAAAGGaGcGGa
```

The Frequent Words Problem is not going to help us, since **AAAAAAAAGGGGGGG** does not even appear in the sequences above. Perhaps, then, we could apply our solution to the Frequent Words with Mismatches Problem. However, in Chapter 1, we implemented an algorithm for the Frequent Words with Mismatches Problem aimed at finding hidden messages with a small number of mismatches and a small *k*-mer size (e.g., one or two mismatches for *DnaA* boxes of length 9). This algorithm will become too slow when searching for the implanted motif above, which is longer and has more mutations.

Furthermore, concatenating all the sequences into a single string is inadequate because it does not correctly model the biological problem of motif finding. A *DnaA* box is a pattern that clumps, or appears frequently, within a relatively short interval of the genome. In contrast, a regulatory motif is a pattern that appears at least once (perhaps with variation) in each of many different regions that are scattered throughout the genome.

A brute force algorithm for motif finding

Given a collection of strings *Dna* and an integer d, a k-mer is a **(k, d)-motif** if it appears in every string from *Dna* with at most d mismatches. For example, the implanted 15-mer in the strings above represents a $(15, 4)$-motif.

Implanted Motif Problem:
Find all (k, d)-motifs in a collection of strings.

> **Input**: A collection of strings *Dna*, and integers k and d.
> **Output**: All (k, d)-motifs in *Dna*.

Brute force search (also known as **exhaustive search**) is a general problem-solving technique that explores all possible candidate solutions and checks whether each candidate solves the problem. Such algorithms require little effort to design and are guaranteed to produce a correct solution, but they may take an enormous amount of time, and the number of candidates may be too large to check.

A brute force approach for solving the Implanted Motif Problem is based on the observation that any (k, d)-motif must be at most d mismatches apart from some k-mer appearing in the first string in *Dna*. Therefore, we can generate all such k-mers and then check which of them are (k, d)-motifs. If you have forgotten how to generate these k-mers, recall **Charging Station: Generating the Neighborhood of a String**.

```
MOTIFENUMERATION(Dna, k, d)
    Patterns ← an empty set
    for each k-mer Pattern in the first string in Dna
        for each k-mer Pattern′ differing from Pattern by at most d mismatches
            if Pattern′ appears in each string from Dna with at most d mismatches
                add Pattern′ to Patterns
    remove duplicates from Patterns
    return Patterns
```

MOTIFENUMERATION is unfortunately rather slow for large values of k and d, and so we will try a different approach instead. Maybe we can detect an implanted pattern simply by identifying the two most similar k-mers between each pair of strings in *Dna*? However, consider the implanted 15-mers **AgAAgAAAGGttGGG** and **cAAtAAAAcGGGGcG**, each of which differs from **AAAAAAAAGGGGGGG** by four mismatches. Although these 15-mers look similar to the correct motif **AAAAAAAAGGGGGGG**, they are not so similar when compared to each other, having eight mismatches:

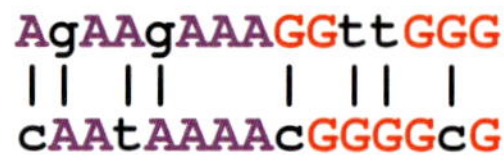

Since these two implanted patterns are so different, we should be concerned whether we will be able to find them by searching for the most similar k-mers among pairs of strings in *Dna*.

In the rest of the chapter, we will benchmark our motif finding algorithms by using a particularly challenging instance of the Implanted Motif Problem. The **Subtle Motif Problem** refers to implanting a 15-mer with four random mutations in ten randomly generated 600 nucleotide-long strings (the typical length of many upstream regulatory regions). The instance of the Subtle Motif Problem that we will use has the implanted 15-mer **AAAAAAAAGGGGGGG**.

It turns out that thousands of pairs of randomly occurring 15-mers in our dataset for the Subtle Motif Problem are fewer than 8 nucleotides apart from each other, preventing us from identifying the true implanted motifs by pairwise comparisons.

Scoring Motifs

From motifs to profile matrices and consensus strings

Although the Implanted Motif Problem offers a useful abstraction of the biological problem of motif finding, it has some limitations. For example, when Steve Kay used a DNA array to infer the set of circadian genes in plants, he did not expect that *all* genes in the resulting set would have the evening element (or its variants) in their upstream regions. Similarly, biologists do not expect that all genes with an elevated expression level in infected flies must be regulated by NF-ϰB. DNA array experiments are inherently noisy, and some genes identified by these experiments have nothing to do with the circadian clock in plants or immunity genes in flies. For such noisy datasets, any algorithm for the Implanted Motif Problem would fail, because as long as a single sequence does not contain the transcription factor binding site, a (k, d)-motif does not exist!

A more appropriate problem formulation would score individual instances of motifs depending on how similar they are to an "ideal" motif (i.e., a transcription factor binding site that binds the best to the transcription factor). However, since the ideal motif is unknown, we attempt to select a k-mer from each string and score these k-mers depending on how similar they are *to each other*.

To define scoring, consider t DNA strings, each of length n, and select a k-mer from each string to form a collection *Motifs*, which we represent as a $t \times k$ **motif matrix**. In Figure 2.2, which shows the motif matrix for the NF-ϰB binding sites from Figure 2.1, we indicate the most frequent nucleotide in each column of the motif matrix by upper case letters. If there are multiple most popular nucleotides in a column, then we arbitrarily select one of them to break the tie. Note that positions 2 and 3 are the most conserved (nucleotide **G** is completely conserved in these positions), whereas position 10 is the least conserved.

By varying the choice of k-mers in each string, we can construct a large number of different motif matrices from a given sample of DNA strings. Our goal is to select k-mers resulting in the most "conserved" motif matrix, meaning the matrix with the most upper case letters (and thus the fewest number of lower case letters). Leaving aside the question of how we select such k-mers, we will first focus on how to score the resulting motif matrices, defining SCORE(*Motifs*) as the number of unpopular (lower case) letters in the motif matrix *Motifs*. Our goal is to find a collection of k-mers that *minimizes* this score.

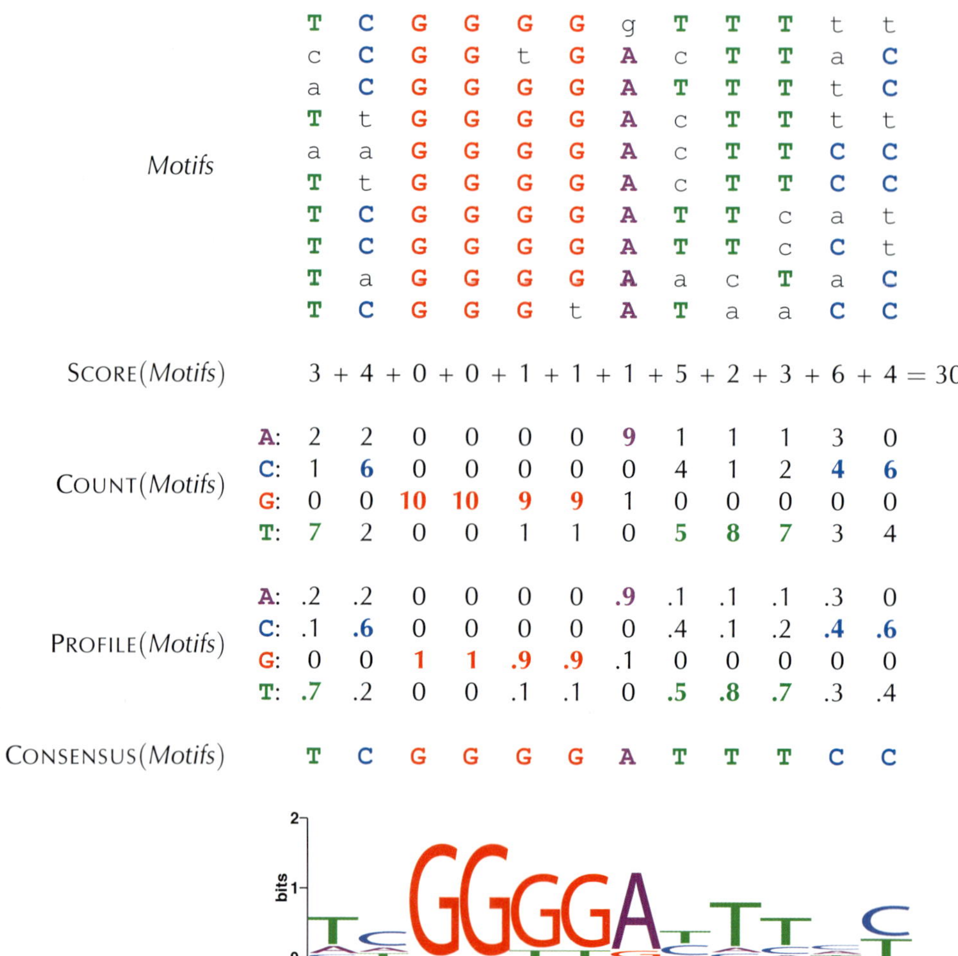

Motifs	T	C	G	G	G	G	g	T	T	T	t	t
	c	C	G	G	t	G	A	c	T	T	a	C
	a	C	G	G	G	G	A	T	T	T	t	C
	T	t	G	G	G	G	A	c	T	T	t	t
	a	a	G	G	G	G	A	c	T	T	C	C
	T	t	G	G	G	G	A	c	T	T	C	C
	T	C	G	G	G	G	A	T	T	c	a	t
	T	C	G	G	G	G	A	T	T	c	C	t
	T	a	G	G	G	G	A	a	c	T	a	C
	T	C	G	G	G	t	A	T	a	a	C	C

SCORE(*Motifs*) $3 + 4 + 0 + 0 + 1 + 1 + 1 + 5 + 2 + 3 + 6 + 4 = 30$

COUNT(*Motifs*)													
	A:	2	2	0	0	0	0	9	1	1	1	3	0
	C:	1	6	0	0	0	0	0	4	1	2	4	6
	G:	0	0	10	10	9	9	1	0	0	0	0	0
	T:	7	2	0	0	1	1	0	5	8	7	3	4

PROFILE(*Motifs*)													
	A:	.2	.2	0	0	0	0	.9	.1	.1	.1	.3	0
	C:	.1	.6	0	0	0	0	0	.4	.1	.2	.4	.6
	G:	0	0	1	1	.9	.9	.1	0	0	0	0	0
	T:	.7	.2	0	0	.1	.1	0	.5	.8	.7	.3	.4

CONSENSUS(*Motifs*) T C G G G G A T T T C C

FIGURE 2.2 From motif matrix to count matrix to profile matrix to consensus string to motif logo. The NF-κB binding sites form a 10×12 motif matrix, with the most frequent nucleotide in each column shown in upper case letters and all other nucleotides shown in lower case letters. SCORE(*Motifs*) counts the total number of unpopular (lower case) symbols in the motif matrix. The motif matrix results in a 4×12 count matrix holding the nucleotide counts in every column of the motif matrix; a profile matrix holding the frequencies of nucleotides in every column of the motif matrix; and a consensus string formed by the most frequent nucleotide in each column of the count matrix. Finally, the motif logo is a common way to visualize the conservation of various positions within a motif. The total height of the letters depicts the information content of the position.

Exercise Break: The minimum possible value of SCORE(*Motifs*) is 0 (if all rows in the $t \times k$ matrix *Motifs* are the same). What is the maximum possible value of SCORE(*Motifs*) in terms of t and k?

We can construct the $4 \times k$ **count matrix** COUNT(*Motifs*) counting the number of occurrences of each nucleotide in each column of the motif matrix; the (i, j)-th element of COUNT(*Motifs*) stores the number of times that nucleotide i appears in column j of *Motifs*. We will further divide all of the elements in the count matrix by t, the number of rows in *Motifs*. This results in a **profile matrix** $P =$ PROFILE(*Motifs*) for which $P_{i,j}$ is the *frequency* of the i-th nucleotide in the j-th column of the motif matrix. Note that the elements of any column of the profile matrix sum to 1.

Finally, we form a **consensus string**, denoted CONSENSUS(*Motifs*), from the most popular nucleotides in each column of the motif matrix (ties are broken arbitrarily). If we select *Motifs* correctly from the collection of upstream regions, then CONSENSUS(*Motifs*) provides an ideal candidate regulatory motif for these regions. For example, the consensus string for the NF-κB binding sites in Figure 2.2 is **TCGGGGATTTCC**.

Towards a more adequate motif scoring function

Consider the second column (containing 6 C, 2 A, and 2 T) and the final column (containing 6 C and 4 T) in the motif matrix from Figure 2.2. Both of these columns contribute 4 to SCORE(*Motifs*).

STOP and Think: Does scoring these two columns equally make sense biologically?

For many biological motifs, certain positions feature two nucleotides with roughly the same ability to bind to a transcription factor. For example, the sixteen nucleotide-long CSRE transcription factor binding site in the yeast *S. cerevisiae* consists of five strongly conserved positions in addition to eleven weakly conserved positions, each of which features two nucleotides with similar frequencies (Figure 2.3).

Following this example, a more appropriate representation of the consensus string TCGGGGATTTCC for the NF-κB binding sites should include viable alternatives to the most popular nucleotides in each column (see Figure 2.4). In this sense, the last column (6 C, 4 T) in the motif matrix from Figure 2.2 is "more conserved" than the second column (6 C, 2 A, 2 T) and should receive a lower score.

1	2	3	4	5	6	7	8	9	10	11	12	13	14	15	16
C	G/C	G/T	T/A	C/T	G/C	C/G	A	T	G/T	C/G	A	T	C/T	C/T	G/T

FIGURE 2.3 The CSRE transcription factor binding site in *S. cerevisiae* is 16 nucleotides long, but only five of these positions (1, 8, 9, 12, 13) are strongly conserved. The remaining 11 positions can take one of two different nucleotides.

1	2	3	4	5	6	7	8	9	10	11	12
T	C	G	G	G	G	A	T/C	T	T	C	C/T

FIGURE 2.4 Taking nucleotides in each column of the NF-κB binding site motif matrix from Figure 2.2 with frequency at least 0.4 yields a representation of the NF-κB binding sites with ten positions represented by a single consensus nucleotide and two positions (8 and 12) represented by a pair of frequent nucleotides.

Entropy and the motif logo

Every column of PROFILE(*Motifs*) corresponds to a **probability distribution**, or a collection of nonnegative numbers that sum to 1. For example, the second column in Figure 2.2 corresponds to the probabilities 0.2, 0.6, 0.0, and 0.2 for A, C, G, and T, respectively.

Entropy is a measure of the uncertainty of a probability distribution $(p_1, \ldots, p_N)$, and is defined as

$$H(p_1, \ldots, p_N) = -\sum_{i=1}^{N} p_i \cdot \log_2(p_i).$$

For example, the entropy of the probability distribution (0.2, 0.6, 0.0, 0.2) corresponding to the second column of the profile matrix in Figure 2.2 is

$$-(0.2\log_2 0.2 + 0.6\log_2 0.6 + 0.0\log_2 0.0 + 0.2\log_2 0.2) \approx 1.371,$$

whereas the entropy of the more conserved final column $(0.0, 0.6, 0.0, 0.4)$ is

$$-(0.0\log_2 0.0 + 0.6\log_2 0.6 + 0.0\log_2 0.0 + 0.4\log_2 0.4) \approx 0.971,$$

and the entropy of the very conserved 5th column $(0.0, 0.0, 0.9, 0.1)$ is

$$-(0.0\log_2 0.0 + 0.0\log_2 0.0 + 0.9\log_2 0.9 + 0.1\log_2 0.1) \approx 0.467.$$

Note that technically, $\log_2 0$ is not defined, but in the computation of entropy, we assume that $0 \cdot \log_2 0$ is equal to 0.

STOP and Think: What are the maximum and minimum possible values for the entropy of a probability distribution containing four values?

The entropy of the completely conserved third column of the profile matrix in Figure 2.2 is 0, which is the minimum possible entropy. On the other hand, a column with equally-likely nucleotides (all probabilities equal to 1/4) has maximum possible entropy $-4 \cdot 1/4 \cdot \log_2(1/4) = 2$. In general, the more conserved the column, the smaller its entropy. Thus, entropy offers an improved method of scoring motif matrices: the entropy of a motif matrix is defined as the sum of the entropies of its columns. In this book, we will continue to use SCORE(*Motifs*) for simplicity, but the entropy score is used more often in practice.

Exercise Break: Compute the entropy of the NF-κB motif matrix from Figure 2.2.

Another application of entropy is the **motif logo**, a diagram for visualizing motif conservation that consists of a stack of letters at each position (see the bottom of Figure 2.2). The relative sizes of letters indicate their frequency in the column. The total height of the letters in each column is based on the **information content** of the column, which is defined as $2 - H(p_1, \ldots, p_N)$. The lower the entropy, the higher the information content, meaning that tall columns in the motif logo are highly conserved.

From Motif Finding to Finding a Median String

The Motif Finding Problem

Now that we have a good grasp of scoring a collection of k-mers, we are ready to formulate the Motif Finding Problem.

Motif Finding Problem:
Given a collection of strings, find a set of k-mers, one from each string, that minimizes the score of the resulting motif.

> **Input**: A collection of strings *Dna* and an integer k.
> **Output**: A collection *Motifs* of k-mers, one from each string in *Dna*, minimizing SCORE(*Motifs*) among all possible choices of k-mers.

A brute force algorithm for the Motif Finding Problem, **BRUTEFORCEMOTIFSEARCH**, considers every possible choice of k-mers *Motifs* from *Dna* (one k-mer from each string of n nucleotides) and returns the collection *Motifs* having minimum score. Because there are $n - k + 1$ choices of k-mers in each of t sequences, there are $(n - k + 1)^t$ different ways to form *Motifs*. For each choice of *Motifs*, the algorithm calculates SCORE(*Motifs*), which requires $k \cdot t$ steps. Thus, assuming that k is much smaller than n, the overall running time of the algorithm is $\mathcal{O}(n^t \cdot k \cdot t)$. We need to come up with a faster algorithm!

Reformulating the Motif Finding Problem

Because **BRUTEFORCEMOTIFSEARCH** is inefficient, we will think about motif finding in a different way. Instead of exploring all *Motifs* in *Dna* and deriving the consensus string from *Motifs* afterwards,

$$Motifs \rightarrow \text{CONSENSUS}(Motifs)\,,$$

we will explore all potential k-mer consensus strings first and then find the best possible collection *Motifs* for each consensus string,

$$\text{CONSENSUS}(Motifs) \rightarrow Motifs\,.$$

To reformulate the Motif Finding Problem, we need to devise an alternative way of computing SCORE(*Motifs*). Until now, we have computed SCORE(*Motifs*), the number of lower case letters in the motif matrix, column-by-column. For example, in Figure 2.2, we computed SCORE(*Motifs*) for the NF-κB motif matrix as

$$3+4+0+0+1+1+1+5+2+3+6+4=30\,.$$

Figure 2.5 illustrates that SCORE(*Motifs*) can just as easily be computed row-by-row as

$$3+4+2+4+3+2+3+2+4+3=30\,.$$

Note that each element in the latter sum represents the number of mismatches between the consensus string **TCGGGGATTTCC** and a motif in the corresponding row of the motif matrix, i.e., the Hamming distance between these strings. For the first row of the motif matrix in Figure 2.5, d(**TCGGGGATTTCC**, **TCGGGG**g**TTT**tt) = 3.

	T	C	G	G	G	G	g	T	T	T	t	t	3
	c	C	G	G	t	G	A	c	T	T	a	C	4
	a	C	G	G	G	G	A	T	T	T	t	C	2
	T	t	G	G	G	G	A	c	T	T	t	t	4
Motifs	a	a	G	G	G	G	A	c	T	T	C	C	3
	T	t	G	G	G	G	A	c	T	T	C	C	2
	T	C	G	G	G	G	A	T	T	c	a	t	3
	T	C	G	G	G	G	A	T	T	c	C	t	2
	T	a	G	G	G	G	A	a	c	T	a	C	4
	T	C	G	G	G	t	A	T	a	a	C	C	+ 3
SCORE(*Motifs*)	3 +	4 +	0 +	0 +	1 +	1 +	1 +	5 +	2 +	3 +	6 +	4 =	30
CONSENSUS(*Motifs*)	T	C	G	G	G	G	A	T	T	T	C	C	

FIGURE 2.5 The motif and score matrices in addition to the consensus string for the NF-κB binding sites, reproduced from Figure 2.2. Rather than count the non-consensus elements (i.e., lower case nucleotides) column-by-column, we can count them row-by-row, as highlighted on the right of the motifs matrix. Each value at the end of a row corresponds to the Hamming distance between that row and the consensus string.

Given a collection of k-mers $Motifs = \{Motif_1, ..., Motif_t\}$ and a k-mer *Pattern*, we now define $d(Pattern, Motifs)$ as the sum of Hamming distances between *Pattern* and each $Motif_i$,

$$d(Pattern, Motifs) = \sum_{i=1}^{t} \text{HAMMINGDISTANCE}(Pattern, Motif_i)\,.$$

Because SCORE(*Motifs*) corresponds to counting the lower case elements of *Motifs* column-by-column and d(CONSENSUS(*Motifs*), *Motifs*) corresponds to counting these elements row-by-row, we obtain that

$$\text{SCORE}(Motifs) = d(\text{CONSENSUS}(Motifs), Motifs)\,.$$

This equation gives us an idea. Instead of searching for a collection of k-mers *Motifs* minimizing SCORE(*Motifs*), let's instead search for a potential consensus string *Pattern* minimizing $d(Pattern, Motifs)$ among all possible k-mers *Pattern* and all possible choices of k-mers *Motifs* in *Dna*. This problem is equivalent to the Motif Finding Problem.

Equivalent Motif Finding Problem:
Given a collection of strings, find a pattern and a collection of k-mers (one from each string) that minimizes the distance between all possible patterns and all possible collections of k-mers.

> **Input**: A collection of strings *Dna* and an integer k.
> **Output**: A k-mer *Pattern* and a collection of k-mers *Motifs*, one from each string in *Dna*, minimizing $d(Pattern, Motifs)$ among all possible choices of *Pattern* and *Motifs*.

The Median String Problem

But wait a second — have we not just made our task more difficult? Instead of having to search for all *Motifs*, we now have to search all *Motifs* as well as all k-mers *Pattern*. The key observation for solving the Equivalent Motif Finding Problem is that, given *Pattern*, we don't need to explore all possible collections *Motifs* in order to minimize $d(Pattern, Motifs)$.

To explain how this can be done, we define MOTIFS(*Pattern*, *Dna*) as a collection of k-mers that minimizes $d(Pattern, Motifs)$ for a given *Pattern* and all possible sets of k-mers *Motifs* in *Dna*. For example, for the strings *Dna* shown below, the five colored 3-mers represent MOTIFS(**AAA**, *Dna*).

```
      ttaccttAAC
      gATAtctgtc
Dna   ACGgcgttcg
      ccctAAAgag
      cgtcAGAggt
```

STOP and Think: Given a collection of strings *Dna* and a k-mer *Pattern*, design a fast algorithm for generating MOTIFS(*Pattern*, *Dna*).

The reason why we don't need to consider all possible collections *Motifs* in $Dna = \{Dna_1, ..., Dna_t\}$ is that we can generate the k-mers in MOTIFS(*Pattern*, *Dna*) one at a time; that is, we can select a k-mer in Dna_i independently of selecting k-mers in all other strings in *Dna*. Given a k-mer *Pattern* and a longer string *Text*, we use $d(Pattern, Text)$ to denote the minimum Hamming distance between *Pattern* and any k-mer in *Text*,

$$d(Pattern, Text) = \min_{\text{all } k\text{-mers } Pattern' \text{ in } Text} \text{HAMMINGDISTANCE}(Pattern, Pattern')\,.$$

For example,

$$d(\mathbf{GATTCTCA}, \texttt{gcaaa}\mathbf{GACGCTGA}\texttt{ccaa}) = \mathbf{3}\,.$$

A k-mer in *Text* that achieves the minimum Hamming distance with *Pattern* is denoted MOTIF(*Pattern*, *Text*). For the above example,

$$\text{MOTIF}(\mathbf{GATTCTCA}, \texttt{gcaaa}\mathbf{GACGCTGA}\texttt{ccaa}) = \mathbf{GACGCTGA}\,.$$

We note that the notation MOTIF(*Pattern*, *Text*) is ambiguous because there may be multiple k-mers in *Text* that achieve the minimum Hamming distance with *Pattern*. For example, MOTIF(**AAG**, gc**AAT**cct**CAG**c) could be either **AAT** or **CAG**. If there are multiple k-mers in *Text* having minimum Hamming distance with *Pattern*, we select the first such k-mer in *Text* as MOTIF(*Pattern*, *Text*).

Given a k-mer *Pattern* and a set of strings $Dna = \{Dna_1, ..., Dna_t\}$, we define $d(Pattern, Dna)$ as the sum of distances between *Pattern* and all strings in *Dna*,

$$d(Pattern, Dna) = \sum_{i=1}^{t} d(Pattern, Dna_i).$$

For example, for the strings *Dna* shown below, $d(\mathbf{AAA}, Dna) = \mathbf{1} + \mathbf{1} + \mathbf{2} + \mathbf{0} + \mathbf{1} = \mathbf{5}$.

	ttacctt**AAC**	**1**
	g**ATA**tctgtc	**1**
Dna	**ACG**gcgttcg	**2**
	ccct**AAA**gag	**0**
	cgtc**AGA**ggt	**1**

Our goal is to find a k-mer *Pattern* that minimizes $d(Pattern, Dna)$ over all k-mers *Pattern*, the same task that the Equivalent Motif Finding Problem is trying to achieve. We call such a k-mer a **median string** for *Dna*.

2B

Median String Problem:
Find a median string.

> **Input**: A collection of strings *Dna* and an integer k.
> **Output**: A k-mer *Pattern* minimizing $d(Pattern, Dna)$ among all possible choices of k-mers.

Notice that finding a median string requires solving a double minimization problem. We must find a k-mer *Pattern* that minimizes $d(Pattern, Dna)$, where this function is itself computed by taking a minimum over all choices of k-mers from each string in *Dna*. The pseudocode for a brute-force algorithm, **MEDIANSTRING**, is given below.

```
MEDIANSTRING(Dna, k)
    distance ← ∞
    for each k-mer Pattern from AA...AA to TT...TT
        if distance > d(Pattern, Dna)
            distance ← d(Pattern, Dna)
            Median ← Pattern
    return Median
```

Charging Station (Solving the Median String Problem): Although this pseudocode is short, it is not without potential pitfalls. Check out this Charging Station if you fall into one of them.

STOP and Think: Instead of making a time-consuming search through all possible k-mers in **MEDIANSTRING**, can you only search through all k-mers that appear in *Dna*?

Why have we reformulated the Motif Finding Problem?

To see why we reformulated the Motif Finding Problem as the equivalent Median String Problem, consider the runtimes of **MEDIANSTRING** and **BRUTEFORCEMOTIFS**. The former algorithm computes $d(Pattern, Dna)$ for each of the 4^k k-mers *Pattern*. Each computation of $d(Pattern, Dna)$ requires a single pass over each string in *Dna*, which requires approximately $k \cdot n \cdot t$ operations for t strings of length n in *Dna*. Therefore, **MEDIANSTRING** has a running time of $\mathcal{O}\left(4^k \cdot n \cdot k \cdot t\right)$, which in practice compares favorably with the $\mathcal{O}(n^t \cdot k \cdot t)$ running time of **BRUTEFORCEMOTIFSEARCH** because the length of a motif (k) typically does not exceed 20 nucleotides, whereas t is measured in the thousands.

The Median String Problem teaches an important lesson, which is that sometimes rethinking how a problem is formulated can lead to dramatic improvements in the runtime required to solve it. In this case, our simple observation that SCORE(*Motifs*) could just as easily be computed row-by-row as column-by-column produced the faster **MEDIANSTRING** algorithm.

Of course, the ultimate test of a bioinformatics algorithm is how it performs in practice. Unfortunately, since **MEDIANSTRING** has to consider 4^k k-mers, it becomes too slow for the Subtle Motif Problem, for which $k = 15$. We will run **MEDIANSTRING** with $k = 13$ in the hope that it will capture a substring of the correct 15-mer motif. The algorithm still requires half a day to run on our computer and returns the median string AAAAAtAGaGGGG (with distance 29). This 13-mer is not a substring of the implanted pattern AAAAAAAAGGGGGGG, but it does come close.

STOP and Think: How can a slightly incorrect median string of length 13 help us find the correct median string of length 15?

STOP

We have thus far assumed that the value of k is known in advance, which is not the case in practice. As a result, we are forced to run our motif finding algorithms for different values of k and then try to deduce the correct motif length. Since some regulatory motifs are rather long — later in the chapter, we will search for a biologically important motif of length 20 — **MEDIANSTRING** may be too slow to find them.

Greedy Motif Search

Using the profile matrix to roll dice

Many algorithms are iterative procedures that must choose among many alternatives at each iteration. Some of these alternatives may lead to correct solutions, whereas others may not. **Greedy algorithms** select the "most attractive" alternative at each iteration.

For example, a greedy algorithm in chess might attempt to capture an opponent's most valuable piece at every move. Yet anyone who has played chess knows that this strategy of looking only one move ahead will likely produce disastrous results. In general, most greedy algorithms typically fail to find an exact solution of the problem; instead, they are often fast **heuristics** that trade accuracy for speed in order to find an *approximate* solution. Nevertheless, for many biological problems that we will study in this book, greedy algorithms will prove quite useful.

In this section, we will explore a greedy approach to motif finding. Again, let *Motifs* be a collection of k-mers taken from t strings *Dna*. Recall from our discussion of entropy that we can view each column of PROFILE(*Motifs*) as a four-sided biased die. Thus, a profile matrix with k columns can be viewed as a collection of k dice, which we will roll to randomly generate a k-mer. For example, if the first column of the profile matrix is $(0.2, 0.1, 0.0, 0.7)$, then we generate **A** as the first nucleotide with probability 0.2, **C** with probability 0.1, **G** with probability 0.0, and **T** with probability 0.7.

In Figure 2.6, we reproduce the profile matrix for the NF-κB binding sites from Figure 2.2, where the lone colored entry in the i-th column corresponds to the i-th nucleotide in **ACGGGGATTACC**. The probability Pr(**ACGGGGATTACC**|*Profile*) that *Profile* generates **ACGGGGATTACC** is computed by simply multiplying the highlighted entries in the profile matrix.

Profile

A:	**.2**	.2	.0	.0	.0	.0	**.9**	.1	.1	**.1**	.3	.0
C:	.1	**.6**	.0	.0	.0	.0	.0	.4	.1	.2	**.4**	**.6**
G:	.0	.0	**1**	**1**	**.9**	**.9**	.1	.0	.0	.0	.0	.0
T:	.7	.2	.0	.0	.1	.1	.0	**.5**	**.8**	.7	.3	.4

$$\Pr(\mathbf{ACGGGGATTACC}|\mathit{Profile}) = .2 \cdot .6 \cdot 1 \cdot 1 \cdot .9 \cdot .9 \cdot .9 \cdot .5 \cdot .8 \cdot .1 \cdot .4 \cdot .6 = 0.000839808$$

FIGURE 2.6 We can generate a random string based on a profile matrix by selecting the i-th nucleotide in the string with the probability corresponding to that nucleotide in the i-th column of the profile matrix. The probability that a profile matrix will produce a given string is given by the product of individual nucleotide probabilities.

A k-mer tends to have a higher probability when it is more similar to the consensus string of a profile. For example, for the NF-κB profile matrix shown in Figure 2.6 and its consensus string **TCGGGGATTTCC**,

$$\begin{aligned}\Pr(\mathbf{TCGGGGATTTCC}|\mathit{Profile}) &= .7 \cdot .6 \cdot 1 \cdot 1 \cdot .9 \cdot .9 \cdot .9 \cdot .5 \cdot .8 \cdot .7 \cdot .4 \cdot .6 \\ &= 0.0205753\,,\end{aligned}$$

which is larger than the value of Pr(ACGGGGATTACC|*Profile*) = 0.000839808 that we computed in Figure 2.6.

Exercise Break: Compute Pr(TCGTGGATTTCC|*Profile*), where *Profile* is the matrix shown in Figure 2.6.

Given a profile matrix *Profile*, we can evaluate the probability of every k-mer in a string *Text* and find a ***Profile*-most probable** k-mer in *Text*, i.e., a k-mer that was most likely to have been generated by *Profile* among all k-mers in *Text*. For the NF-κB profile matrix, ACGGGGATTACC is the *Profile*-most probable 12-mer in ggtACGGGGATTACCt. Indeed, every other 12-mer in this string has probability 0. In general, if there are multiple *Profile*-most probable k-mers in *Text*, then we select the first such k-mer occurring in *Text*.

***Profile*-most Probable k-mer Problem**:
Find a Profile-most probable k-mer in a string.

Input: A string *Text*, an integer k, and a $4 \times k$ matrix *Profile*.
Output: A *Profile*-most probable k-mer in *Text*.

Our proposed greedy motif search algorithm, GREEDYMOTIFSEARCH, starts by forming a motif matrix from arbitrarily selected k-mers in each string from *Dna* (which in our specific implementation is the first k-mer in each string). It then attempts to improve this initial motif matrix by trying each of the k-mers in Dna_1 as the first motif. For a given choice of k-mer $Motif_1$ in Dna_1, it builds a profile matrix *Profile* for this lone k-mer, and sets $Motif_2$ equal to the *Profile*-most probable k-mer in Dna_2. It then iterates by updating *Profile* as the profile matrix formed from $Motif_1$ and $Motif_2$, and sets $Motif_3$ equal to the *Profile*-most probable k-mer in Dna_3. In general, after finding $i-1$ k-mers *Motifs* in the first $i-1$ strings of *Dna*, GREEDYMOTIFSEARCH constructs *Profile*(*Motifs*) and selects the *Profile*-most probable k-mer from Dna_i based on this profile matrix. After obtaining a k-mer from each string to generate a collection *Motifs*, GREEDYMOTIFSEARCH tests to see whether *Motifs* outscores the current best scoring collection of motifs and then moves $Motif_1$ one symbol over in Dna_1, beginning the entire process of generating *Motifs* again.

```
GREEDYMOTIFSEARCH(Dna, k, t)
    BestMotifs ← motif matrix formed by first k-mers in each string from Dna
    for each k-mer Motif in the first string from Dna
        Motif1 ← Motif
        for i = 2 to t
            form Profile from motifs Motif1, ..., Motif(i-1)
            Motif_i ← Profile-most probable k-mer in the i-th string in Dna
        Motifs ← (Motif1, ..., Motif_t)
        if SCORE(Motifs) < SCORE(BestMotifs)
            BestMotifs ← Motifs
    return BestMotifs
```

If you are not satisfied with the performance of **GREEDYMOTIFSEARCH** — even if you implemented it correctly — then wait until we discuss this algorithm in the next section.

Analyzing greedy motif finding

In contrast to **MEDIANSTRING**, **GREEDYMOTIFSEARCH** is fast and can be run with $k = 15$ to solve the Subtle Motif Problem (recall that we settled for $k = 13$ in the case of **MEDIANSTRING**). However, it trades speed for accuracy and returns the 15-mer **gtAAAtAgaGatGtG** (total distance: 58), which is very different from the true implanted motif **AAAAAAAAGGGGGGG**.

STOP and Think: Why does **GREEDYMOTIFSEARCH** perform so poorly?

At first glance, **GREEDYMOTIFSEARCH** may seem like a reasonable algorithm, but it is not! Let's see whether **GREEDYMOTIFSEARCH** will find the (4, 1)-motif **ACGT** implanted in the following strings *Dna*:

```
ttACCTtaac
gATGTctgtc
acgGCGTtag
ccctaACGAg
cgtcagAGGT
```

We will assume an optimistic scenario in which the algorithm has already correctly chosen the implanted 4-mer **ACCT** from the first string in *Dna* and constructed the corresponding *Profile*:

A:	**1**	0	0	0
C:	0	**1**	**1**	0
G:	0	0	0	0
T:	0	0	0	**1**

The algorithm is now ready to search for a *Profile*-most probable 4-mer in the second sequence. The issue, however, is that there are so many zeroes in the profile matrix that the probability of every 4-mer but **ACCT** is zero! Thus, unless **ACCT** is present in every string in *Dna*, there is little chance that **GREEDYMOTIFSEARCH** will find the implanted motif. Zeroes in the profile matrix are not just a minor annoyance but rather a persistent problem that we must address.

Motif Finding Meets Oliver Cromwell

What is the probability that the sun will not rise tomorrow?

In 1650, after the Scots proclaimed Charles II as king during the English Civil War, Oliver Cromwell made a famous appeal to the Church of Scotland. Urging them to see the error of their royal alliance, he pleaded,

> *I beseech you, in the bowels of Christ, think it possible that you may be mistaken.*

The Scots rejected the appeal, and Cromwell invaded Scotland in response. His quotation later inspired the statistical maxim called **Cromwell's rule**, which states that we should not use probabilities of 0 or 1 unless we are talking about logical statements that can only be true or false. In other words, we should allow a small probability for extremely unlikely events, such as "this book was written by aliens" or "the sun will not rise tomorrow". We cannot speak to the likelihood of the former event, but in the 18th Century, the French mathematician Pierre-Simon Laplace actually estimated the probability that the sun will not rise tomorrow (1/1826251), given that it has risen every day for the past 5000 years. Although this estimate was ridiculed by his contemporaries, Laplace's approach to this question now plays an important role in statistics.

In any observed data set, there is the possibility, especially with low-probability events or small data sets, that an event with nonzero probability does not occur. Its observed frequency is therefore zero; however, setting the empirical probability of the event equal to zero represents an inaccurate oversimplification that may cause problems. By artificially adjusting the probability of rare events, these problems can be mitigated.

Laplace's Rule of Succession

Cromwell's rule is relevant to the calculation of the probability of a string based on a profile matrix. For example, consider the following *Profile*:

Profile													
A:	.2	.2	.0	.0	.0	.0	**.9**	.1	.1	.1	.3	.0
C:	.1	**.6**	.0	.0	.0	.0	.0	.4	.1	.2	**.4**	**.6**
G:	.0	.0	**1**	1	**.9**	**.9**	.1	.0	.0	.0	.0	.0
T:	**.7**	.2	.0	**.0**	.1	.1	.0	**.5**	**.8**	**.7**	.3	.4

$$\Pr(\mathtt{TCGTGGATTTCC}|\mathit{Profile}) = .7 \cdot .6 \cdot 1 \cdot .0 \cdot .9 \cdot .9 \cdot .9 \cdot .5 \cdot .8 \cdot .7 \cdot .4 \cdot .6 = 0$$

The fourth symbol of **TCGTGGATTTCC** causes Pr(**TCGTGGATTTCC**|*Profile*) to equal zero. As a result, the entire string is assigned a zero probability, even though **TCGTGGATTTCC** differs from the consensus string at only one position. For that matter, **TCGTGGATTTCC** has the same low probability as **AAATCTTGGAA**, which differs from the consensus string at every position.

To improve this unfair scoring, bioinformaticians often substitute zeroes with small numbers called **pseudocounts**. The simplest approach to introducing pseudocounts, called **Laplace's Rule of Succession**, is similar to the principle that Laplace used to calculate the probability that the sun will not rise tomorrow. In the case of motifs, pseudocounts often amount to adding 1 (or some other small number) to each element of COUNT(*Motifs*). Say that we have the following motif, count, and profile matrices:

Motifs

```
T A A C
G T C T
A C T A
A G G T
```

COUNT(*Motifs*)

A:	2	1	1	1
C:	0	1	1	1
G:	1	1	1	0
T:	1	1	1	2

PROFILE(*Motifs*)

2/4	1/4	1/4	1/4
0	1/4	1/4	1/4
1/4	1/4	1/4	0
1/4	1/4	1/4	2/4

Laplace's Rule of Succession adds 1 to each element of COUNT(*Motifs*), updating the two matrices to the following:

COUNT(*Motifs*)	A:	2+1	1+1	1+1	1+1	
	C:	0+1	1+1	1+1	1+1	
	G:	1+1	1+1	1+1	0+1	
	T:	1+1	1+1	1+1	2+1	

PROFILE(*Motifs*)	3/8	2/8	2/8	2/8
	1/8	2/8	2/8	2/8
	2/8	2/8	2/8	1/8
	2/8	2/8	2/8	3/8

STOP and Think: How would you use Laplace's Rule of Succession to address the shortcomings of **GREEDYMOTIFSEARCH**?

An improved greedy motif search

The only change we need to introduce to **GREEDYMOTIFSEARCH** in order to eliminate zeroes from the profile matrices that it constructs is to replace line 6 of the pseudocode for **GREEDYMOTIFSEARCH**:

```
form Profile from motifs Motif_1, ... Motif_{i-1}
```

with the following line:

```
apply Laplace's Rule of Succession to form Profile from motifs Motif_1, ... Motif_{i-1}
```

We now will apply Laplace's Rule of Succession to search for the (4, 1)-motif **ACGT** implanted in the following strings *Dna*:

```
     ttACCTtaac
     gATGTctgtc
Dna  acgGCGTtag
     ccctaACGAg
     cgtcagAGGT
```

Again, let's assume that the algorithm has already chosen the implanted 4-mer **ACCT** from the first sequence. We can construct the corresponding count and profile matrices using Laplace's Rule of Succession:

Motifs **ACCT**

COUNT(*Motifs*)					PROFILE(*Motifs*)			
A:	1+1	0+1	0+1	0+1	2/5	1/5	1/5	1/5
C:	0+1	1+1	1+1	0+1	1/5	2/5	2/5	1/5
G:	0+1	0+1	0+1	0+1	1/5	1/5	1/5	1/5
T:	0+1	0+1	0+1	1+1	1/5	1/5	1/5	2/5

We use this profile matrix to compute the probabilities of all 4-mers in the second string from *Dna*:

gATG	ATGT	TGTc	GTct	Tctg	ctgt	tgtc
$1/5^4$	$4/5^4$	$1/5^4$	$4/5^4$	$2/5^4$	$2/5^4$	$1/5^4$

There are two *Profile*-most probable 4-mers in the second sequence (**ATGT** and **GT**ct); let's assume that we get lucky again and choose the implanted 4-mer **ATGT**. We now have the following motif, count, and profile matrices:

Motifs
ACCT
ATGT

COUNT(*Motifs*)					PROFILE(*Motifs*)			
A:	2+1	0+1	0+1	0+1	3/6	1/6	1/6	1/6
C:	0+1	1+1	1+1	0+1	1/6	2/6	2/6	1/6
G:	0+1	0+1	1+1	0+1	1/6	1/6	2/6	1/6
T:	0+1	1+1	0+1	2+1	1/6	2/6	1/6	3/6

We use this profile matrix to compute the probabilities of all 4-mers in the third string from *Dna*:

acgG	cgGC	gGCG	GCGT	CGTt	GTta	Ttag
$12/6^4$	$2/6^4$	$2/6^4$	$12/6^4$	$3/6^4$	$2/6^4$	$2/6^4$

Again, there are two *Profile*-most probable 4-mers in the second sequence (acg**G** and **GCGT**). This time, we will assume that acg**G** is selected instead of the implanted 4-mer **GCGT**. We now have the following motif, count, and profile matrices:

Motifs	**ACCT** **ATGT** acg**G**

COUNT(*Motifs*)

A:	3+1	0+1	0+1	1+1
C:	0+1	2+1	1+1	0+1
G:	0+1	0+1	2+1	1+1
T:	0+1	1+1	0+1	2+1

PROFILE(*Motifs*)

4/7	1/7	1/7	1/7
1/7	3/7	2/7	1/7
1/7	1/7	3/7	2/7
1/7	2/7	1/7	3/7

We use this profile matrix to compute probabilities of all 4-mers in the fourth string from *Dna*:

cccct	ccta	cta**A**	ta**AC**	a**ACG**	**ACGA**	**CGA**g
$18/7^4$	$3/7^4$	$2/7^4$	$1/7^4$	$16/7^4$	$36/7^4$	$2/7^4$

Despite the fact that we missed the implanted 4-mer in the third sequence, we have now found the implanted 4-mer in the fourth string in *Dna* as the *Profile*-most probable 4-mer **ACGA**. This provides us with the following motif, count, and profile matrices:

Motifs	**ACCT** **ATGT** acg**G** **ACGA**

COUNT(*Motifs*)

A:	4+1	0+1	0+1	0+1
C:	0+1	3+1	1+1	0+1
G:	0+1	0+1	3+1	1+1
T:	0+1	1+1	0+1	2+1

PROFILE(*Motifs*)

5/8	1/8	1/8	2/8
1/8	4/8	2/8	1/8
1/8	1/8	4/8	2/8
1/8	2/8	1/8	3/8

We now use this profile to compute the probabilities of all 4-mers in the fifth string in *Dna*:

cgtc	gtca	tcag	cag**A**	ag**AG**	g**AGG**	**AGGT**
$1/8^4$	$8/8^4$	$8/8^4$	$8/8^4$	$10/8^4$	$8/8^4$	$60/8^4$

The *Profile*-most probable 4-mer of the fifth string in *Dna* is **AGGT**, the implanted 4-mer. As a result, **GREEDYMOTIFSEARCH** has produced the following motif matrix, which implies the correct consensus string **ACGT**:

	ACCT
	ATGT
Motifs	acgG
	ACGA
	AGGT
CONSENSUS(*Motifs*)	ACGT

You have now seen the power of pseudocounts illustrated on a small example. Running **GREEDYMOTIFSEARCH** with pseudocounts to solve the Subtle Motif Problem returns a collection of 15-mers *Motifs* with SCORE(*Motifs*) = 41 and CONSENSUS(*Motifs*) = **AAAAAtAgaGGGGtt**. Thus, Laplace's Rule of Succession has provided a great improvement over the original **GREEDYMOTIFSEARCH**, which returned the consensus string **gTtAAAtAgaGatGtG** with SCORE(*Motifs*) = 58.

You may be satisfied with the performance of **GREEDYMOTIFSEARCH**, but you should know by now that your authors are never satisfied. Can we design an even more accurate motif finding algorithm?

Randomized Motif Search

Rolling dice to find motifs

We will now turn to **randomized algorithms** that flip coins and roll dice in order to search for motifs. Making random algorithmic decisions may sound like a disastrous idea — just imagine a chess game in which every move would be decided by rolling a die. However, an 18th Century French mathematician and naturalist, Comte de Buffon, first proved that randomized algorithms are useful by randomly dropping needles onto parallel strips of wood and using the results of this experiment to accurately approximate the constant π (see **DETOUR: Buffon's Needle**).

Randomized algorithms may be nonintuitive because they lack the control of traditional algorithms. Some randomized algorithms are **Las Vegas algorithms**, which deliver solutions that are guaranteed to be exact, despite the fact that they rely on making random decisions. Yet most randomized algorithms, including the motif finding algorithms that we will consider in this chapter, are **Monte Carlo algorithms**. These algorithms are not guaranteed to return exact solutions, but they do quickly find *approximate* solutions. Because of their speed, they can be run many times, allowing us to choose the best approximation from thousands of runs.

We previously defined PROFILE(*Motifs*) as the profile matrix constructed from a collection of k-mers *Motifs* in *Dna*. Now, given a collection of strings *Dna* and an arbitrary $4 \times k$ matrix *Profile*, we define MOTIFS(*Profile*, *Dna*) as the collection of k-mers formed by the *Profile*-most probable k-mers in each sequence from *Dna*. For example, consider the following *Profile* and *Dna*:

Profile

A:	4/5	0	0	1/5
C:	0	3/5	1/5	0
G:	1/5	1/5	4/5	0
T:	0	1/5	0	4/5

Dna

```
ttaccttaac
gatgtctgtc
acggcgttag
ccctaacgag
cgtcagaggt
```

Taking the *Profile*-most probable 4-mer from each row of *Dna* produces the following 4-mers (shown in red):

MOTIFS(*Profile*, *Dna*)

tt**acct**taac
g**atgt**ctgtc
acg**gcgt**tag
cccta**acga**g
cgtcag**aggt**

In general, we can begin from a collection of randomly chosen k-mers *Motifs* in *Dna*, construct PROFILE(*Motifs*), and use this profile to generate a new collection of k-mers:

MOTIFS(PROFILE(*Motifs*),*Dna*)

Why would we do this? Because our hope is that MOTIFS(PROFILE(*Motifs*), *Dna*) has a better score than the original collection of k-mers *Motifs*. We can then form the profile matrix of these k-mers,

PROFILE(MOTIFS(PROFILE(*Motifs*), *Dna*)) ,

and use it to form the most probable k-mers,

MOTIFS(PROFILE(MOTIFS(PROFILE(*Motifs*), *Dna*)), *Dna*) .

We can continue to iterate...

...PROFILE(MOTIFS(PROFILE(MOTIFS(PROFILE(*Motifs*), *Dna*)), *Dna*))...

for as long as the score of the constructed motifs keeps improving, which is exactly what **RANDOMIZEDMOTIFSEARCH** does. To implement this algorithm, you will need to randomly select the initial collection of k-mers that form the motif matrix *Motifs*. To do so, you will need a **random number generator** (denoted RANDOMNUMBER(N)) that is equally likely to return any integer from 1 to N. You might like to think about this random number generator as an unbiased N-sided die.

2F

```
RANDOMIZEDMOTIFSEARCH(Dna, k, t)
    randomly select k-mers Motifs = (Motif_1, ..., Motif_t) in each string from Dna
    BestMotifs ← Motifs
    while forever
        Profile ← PROFILE(Motifs)
        Motifs ← MOTIFS(Profile, Dna)
        if SCORE(Motifs) < SCORE(BestMotifs)
            BestMotifs ← Motifs
        else
            return BestMotifs
```

Exercise Break: Prove that **RANDOMIZEDMOTIFSEARCH** will eventually terminate.

Since a single run of **RANDOMIZEDMOTIFSEARCH** may generate a rather poor set of motifs, bioinformaticians usually run this algorithm thousands of times. On each run, they begin from a new randomly selected set of k-mers, selecting the best set of k-mers found in all these runs.

Why randomized motif search works

At first glance, **RANDOMIZEDMOTIFSEARCH** appears to be doomed. How can this algorithm, which starts from a random guess, possibly find anything useful? To explore **RANDOMIZEDMOTIFSEARCH**, let's run it on five short strings with the implanted $(4,1)$-motif `ACGT` (shown in upper case letters below) and imagine that it chooses the following 4-mers *Motifs* (shown in red) at the first iteration. As expected, it misses the implanted motif in nearly every string.

```
       ttACCTtaac
       gATGTctgtc
Dna    ccgGCGTtag
       cactaACGAg
       cgtcagAGGT
```

Below, we construct the profile matrix PROFILE(*Motifs*) of the chosen 4-mers.

Motifs					PROFILE(*Motifs*)			
t	a	a	c	A:	0.4	0.2	0.2	0.2
G	T	c	t	C:	0.2	0.4	0.2	0.2
c	c	g	G	G:	0.2	0.2	0.4	0.2
a	c	t	a	T:	0.2	0.2	0.2	0.4
A	G	G	T					

We can now compute the probabilities of every 4-mer in *Dna* based on this profile matrix. For example, the probability of the first 4-mer in the first string of *Dna* is PR(ttAC|*Profile*) $= 0.2 \cdot 0.2 \cdot 0.2 \cdot 0.2 = 0.0016$. The maximum probabilities in every row are shown in red below.

ttAC	tACC	ACCT	CCTt	CTta	Ttaa	taac
.0016	.0016	**.0128**	.0064	.0016	.0016	.0016
gATG	ATGT	TGTc	GTct	Tctg	ctgt	tgtc
.0016	**.0128**	.0016	.0032	.0032	.0032	.0016
ccgG	cgGC	gGCG	GCGT	CGTt	GTta	Ttag
.0064	.0036	.0016	**.0128**	.0032	.0016	.0016
cact	acta	ctaA	taAC	aACG	ACGA	CGAg
.0032	.0064	.0016	.0016	.0032	**.0128**	.0016
cgtc	gtca	tcag	cagA	agAG	gAGG	AGGT
.0016	.0016	.0016	.0032	.0032	.0032	**.0128**

We select the most probable 4-mer in each row above as our new collection *Motifs* (shown below). Notice that this collection has captured all five implanted motifs in *Dna*!

	ttACCTtaac
	gATGTctgtc
Dna	ccgGCGTtag
	cactaACGAg
	cgtcagAGGT

STOP and Think: How is it possible that randomly chosen k-mers have led us to the correct implanted k-mer? If you think we manufactured this example, select your own initial 4-mers and see what happens.

For the Subtle Motif Problem with implanted 15-mer AAAAAAAAGGGGGGG, when we run RANDOMIZEDMOTIFSEARCH 100,000 times (each time with new randomly selected k-mers), it returns the 15-mers shown in Figure 2.7 as the lowest scoring collection *Motifs* across all iterations, resulting in the consensus string AAAAAAAAacaGGGG with score 43. These strings are only slightly less conserved than the collection of implanted (15, 4)-motifs with score 40 (or the motif returned by GREEDYMOTIFSEARCH with score 41), and it largely captures the implanted motif. Furthermore, unlike GREEDYMOTIFSEARCH, RANDOMIZEDMOTIFSEARCH can be run for a larger number of iterations to discover better and better motifs.

		Score
	AAAtAcAgACAGcGt	5
	AAAAAAtAgCAGGGt	3
	tAAAAtAAACAGcGG	3
	AcAgAAAAAaAGGGG	3
Motifs	AAAAtAAAACtGcGa	4
	AtAgAcgAACAcGGt	6
	cAAAAAgAgaAGGGG	4
	AtAgAAAAggAaGGG	5
	AAgAAAAAAgAGaGG	3
	cAtAAtgAACtGtGa	7
CONSENSUS(*Motifs*)	AAAAAAAAACAGGGG	**43**

FIGURE 2.7 The lowest scoring collection of strings *Motifs* produced by 100,000 runs of **RANDOMIZEDMOTIFSEARCH**, along with their consensus string and score for the Subtle Motif Problem.

STOP and Think: Does your run of **RANDOMIZEDMOTIFSEARCH** return a similar consensus string? How many times do you need to run **RANDOMIZEDMOTIFSEARCH** to obtain the implanted (15, 4)-motif with distance 40?

Although the motifs returned by **RANDOMIZEDMOTIFSEARCH** are slightly less conserved than the motifs returned by **MEDIANSTRING**, **RANDOMIZEDMOTIFSEARCH** has the advantage of being able to find longer motifs (since **MEDIANSTRING** becomes too slow for longer motifs). In the epilogue, we will see that this feature is important in practice.

How Can a Randomized Algorithm Perform So Well?

In the previous section, we began with a collection of implanted motifs (with consensus `ACGT`) that resulted in the following profile matrix.

A:	**0.8**	0.0	0.0	0.2
C:	0.0	**0.6**	0.2	0.0
G:	0.2	0.2	**0.8**	0.0
T:	0.0	0.2	0.0	**0.8**

If the strings in *Dna* were truly random, then we would expect that all nucleotides in the selected k-mers would be equally likely, resulting in an expected *Profile* in which every entry is approximately 0.25:

A:	0.25	0.25	0.25	0.25
C:	0.25	0.25	0.25	0.25
G:	0.25	0.25	0.25	0.25
T:	0.25	0.25	0.25	0.25

Such a **uniform profile** is essentially useless for motif finding because no string is more probable than any other according to this profile and because it does not provide any clues on what an implanted motif looks like.

At the opposite end of the spectrum, if we were incredibly lucky, we would choose the implanted k-mers *Motifs* from the very beginning, resulting in the first of the two profile matrices above. In practice, we are likely to obtain a profile somewhere in between these two extremes, such as the following:

A:	**0.4**	0.2	0.2	0.2
C:	0.2	**0.4**	0.2	0.2
G:	0.2	0.2	**0.4**	0.2
T:	0.2	0.2	0.2	**0.4**

This profile matrix has already started to point us toward the implanted motif `ACGT`, i.e., `ACGT` is the most likely 4-mer that can be generated by this profile. Fortunately, **RANDOMIZEDMOTIFSEARCH** is designed so that subsequent steps have a good chance of leading us toward this implanted motif (although it is not certain).

If you still doubt the efficacy of randomized algorithms, consider the following argument. We have already noticed that if the strings in *Dna* were random, then **RANDOMIZEDMOTIFSEARCH** would start from a nearly uniform profile, and there would be nothing to work with. However, the key observation is that the strings in *Dna* are not random because they include the implanted motif! These multiple occurrences of the same motif may direct the profile matrix away from the uniform profile and toward the implanted motif. For example, consider again the original randomly selected *k*-mers *Motifs* (shown in red):

```
      ttACCTtaac
      gATGTctgtc
Dna   ccgGCGTtag
      cactaACGAg
      cgtcagAGGT
```

You will see that the 4-mer **AGGT** in the last string happened to capture the implanted motif simply by chance. In fact, the profile formed from the remaining 4-mers (**taac**, **GTct**, **ccgG**, and **acta**) is uniform. Note that only completely captured motifs (like **AGGT**) rather than partially captured motifs (like **GTct** or **ccgG**) contribute to the statistical bias in the profile matrix.

Exercise Break: Compute the probability that ten randomly selected 15-mers from ten 600-nucleotide long strings (such as in the Subtle Motif Problem) capture at least one implanted 15-mer.

Although the probability that randomly selected *k*-mers match *all* implanted motifs is negligible, the probability that they capture *at least one* implanted motif is significant. Even in the case of difficult motif finding problems for which this probability is small, we can run **RANDOMIZEDMOTIFSEARCH** many times, so that it will almost certainly catch at least one implanted motif, thus creating a statistical bias pointing toward the correct motif.

Unfortunately, capturing a single implanted motif is often insufficient to steer **RANDOMIZEDMOTIFSEARCH** to an optimal solution. Therefore, since the number of starting positions of k-mers is huge, the strategy of randomly selecting motifs is often not as successful as in the simple example above. The chance that these randomly selected k-mers will be able to guide us to the optimal solution is relatively small.

Exercise Break: Compute the probability that ten randomly selected 15-mers from the ten 600-nucleotide long strings in the Subtle Motif Problem capture at least two implanted 15-mers.

Gibbs Sampling

Note that **RANDOMIZEDMOTIFSEARCH** may change all t strings in *Motifs* in a single iteration. This strategy may prove reckless, since some correct motifs (captured in *Motifs*) may potentially be discarded at the next iteration. **GIBBSSAMPLER** is a more cautious iterative algorithm that discards a single k-mer from the current set of motifs at each iteration and decides to either keep it or replace it with a new one. This algorithm thus moves with more caution in the space of all motifs, as illustrated below.

```
ttaccttaac      ttaccttaac      ttaccttaac      ttaccttaac
gatatctgtc      gatatctgtc      gatatctgtc      gatatctgtc
acggcgttcg  →   acggcgttcg      acggcgttcg  →   acggcgttcg
ccctaaagag      ccctaaagag      ccctaaagag      ccctaaagag
cgtcagaggt      cgtcagaggt      cgtcagaggt      cgtcagaggt
```

RANDOMIZEDMOTIFSEARCH
(may change all k-mers in one step)

GIBBSSAMPLER
(changes one k-mer in one step)

Like **RANDOMIZEDMOTIFSEARCH**, **GIBBSSAMPLER** starts with randomly chosen k-mers in each of t DNA sequences, but it makes a random rather than a deterministic choice at each iteration. It uses randomly selected k-mers $Motifs = (Motif_1, \ldots, Motif_t)$ to come up with another (hopefully better scoring) set of k-mers. In contrast with **RANDOMIZEDMOTIFSEARCH**, which deterministically defines new motifs as

$$\text{MOTIFS}(\text{PROFILE}(Motifs), Dna),$$

GIBBSSAMPLER randomly selects an integer i between 1 and t and then randomly changes a single k-mer $Motif_i$.

To describe how **GIBBSSAMPLER** updates *Motifs*, we will need a slightly more advanced random number generator. Given a probability distribution $(p_1, \ldots, p_n)$, this random number generator, denoted $\text{RANDOM}(p_1, \ldots, p_n)$, models an n-sided biased die and returns integer i with probability p_i. For example, the standard six-sided fair die represents the random number generator

$$\text{RANDOM}(1/6, 1/6, 1/6, 1/6, 1/6, 1/6)\,,$$

whereas a biased die might represent the random number generator

$$\text{RANDOM}(0.1, 0.2, 0.3, 0.05, 0.1, 0.25)\,.$$

GIBBSSAMPLER further generalizes the random number generator by using the function $\text{RANDOM}(p_1, \ldots, p_n)$ defined for any set of non-negative numbers, i.e., not necessarily satisfying the condition $\sum_{i=1}^{n} p_i = 1$. Specifically, if $\sum_{i=1}^{n} p_i = C > 0$, then $\text{RANDOM}(p_1, \ldots, p_n)$ is defined as $\text{RANDOM}(p_1/C, \ldots, p_n/C)$, where $(p_1/C, \ldots, p_n/C)$ is a probability distribution. For example, given the values $(p_1, p_2, p_3) = (0.1, 0.2, 0.3)$ with $0.1 + 0.2 + 0.3 = 0.6$,

$$\begin{aligned}\text{RANDOM}(0.1, 0.2, 0.3) &= \text{RANDOM}(0.1/0.6, 0.2/0.6, 0.3/0.6)\\ &= \text{RANDOM}(1/6, 1/3, 1/2)\,.\end{aligned}$$

STOP and Think: Implement $\text{RANDOM}(p_1, \ldots, p_n)$ so that it uses $\text{RANDOMNUMBER}(X)$ (for an appropriately chosen integer X) as a subroutine.

We have previously defined the notion of a *Profile*-most probable k-mer in a string. We now define a ***Profile*-randomly generated k-mer** in a string *Text*. For each k-mer *Pattern* in *Text*, compute the probability $\Pr(Pattern|Profile)$, resulting in $n = |Text| - k + 1$ probabilities $(p_1, \ldots, p_n)$. These probabilities do not necessarily sum to 1, but we can still form the random number generator $\text{RANDOM}(p_1, \ldots, p_n)$ based on them. **GIBBSSAMPLER** uses this random number generator to select a *Profile*-randomly generated k-mer at each step: if the die rolls the number i, then we define the *Profile*-randomly generated k-mer as the i-th k-mer in *Text*. While the pseudocode below repeats this procedure N times, in practice **GIBBSSAMPLER** depends on various stopping rules that are beyond the scope of this chapter.

```
GIBBSSAMPLER(Dna, k, t, N)
    randomly select k-mers Motifs = (Motif_1, ..., Motif_t) in each string from Dna
    BestMotifs ← Motifs
    for j ← 1 to N
        i ← RANDOM(t)
        Profile ← profile matrix formed from all strings in Motifs except for Motif_i
        Motif_i ← Profile-randomly generated k-mer in the i-th sequence
        if SCORE(Motifs) < SCORE(BestMotifs)
            BestMotifs ← Motifs
    return BestMotifs
```

STOP and Think: Note that in contrast to **RANDOMIZEDMOTIFSEARCH**, which always moves from higher to lower scoring *Motifs*, **GIBBSSAMPLER** may move from lower to higher scoring *Motifs*. Why is this reasonable?

Gibbs Sampling in Action

We illustrate how **GIBBSSAMPLER** works on the same strings *Dna* that we considered before. Imagine that, at the initial step, the algorithm has chosen the following 4-mers (shown in red) and has randomly selected the third string for removal. To be more precise, **GIBBSSAMPLER** does not really remove the third string; it ignores it at this particular step and may analyze it again in subsequent steps.

```
     ttACCTtaac        ttACCTtaac
     gATGTctgtc        gATGTctgtc
Dna  ccgGCGTtag   -->   ----------
     cactaACGAg        cactaACGAg
     cgtcagAGGT        cgtcagAGGT
```

This results in the following motif, count, and profile matrices.

$$\textit{Motifs}\quad \begin{matrix} \texttt{t} & \texttt{a} & \texttt{a} & \texttt{c} \\ \texttt{G} & \texttt{T} & \texttt{c} & \texttt{t} \\ \texttt{a} & \texttt{c} & \texttt{t} & \texttt{a} \\ \texttt{A} & \texttt{G} & \texttt{G} & \texttt{T} \end{matrix}$$

$$\textsc{Count}(\textit{Motifs})\quad \begin{matrix} \texttt{A:} & 2 & 1 & 1 & 1 \\ \texttt{C:} & 0 & 1 & 1 & 1 \\ \texttt{G:} & 1 & 1 & 1 & 0 \\ \texttt{T:} & 1 & 1 & 1 & 2 \end{matrix} \qquad \textsc{Profile}(\textit{Motifs})\quad \begin{matrix} \texttt{A:} & 2/4 & 1/4 & 1/4 & 1/4 \\ \texttt{C:} & 0 & 1/4 & 1/4 & 1/4 \\ \texttt{G:} & 1/4 & 1/4 & 1/4 & 0 \\ \texttt{T:} & 1/4 & 1/4 & 1/4 & 2/4 \end{matrix}$$

Note that the profile matrix is only slightly more conserved than the uniform profile, making us wonder whether we have any chance to be steered toward the implanted motif. We now use this profile matrix to compute the probabilities of all 4-mers in the deleted string `ccgGCGTtag`:

`ccgG`	`cgGC`	`gGCG`	`GCGT`	`CGTt`	`GTta`	`Ttag`
0	0	0	1/128	0	1/256	0

Note that all but two of these probabilities are zero. This situation is similar to the one we encountered with **GREEDYMOTIFSEARCH**, and as before, we need to augment zero probabilities with small pseudocounts to avoid disastrous results.

Application of Laplace's Rule of Succession to the count matrix above yields the following updated count and profile matrices:

$$\textsc{Count}(\textit{Motifs})\quad \begin{matrix} \texttt{A:} & 3 & 2 & 2 & 2 \\ \texttt{C:} & 1 & 2 & 2 & 2 \\ \texttt{G:} & 2 & 2 & 2 & 1 \\ \texttt{T:} & 2 & 2 & 2 & 3 \end{matrix} \qquad \textsc{Profile}(\textit{Motifs})\quad \begin{matrix} \texttt{A:} & 3/8 & 2/8 & 2/8 & 2/8 \\ \texttt{C:} & 1/8 & 2/8 & 2/8 & 2/8 \\ \texttt{G:} & 2/8 & 2/8 & 2/8 & 1/8 \\ \texttt{T:} & 2/8 & 2/8 & 2/8 & 3/8 \end{matrix}$$

After adding pseudocounts, the 4-mer probabilities in the deleted string `ccgGCGTtag` are recomputed as follows:

`ccgG`	`cgGC`	`gGCG`	`GCGT`	`CGTt`	`GTta`	`Ttag`
$4/8^4$	$8/8^4$	$8/8^4$	$24/8^4$	$12/8^4$	$16/8^4$	$8/8^4$

Since these probabilities sum to $C = 80/8^4$, our hypothetical seven-sided die is represented by the random number generator

$$\textsc{Random}\left(\frac{4/8^4}{80/8^4}, \frac{8/8^4}{80/8^4}, \frac{8/8^4}{80/8^4}, \frac{24/8^4}{80/8^4}, \frac{12/8^4}{80/8^4}, \frac{16/8^4}{80/8^4}, \frac{8/8^4}{80/8^4}\right)$$
$$= \textsc{Random}\left(\frac{4}{80}, \frac{8}{80}, \frac{8}{80}, \frac{24}{80}, \frac{12}{80}, \frac{16}{80}, \frac{8}{80}\right).$$

Let's assume that after rolling this seven-sided die, we arrive at the *Profile*-randomly generated 4-mer GCGT (the fourth 4-mer in the deleted sequence). The deleted string ccgGCGTtag is now added back to the collection of motifs, and **GCGT** substitutes the previously chosen ccgG in the third string in *Dna*, as shown below. We then roll a fair five-sided die and randomly select the first string from *Dna* for removal.

	ttACCT**taac**		----------
	gAT**GTct**gtc		gAT**GTct**gtc
Dna	ccg**GCGT**tag	$\longrightarrow$	ccg**GCGT**tag
	c**acta**ACGAg		c**acta**ACGAg
	cgtcag**AGGT**		cgtcag**AGGT**

After constructing the motif and profile matrices, we obtain the following:

Motifs					PROFILE(*Motifs*)				
G	T	c	t		A:	2/4	0	0	1/4
G	C	G	T		C:	0	2/4	1/4	0
a	c	t	a		G:	2/4	1/4	2/4	0
A	G	G	T		T:	0	1/4	1/4	3/4

Note that the profile matrix looks more biased toward the implanted motif than the previous profile matrix did. We update the count and profile matrices with pseudocounts:

COUNT(*Motifs*)					PROFILE(*Motifs*)				
A:	3	1	1	2	A:	3/8	1/8	1/8	2/8
C:	1	3	2	1	C:	1/8	3/8	2/8	1/8
G:	3	2	3	1	G:	3/8	2/8	3/8	1/8
T:	1	2	2	4	T:	1/8	2/8	2/8	4/8

Then, we compute the probabilities of all 4-mers in the deleted string ttACCTtaac:

ttAC	tACC	ACCT	CCTt	CTta	Ttaa	taac
$2/8^4$	$2/8^4$	$72/8^4$	$24/8^4$	$8/8^4$	$4/8^4$	$1/8^4$

When we roll a seven-sided die, we arrive at the *Profile*-randomly generated *k*-mer **ACCT**, which we add to the collection *Motifs*. After rolling the five-sided die once again, we randomly select the fourth string for removal.

	tt**ACCT**taac		tt**ACCT**taac
	gAT**GTct**gtc		gAT**GTct**gtc
Dna	ccg**GCGT**tag	$\longrightarrow$	ccg**GCGT**tag
	c**acta**ACGAg		----------
	cgtcag**AGGT**		cgtcag**AGGT**

We further add pseudocounts and construct the resulting count and profile matrices:

```
        A C C T
Motifs  G T c t
        G C G T
        A G G T
```

COUNT(*Motifs*)	A:	3	1	1	1	PROFILE(*Motifs*)	A:	3/8	1/8	1/8	1/8
	C:	1	3	3	1		C:	1/8	3/8	3/8	1/8
	G:	3	2	3	1		G:	3/8	2/8	3/8	1/8
	T:	1	2	1	5		T:	1/8	2/8	1/8	5/8

We now compute the probabilities of all 4-mers in the deleted string `cactaACGAg`:

cact	acta	ctaA	taAC	aACG	ACGA	CGAg
$15/8^4$	$9/8^4$	$2/8^4$	$1/8^4$	$9/8^4$	$27/8^4$	$2/8^4$

We need to roll a seven-sided die to produce a *Profile*-randomly generated 4-mer. Assuming the most probable scenario in which we select **ACGA**, we update the selected 4-mers as follows:

```
     tt**ACCT**taac
     gAT**GTct**gtc
Dna  ccg**GCGT**tag
     cacta**ACGA**g
     cgtcag**AGGT**
```

You can see that the algorithm is beginning to converge. Rest assured that a subsequent iteration will produce all implanted motifs after we select the second string in *Dna* (when the incorrect 4-mer `GTct` will likely change into the implanted (4, 1)-motif `ATGT`).

STOP and Think: Run **GIBBSSAMPLER** on the Subtle Motif Problem. What do you find?

Although **GIBBSSAMPLER** performs well in many cases, it may converge to a suboptimal solution, particularly for difficult search problems with elusive motifs. A **local optimum** is a solution that is optimal within a small neighboring set of solutions, which is in contrast to a **global optimum**, or the optimal solution among all possible solutions. Since **GIBBSSAMPLER** explores just a small subset of solutions, it may "get stuck" in a local optimum. For this reason, similarly to **RANDOMIZEDMOTIFSEARCH**, it should be run many times with the hope that one of these runs will produce the best-scoring motifs. Yet convergence to a local optimum is just one of many issues we must consider in motif finding; see **DETOUR: Complications in Motif Finding** for some other challenges.

When we run **GIBBSSAMPLER** 2,000 times on the Subtle Motif Problem with implanted 15-mer **AAAAAAAAGGGGGGG** (each time with new randomly selected *k*-mers for $N = 200$ iterations), it returns a collection *Motifs* with consensus **AAAAAAgAGGGGGGt** and SCORE(*Motifs*) equal to 38. This score is even lower than the score of 40 expected from the implanted motifs!

Epilogue: How Does Tuberculosis Hibernate to Hide from Antibiotics?

Tuberculosis (TB) is an infectious disease that is caused by the *Mycobacterium tuberculosis* bacterium (MTB) and is responsible for over a million deaths each year. Although the spread of TB has been greatly reduced due to antibiotics, strains that resist all available treatments are now emerging. MTB is successful as a pathogen because it can persist in humans for decades without causing disease; in fact, one-third of the world population has **latent MTB infections**, in which MTB lies dormant within the host's body and may or may not reactivate at a later time. The widespread prevalence of latent infections makes it difficult to control TB epidemics. Biologists are therefore interested in finding out what makes the disease latent and how MTB activates itself within a host.

It remains unclear why MTB can stay latent for so long and how it survives during latency. The resistance of latent TB to antibiotics implies that MTB may have an ability to shut down expression of most genes and stay dormant, not unlike bears hibernating in the winter. Hibernation in bacteria is called **sporulation** because many bacteria form protective and metabolically dormant **spores** that can survive in tough conditions, allowing the bacteria to persist in the environment until conditions improve.

Hypoxia, or oxygen shortage, is often associated with latent forms of TB. Biologists have found that MTB becomes dormant in low-oxygen environments, presumably with the idea that the host's lungs will recover enough to potentially spread the disease in the future. Since MTB shows a remarkable ability to survive for years without oxygen, it is important to identify MTB genes responsible for the development of the latent state under hypoxic conditions. Biologists are interested in finding a **transcription factor** that "senses" the shortage of oxygen and starts a genetic program that affects the expression of many genes, allowing MTB to adapt to hypoxia.

In 2003, biologists found the **dormancy survival regulator(DosR)**, a transcription factor that regulates many genes whose expression dramatically changes under hypoxic conditions. However, it remained unclear how DosR regulates these genes, and its transcription factor binding site remained unknown. In an attempt to resolve this puzzle, biologists performed a DNA array experiment and found 25 genes whose expression levels significantly changed in hypoxic conditions. Given the upstream regions of these genes, each of which is 250 nucleotides long, we would like to discover the "hidden message" that DosR uses to control the expression of these genes.

To simplify the problem a bit, we have selected just 10 of the 25 genes, resulting in the **DosR dataset**. We will try to identify motifs in this dataset using the arsenal of motif finding tools that we have developed. However, we will not give you a hint about the DosR motif.

What *k*-mer size should we choose in order to analyze the DosR dataset using MEDIANSTRING and RANDOMIZEDMOTIFSEARCH? Taking a wild guess and running these algorithms for *k* from 8 to 12 returns the consensus strings shown below.

	MEDIANSTRING			RANDOMIZEDMOTIFSEARCH	
k	**Consensus**	**Score**	***k***	**Consensus**	**Score**
8	CATCGGCC	11	8	CCGACGGG	13
9	GGCGGGGAC	16	9	CCATCGGCC	16
10	GGTGGCCACC	19	10	CCATCGGCCC	21
11	GGACTTCCGGC	20	11	ACCTTCGGCCC	25
12	GGACTTCCGGCC	23	12	GGACCAACGGCC	28

STOP and Think: Can you infer the DosR binding site from these median strings? What do you think is the length of the binding site?

Note that although the consensus strings returned by RANDOMIZEDMOTIFSEARCH generally deviate from the median strings, the consensus 12-mer (GGACCAACGGCC, with score 28) is very similar to the median string (GGACTTCCGGCC, with score 23).

While the motifs returned by RANDOMIZEDMOTIFSEARCH are slightly less conserved than the motifs returned by MEDIANSTRING, the former algorithm has the advantage of being able to find longer motifs (since MEDIANSTRING becomes too slow for longer motifs). The motif of length 20 returned by RANDOMIZEDMOTIFSEARCH is CGGGACCTACGTCCCTAGCC (with score 57). As shown below, the consensus strings of length 12 found by RANDOMIZEDMOTIFSEARCH and MEDIANSTRING are "embedded" with small variations in the longer motif of length 20:

```
   GGACCAACGGCC
CGGGACCTACGTCCCTAGCC
   GGACTTCCGGCC
```

Finally, in 2,000 runs with $N = 200$, **GIBBSSAMPLER** returned the same consensus string of length 20 for the DosR dataset as **RANDOMIZEDMOTIFSEARCH** but generated a different collection of motifs with a smaller score of 55.

As you have seen in this chapter, different motif finding algorithms generate somewhat different results, and it remains unclear how to identify the DosR motif in MTB. Try to answer this question and find all putative DosR motifs in MTB as well as all genes that they regulate. We will provide the upstream regions of all 25 genes identified in the DosR study to help you address the following problem.

Challenge Problem: Infer the profile of the DosR motif and find all its putative occurrences in *Mycobacterium tuberculosis.*

Charging Stations

Solving the Median String Problem

The first potential issue with implementing **MEDIANSTRING** is writing a function to compute $d(Pattern, Dna) = \sum_{i=1}^{t} d(Pattern, Dna_i)$, the sum of distances between *Pattern* and each string in $Dna = \{Dna_1, \ldots, Dna_t\}$. This task is achieved by the following pseudocode.

```
DISTANCEBETWEENPATTERNANDSTRINGS(Pattern, Dna)
    k ← |Pattern|
    distance ← 0
    for each string Text in Dna
        HammingDistance ← ∞
        for each k-mer Pattern' in Text
            if HammingDistance > HAMMINGDISTANCE(Pattern, Pattern')
                HammingDistance ← HAMMINGDISTANCE(Pattern, Pattern')
        distance ← distance + HammingDistance
    return distance
```

To solve the Median String Problem, we need to iterate through all possible 4^k k-mers *Pattern* before computing $d(Pattern, Dna)$. The pseudocode below is a modification of **MEDIANSTRING** using the function NUMBERTOPATTERN (implemented in **Charging Station: Converting Patterns Into Numbers and Vice-Versa**), which is applied to convert all integers from 0 to $4^k - 1$ into all possible k-mers.

```
MEDIANSTRING(Dna, k)
    distance ← ∞
    for i ← 0 to 4^k − 1
        Pattern ← NUMBERTOPATTERN(i, k)
        if distance > DISTANCEBETWEENPATTERNANDSTRINGS(Pattern, Dna)
            distance ← DISTANCEBETWEENPATTERNANDSTRINGS(Pattern, Dna)
            Median ← Pattern
    return Median
```

Detours

Gene expression

Genes encode proteins, and proteins dictate cell function. To respond to changes in their environment, cells must therefore control their protein levels. The flow of information from DNA to RNA to protein means that the cell can adjust the amount of proteins that it produces during both transcription (DNA to RNA) and translation (RNA to protein).

Transcription begins when an RNA polymerase binds to a **promoter sequence** on the DNA molecule, which is often located just upstream from the starting point for transcription. The initiation of transcription is a convenient control point for the cell to regulate gene expression since it is at the very beginning of the protein production process. The genes transcribed in a cell are controlled by various transcription regulators that may increase or suppress transcription.

DNA arrays

A **DNA array** is a collection of DNA molecules attached to a solid surface. Each spot on the array is assigned a unique DNA sequence called a **probe** that measures the expression level of a specific gene, known as a **target**. In most arrays, probes are synthesized and then attached to a glass or silicon chip (Figure 2.8).

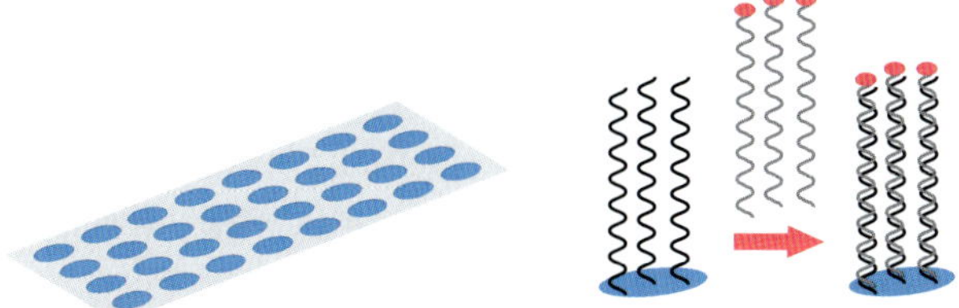

FIGURE 2.8 Fluorescently labeled DNA binds to a complementary probe on a DNA array.

Fluorescently labeled targets then bind to their corresponding probe (e.g., when their sequences are complementary), generating a fluorescent signal. The strength of this signal depends upon the amount of target sample that binds to the probe at a given spot. Thus, the higher the expression level of a gene, the higher the intensity of its fluorescent signal on the array. Since an array may contain millions of probes, biologists can measure the expression of many genes in a single array experiment. The DNA array experiment that identified the evening element in *Arabidopsis thaliana* measured the expression of 8,000 genes.

Buffon's needle

Comte de Buffon was a prolific 18th Century naturalist whose writings on natural history were popular at the time. However, his first paper was in mathematics; in 1733, he wrote an essay on a Medieval French game called "Le jeu de franc carreau". In this game, a single player flips a coin into the air, and the coin lands on a checkerboard. The player wins if the coin lands completely within one of the squares on the board, and loses otherwise (Figure 2.9 (left)). Buffon asked a natural question: what is the probability that the player wins?

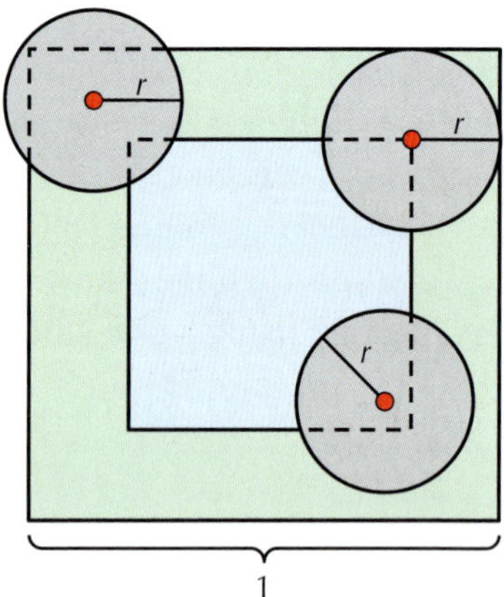

FIGURE 2.9 (Left) A game of "franc carreau" with four coins. Two of the coins have landed within one of the squares of the checkerboard and are considered winners, whereas the other two have landed on a boundary and are considered losers. (Right) Three coins shown on a single square of the checkerboard (the green outside square); one coin is a loser, another is a winner, and the third corresponds to a boundary case. You can see that if the coin has radius r, then the probability of winning the game corresponds to the probability that the center of the coin (shown as a red dot) lands within the blue square, which has side length $1 - 2r$. This probability is the ratio of the squares' areas, which is $(1 - 2r)^2$.

Let's assume that the checkerboard consists of just a single square with side length 1, that the coin has radius $r < 1/2$, and that the center of the coin always lands within the square. Then the player can only win if the center of the circle falls within an imaginary central square of side length $1 - 2r$ (Figure 2.9 (right)). Assuming that the coin lands anywhere on the larger square with uniform probability, then the probability that the coin falls completely within the smaller square is given by the ratio of the areas of the two squares, or $(1 - 2r)^2$.

Four decades later, Buffon published a paper describing a similar game in which the player uniformly drops a needle onto a floor covered by long wooden panels of equal width. In this game, which has become known as **Buffon's needle**, the player wins if the needle falls entirely within one of the panels. Note that computing the probability of a win is now complicated by the fact that the needle is described by an orientation in addition to its position. Nevertheless, the first game gives us an idea for how to solve this problem: once we fix a position for the center of the needle, its collection of different possible orientations sweep out a circle (Figure 2.10 (left)).

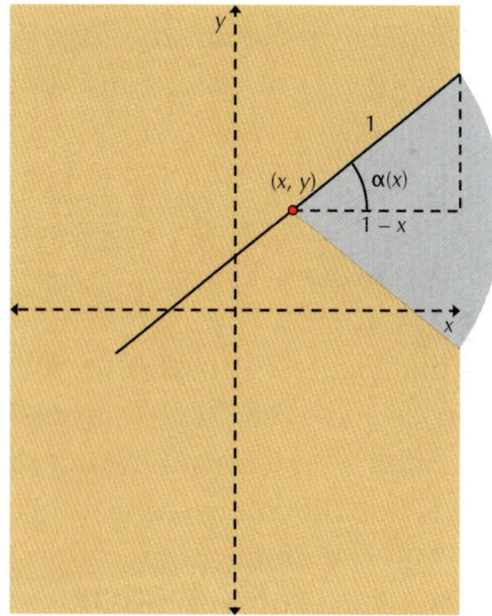

FIGURE 2.10 (Left) Once we fix a point for the center of the needle (shown as a red dot), its collection of possible orientations sweep out a circle. In the circle on the left, the needle will always lie within the dark brown panel, regardless of its orientation. In the circle on the right, one of the two needles lies within the dark brown panel, whereas another is shown crossing over into the adjacent panel. (Right) Once we fix a point (x, y) for the center of the needle, there is a critical angle $\alpha(x)$ such that all angles between $-\alpha(x)$ and $\alpha(x)$ will cause the needle to cross over into the next panel. In this figure, the length of the needle is equal to the width of the panel.

The probability that the player wins depends on the length of the needle with respect to the distance between wooden panels. We will assume that both of these lengths are equal to 2, and we will find the probability of a *loss* instead of a win. To this end, we first ask a simpler question: if the center of the needle were to land in the same place every time, then what is the likelihood that the needle crosses a panel?

To address this question, let's map the panel into which the needle falls onto a coordinate plane, with the y-axis dividing the panel into two smaller panels of width 1 (Figure 2.10 (right)). If the center of the needle lands at position (x, y) with $x > 0$, then its orientation can be described by an angle θ, where θ ranges from $-\pi/2$ to $\pi/2$ radians. If $\theta = 0$, then the needle will cross the vertical line $x = 1$; if $\theta = \pi/2$, then the needle will not cross the line $x = 1$. Yet more importantly, since the needle's center position is fixed, there must be some critical angle $\alpha(x)$ such that the needle always touches this line if $-\alpha(x) \leq \theta \leq \alpha(x)$. If the needle is dropped randomly, then any value of θ between $-\pi/2$ and $\pi/2$ is equally likely, and so we obtain that the probability of a loss given this position of the needle is equal to $2 \cdot \alpha(x)/\pi$.

Following the same reasoning, the needle can take any position x with equal probability. To find the probability of a loss, Pr(loss), we must therefore compute an "average" of the values $2 \cdot \alpha(x)/\pi$ as x continuously ranges from -1 to 1. This average can be represented using an integral,

$$\Pr(\text{loss}) = \frac{\int_{-1}^{1} \frac{2 \cdot \alpha(x)}{\pi} \, dx}{1 - (-1)} = \int_{-1}^{1} \frac{\alpha(x)}{\pi} \, dx = 2 \int_{0}^{1} \frac{\alpha(x)}{\pi} \, dx \, .$$

Revisiting Figure 2.10 (right), applying some basic trigonometry tells us that $\cos \alpha(x)$ is equal to $1 - x$, so that $\alpha(x) = \arccos(1 - x)$. After making this substitution into the above equation — and consulting our dusty calculus textbook — Pr(loss) must be equal to $2/\pi$. It is not difficult to see that this probability will be the same when the needle is dropped onto any number of wooden panels.

But what does Buffon's needle have to do with randomized algorithms? In 1812, none other than Laplace pointed out that Buffon's needle could be used to approximate π, and the world's first Monte Carlo algorithm was born. Specifically, we can approximate the probability P_e of a loss *empirically* by dully flipping a needle into the air thousands of times (or asking a computer to do it for us). Once we have computed this empirical probability, we can conclude that P_e is approximately equal to $2/\pi$, and thus

$$\pi \approx \frac{2}{P_e} \, .$$

STOP and Think: How does this approximation change in the following cases?

1. The needle is shorter than the width between panels.
2. The needle is longer than the width between panels.

Complications in motif finding

Motif finding becomes difficult if the **background nucleotide distribution** in the sample is skewed. In this case, searching for k-mers with the minimum score or entropy may lead to a biologically irrelevant motif composed from the most frequent nucleotides in the sample. For example, if A has frequency 85% and T, G, and C have frequencies of 5%, then k-mer AA...AA may represent a motif with minimum score, thus disguising biologically relevant motifs. For example, the relevant motif GCCG with score 5 in the example below loses out to the motif **aaaa** with score 1.

```
      taaaaGTCGa
      acGCTGaaaa
Dna   aaaaGCCTat
      aCCCGaataa
      agaaaaGGCG
```

To find biologically relevant motifs in samples with biased nucleotide frequencies, you may therefore want to use a generalization of entropy called "relative entropy" (see **DETOUR: Relative Entropy**).

Another complication in motif finding is that many motifs are best represented in a different alphabet than the alphabet of 4 nucleotides. Let W denote either A or T, S denote either G or C, K denote either G or T, and Y denote either C or T. Now, consider the motif CSKWYWWATKWATYYK, which represents the CSRE motif in yeast (recall Figure 2.3 from page 76). This strong motif in a hybrid alphabet corresponds to 2^{11} different motifs in the standard 4-letter alphabet of nucleotides. However, each of these 2^{11} motifs is too weak to be found by the algorithms we have considered in this chapter.

Relative entropy

Given a collection of strings *Dna*, the **relative entropy** of a $4 \times k$ profile matrix $P = (p_{r,j})$ is defined as

$$\sum_{j=1}^{k} \sum_{r\in\{\texttt{A,C,G,T}\}} p_{r,j} \cdot \log_2(p_{r,j}/b_r) =$$

$$\sum_{j=1}^{k} \sum_{r\in\{\texttt{A,C,G,T}\}} p_{r,j} \cdot \log_2(p_{r,j}) - \sum_{j=1}^{k} \sum_{r\in\{\texttt{A,C,G,T}\}} p_{r,j} \cdot \log_2(b_r),$$

where b_r is the frequency of nucleotide r in *Dna*. Note that the sum in the expression for entropy is preceded by a negative sign ($-\sum_{j=1}^{k}\sum_{r\in\{\texttt{A,C,G,T}\}} p_{r,j} \cdot \log_2(p_{r,j})$), whereas the sum on the left side of the *relative* entropy equation does not have this negative sign. Therefore, although we minimized the entropy of a motif matrix, we will now attempt to maximize the relative entropy.

The term $-\sum_{j=1}^{k}\sum_{r\in\{\texttt{A,C,G,T}\}} p_{r,j} \cdot \log_2(b_r)$ is called the **cross-entropy** of the profile matrix P; note that the relative entropy of a profile matrix is simply the difference between the profile's cross-entropy and its entropy. For example, the relative entropy for the motif GCCG in the example from **DETOUR: Complications in Motif Finding** is equal to $9.85 - 3.53 = 6.32$, as shown below. In this example, $b_{\texttt{A}} = 0.5$, $b_{\texttt{C}} = 0.18$, $b_{\texttt{G}} = 0.2$, and $b_{\texttt{T}} = 0.12$.

PAGE 113

		G	T	C	G
		G	C	T	G
Motifs		G	C	C	T
		c	C	C	G
		G	G	C	G
	A:	0.0	0.0	0.0	0.0
	C:	0.2	0.6	0.8	0.0
PROFILE(*Motifs*)	G:	0.8	0.2	0.0	0.8
	T:	0.0	0.2	0.2	0.2

Entropy	0.72 + 1.37 + 0.72 + 0.72 = 3.53
Cross-entropy	2.35 + 2.56 + 2.47 + 2.47 = 9.85

For the more conserved but irrelevant motif aaaa, the relative entropy is equal to $4.18 - 0.72 = 3.46$, as shown below. Thus, GCCG loses to aaaa with respect to entropy but wins with respect to relative entropy.

		a	a	a	a
		a	a	a	a
Motifs		a	a	a	a
		a	t	a	a
		a	a	a	a
	A:	1.0	0.8	1.0	1.0
PROFILE(*Motifs*)	C:	0.0	0.0	0.0	0.0
	G:	0.0	0.0	0.0	0.0
	T:	0.0	0.2	0.0	0.0

$$\textbf{Entropy} \quad 0.0 + 0.72 + 0.0 + 0.0 = 0.72$$
$$\textbf{Cross-entropy} \quad 0.94 + 1.36 + 0.94 + 0.94 = 4.18$$

Bibliography Notes

Konopka and Benzer, 1971 bred flies with abnormally short (19 hours) and long (28 hours) circadian rhythms and then traced these abnormalities to a single gene. Harmer et al., 2000 discovered the evening transcription factor binding site that orchestrates the circadian clock in plants. Excellent coverage of this discovery is given by Cristianini and Hahn, 2006. Park et al., 2003 found a transcription factor that mediates the hypoxic response of *Mycobacterium tuberculosis*.

Hertz and Stormo, 1999 described the first greedy algorithm for motif finding. The general framework for Gibbs sampling was described by Geman and Geman, 1984 and was named Gibbs sampling in reference to its similarities with some approaches in statistical mechanics (Josiah Willard Gibbs was one of the founders of statistical mechanics). Lawrence et al., 1993 adapted Gibbs sampling for motif finding.

HOW DO WE ASSEMBLE GENOMES?
Graph Algorithms
1735
1857
1946
2000
BIOINFORMATICS TIME MACHINE

Exploding Newspapers

Imagine that we stack a hundred copies of the June 27, 2000 edition of the *New York Times* on a pile of dynamite, and then we light the fuse. We ask you to further suspend your disbelief and assume that the newspapers are not all incinerated but instead explode cartoonishly into smoldering pieces of confetti. How could we use the tiny snippets of newspaper to figure out what the news was on June 27, 2000? We will call this crazy conundrum the **Newspaper Problem** (see Figure 3.1).

FIGURE 3.1 Don't try this at home! Crazy as it may seem, the Newspaper Problem serves as an analogy for the computational framework of genome assembly.

The Newspaper Problem is even more difficult than it may seem. Because we had multiple copies of the same edition of the newspaper, and because we undoubtedly lost some information in the blast, we cannot simply glue together one of the newspaper copies in the same way that we would assemble a jigsaw puzzle. Instead, we need to use *overlapping* fragments from different copies of the newspaper to reconstruct the day's news, as shown in Figure 3.2.

Fine, you ask, *but what do exploding newspapers have to do with biology?* Determining the order of nucleotides in a genome, or **genome sequencing**, presents a fundamental task in bioinformatics. Genomes vary in length; your own genome is roughly 3 billion

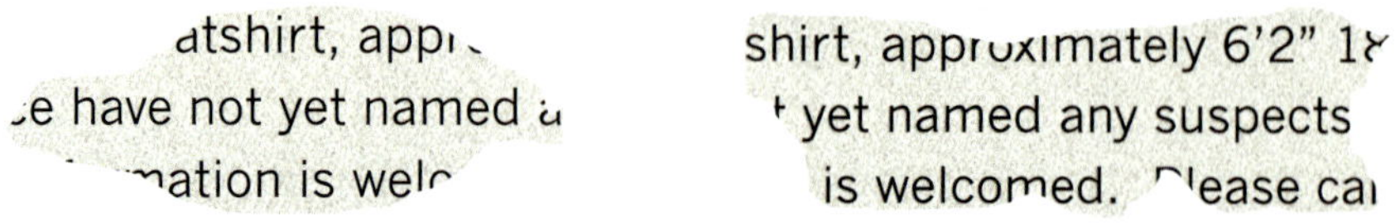
atshirt, app
e have not yet named
nation is wel
shirt, appr ximately 6'2" 1
yet named any suspects
is welcomed. 'ease ca

FIGURE 3.2 In the Newspaper Problem, we need to use overlapping shreds of paper to figure out the news.

nucleotides long, whereas the genome of *Amoeba dubia*, an amorphous unicellular organism, is approximately 200 times longer! This unicellular organism competes with the rare Japanese flower *Paris japonica* for the title of species with the longest genome.

The first sequenced genome, belonging to a φX174 bacterial phage (i.e., a virus that preys on bacteria), had only 5,386 nucleotides and was completed in 1977 by Frederick Sanger. Four decades after this Nobel Prize-winning discovery, genome sequencing has raced to the forefront of bioinformatics research, as the cost of genome sequencing plummeted. Because of the decreasing cost of sequencing, we now have thousands of sequenced genomes, including those of many mammals (Figure 3.3).

2010
bat
panda
neanderthal
2009
cow
horse
elephant
2007
cat
opossum
2006
macaque
2005
dog
chimpanzee
2004
rat
2002
mouse
2000
human

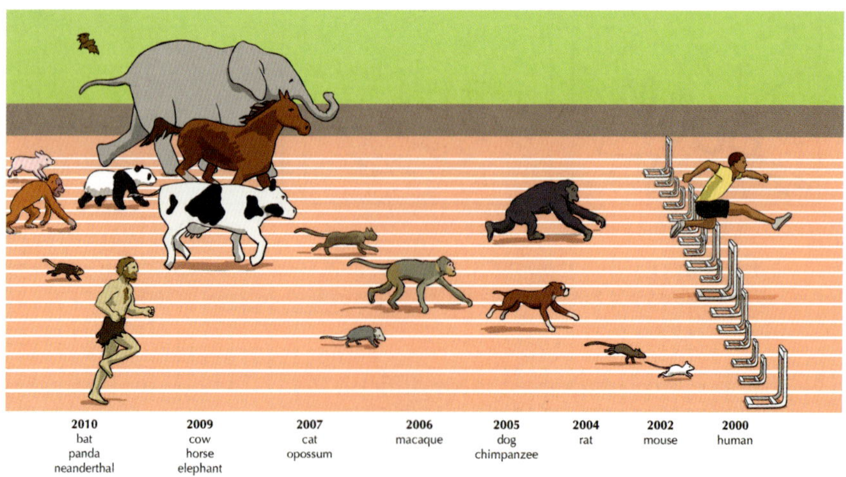

FIGURE 3.3 The first mammals with sequenced genomes.

To sequence a genome, we must clear some practical hurdles. The largest obstacle is the fact that biologists still lack the technology to read the nucleotides of a genome from beginning to end in the same way that you would read a book. The best they can do

is sequence much shorter DNA fragments called **reads**. The reasons why researchers can sequence small pieces of DNA but not long genomes warrant their own discussion in **DETOUR: A Short History of DNA Sequencing Technologies**. In this chapter, our aim is to turn an apparent handicap into a useful tool for assembling the genome back together.

PAGE 171

The traditional method for sequencing genomes is described as follows. Researchers take a small tissue or blood sample containing millions of cells with identical DNA, use biochemical methods to break the DNA into fragments, and then sequence these fragments to produce reads (Figure 3.4). The difficulty is that researchers do not know where in the genome these reads came from, and so they must use overlapping reads to reconstruct the genome. Thus, putting a genome back together from its reads, or **genome assembly**, is just like the Newspaper Problem.

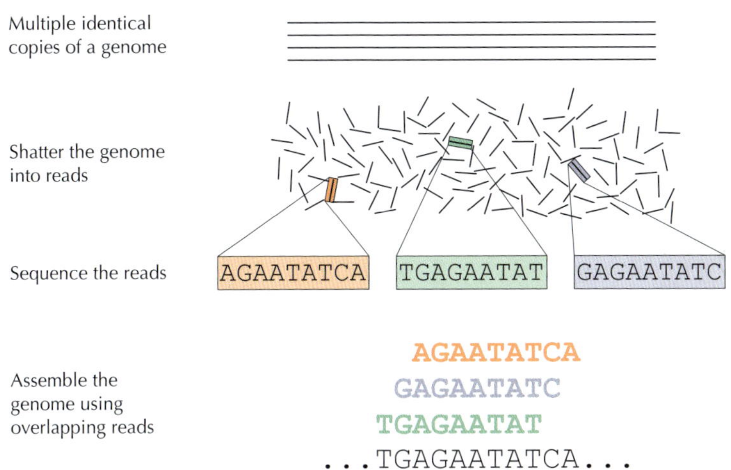

FIGURE 3.4 In DNA sequencing, many identical copies of a genome are broken in random locations to generate short reads, which are then sequenced and assembled into the nucleotide sequence of the genome.

Even though researchers have sequenced many genomes, a giant genome like that of *Amoeba dubia* still remains beyond the reach of modern sequencing technologies. You might guess that the barrier to sequence such a genome would be experimental, but that is not true; biologists can easily generate enough reads to analyze a large genome, but assembling these reads still presents a major computational challenge.

The String Reconstruction Problem

Genome assembly is more difficult than you think

Before we introduce a computational problem modeling genome assembly, we will take a moment to discuss a few practical complications that make genome assembly more difficult than the Newspaper Problem.

First, DNA is double-stranded, and we have no way of knowing *a priori* which strand a given read derives from, meaning that we will not know whether to use a read or its reverse complement when assembling a particular strand of a genome. Second, modern sequencing machines are not perfect, and the reads that they generate often contain errors. Sequencing errors complicate genome assembly because they prevent us from identifying all overlapping reads. Third, some regions of the genome may not be covered by any reads, making it impossible to reconstruct the entire genome.

Since the reads generated by modern sequencers often have the same length, we may safely assume that reads are all k-mers for some value of k. The first part of this chapter will assume an ideal — and unrealistic — situation in which all reads come from the same strand, have no errors, and exhibit **perfect coverage**, so that every k-mer substring of the genome is generated as a read. Later, we will show how to relax these assumptions for more realistic datasets.

Reconstructing strings from k-mers

We are now ready to define a computational problem modeling genome assembly. Given a string *Text*, its **k-mer composition** $\text{COMPOSITION}_k(Text)$ is the collection of all k-mer substrings of *Text* (including repeated k-mers). For example,

$$\text{COMPOSITION}_3(\texttt{TATGGGGTGC}) = \{\texttt{ATG}, \texttt{GGG}, \texttt{GGG}, \texttt{GGT}, \texttt{GTG}, \texttt{TAT}, \texttt{TGC}, \texttt{TGG}\}.$$

Note that we have listed k-mers in **lexicographic order** (i.e., how they would appear in a dictionary) rather than in the order of their appearance in TATGGGGTGC. We have done this because the correct ordering of reads is unknown when they are generated.

String Composition Problem:
Generate the k-mer composition of a string.

Input: A string *Text* and an integer k.
Output: $\text{COMPOSITION}_k(Text)$, where the k-mers are arranged in lexicographic order.

Solving the String Composition Problem is a straightforward exercise, but in order to model genome assembly, we need to solve its inverse problem.

String Reconstruction Problem:
Reconstruct a string from its k-mer composition.

> **Input**: An integer k and a collection *Patterns* of k-mers.
> **Output**: A string *Text* with k-mer composition equal to *Patterns* (if such a string exists).

Before we ask you to solve the String Reconstruction Problem, let's consider the following example of a 3-mer composition:

```
AAT ATG GTT TAA TGT
```

The most natural way to solve the String Reconstruction Problem is to mimic the solution of the Newspaper Problem and "connect" a pair of k-mers if they overlap in $k-1$ symbols. For the above example, it is easy to see that the string should start with TAA because there is no 3-mer ending in TA. This implies that the next 3-mer in the string should start with AA. There is only one 3-mer satisfying this condition, AAT:

```
TAA
 AAT
```

In turn, AAT can only be extended by ATG, which can only be extended by TGT, and so on, leading us to reconstruct **TAATGTT**:

```
TAA
 AAT
  ATG
   TGT
    GTT
TAATGTT
```

It looks like we are finished with the String Reconstruction Problem and can let you move on to the next chapter. To be sure, let's consider another 3-mer composition:

```
AAT ATG ATG ATG CAT CCA GAT GCC GGA GGG GTT TAA TGC TGG TGT
```

Exercise Break: Reconstruct a string with this 3-mer composition.

If we start again with TAA, then the next 3-mer in the string should start with AA, and there is only one such 3-mer, AAT. In turn, AAT can only be extended by ATG:

```
TAA
 AAT
  ATG
TAATG
```

ATG can be extended either by TGC, or TGG, or TGT. Now we must decide which of these 3-mers to choose. Let's select TGT:

```
TAA
 AAT
  ATG
   TGT
TAATGT
```

After TGT, our only choice is GTT:

```
TAA
 AAT
  ATG
   TGT
    GTT
TAATGTT
```

Unfortunately, now we are stuck at GTT because no 3-mers in the composition start with TT! We could try to extend TAA to the left, but no 3-mers in the composition end with TA.

You may have found this trap on your own and already discovered how to escape it. Like a good chess player, if you think a few steps ahead, then you would never extend ATG by TGT until reaching the end of the genome. With this thought in mind, let's take a step back, extending ATG by TGC instead:

```
TAA
 AAT
  ATG
   TGC
TAATGC
```

Continuing the process, we obtain the following assembly:

```
TAA
 AAT
  ATG
   TGC
    GCC
     CCA
      CAT
       ATG
        TGG
         GGA
          GAT
           ATG
            TGT
             GTT
TAATGCCATGGATGTT
```

Yet this assembly is incorrect because we have only used fourteen of the fifteen 3-mers in the composition (we omitted GGG), making our reconstructed genome one nucleotide too short.

Repeats complicate genome assembly

The difficulty in assembling this genome arises because ATG is *repeated* three times in the 3-mer composition, which causes us to have the three choices TGG, TGC, and TGT by which to extend ATG. Repeated substrings in the genome are not a serious problem when we have just fifteen reads, but with millions of reads, repeats make it much more difficult to "look ahead" and construct the correct assembly.

If you followed **DETOUR: Probabilities of Patterns in a String** from Chapter 1, you know how unlikely it is to witness a long repeat in a randomly generated sequence of nucleotides. You also know that real genomes are anything but random. Indeed, approximately 50% of the human genome is made up of repeats, e.g., the approximately 300 nucleotide-long **Alu sequence** is repeated over a million times, with only a few nucleotides inserted/deleted/substituted each time (see **DETOUR: Repeats in the Human Genome**).

PAGE 53

PAGE 173

An analogy illustrating the difficulty of assembling a genome with many repeats is the Triazzle® jigsaw puzzle (Figure 3.5). People usually put together jigsaw puzzles by connecting matching pieces. However, every piece in the Triazzle matches more than one other piece; in Figure 3.5, each frog appears several times. If you proceed carelessly, then you will likely match most of the pieces but fail to fit the remaining ones. And yet the Triazzle has only 16 pieces, which should give us pause about assembling a genome from millions of reads.

FIGURE 3.5 Each Triazzle has only sixteen pieces but carries a warning: "It's Harder than it Looks!"

Exercise Break: Design a strategy for assembling the Triazzle puzzle.

String Reconstruction as a Walk in the Overlap Graph

From a string to a graph

Repeats in a genome necessitate some way of looking ahead to see the correct assembly in advance. Returning to our previous example, you may have already found that **TAATGCCATGGGATGTT** is a solution to the String Reconstruction Problem for the collection of fifteen 3-mers in the last section, as illustrated below. Note that we use a different color for each interval of the string between occurrences of **ATG**.

```
TAA
 AAT
  ATG
   TGC
    GCC
     CCA
      CAT
       ATG
        TGG
         GGG
          GGA
           GAT
            ATG
             TGT
              GTT
TAATGCCATGGGATGTT
```

STOP and Think: Is this the only solution to the String Reconstruction Problem for this collection of 3-mers?

In Figure 3.6, consecutive 3-mers in **TAATGCCATGGGATGTT** are linked together to form the **genome path**.

FIGURE 3.6 The fifteen color-coded 3-mers making up **TAATGCCATGGGATGTT** are joined into the genome path according to their order in the genome.

String Spelled by a Genome Path Problem:
Reconstruct a string from its genome path.

Input: A sequence of k-mers $Pattern_1, \ldots, Pattern_n$ such that the last $k-1$ symbols of $Pattern_i$ are equal to the first $k-1$ symbols of $Pattern_{i+1}$ for $1 \leq i \leq n-1$.
Output: A string *Text* of length $k+n-1$ such that the i-th k-mer in *Text* is equal to $Pattern_i$ (for $1 \leq i \leq n$).

Reconstructing a genome from its genome path is easy: as we proceed from left to right, the 3-mers "spell" out **TAATGCCATGGGATGTT**, adding one new symbol to the genome at each new 3-mer. Unfortunately, constructing this string's genome path requires us to know the genome in advance.

STOP and Think: Could you construct the genome path if you knew only the genome's 3-mer composition?

In this chapter, we will use the terms **prefix** and **suffix** to refer to the first $k - 1$ nucleotides and last $k - 1$ nucleotides of a k-mer, respectively. For example, PREFIX(**TAA**) = **TA** and SUFFIX(**TAA**) = **AA**. We note that the suffix of a 3-mer in the genome path is equal to the prefix of the following 3-mer in the path. For example, SUFFIX(**TAA**) = PREFIX(**AAT**) = **AA** in the genome path for **TAATGCCATGGGATGTT**.

This observation suggests a method of constructing a string's genome path from its k-mer composition: we will use an arrow to connect any k-mer *Pattern* to a k-mer *Pattern'* if the suffix of *Pattern* is equal to the prefix of *Pattern'*.

STOP and Think: Apply the rule we just described to the 3-mer composition of **TAATGCCATGGGATGTT**. Are you able to reconstruct the genome path of **TAATGCCATGGGATGTT**?

If we follow the rule of connecting two 3-mers with an arrow every time the suffix of one is equal to the prefix of the other, then we will connect all consecutive 3-mers in **TAATGCCATGGGATGTT** as in Figure 3.6. However, because we don't know this genome in advance, we wind up having to connect many other pairs of 3-mers as well. For example, each of the three occurrences of **ATG** should be connected to **TGC**, **TGG**, and **TGT**, as shown in Figure 3.7.

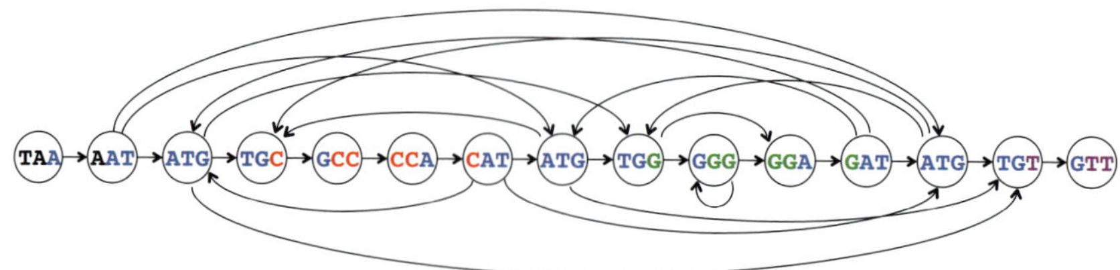

FIGURE 3.7 The graph showing all connections between nodes representing the 3-mer composition of **TAATGCCATGGGATGTT**. This graph has fifteen nodes and 28 edges. Note that the genome can still be spelled out by walking along the horizontal edges from **TAA** to **GTT**.

Figure 3.7 presents an example of a **graph**, or a network of **nodes** connected by **edges**. This particular graph is an example of a **directed graph**, whose edges have a direction and are represented by arrows (as opposed to an **undirected graph** whose edges do not have directions). If you are unfamiliar with graphs, see DETOUR: Graphs.

PAGE 174

The genome vanishes

The genome can still be traced out in the graph in Figure 3.7 by following the horizontal path from **TAA** to **GTT**. But in genome sequencing, we do not know in advance how to correctly order reads. Therefore we will arrange the 3-mers lexicographically, which produces the **overlap graph** shown in Figure 3.8. The genome path has disappeared!

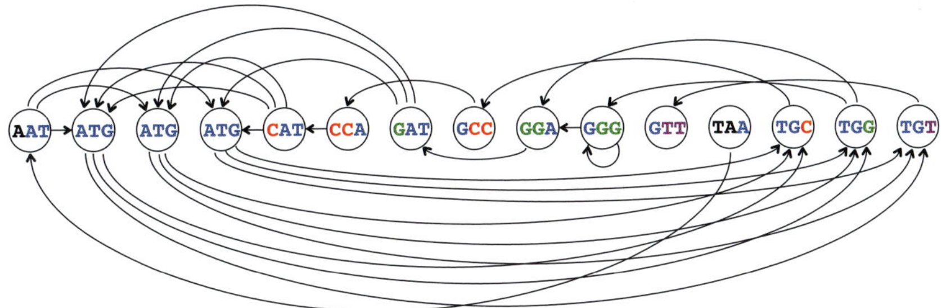

FIGURE 3.8 The same graph as the one in Figure 3.7 with 3-mers ordered lexicographically. The path through the graph representing the correct assembly is now harder to see.

The genome path may have disappeared to the naked eye, but it must still be there, since we have simply rearranged the nodes of the graph. Indeed, Figure 3.9 (top) highlights the genome path spelling out **TAATGCCATGGGATGTT**. However, if we had given you this graph to begin with, you would have needed to find a path through the graph visiting each node exactly once; such a path "explains" all the 3-mers in the 3-mer composition of the genome. Although finding such a path is currently just as difficult as trying to assemble the genome by hand, the graph nevertheless gives us a nice way of visualizing the overlap relationships between reads.

STOP and Think: Can any other strings be reconstructed by following a path visiting all the nodes in Figure 3.8?

To generalize the construction of the graph in Figure 3.8 to an arbitrary collection of k-mers *Patterns*, we form a node for each k-mer in *Patterns* and connect k-mers *Pattern* and *Pattern'* by a directed edge if SUFFIX(*Pattern*) = PREFIX(*Pattern'*). The resulting graph is called the **overlap graph** on these k-mers, denoted OVERLAP(*Patterns*).

3C **Overlap Graph Problem**:
Construct the overlap graph of a collection of k-mers.

> **Input**: A collection *Patterns* of k-mers.
> **Output**: The overlap graph OVERLAP(*Patterns*).

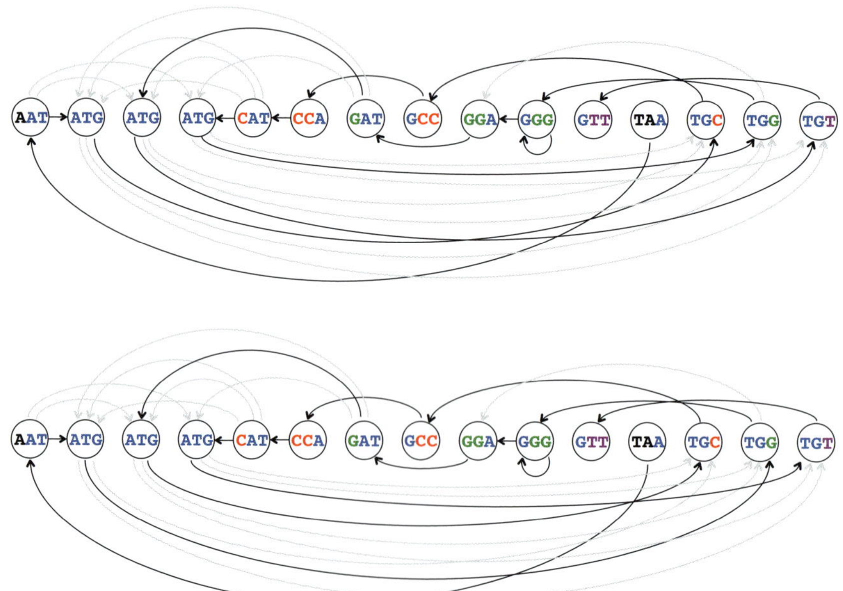

FIGURE 3.9 (Top) The genome path spelling out **TAATGCCATGGGATGTT**, highlighted in the overlap graph. (Bottom) Another Hamiltonian path in the overlap graph spells the genome **TAATGGGATGCCATGTT**. These two genomes differ by exchanging the positions of **CC** and **GG** but have the same 3-mer composition.

Two graph representations

If you have never worked with graphs before, you may be wondering how to represent graphs in your programs. To make a brief digression from our discussion of genome assembly, consider the graph in Figure 3.10 (top left); we can move around this graph's nodes without changing the graph. (For another example, the graphs in Figure 3.7 and Figure 3.8 are the same). As a result, when we are representing a graph computationally, the only information we need to store is the pair of nodes that each edge connects.

There are two standard ways of representing a graph. For a directed graph with n nodes, the $n \times n$ **adjacency matrix** $(A_{i,j})$ is defined by the following rule: $A_{i,j} = 1$ if a directed edge connects node i to node j, and $A_{i,j} = 0$ otherwise. Another, more memory-efficient way of representing a graph is to use an **adjacency list**, for which we simply list all nodes connected to each node (see Figure 3.10).

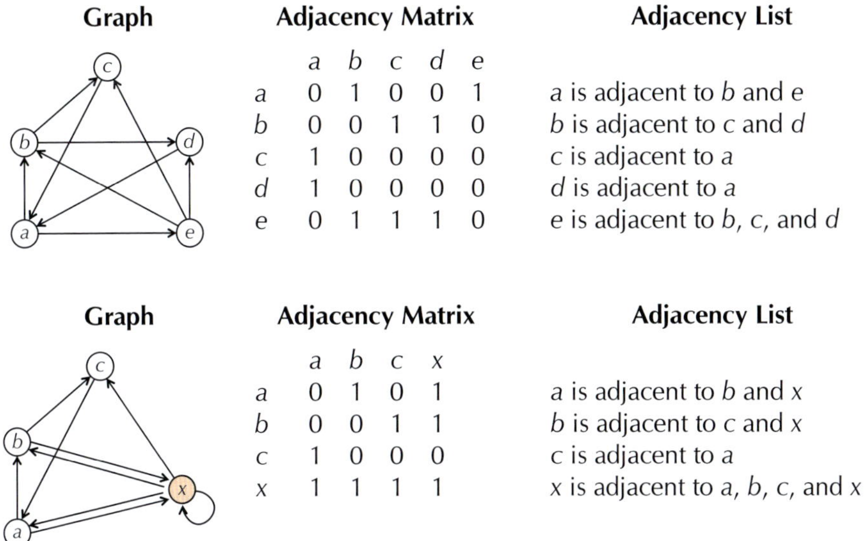

FIGURE 3.10 (Top) A graph with five nodes and nine edges, followed by its adjacency matrix and adjacency list. (Bottom) The graph produced by gluing nodes *d* and e into a single node x, along with the new graph's adjacency matrix and adjacency list.

Hamiltonian paths and universal strings

We now know that to solve the String Reconstruction Problem, we are looking for a path in the overlap graph that visits every node exactly once. A path in a graph visiting every node once is called a **Hamiltonian path**, in honor of the Irish mathematician William Hamilton (see **DETOUR: The Icosian Game**). As Figure 3.9 illustrates, a graph may have more than one Hamiltonian path.

Hamiltonian Path Problem:
Construct a Hamiltonian path in a graph.

> **Input**: A directed graph.
> **Output**: A path visiting every node in the graph exactly once (if such a path exists).

We do not ask you to solve the Hamiltonian Path Problem yet, since it is not clear how we could design an efficient algorithm for it. Instead, we want you to meet Nicolaas de Bruijn, a Dutch mathematician. In 1946, de Bruijn was interested in solving a purely theoretical problem, described as follows. A **binary string** is a string composed only of 0's and 1's; a binary string is ***k*-universal** if it contains every binary *k*-mer exactly once. For example, **0001110100** is a 3-universal string, as it contains each of the eight binary 3-mers (**000**, **001**, **011**, **111**, **110**, **101**, **010**, and **100**) exactly once.

Finding a *k*-universal string can be reduced to solving the String Reconstruction Problem when the *k*-mer composition is the collection of all binary *k*-mers. Thus, finding a *k*-universal string is equivalent to finding a Hamiltonian path in the overlap graph formed on all binary *k*-mers (Figure 3.11). Although the Hamiltonian path in Figure 3.11 can easily be found by hand, de Bruijn was interested in constructing *k*-universal strings for arbitrary values of *k*. For example, to find a 20-universal string, you would have to consider a graph with over a million nodes. It is absolutely unclear how to find a Hamiltonian path in such a huge graph, or even whether such a path exists!

Instead of searching for Hamiltonian paths in huge graphs, de Bruijn developed a completely different (and somewhat non-intuitive) way of representing a *k*-mer composition using a graph. Later in this chapter, we will learn how he used this method to construct universal strings.

Exercise Break: Construct a 4-universal string.

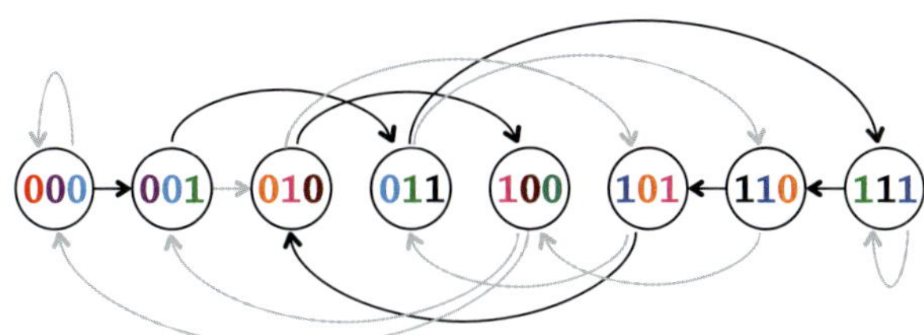

FIGURE 3.11 A Hamiltonian path (connecting node 000 to 100) highlighted in the overlap graph of all binary 3-mers. This path spells out the 3-universal binary string 0001110100.

Another Graph for String Reconstruction

Gluing nodes and de Bruijn graphs

Let's again represent the genome **TAATGCCATGGGATGTT** as a sequence of its 3-mers:

TAA AAT ATG TGC GCC CCA CAT ATG TGG GGG GGA GAT ATG TGT GTT

This time, instead of assigning these 3-mers to *nodes*, we will assign them to *edges*, as shown in Figure 3.12. You can once again reconstruct the genome by following this path from left to right, adding one new nucleotide at each step. Since each pair of consecutive edges represent consecutive 3-mers that overlap in two nucleotides, we will label each node of this graph with a 2-mer representing the overlapping nucleotides shared by the edges on either side of the node. For example, the node with incoming edge **CAT** and outgoing edge **ATG** is labeled **AT**.

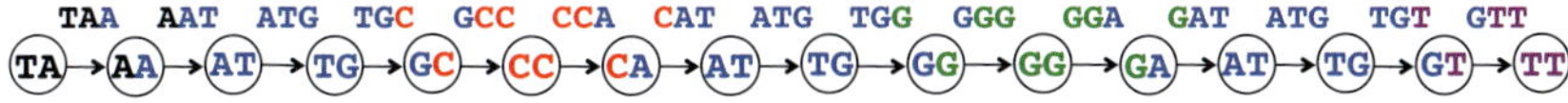

FIGURE 3.12 Genome **TAATGCCATGGGATGTT** represented as a path with edges (rather than nodes) labeled by 3-mers and nodes labeled by 2-mers.

Nothing seems new here until we start **gluing** identically labeled nodes. In Figure 3.13 (top panels), we bring the three **AT** nodes closer and closer to each other until they have been glued into a single node. Note that there are also three nodes labeled by

TG, which we glue together in Figure 3.13 (middle panels). Finally, we glue together the two nodes labeled GG (**GG** and **GG**), as shown in Figure 3.13 (bottom panels), which produces a special type of edge called a **loop** connecting GG to itself.

The number of nodes in the resulting graph (Figure 3.13 (bottom right)) has reduced from 16 to 11, while the number of edges stayed the same. This graph is called the **de Bruijn graph** of **TAATGCCATGGGATGTT**, denoted DEBRUIJN$_3$(**TAATGCCATGGGATGTT**). Note that this de Bruijn graph has three different edges connecting **AT** to **TG**, representing three copies of the repeat **ATG**.

In general, given a genome *Text*, PATHGRAPH$_k$(*Text*) is the path consisting of $|Text| - k + 1$ edges, where the i-th edge of this path is labeled by the i-th k-mer in *Text* and the i-th node of the path is labeled by the i-th $(k-1)$-mer in *Text*. The de Bruijn graph DEBRUIJN$_k$(*Text*) is formed by gluing identically labeled nodes in PATHGRAPH$_k$(*Text*).

De Bruijn Graph from a String Problem:
Construct the de Bruijn graph of a string.

Input: A string *Text* and an integer k.
Output: DEBRUIJN$_k$(*Text*).

STOP and Think: Consider the following questions.

1. If we gave you the de Bruijn graph DEBRUIJN$_k$(*Text*) without giving you *Text*, could you reconstruct *Text*?
2. Construct the de Bruijn graphs DEBRUIJN$_2$(*Text*), DEBRUIJN$_3$(*Text*), and DEBRUIJN$_4$(*Text*) for *Text* = **TAATGCCATGGGATGTT**. What do you notice?
3. How does the graph DEBRUIJN$_3$(**TAATGCCATGGGATGTT**) compare to DEBRUIJN$_3$(**TAATGGGATGCCATGTT**)?

PAGE 165

Charging Station (The Effect of Gluing on the Adjacency Matrix): Figure 3.10 (bottom) shows how the gluing operation affects the adjacency matrix and adjacency list of a graph. Check out this Charging Station to see how gluing works for a de Bruijn graph.

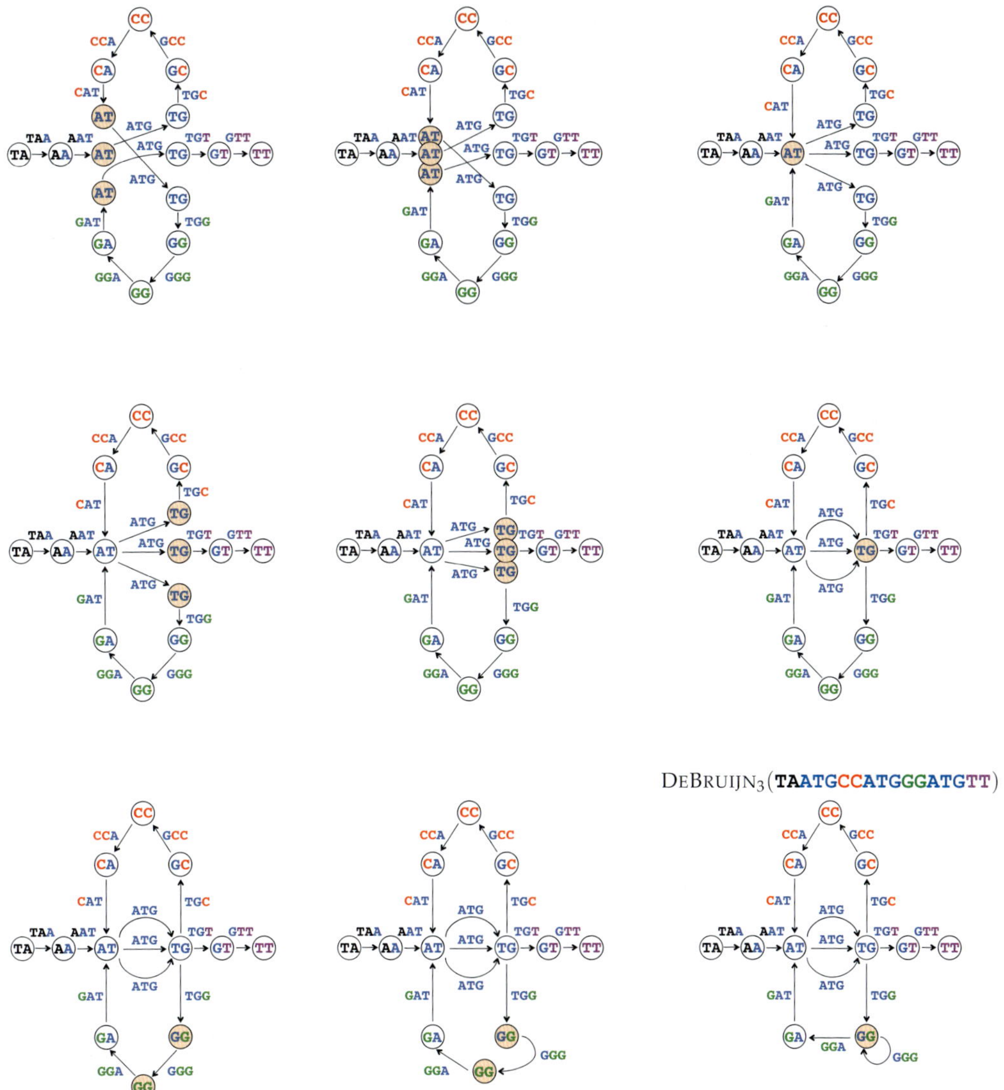

FIGURE 3.13 (Top panels) Bringing the three nodes labeled **AT** in Figure 3.12 closer (left) and closer (middle) to each other to eventually glue them into a single node (right). (Middle panels) Bringing the three nodes labeled **TG** closer (left) and closer (middle) to each other to eventually glue them into a single node (right). (Bottom panels) Bringing the two nodes labeled GG closer (left) and closer (middle) to each other to eventually glue them into a single node (right). The path with 16 nodes from Figure 3.12 has been transformed into the graph DEBRUIJN3(**TAATGCCATGGGATGTT**) with eleven nodes.

Walking in the de Bruijn Graph

Eulerian paths

Even though we have glued together nodes to form the de Bruijn graph, we have not changed its edges, and so the path from **TA** to **TT** reconstructing the genome is still hiding in DEBRUIJN3(**TAATGCCATGGGATGTT**) (Figure 3.14), although this path has become "tangled" after gluing. Therefore, solving the String Reconstruction Problem reduces to finding a path in the de Bruijn graph that visits every *edge* exactly once. Such a path is called an **Eulerian Path** in honor of the great mathematician Leonhard Euler (pronounced "oiler").

Eulerian Path Problem:
Construct an Eulerian path in a graph.

> **Input**: A directed graph.
> **Output**: A path visiting every edge in the graph exactly once (if such a path exists).

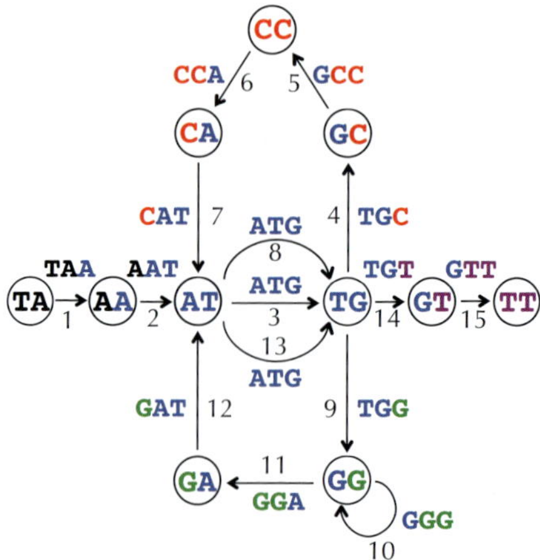

FIGURE 3.14 The path from **TA** to **TT** spelling out the genome **TAATGCCATGGGATGTT** has become "tangled" in the de Bruijn graph. The numbering of the fifteen edges of the path indicates an Eulerian path reconstructing the genome.

We now have an alternative way of solving the String Reconstruction Problem that amounts to finding an Eulerian path in the de Bruijn graph. But wait — to construct the de Bruijn graph of a genome, we glued together nodes of PathGraph$_k$(*Text*). However, constructing this graph requires us to know the correct ordering of the k-mers in *Text*!

STOP and Think: Can you construct DeBruijn$_k$(*Text*) if you don't know *Text* but you do know its k-mer composition?

Another way to construct de Bruijn graphs

Figure 3.15 (top) represents the 3-mer composition of **TAATGCCATGGGATGTT** as a composition graph CompositionGraph3(**TAATGCCATGGGATGTT**). As with the de Bruijn graph, each 3-mer is assigned to a directed edge, with its prefix labeling the first node of the edge and its suffix labeling the second node of the edge. However, the edges of this graph are **isolated**, meaning that no two edges share a node.

STOP and Think: Given *Text* = **TAATGCCATGGGATGTT**, glue identically labeled nodes in CompositionGraph3(*Text*). How does the resulting graph differ from DeBruijn3(*Text*) obtained by gluing the identically labeled nodes in PathGraph3(*Text*)?

Figure 3.15 shows how CompositionGraph3(*Text*) changes after gluing nodes with the same label, for *Text* = **TAATGCCATGGGATGTT**. These operations glue the fifteen isolated edges in CompositionGraph3(*Text*) into the path PathGraph3(*Text*). Follow-up gluing operations proceed in exactly the same way as when we glued nodes of PathGraph3(*Text*), which results in DeBruijn3(*Text*). Thus, we can construct the de Bruijn graph from this genome's 3-mer composition without knowing the genome!

STOP and Think: In Figure 3.15, we identified **ATG** and **TGC** as overlapping 3-mers. In reality, since the genome is unknown, we don't know whether this **ATG** is followed by **TGC**, **TGG**, or **TGT**. What would happen if we had identified **ATG** and **TGG** as overlapping 3-mers instead? Verify that the final result will be the same graph shown in Figure 3.14.

For an arbitrary string *Text*, we define CompositionGraph$_k$(*Text*) as the graph consisting of $|Text| - k + 1$ isolated edges, where edges are labeled by k-mers in *Text*; every edge labeled by a k-mer edge connects nodes labeled by the prefix and suffix of this

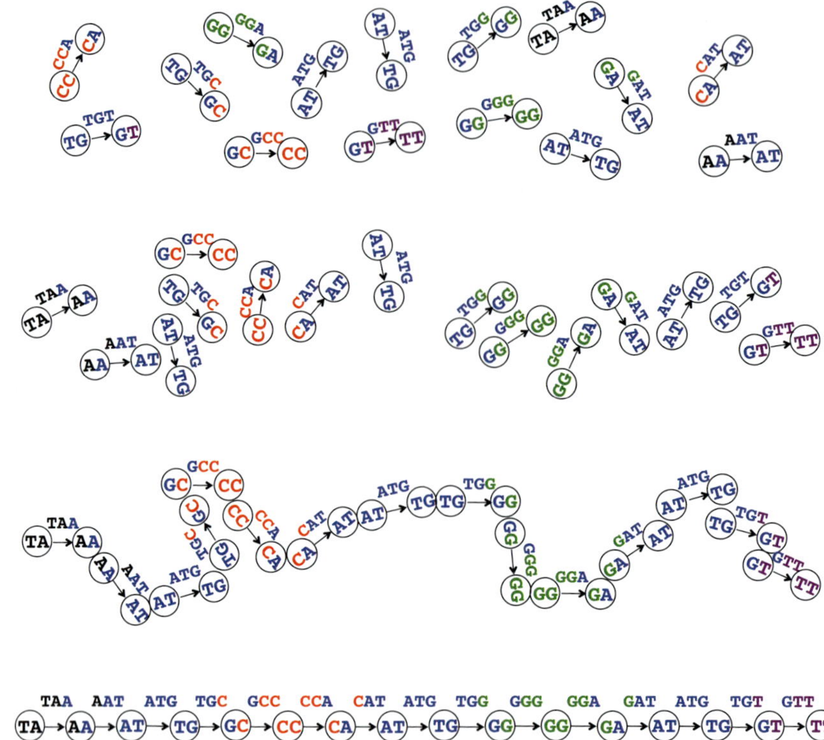

FIGURE 3.15 Gluing some identically labeled nodes transforms the graph COMPOSITIONGRAPH3(**TAATGCCATGGGATGTT**) (top) into the graph PATHGRAPH3(**TAATGCCATGGGATGTT**) (bottom). Gluing all identically labeled nodes produces DEBRUIJN3(**TAATGCCATGGGATGTT**) from Figure 3.14.

k-mer. The graph COMPOSITIONGRAPH$_k$(*Text*) is just a collection of isolated edges representing the k-mers in the k-mer composition of *Text*, meaning that we can construct COMPOSITIONGRAPH$_k$(*Text*) from the k-mer composition of *Text*. Gluing nodes with the same label in COMPOSITIONGRAPH$_k$(*Text*) produces DEBRUIJN$_k$(*Text*).

Given an arbitrary collection of k-mers *Patterns* (where some k-mers may appear multiple times), we define COMPOSITIONGRAPH(*Patterns*) as a graph with |*Patterns*| isolated edges. Every edge is labeled by a k-mer from *Patterns*, and the starting and ending nodes of an edge are labeled by the prefix and suffix of the k-mer labeling that edge. We then define DEBRUIJN(*Patterns*) by gluing identically labeled nodes in COMPOSITIONGRAPH(*Patterns*), which yields the following algorithm.

DEBRUIJN(*Patterns*)
  represent every *k*-mer in *Patterns* as an isolated edge between its prefix and suffix
  glue all nodes with identical labels, yielding the graph DEBRUIJN(*Patterns*)
  return DEBRUIJN(*Patterns*)

Constructing de Bruijn graphs from k-mer composition

Constructing the de Bruijn graph by gluing identically labeled nodes will help us later when we generalize the notion of de Bruijn graph for other applications. We will now describe another useful way to construct de Bruijn graphs without gluing.

Given a collection of k-mers *Patterns*, the nodes of $\text{DEBRUIJN}_k(Patterns)$ are simply all unique $(k-1)$-mers occurring as a prefix or suffix of 3-mers in *Patterns*. For example, say we are given the following collection of 3-mers:

AAT ATG ATG ATG CAT CCA GAT GCC GGA GGG GTT TAA TGC TGG TGT

Then the set of eleven *unique* 2-mers occurring as a prefix or suffix in this collection is as follows:

AA AT CA CC GA GC GG GT TA TG TT

For every k-mer in *Patterns*, we connect its prefix node to its suffix node by a directed edge in order to produce DEBRUIJN(*Patterns*). You can verify that this process produces the same de Bruijn graph that we have been working with (Figure 3.16).

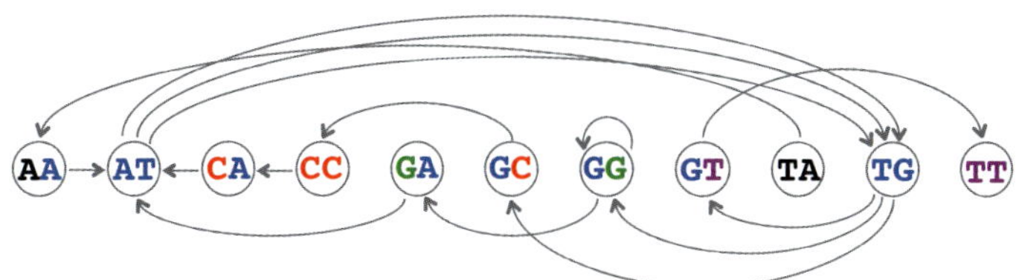

FIGURE 3.16 The de Bruijn graph above is the same as the graph in Figure 3.14, although it has been drawn differently.

De Bruijn Graph from *k*-mers Problem:
Construct the de Bruijn graph of a collection of k-mers.

Input: A collection of *k*-mers *Patterns*.
Output: The de Bruijn graph DEBRUIJN(*Patterns*).

De Bruijn graphs versus overlap graphs

We now have two ways of solving the String Reconstruction Problem. We can either find a Hamiltonian path in the overlap graph or find an Eulerian path in the de Bruijn graph (Figure 3.17). Your inner voice may have already started complaining: *was it really worth my time to learn two slightly different ways of solving the same problem*? After all, we have only changed a single word in the statements of the Hamiltonian and Eulerian Path Problems, from finding a path visiting every *node* exactly once to finding a path visiting every *edge* exactly once.

STOP and Think: Which graph would you rather work with, the overlap graph or the de Bruijn graph?

Our guess is that you would probably prefer working with the de Bruijn graph, since it is smaller. However, this would be the wrong reason to choose one graph over the other. In the case of real assembly problems, both graphs will have millions of nodes, and so all that matters is finding an efficient algorithm for reconstructing the genome. If we can find an *efficient algorithm* for the Hamiltonian Path Problem, but not for the Eulerian path Problem, then you should select the overlap graph even though it looks more complex.

The choice between these two graphs is the pivotal decision of this chapter. To help you make this decision, we will ask you to hop onboard our bioinformatics time machine for a field trip to the 18th Century.

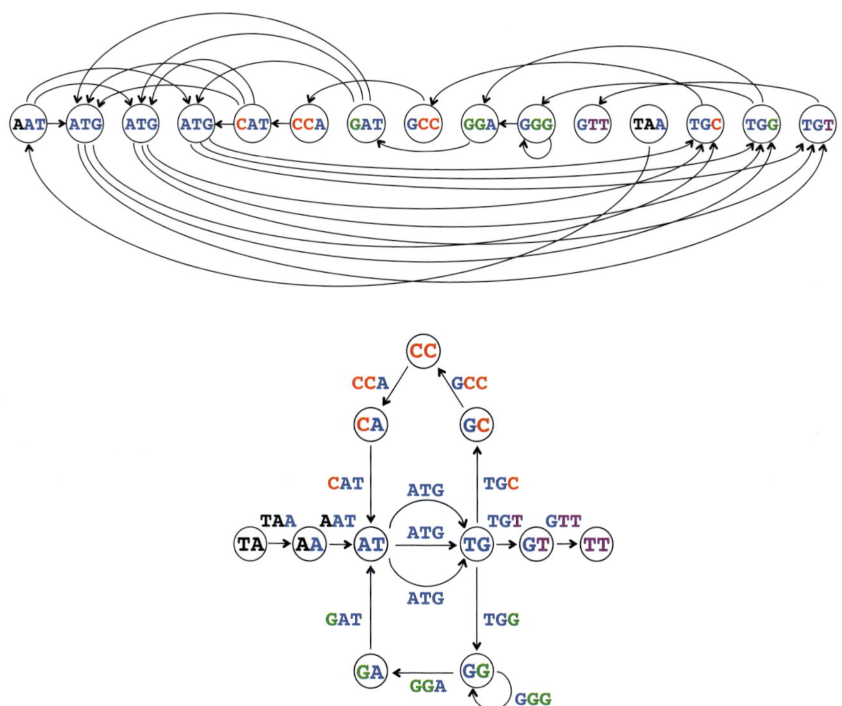

FIGURE 3.17 The overlap graph (top) and de Bruijn graph (bottom) for the same collection of 3-mers.

The Seven Bridges of Königsberg

Our destination is 1735 and the Prussian city of Königsberg. This city, which today is Kaliningrad, Russia, comprised both banks of the Pregel River as well as two river islands; seven bridges connected these four different parts of the city, as illustrated in Figure 3.18 (top). Königsberg's residents enjoyed taking walks, and they asked a simple question: *Is it possible to set out from my house, cross each bridge exactly once, and return home?* Their question became known as the **Bridges of Königsberg Problem**.

Exercise Break: Does the Bridges of Königsberg Problem have a solution?

In 1735, Leonhard Euler drew the graph in Figure 3.18 (bottom), which we call *Königsberg*; this graph's nodes represent the four sectors of the city, and its edges represent the seven bridges connecting different sectors. Note that the edges of *Königsberg* are undirected, meaning that they can be traversed in either direction.

STOP and Think: Redefine the Bridges of Königsberg Problem as a question about the graph *Königsberg*.

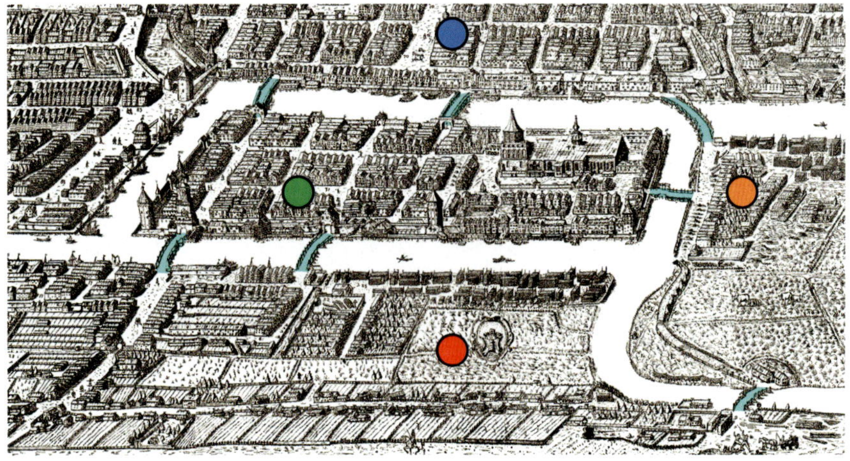

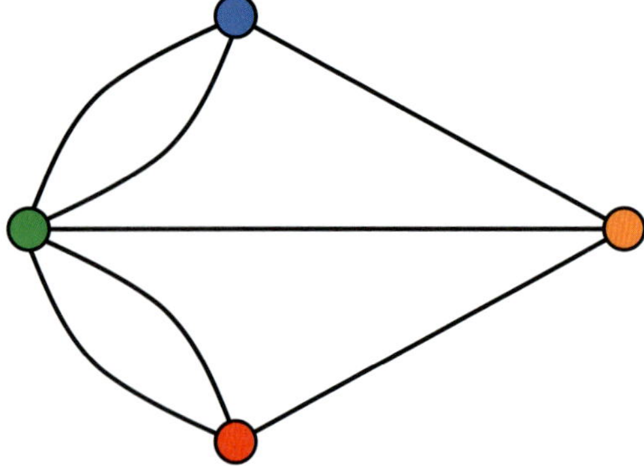

FIGURE 3.18 (Top) A map of Königsberg, adapted from Joachim Bering's 1613 illustration. The city was made up of four sectors represented by the blue, red, yellow, and green dots. The seven bridges connecting the different parts of the city have been highlighted to make them easier to see. (Bottom) The graph *Königsberg*.

We have already defined an Eulerian path as a path in a graph traversing each edge of a graph exactly once. A cycle that traverses each edge of a graph exactly once is called an **Eulerian cycle**, and we say that a graph containing such a cycle is **Eulerian**. Note that an Eulerian cycle in *Königsberg* would immediately provide the residents of the city with the walk they had wanted. We now can redefine the Bridges of Königsberg Problem as an instance of the following more general problem.

Eulerian Cycle Problem:
Find an Eulerian cycle in a graph.

Input: A graph.
Output: An Eulerian cycle in this graph, if one exists.

Euler solved the Bridges of Königsberg Problem, showing that *no* walk can cross each bridge exactly once (i.e., the graph *Königsberg* is not Eulerian), which you may have already figured out for yourself. Yet his real contribution, and the reason why he is viewed as the founder of **graph theory**, a field of study that still flourishes today, is that he proved a theorem dictating when a graph will have an Eulerian cycle. His theorem immediately implies an efficient algorithm for constructing an Eulerian cycle in any Eulerian graph, even one having millions of edges. Furthermore, this algorithm can easily be extended into an algorithm constructing an Eulerian *path* (in a graph having such a path), which will allow us to solve the String Reconstruction Problem by using the de Bruijn graph.

On the other hand, it turns out that no one has ever been able to find an efficient algorithm solving the Hamiltonian Path Problem. The search for such an algorithm, or for a proof that an efficient algorithm does not exist for this problem, is at the heart of one of the most fundamental unanswered questions in computer science. Computer scientists classify an algorithm as **polynomial** if its runtime can be bounded by a polynomial in the length of the input data. On the other hand, an algorithm is **exponential** if its runtime on some datasets is exponential in the length of the input data.

Exercise Break: Classify the algorithms that we encountered in Chapter 1 as polynomial or exponential.

Although Euler's algorithm is polynomial, the Hamiltonian Path Problem belongs to a special class of problems for which all attempts to develop a polynomial algorithm have failed (see **DETOUR: Tractable and Intractable Problems**). Yet instead of trying to solve a problem that has stumped computer scientists for decades, we will set aside the overlap graph and instead focus on the de Bruijn graph approach to genome assembly.

PAGE 178

For the first two decades following the invention of DNA sequencing methods, biologists assembled genomes using overlap graphs, since they failed to realize that the Bridges of Königsberg held the key to DNA assembly (see **DETOUR: From Euler to Hamilton to de Bruijn**). Indeed, overlap graphs were used to assemble the human genome. It took bioinformaticians some time to figure out that the de Bruijn graph, first constructed to solve a completely theoretical problem, was relevant to genome assembly. Moreover, when the de Bruijn graph was brought to bioinformatics, it was considered an exotic mathematical concept with limited practical applications. Today, the de Bruijn graph has become the dominant approach for genome assembly.

Euler's Theorem

We will now explore Euler's method for solving the Eulerian Cycle Problem. Euler worked with undirected graphs like *Königsberg*, but we will consider an analogue of his algorithm for directed graphs so that his method will apply to genome assembly.

Consider an ant, whom we will call Leo, walking along the edges of an Eulerian cycle. Every time Leo enters a node of this graph by an edge, he is able to leave this node by another, unused edge. Thus, in order for a graph to be Eulerian, the number of incoming edges at any node must be equal to the number of outgoing edges at that node. We define the **indegree** and **outdegree** of a node v (denoted $\text{IN}(v)$ and $\text{OUT}(v)$, respectively) as the number of edges leading into and out of v. A node v is **balanced** if $\text{IN}(v) = \text{OUT}(v)$, and a graph is **balanced** if all its nodes are balanced. Because Leo must always be able to leave a node by an unused edge, any Eulerian graph must be balanced. Figure 3.19 shows a balanced graph and an unbalanced graph.

STOP and Think: We now know that every Eulerian graph is balanced; is every balanced graph Eulerian?

The graph in Figure 3.20 is balanced but not Eulerian because it is **disconnected**, meaning that some nodes cannot be reached from other nodes. In any disconnected graph, it is impossible to find an Eulerian cycle. In contrast, we say that a directed graph is **strongly connected** if it is possible to reach any node from every other node.

We now know that an Eulerian graph must be both balanced and strongly connected. Euler's Theorem states that these two conditions are sufficient to guarantee that an arbitrary graph is Eulerian. As a result, it implies that we can determine whether a graph is Eulerian without ever having to draw any cycles.

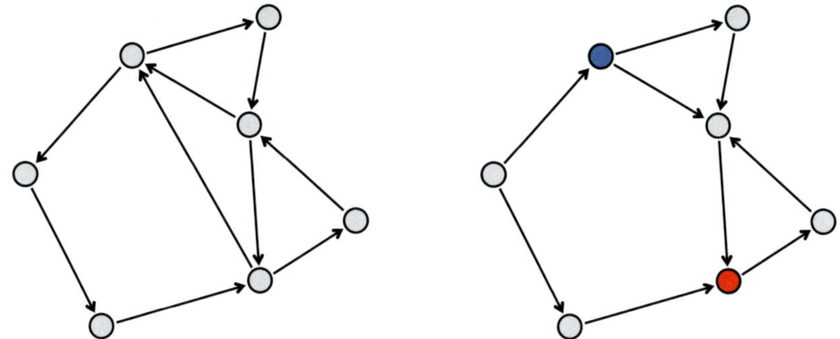

FIGURE 3.19 Balanced (left) and unbalanced (right) directed graphs. For the (unbalanced) blue node v, $\text{IN}(v) = 1$ and $\text{OUT}(v) = 2$, whereas for the (unbalanced) red node w, $\text{IN}(w) = 2$ and $\text{OUT}(w) = 1$.

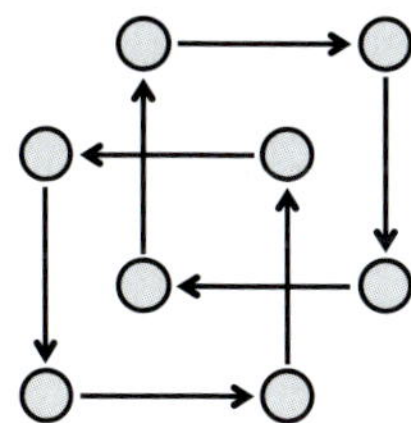

FIGURE 3.20 A balanced, disconnected graph.

Euler's Theorem: *Every balanced, strongly connected directed graph is Eulerian.*

Proof. Let *Graph* be an arbitrary balanced and strongly connected directed graph. To prove that *Graph* has an Eulerian cycle, place Leo at any node v_0 of *Graph* (the green node in Figure 3.21), and let him randomly walk through the graph under the condition that he cannot traverse the same edge twice.

If Leo were incredibly lucky — or a genius — then he would traverse each edge exactly once and return back to v_0. However, odds are that he will "get stuck" somewhere before he can complete an Eulerian cycle, meaning that he reaches a node and finds no unused edges leaving that node.

STOP and Think: Where is Leo when he gets stuck? Can he get stuck in any node of the graph or only in certain nodes?

STOP

It turns out that the only node where Leo can get stuck is the starting node v_0! The reason why is that *Graph* is balanced: if Leo walks into any node other than v_0 (through

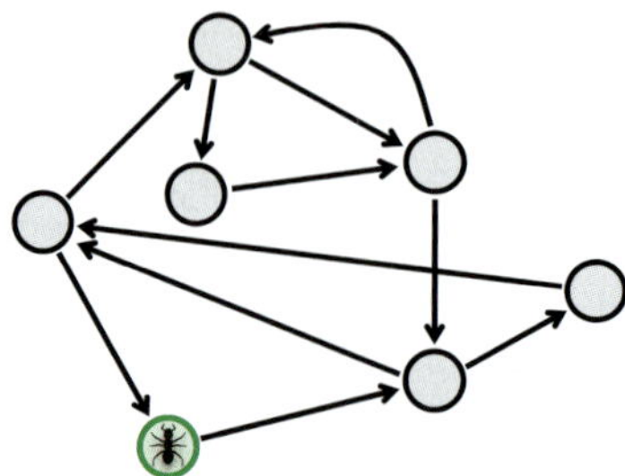

FIGURE 3.21 Leo starts at the green node v_0 and walks through a balanced and strongly connected graph.

an incoming edge), then he will always be able to escape via an unused outgoing edge. The only exception to this rule is the starting node v_0, since Leo used up one of the outgoing edges of v_0 on his first move. Now, because Leo has returned to v_0, the result of his walk was a cycle, which we call *Cycle*$_0$ (Figure 3.22 (left)).

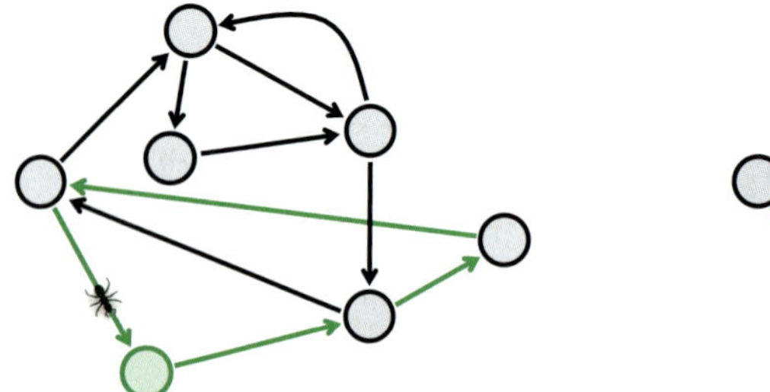

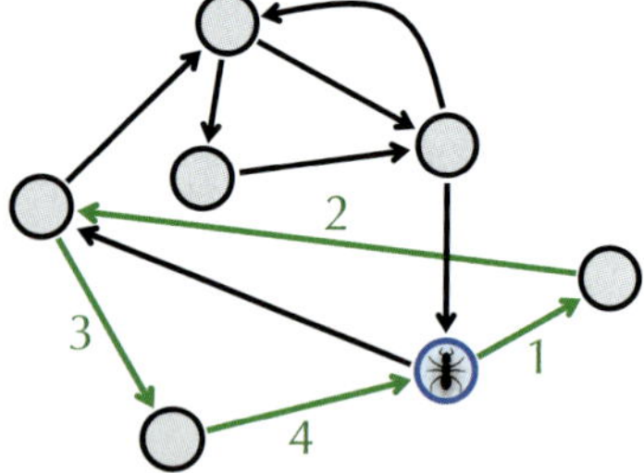

FIGURE 3.22 (Left) Leo produces a cycle *Cycle*$_0$ (formed by green edges) when he gets stuck at the green node v_0. In this case, he has not yet visited every edge in the graph. (Right) Starting at a new node v_1 (shown in blue), Leo first travels along *Cycle*$_0$, returning to v_1. Note that the blue node v_1, unlike the green node v_0, has unused outgoing and incoming edges.

STOP and Think: Is there a way to give Leo different instructions so that he selects a longer walk through the graph before he gets stuck?

As we mentioned, if *Cycle*$_0$ is Eulerian, then we are finished. Otherwise, because *Graph* is strongly connected, some node on *Cycle*$_0$ must have unused edges entering it and leaving it (why?). Naming this node v_1, we ask Leo to start at v_1 instead of v_0 and traverse *Cycle*$_0$ (thus returning to v_1), as shown in Figure 3.22 (right).

Leo is probably annoyed that we have asked him to travel along the exact same cycle, since as before, he will eventually return to v_1, the node where he started. However, now there are unused edges starting at this node, and so he can continue walking from v_1, using a new edge each time. The same argument as the one that we used before implies that Leo must eventually get stuck at v_1. The result of Leo's walk is a new cycle, $Cycle_1$ (Figure 3.23), which is larger than $Cycle_0$.

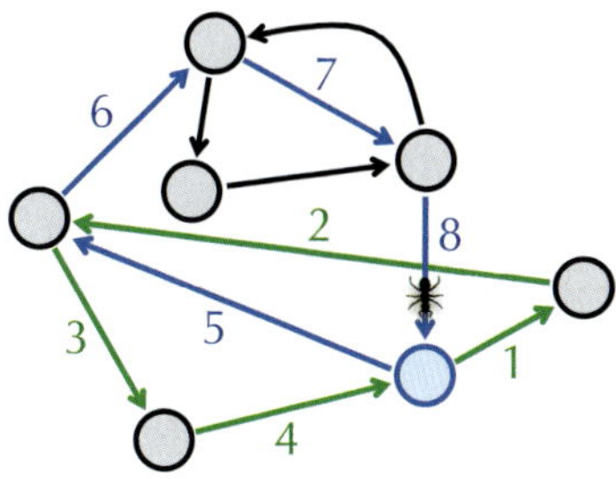

FIGURE 3.23 After traversing the previously constructed green cycle $Cycle_0$, Leo continues walking and eventually produces a larger cycle $Cycle_1$ formed of both the green and the blue cycles put together into a single cycle.

If $Cycle_1$ is an Eulerian cycle, then Leo has completed his job. Otherwise, we select a node v_2 in $Cycle_1$ that has unused edges entering it and leaving it (the red node in Figure 3.24 (left)). Starting at v_2, we ask Leo to traverse $Cycle_1$, returning to v_2, as shown in Figure 3.24 (left). Afterwards, he will randomly walk until he gets stuck at v_2, creating an even larger cycle that we name $Cycle_2$.

In Figure 3.24 (right), $Cycle_2$ happens to be Eulerian, although this is certainly not the case for an arbitrary graph. In general, Leo generates larger and larger cycles at each iteration, and so we are guaranteed that sooner or later some $Cycle_m$ will traverse all the edges in *Graph*. This cycle must be Eulerian, and so we (and Leo) are finished. □

STOP and Think: Formulate and prove an analogue of Euler's Theorem for undirected graphs.

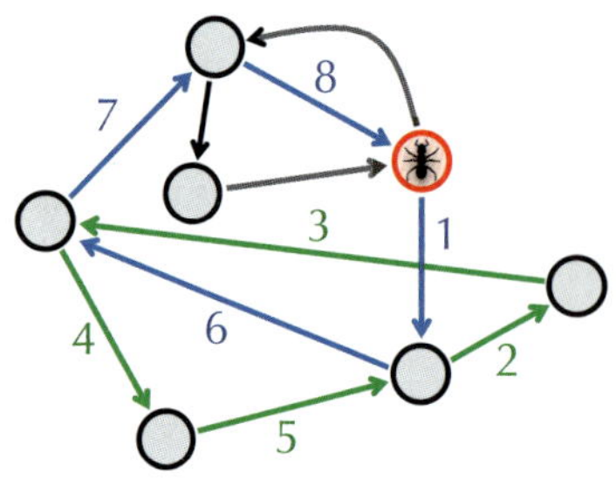

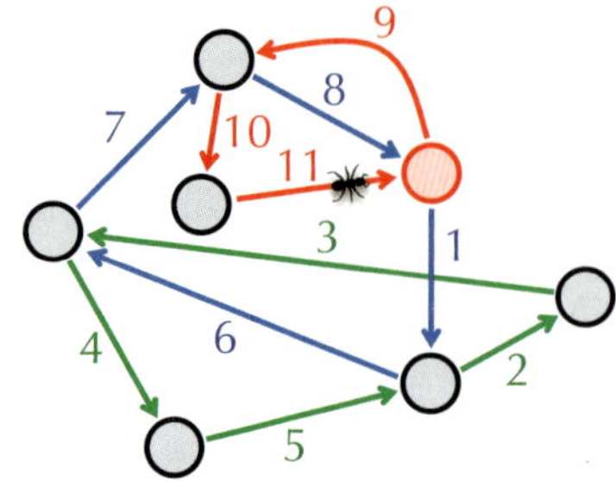

FIGURE 3.24 (Left) Starting at a new node v_2 (shown in red), Leo first travels along the previously constructed $Cycle_1$ (shown as green and blue edges). (Right) After completing the walk through $Cycle_1$, Leo continues randomly walking through the graph and finally produces an Eulerian cycle.

From Euler's Theorem to an Algorithm for Finding Eulerian Cycles

Constructing Eulerian cycles

The proof of Euler's Theorem offers an example of what mathematicians call a **constructive proof**, which not only proves the desired result, but also provides us with a method for constructing the object we need. In short, we track Leo's movements until he inevitably produces an Eulerian cycle in a balanced and strongly connected graph *Graph*, as summarized in the following pseudocode.

```
EULERIANCYCLE(Graph)
    form a cycle Cycle by randomly walking in Graph (don't visit the same edge twice!)
    while there are unexplored edges in Graph
        select a node newStart in Cycle with still unexplored edges
        form Cycle' by traversing Cycle (starting at newStart) and randomly walking
        Cycle ← Cycle'
    return Cycle
```

It may not be obvious, but a good implementation of EULERIANCYCLE will work in linear time. To achieve this runtime speedup, you would need to use an efficient data structure in order to maintain the current cycle that Leo is building as well as the list of unused edges incident to each node and the list of nodes on the current cycle that have unused edges.

From Eulerian cycles to Eulerian paths

We can now check if a directed graph has an Eulerian *cycle*, but what about an Eulerian *path*? Consider the de Bruijn graph in Figure 3.25 (left), which we already know has an Eulerian path, but which does not have an Eulerian cycle because nodes **TA** and **TT** are not balanced. However, we can transform this Eulerian path into an Eulerian cycle by adding a single edge connecting **TT** to **TA**, as shown in Figure 3.25 (right).

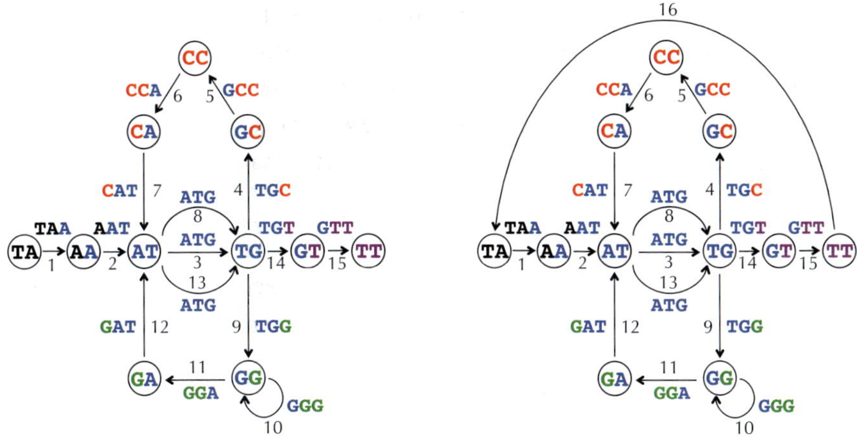

FIGURE 3.25 Transforming an Eulerian path (left) into an Eulerian cycle (right) by adding an edge.

STOP and Think: How many unbalanced nodes does a graph with an Eulerian path have?

More generally, consider a graph that does not have an Eulerian cycle but does have an Eulerian path. If an Eulerian path in this graph connects a node v to a different node w, then the graph is **nearly balanced**, meaning that all its nodes except v and w are balanced. In this case, adding an extra edge from w to v transforms the Eulerian path into an Eulerian cycle. Thus, a nearly balanced graph has an Eulerian path if and only if adding an edge between its unbalanced nodes makes the graph balanced and strongly connected.

You now have a method to assemble a genome, since the String Reconstruction Problem reduces to finding an Eulerian path in the de Bruijn graph generated from reads.

Exercise Break: Find an analogue of the nearly balanced condition that will determine when an *undirected* graph has an Eulerian path.

The analogue of Euler's theorem for undirected graphs immediately implies that there is no Eulerian path in 18th Century Königsberg, but the story is different in modern-day Kaliningrad (see **DETOUR: The Seven Bridges of Kaliningrad**).

PAGE 179

Constructing universal strings

Now that you know how to use the de Bruijn graph to solve the String Reconstruction Problem, you can also construct a k-universal string for any value of k. We should note that de Bruijn was interested in constructing k-universal *circular* strings. For example, **00011101** is a 3-universal circular string, as it contains each of the eight binary 3-mers exactly once (Figure 3.26).

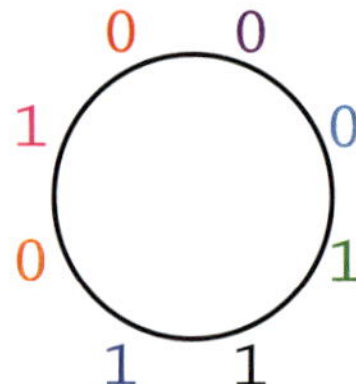

FIGURE 3.26 The circular 3-universal string **00011101** contains each of the binary 3-mers (**000**, **001**, **011**, **111**, **110**, **101**, **010**, and **100**) exactly once.

***k*-Universal Circular String Problem**:
Find a k-universal circular string.

Input: An integer k.
Output: A k-universal circular string.

Like its analogue for linear strings, the k-Universal Circular String Problem is just a specific case of a more general problem, which requires us to reconstruct a circular string given its k-mer composition. This problem models the assembly of a circular genome containing a single chromosome, like the genomes of most bacteria. We know that we can reconstruct a circular string from its k-mer composition by finding an Eulerian cycle in the de Bruijn graph constructed from these k-mers. Therefore, we can construct a k-universal circular binary string by finding an Eulerian cycle in the de Bruijn graph constructed from the collection of all binary k-mers (Figure 3.27).

Exercise Break: How many 3-universal circular strings are there?

Even though finding a 20-universal circular string amounts to finding an Eulerian cycle in a graph with over a million edges, we now have a fast algorithm for solving this problem. Let $BinaryStrings_k$ be the set of all 2^k binary k-mers. The only thing we need to do is to solve the k-Universal Circular String Problem is to find an Eulerian cycle in DEBRUIJN($BinaryStrings_k$). Note that the nodes of this graph represent all possible binary $(k-1)$-mers. A directed edge connects $(k-1)$-mer *Pattern* to $(k-1)$-mer *Pattern'* in this graph if there exists a k-mer whose prefix is *Pattern* and whose suffix is *Pattern'*.

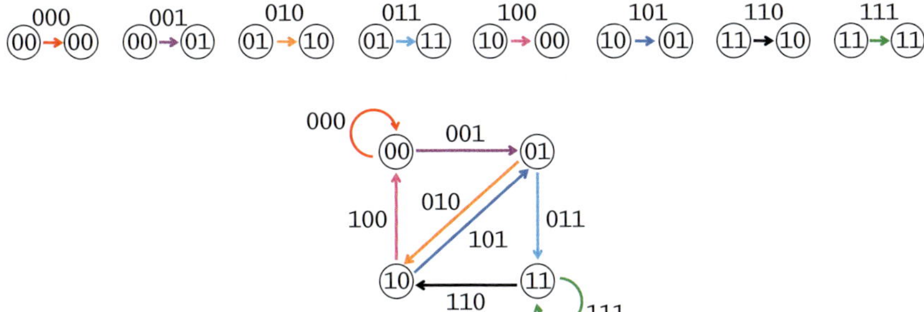

FIGURE 3.27 (Top) A graph consisting of eight isolated directed edges, one for each binary 3-mer. The nodes of each edge correspond to the 3-mer's prefix and suffix. (Bottom) Gluing identically labeled nodes in the graph on top results in a de Bruijn graph containing four nodes. An Eulerian cycle through the edges 000 → 001 → 011 → 111 → 110 → 101 → 010 → 100 → 000 yields the 3-universal circular string 00011101.

STOP and Think: Figure 3.28 illustrates that DEBRUIJN($BinaryStrings_4$) is balanced and strongly connected and is thus Eulerian. Can you prove that for any k, DEBRUIJN($BinaryStrings_k$) is Eulerian?

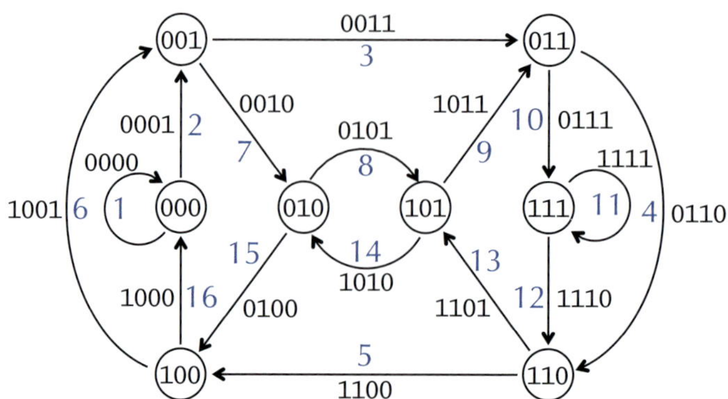

FIGURE 3.28 An Eulerian cycle spelling the cyclic 4-universal string 0000110010111101 in DEBRUIJN(*BinaryStrings*$_4$).

Assembling Genomes from Read-Pairs

From reads to read-pairs

Previously, we described an idealized form of genome assembly in order to build up your intuition about de Bruijn graphs. In the rest of the chapter, we will discuss a number of practically motivated topics that will help you appreciate the advanced methods used by modern assemblers.

We have already mentioned that assembling reads sampled from a randomly generated text is a trivial problem, since random strings are not expected to have long repeats. Moreover, de Bruijn graphs become less and less tangled when read length increases (Figure 3.29). As soon as read length exceeds the length of all repeats in a genome (provided the reads have no errors), the de Bruijn graph turns into a path. However, despite many attempts, biologists have not yet figured out how to generate *long* and *accurate* reads. The most accurate sequencing technologies available today generate reads that are only about 300 nucleotides long, which is too short to span most repeats, even in short bacterial genomes.

We saw earlier that the string **TAATGCCATGGGATGTT** cannot be uniquely reconstructed from its 3-mer composition since another string (**TAATGGGATGCCATGTT**) has the same 3-mer composition.

STOP and Think: What additional experimental information would allow you to uniquely reconstruct the string **TAATGCCATGGGATGTT** ?

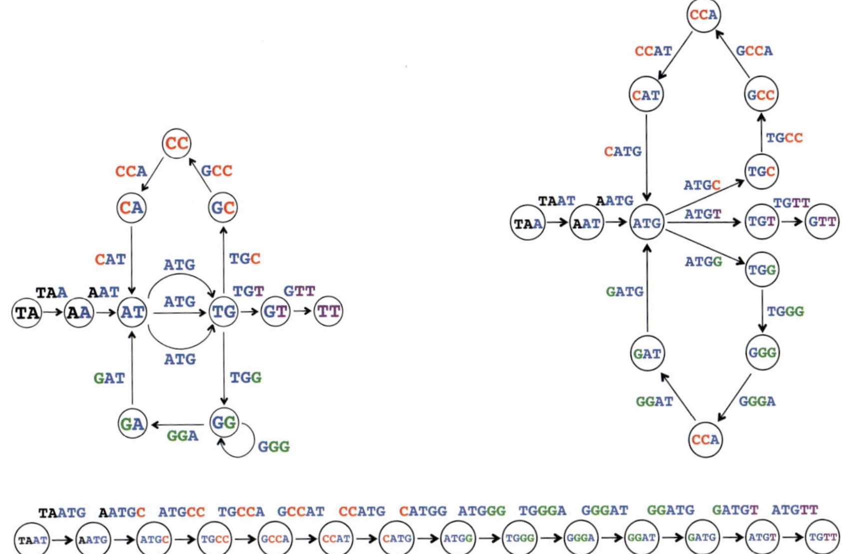

FIGURE 3.29 The graph DEBRUIJN$_4$(**TAATGCCATGGGATGTT**) (top right) is less tangled than the graph DEBRUIJN$_3$(**TAATGCCATGGGATGTT**) (top left). The graph DEBRUIJN$_5$(**TAATGCCATGGGATGTT**) (bottom) is a path.

Increasing read length would help identify the correct assembly, but since increasing read length presents a difficult experimental problem, biologists have devised an ingenious experimental approach to increase read length by generating **read-pairs**, which are pairs of reads separated by a fixed distance d in the genome (Figure 3.30). You can

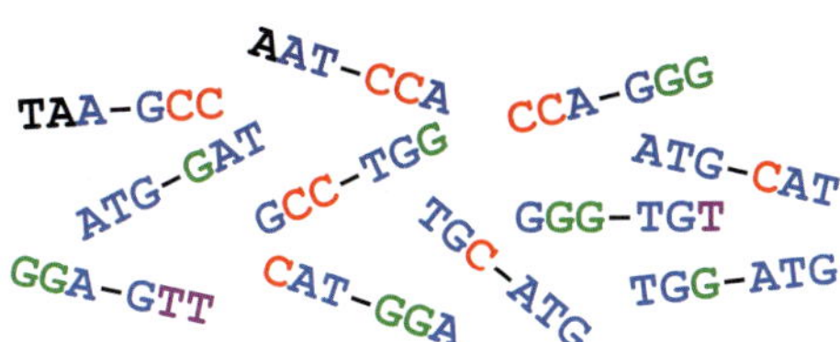

FIGURE 3.30 Read-pairs sampled from **TAATGCCATGGGATGTT** and formed by reads of length 3 separated by a gap of length 1. A simple but inefficient way to assemble these read-pairs is to construct the de Bruijn graph of individual reads (3-mers) within the read-pairs.

think about a read-pair as a long "gapped" read of length $k + d + k$ whose first and last k-mers are known but whose middle segment of length d is unknown. Nevertheless, read-pairs contain more information than k-mers alone, and so we should be able to use them to improve our assemblies. If only you could infer the nucleotides in the middle segment of a read-pair, you would immediately increase the read length from k to $2 \cdot k + d$.

Transforming read-pairs into long virtual reads

Let *Reads* be the collection of all $2N$ k-mer reads taken from N read-pairs. Note that a read-pair formed by k-mer reads $Read_1$ and $Read_2$ corresponds to two edges in the de Bruijn graph DEBRUIJN$_k$(*Reads*). Since these reads are separated by distance d in the genome, there must be a path of length $k + d + 1$ in DEBRUIJN$_k$(*Reads*) connecting the node at the beginning of the edge corresponding to $Read_1$ with the node at the end of the edge corresponding to $Read_2$, as shown in Figure 3.31. If there is only one path of length $k + d + 1$ connecting these nodes, or if all such paths spell out the same string, then we can transform a read-pair formed by reads $Read_1$ and $Read_2$ into a virtual read of length $2 \cdot k + d$ that starts as $Read_1$, spells out this path, and ends with $Read_2$.

For example, consider the de Bruijn graph in Figure 3.31, which is generated from all reads present in the read-pairs in Figure 3.30. There is a unique string spelled by

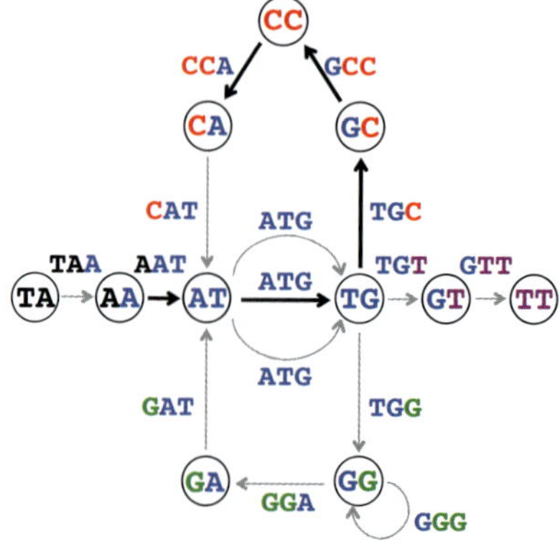

FIGURE 3.31 The highlighted path of length $k + d + 1 = 3 + 1 + 1 = 5$ (connecting node **AAT** to node **CCA**) spells out **AATGCCA**. (There are three such paths because there are three possible choices of edges labeled **ATG**.) Thus, the gapped read **AAT–CCA** can be transformed into a long virtual read **AATGCCA**.

paths of length $k + d + 1 = 5$ between edges labeled **AAT** and **CCA** within a read-pair represented by the gapped read **AAT–CCA**. Thus, from two short reads of length k, we have generated a long virtual read of length $2 \cdot k + d$, achieving computationally what researchers still cannot achieve experimentally! After preprocessing the de Bruijn graph to produce long virtual reads, we can simply construct the de Bruijn graph from these long reads and use it for genome assembly.

Although the idea of transforming read-pairs into long virtual reads is used in many assembly programs, we have made an optimistic assumption: *"If there is only one path of length $k + d + 1$ connecting these nodes, or if all such paths spell out the same string ..."*. In practice, this assumption limits the application of the long virtual read approach to assembling read-pairs because highly repetitive genomic regions often contain multiple paths of the same length between two edges, and these paths often spell different strings (Figure 3.32). If this is the case, then we cannot reliably transform a read-pair into a long read. Instead, we will describe an alternative approach to analyzing read-pairs.

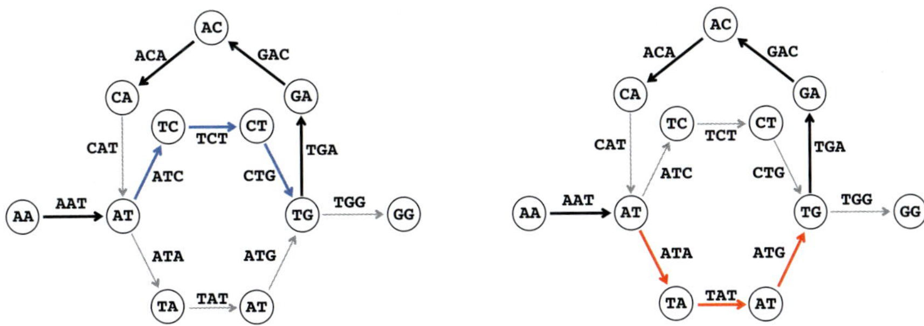

FIGURE 3.32 (Left) The highlighted path in DEBRUIJN$_3$(AATCTGACATATGG) spells out the long virtual read AATCTGACA, which is a substring of AATCTGACATATGG. (Right) The highlighted path in the same graph spells out the long virtual read AATATGACA, which does not occur in AATCTGACATATGG.

From composition to paired composition

Given a string *Text*, a (k, d)-mer is a pair of k-mers in *Text* separated by distance d. We use the notation $(Pattern_1 \mid Pattern_2)$ to refer to a (k, d)-mer whose k-mers are $Pattern_1$ and $Pattern_2$. For example, (**AAT** | **TGG**) is a $(3, 4)$-mer in **TAATGCCATGGGATGTT**. The (k, d)-mer composition of *Text*, denoted PAIREDCOMPOSITION$_{k,d}$(*Text*), is the collection of all (k, d)-mers in *Text* (including repeated (k, d)-mers). For example, here is PAIREDCOMPOSITION$_{3,1}$(**TAATGCCATGGGATGTT**):

```
TAA GCC
 AAT CCA
  ATG CAT
   TGC ATG
    GCC TGG
     CCA GGG
      CAT GGA
       ATG GAT
        TGG ATG
         GGG TGT
          GGA GTT
TAATGCCATGGGATGTT
```

Exercise Break: Generate the (3,2)-mer composition of the string **TAATGCCATGGGATGTT**.

Since the order of (3,1)-mers in PAIREDCOMPOSITION(**TAATGCCATGGGATGTT**) is unknown, we list them according to the lexicographic order of the 6-mers formed by their concatenated 3-mers:

(**AAT** | **CCA**) (**ATG** | **CAT**) (**ATG** | **GAT**) (**CAT** | **GGA**) (**CCA** | **GGG**) (**GCC** | **TGG**)
(**GGA** | **GTT**) (**GGG** | **TGT**) (**TAA** | **GCC**) (**TGC** | **ATG**) (**TGG** | **ATG**)

Note that whereas there are repeated 3-mers in the 3-mer composition of this string, there are no repeated (3,1)-mers in its paired composition. Furthermore, although **TAATGCCATGGGATGTT** and **TAATGGGATGCCATGTT** have the same 3-mer composition, they have different (3,1)-mer compositions. Thus, if we can generate the (3,1)-mer composition of these strings, then we will be able to distinguish between them. But how can we reconstruct a string from its (k,d)-mer composition? And can we adapt the de Bruijn graph approach for this purpose?

String Reconstruction from Read-Pairs Problem:
Reconstruct a string from its paired composition.

Input: A collection of paired k-mers *PairedReads* and an integer d.
Output: A string *Text* with (k,d)-mer composition equal to *PairedReads* (if such a string exists).

Paired de Bruijn graphs

Given a (k,d)-mer $(a_1 \ldots a_k \,|\, b_1, \ldots b_k)$, we define its **prefix** and **suffix** as the following $(k-1, d+1)$-mers:

$$\text{PREFIX}((a_1 \ldots a_k \,|\, b_1, \ldots b_k)) = (a_1 \ldots a_{k-1} \,|\, b_1 \ldots b_{k-1})$$
$$\text{SUFFIX}((a_1 \ldots a_k \,|\, b_1, \ldots b_k)) = (a_2 \ldots a_k \,|\, b_2 \ldots b_k)$$

For example, PREFIX((GAC | TCA)) = (GA | TC) and SUFFIX((GAC | TCA)) = (AC | CA).

Note that for consecutive (k,d)-mers appearing in *Text*, the suffix of the first (k,d)-mer is equal to the prefix of the second (k,d)-mer. For example, for the consecutive (k,d)-mers (**TAA** | **GCC**) and (**AAT** | **CCA**) in **TAATGCCATGGGATGTT**,

SUFFIX((**TAA** | **GCC**)) = PREFIX((**AAT** | **CCA**)) = (**AA** | **CC**) .

Given a string *Text*, we construct a graph PATHGRAPH$_{k,d}$(*Text*) that represents a path formed by $|Text| - (k + d + k) + 1$ edges corresponding to all (k,d)-mers in *Text*. We label edges in this path by (k,d)-mers and label the starting and ending nodes of an edge by its prefix and suffix, respectively (Figure 3.33).

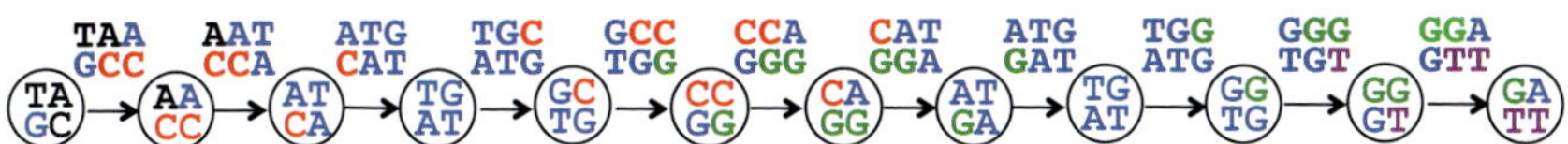

FIGURE 3.33 PATHGRAPH$_{3,1}$(**TAATGCCATGGGATGTT**). Each $(3,1)$-mer has been displayed as a two-line expression to save space.

The **paired de Bruijn graph**, denoted DEBRUIJN$_{k,d}$(*Text*), is formed by gluing identically labeled nodes in PATHGRAPH$_{k,d}$(*Text*) (Figure 3.34). Note that the paired de Bruijn graph is less tangled than the de Bruijn graph constructed from individual reads from Figure 3.13.

STOP and Think: It is easy to construct a paired de Bruijn graph from a string *Text*. But how can we construct the paired de Bruijn graph from the (k,d)-mer composition of *Text*?

We define PAIREDCOMPOSITIONGRAPH$_{k,d}$(*Text*) as the graph consisting of $|Text| - (k + d + k) + 1$ isolated edges that are labeled by the (k,d)-mers in *Text*, and whose nodes are labeled by the prefixes and suffixes of these labels (see Figure 3.35). As you may have

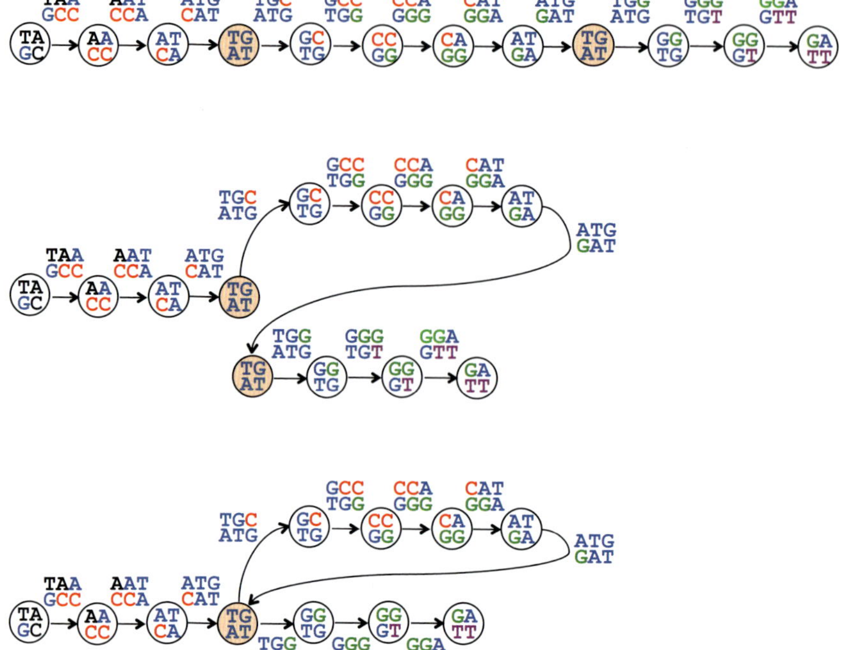

FIGURE 3.34 (Top) PATHGRAPH$_{3,1}$(**TAATGCCATGGGATGTT**) is formed by eleven edges and twelve nodes. Only two of these nodes have the same label, (**TG** | **AT**). (Middle) Bringing the two identically labeled nodes closer to each other in preparation for gluing. (Bottom) The paired de Bruijn graph DEBRUIJN$_{3,1}$(**TAATGCCATGGGATGTT**) is obtained from PATHGRAPH$_{3,1}$(**TAATGCCATGGGATGTT**) by gluing the nodes sharing the label (**TG** | **AT**). This paired de Bruijn graph has a unique Eulerian path, which spells out **TAATGCCATGGGATGTT**.

guessed, gluing identically labeled nodes in PAIREDCOMPOSITIONGRAPH$_{k,d}$(*Text*) results in exactly the same de Bruijn graph as gluing identically labeled nodes in PATHGRAPH$_{k,d}$(*Text*). Of course, in practice, we will not know *Text*; however, we can form PAIREDCOMPOSITIONGRAPH$_{k,d}$(*Text*) directly from the (k, d)-mer composition of *Text*, and the gluing step will result in the paired de Bruijn graph of this composition. The genome can be reconstructed by following an Eulerian path in this de Bruijn graph.

A pitfall of paired de Bruijn graphs

We saw earlier that every solution of the String Reconstruction Problem corresponds to an Eulerian path in the de Bruijn graph constructed from a k-mer composition. Likewise,

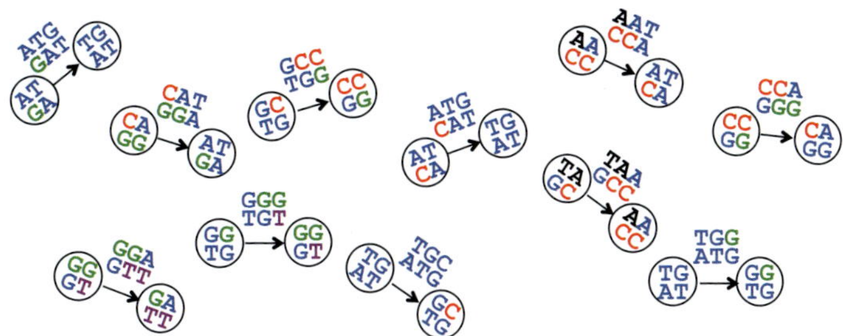

FIGURE 3.35 The graph PAIREDCOMPOSITIONGRAPH$_{3,1}$(**TAATGCCATGGGATGTT**) is a collection of isolated edges. Each edge is labeled by a (3, 1)-mer in **TAATGCCATGGGATGTT**; the starting node of an edge is labeled by the prefix of the edge's (3, 1)-mer, and the ending node of an edge is labeled by the suffix of this (3, 1)-mer. Gluing identically labeled nodes yields the paired de Bruijn graph shown in Figure 3.34 (bottom).

every solution of the String Reconstruction from Read-Pairs Problem corresponds to an Eulerian path in the Paired de Bruijn graph constructed from a (k, d)-mer composition.

Exercise Break: In the paired de Bruijn graph shown in Figure 3.36, reconstruct the genome spelled by the following Eulerian path of (2, 1)-mers: (AG | AG) → (GC | GC) → (CA | CT) → (AG | TG) → (GC | GC) → (CT | CT) → (TG | TG) → (GC | GC) → (CT | CA).

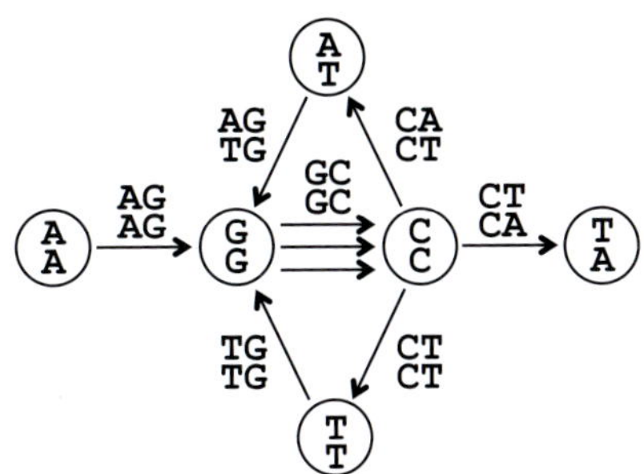

FIGURE 3.36 A paired de Bruijn graph constructed from the collection of nine (2, 1)-mers (AG | AG), (AG | TG), (CA | CT), (CT | CA), (CT | CT), (GC | GC), (GC | GC), (GC | GC), and (TG | TG).

We also saw that every Eulerian path in the de Bruijn graph constructed from a k-mer composition spells out a solution of the String Reconstruction Problem. But is this the case for the paired de Bruijn graph?

STOP and Think: The graph shown in Figure 3.36 has another Eulerian path: (AG|AG) → (GC|GC) → (CT|CT) → (TG|TG) → (GC|GC) → (CA|CT) → (AG|TG) → (GC|GC) → (CT|CA). Can you reconstruct a genome spelled by this path?

If you attempted the preceding question, then you know that not every Eulerian path in the paired de Bruijn graph constructed from a (k, d)-mer composition spells out a solution of the String Reconstruction from Read-Pairs Problem. You are now ready to solve this problem and become a genome assembly expert.

Charging Station (Generating All Eulerian Cycles): You know how to construct a single Eulerian cycle in a graph, but it remains unclear how to find *all possible* Eulerian cycles, which will be helpful when solving the String Reconstruction from Read-Pairs Problem. Check out this Charging Station to see how to generate all Eulerian cycles in a graph.

Charging Station (Reconstructing a String Spelled by a Path in the Paired de Bruijn Graph): To solve the String Reconstruction from Read-Pairs Problem, you will need to reconstruct a string from its path in the paired de Bruijn graph. Check out this Charging Station to see an example of how this can be done.

Epilogue: Genome Assembly Faces Real Sequencing Data

Our discussion of genome assembly has thus far relied upon various assumptions. Accordingly, applying de Bruijn graphs to real sequencing data is not a straightforward procedure. Below, we describe practical challenges introduced by quirks in modern sequencing technologies and some computational techniques that have been devised to address these challenges. In this discussion, we will first assume that reads are generated as contiguous substrings of a genome instead of read-pairs for the sake of simplicity.

Breaking reads into k-mers

Given a k-mer substring of a genome, we define its **coverage** as the number of reads to which this k-mer belongs. We have taken for granted that a sequencing machine can generate all k-mers present in the genome, but this assumption of "perfect k-mer coverage" does not hold in practice. For example, the popular Illumina sequencing technology generates reads that are approximately 300 nucleotides long, but this technology still misses many 300-mers present in the genome (even if the average coverage is very high), and nearly all the reads that it does generate have sequencing errors.

STOP and Think: Given a set of reads having imperfect k-mer coverage, can you find a parameter $l < k$ so that the same reads have perfect l-mer coverage? What is the maximum value of this parameter?

Figure 3.37 (left) shows four 10-mer reads that capture some but not all of the 10-mers in an example genome. However, if we take the counterintuitive step of breaking these reads into *shorter* 5-mers (Figure 3.37, right), then these 5-mers exhibit perfect coverage. This **read breaking** approach, in which we break reads into shorter k-mers, is used by many modern assemblers.

```
ATGCCGTATGGACAACGACT          ATGCCGTATGGACAACGACT
ATGCCGTATG                    ATGCC
  GCCGTATGGA                   TGCCG
     GTATGGACAA                 GCCGT
          GACAACGACT             CCGTA
                                  CGTAT
                                   GTATG
                                    TATGG
                                     ATGGA
                                      TGGAC
                                       GGACA
                                        GACAA
                                         ACAAC
                                          CAACG
                                           AACGA
                                            ACGAC
                                             CGACT
```

FIGURE 3.37 Breaking 10-mer reads (left) into 5-mers results in perfect coverage of a genome by 5-mers (right).

Read breaking must deal with a practical trade-off. On the one hand, the smaller the value of k, the larger the chance that the k-mer coverage is perfect. On the other hand, smaller values of k result in a more tangled de Bruijn graph, making it difficult to infer the genome from this graph.

Splitting the genome into contigs

Even after read breaking, most assemblies still have gaps in k-mer coverage, causing the de Bruijn graph to have missing edges, and so the search for an Eulerian path fails. In this case, biologists often settle on assembling **contigs** (long, contiguous segments of the genome) rather than entire chromosomes. For example, a typical bacterial sequencing project may result in about a hundred contigs, ranging in length from a few thousand to a few hundred thousand nucleotides. For most genomes, the order of these contigs along the genome remains unknown. Needless to say, biologists would prefer to have the entire genomic sequence, but the cost of ordering the contigs into a final assembly and closing the gaps using more expensive experimental methods is often prohibitive.

Fortunately, we can derive contigs from the de Bruijn graph. A path in a graph is called **non-branching** if $\text{IN}(v) = \text{OUT}(v) = 1$ for each intermediate node v of this path, i.e., for each node except possibly the starting and ending node of a path. A **maximal non-branching path** is a non-branching path that cannot be extended into a longer non-branching path. We are interested in these paths because the strings of nucleotides that they spell out must be present in any assembly with a given k-mer composition. For this reason, contigs correspond to strings spelled by maximal non-branching paths in the de Bruijn graph. For example, the de Bruijn graph in Figure 3.38, constructed for the 3-mer composition of **TAATGCCATGGGATGTT**, has nine maximal non-branching paths that spell out the contigs **TAAT**, **TGTT**, **TGCCAT**, **ATG**, **ATG**, **ATG**, **TGG**, **GGG**, and **GGAT**. In practice, biologists have no choice but to break genomes into contigs, even in the case of perfect coverage (like in Figure 3.38), since repeats prevent them from being able to infer a unique Eulerian path.

Contig Generation Problem:
Generate the contigs from a collection of reads (with imperfect coverage).

Input: A collection of k-mers *Patterns*.
Output: All contigs in DEBRUIJN(*Patterns*).

Charging Station (Maximal Non-Branching Paths in a Graph): If you have difficulties finding maximal non-branching paths in a graph, check out this Charging Station.

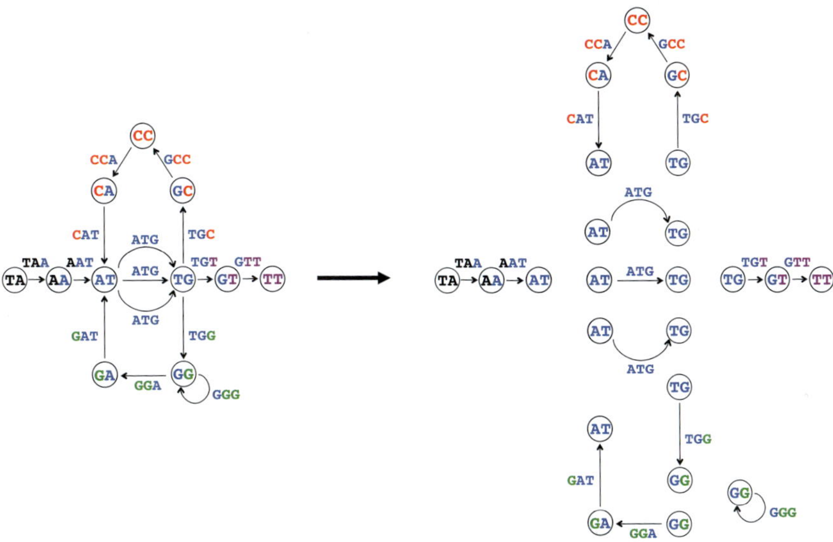

FIGURE 3.38 Breaking the graph DEBRUIJN3(**TAATGCCATGGGATGTT**) into nine maximal non-branching paths representing contigs **TAAT**, **TGTT**, **TGCCAT**, **ATG**, **ATG**, **ATG**, **TGG**, **GGG**, and **GGAT**.

Assembling error-prone reads

Error-prone reads represent yet another barrier to real sequencing projects. Adding the single erroneous read CGTA**C**GGACA (with a single error that misreads T as **C**) to the set of reads in Figure 3.37 results in erroneous 5-mers CGTA**C**, GTA**C**G, TA**C**GG, A**C**GGA, and **C**GGAC after read breaking. These 5-mers result in an erroneous path from node CGTA to node GGAC in the de Bruijn graph (Figure 3.39 (top)), meaning that if the correct read CGTATGGACA is generated as well, then we will have two paths connecting CGTA to GGAC in the de Bruijn graph. This structure is called a **bubble**, which we define as two short disjoint paths (e.g., shorter than some threshold length) connecting the same pair of nodes in the de Bruijn graph.

STOP and Think: What is the size of the bubble (in the de Bruijn graph constructed from *k*-mers) introduced by a single error in a read?

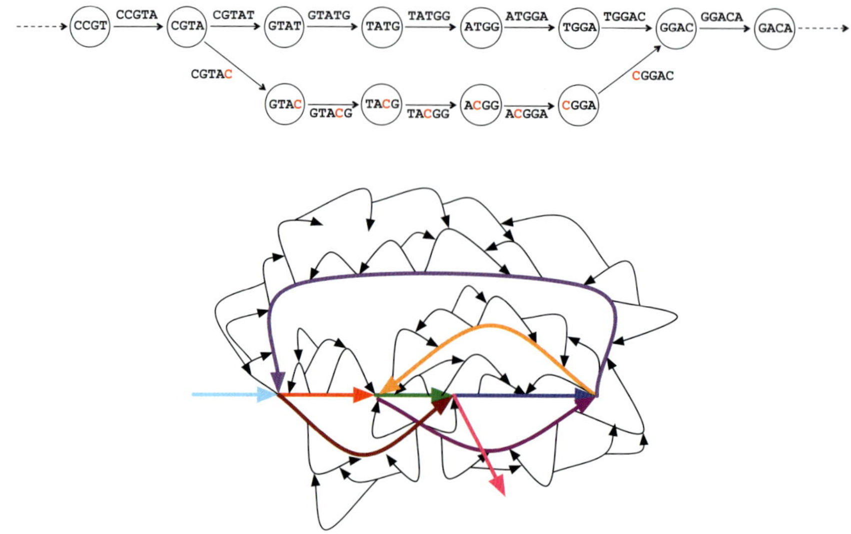

FIGURE 3.39 (Top) A correct path CGTA → GTAT → TATG → ATGG → TGGA → GGAC along with an incorrect path CGTA → GTA**C** → TA**C**G → A**C**GG → **C**GGA → GGAC form a bubble in a de Bruijn graph, making it difficult to identify which path is correct. (Bottom) An illustration of a de Bruijn graph with many bubbles. Bubble removal should leave only the colored paths remaining.

STOP and Think: Design an algorithm for detecting bubbles in de Bruijn graphs. After a bubble is detected, you must decide which of two paths in the bubble to remove. How should you make this decision?

Existing assemblers remove bubbles from de Bruijn graphs. The practical challenge is that, since nearly all reads have errors, de Bruijn graphs have millions of bubbles (Figure 3.39 (bottom)). Bubble removal occasionally removes the correct path, thus introducing errors rather than fixing them. To make matters worse, in a genome having **inexact repeats**, where the repeated regions differ by a single nucleotide or some other small variation, reads from the two repeat copies will also generate bubbles in the de Bruijn graph because one of the copies may appear to be an erroneous version of the other. Applying bubble removal to these regions introduces assembly errors by making repeats appear more similar than they are. Thus, modern genome assemblers attempt to distinguish bubbles caused by sequencing errors (which should be removed) from bubbles caused by variations (which should be retained).

Inferring multiplicities of edges in de Bruijn graphs

Although the de Bruijn graph framework requires that we know the **multiplicity** of each k-mer in the genome (i.e., the number of times the k-mer appears), this information is not readily available from reads. However, the multiplicity of a k-mer in a genome can often be estimated using its coverage. Indeed, k-mers that appear t times in a genome are expected to have approximately t times higher coverage than k-mers that appear just once. Needless to say, coverage varies across the genome, and this condition is often violated. As a result, existing assemblers often assemble repetitive regions in genomes without knowing the exact number of times each k-mer from this region occurs in the genome.

We have introduced you to some of the practical considerations in genome sequencing, but some still remain; for example, how do we deal with the fact that genomes are double-stranded? (See **DETOUR: Pitfalls of Assembling Double-Stranded DNA**.) PAGE 180 In practice, researchers will eventually obtain a collection of contigs from a genome sequencing experiments, but partitioning these contigs into two groups (corresponding to each strand) is a difficult problem that modern assemblers have yet to fully resolve. For example, if an assembler outputs five contigs, it may be that contigs 1 and 3 belong to one strand while contigs 2, 4, and 5 belong to the opposing strand.

However, we will give you a challenge problem that does not encounter these issues. Why? Developing assembly algorithms for large genomes is a formidable challenge because even the seemingly simple problem of constructing the de Bruijn graph from a collection of all k-mers present in millions of reads is nontrivial. To make your life easier, we will give you a small bacterial genome for your first assembly dataset.

Challenge Problem: *Carsonella ruddii* is a bacterium that lives symbiotically inside some insects. Its sheltered life has allowed it to reduce its genome to only about 160,000 base pairs. With only about 200 genes, it lacks some genes necessary for survival, but these genes are supplied by its insect host. In fact, *Carsonella* has such a small genome that biologists have conjectured that it is losing its "bacterial" identity and turning into an **organelle**, which is part of the host's genome. This transition from bacterium to organelle has happened many times during evolutionary history; in fact, the mitochondrion responsible for energy production in human cells was once a free-roaming bacterium that we assimilated in the distant past.

Given a collection of simulated error-free read-pairs, use the paired de Bruijn graph to reconstruct the *Carsonella ruddii* genome. Compare this assembly to the assembly obtained from the classic de Bruijn graph (i.e., when all we know is the reads themselves and do not know the distance between paired reads) in order to better appreciate the benefits of read-pairs. For each k, what is the minimum value of d needed to enable reconstruction of the entire *Carsonella ruddii* genome from its (k, d)-mer composition?

Exercise Break: By the way, one more thing . . . what was the headline of the June 27, 2000 edition of the *New York Times*?

Charging Stations

The effect of gluing on the adjacency matrix

Figure 3.40 uses *Text* = **TAATGCCATGGGATGTT** to illustrate how gluing converts the adjacency matrix of PATHGRAPH$_3$(*Text*) into the adjacency matrix of DEBRUIJN$_3$(*Text*).

	TA	**AA**	**AT$_1$**	**TG$_1$**	**GC**	**CC**	**CA**	**AT$_2$**	**TG$_2$**	**GG$_1$**	**GG$_2$**	**GA**	**AT$_3$**	**TG$_3$**	**GT**	**TT**
TA	0	1	0	0	0	0	0	0	0	0	0	0	0	0	0	0
AA	0	0	1	0	0	0	0	0	0	0	0	0	0	0	0	0
AT$_1$	0	0	0	1	0	0	0	0	0	0	0	0	0	0	0	0
TG$_1$	0	0	0	0	1	0	0	0	0	0	0	0	0	0	0	0
GC	0	0	0	0	0	1	0	0	0	0	0	0	0	0	0	0
CC	0	0	0	0	0	0	1	0	0	0	0	0	0	0	0	0
CA	0	0	0	0	0	0	0	1	0	0	0	0	0	0	0	0
AT$_2$	0	0	0	0	0	0	0	0	1	0	0	0	0	0	0	0
TG$_2$	0	0	0	0	0	0	0	0	0	1	0	0	0	0	0	0
GG$_1$	0	0	0	0	0	0	0	0	0	0	1	0	0	0	0	0
GG$_2$	0	0	0	0	0	0	0	0	0	0	0	1	0	0	0	0
GA	0	0	0	0	0	0	0	0	0	0	0	0	1	0	0	0
AT$_3$	0	0	0	0	0	0	0	0	0	0	0	0	0	1	0	0
TG$_3$	0	0	0	0	0	0	0	0	0	0	0	0	0	0	1	0
GT	0	0	0	0	0	0	0	0	0	0	0	0	0	0	0	1
TT	0	0	0	0	0	0	0	0	0	0	0	0	0	0	0	0

	TA	**AA**	**AT**	**TG**	**GC**	**CC**	**CA**	**GG**	**GA**	**GT**	**TT**
TA	0	1	0	0	0	0	0	0	0	0	0
AA	0	0	1	0	0	0	0	0	0	0	0
AT	0	0	0	3	0	0	0	0	0	0	0
TG	0	0	0	0	1	0	0	1	0	1	0
GC	0	0	0	0	0	1	0	0	0	0	0
CC	0	0	0	0	0	0	1	0	0	0	0
CA	0	0	1	0	0	0	0	0	0	0	0
GG	0	0	0	0	0	0	0	1	1	0	0
GA	0	0	1	0	0	0	0	0	0	0	0
GT	0	0	0	0	0	0	0	0	0	0	1
TT	0	0	0	0	0	0	0	0	0	0	0

FIGURE 3.40 Adjacency matrices. (Top) The 16×16 adjacency matrix of PATHGRAPH$_3$(**TAATGCCATGGGATGTT**). Note that we have used indexing to differentiate multiple occurrences of AT, TG, and GG in PATHGRAPH$_3$(**TAATGCCATGGGATGTT**). (Bottom) The 11×11 adjacency matrix of DEBRUIJN$_3$(**TAATGCCATGGGATGTT**), produced after gluing nodes labeled by identical 2-mers. Note that if there are *m* edges connecting nodes *i* and *j*, the (*i*, *j*)-th element of the adjacency matrix is *m*.

Generating all Eulerian cycles

The inherent difficulty in generating all Eulerian cycles in a graph is keeping track of potentially many different alternatives at any given node. On the opposite end of the spectrum, a **simple directed graph**, a connected graph in which each node has indegree and outdegree equal to 1, offers a trivial case, since there is only one Eulerian cycle.

Our idea, then, is to transform a single labeled directed graph *Graph* containing $n \geq 1$ Eulerian cycles to n different simple directed graphs, each containing a single Eulerian cycle. This transformation has the property that it is easily invertible, i.e., that given the unique Eulerian cycle in one of the simple directed graphs, we can easily reconstruct the original Eulerian cycle in *Graph*.

Given a node v in *Graph* (of indegree greater than 1) with incoming edge (u, v) and outgoing edge (v, w), we will construct a "simpler" **(u, v, w)-bypass graph** in which we remove the edges (u, v) and (v, w) from *Graph*, and add a new node x along with the edges (u, x) and (x, w) (Figure 3.41 (top)). The new edges (u, x) and (x, w) in the bypass graph inherit the labels of the removed edges (u, v) and (v, w), respectively. The critical property of this graph is revealed by the following exercise.

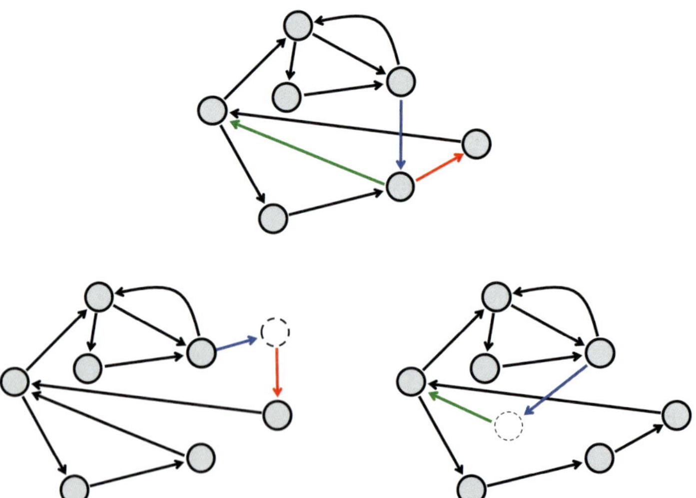

FIGURE 3.41 (Top) An Eulerian graph. (Bottom left) The (u, v, w)-bypass graph (right) constructed for the blue and red edges. (Bottom right) The other bypass graph constructed for the blue and green edges.

Exercise Break: Show that any Eulerian cycle in *Graph* passing through (u, v) and then (v, w) corresponds to an Eulerian cycle (with the same edge labels) in the (u, v, w)-bypass graph passing through (u, x) and then (x, w).

In general, given an incoming edge (u, v) into v along with k outgoing edges (v, w_1), $\ldots$, (v, w_k) from v, we can construct k different bypass graphs (Figure 3.41 (bottom)). Note that since no two bypass graphs have the same Eulerian cycle, the total number of Eulerian cycles in all k of these bypass graphs is equal to the number of Eulerian cycles in the original graph.

Our idea, roughly stated, is to iteratively construct every possible bypass graph for *Graph* until we obtain a large family of simple directed graphs; each one of these graphs will correspond to a distinct Eulerian cycle in *Graph*. This idea is implemented by the pseudocode below.

```
ALLEULERIANCYCLES(Graph)
    AllGraphs ← the set consisting of a single graph Graph
    while there is a non-simple graph G in AllGraphs
        v ← a node with indegree larger than 1 in G
        for each incoming edge (u, v) into v
            for each outgoing edge (v, w) from v
                NewGraph ← (u, v, w)-bypass graph of G
                if NewGraph is connected
                    add NewGraph to AllGraphs
        remove G from AllGraphs
    for each graph G in AllGraphs
        output the (single) Eulerian cycle in G
```

There exists a more elegant approach to constructing all Eulerian cycles in an Eulerian graph that is based on a theorem that de Bruijn had a hand in proving. To learn about this theorem, see DETOUR: The BEST Theorem.

Reconstructing a string spelled by a path in the paired de Bruijn graph

Consider the following Eulerian path formed by nine edges in the paired de Bruijn graph from Figure 3.36.

AG-AG → GC-GC → CA-CT → AG-TG →

GC-GC → CT-CT → TG-TG → GC-GC → CT-CA

We can arrange the (2, 1)-mers in this path into the nine rows shown below, revealing the string **AGCAGCTGCTGCA** spelled by this path:

```
AG-AG
 GC-GC
  CA-CT
   AG-TG
    GC-GC
     CT-CT
      TG-TG
       GC-GC
        CT-CA
AGCAGCTGCTGCA
```

Now, consider another Eulerian path in the paired de Bruijn graph from Figure 3.36:

AG-AG → GC-GC → CT-CT → TG-TG →
GC-GC → CA-CT → AG-TG → GC-GC → CT-CA

An attempt to assemble these (2, 1)-mers reveals that not every column has the same nucleotide (see the two columns shown in red below). This example illustrates that not all Eulerian paths in the paired de Bruijn graph spell solutions of the String Reconstruction from Read-Pairs Problem.

```
AG-AG
 GC-GC
  CT-CT
   TG-TG
    GC-GC
     CA-CT
      AG-TG
       GC-GC
        CT-CA
AGC?GC?GCTGCA
```

String Spelled by a Gapped Genome Path Problem:
Reconstruct a sequence of (k, d)-mers corresponding to a path in a paired de Bruijn graph.

> **Input**: A sequence of (k, d)-mers $(a_1|b_1), \ldots, (a_n|b_n)$ such that $\text{SUFFIX}((a_i|b_i)) = \text{PREFIX}((a_{i+1}|b_{i+1}))$ for $1 \leq i \leq n-1$.
> **Output**: A string *Text* of length $k + d + k + n - 1$ such that the i-th (k, d)-mer of *Text* is equal to $(a_i|b_i)$ for $1 \leq i \leq n$ (if such a string exists).

Our approach to solving this problem will split the given (k, d)-mers $(a_1|b_1), \ldots, (a_n|b_n)$ into their initial k-mers, *FirstPatterns* $= (a_1, \ldots, a_n)$, and their terminal k-mers, *SecondPatterns* = $(b_1, \ldots, b_n)$. Assuming that we have implemented an algorithm solving the String Spelled by a Genome Path Problem (denoted **STRINGSPELLEDBYPATTERNS**), we can assemble *FirstPatterns* and *SecondPatterns* into strings *PrefixString* and *SuffixString*, respectively.

For the first example above, we have that *PrefixString* = AGCAGCTGCT and *SuffixString* = AGCTGCTGCA. These strings perfectly overlap starting at the fourth nucleotide of *PrefixString*:

```
PrefixString = AGCAGCTGCT
SuffixString =    AGCTGCTGCA
      Genome = AGCAGCTGCTGCA
```

However, for the second example above, there is no perfect overlap:

```
PrefixString = AGCTGCAGCT
SuffixString =    AGCTGCTGCA
      Genome = AGC?GC?GCTGCA
```

The following algorithm, **STRINGSPELLEDBYGAPPEDPATTERNS**, generalizes this approach to an arbitrary sequence *GappedPatterns* of (k, d)-mers. It constructs strings *PrefixString* and *SuffixString* as described above, and checks whether they have perfect overlap (i.e., form the prefix and suffix of a reconstructed string). It also assumes that the number of (k, d)-mers in *GappedPatterns* is at least d; otherwise, it is impossible to reconstruct a contiguous string.

```
StringSpelledByGappedPatterns(GappedPatterns, k, d)
    FirstPatterns ← the sequence of initial k-mers from GappedPatterns
    SecondPatterns ← the sequence of terminal k-mers from GappedPatterns
    PrefixString ← StringSpelledByPatterns(FirstPatterns, k)
    SuffixString ← StringSpelledByPatterns(SecondPatterns, k)
    for i = k + d + 1 to |PrefixString|
        if the i-th symbol in PrefixString ≠ the (i − k − d)-th symbol in SuffixString
            return "there is no string spelled by the gapped patterns"
    return PrefixString concatenated with the last k + d symbols of SuffixString
```

Maximal non-branching paths in a graph

A node v in a directed graph *Graph* is called a **1-in-1-out node** if its indegree and outdegree are both equal to 1. We can rephrase the definition of a "maximal non-branching path" from the main text as a path whose internal nodes are 1-in-1-out nodes and whose initial and final nodes are not 1-in-1-out nodes. Also, note that the definition from the main text does not handle the case when *Graph* has a connected component that is an **isolated cycle**, in which all nodes are 1-in-1-out nodes (recall Figure 3.38).

MaximalNonBranchingPaths, shown below, iterates through all nodes of the graph that are not 1-in-1-out nodes and generates all non-branching paths starting at each such node. In a final step, it finds all isolated cycles in the graph.

3M

```
MaximalNonBranchingPaths(Graph)
    Paths ← empty list
    for each node v in Graph
        if v is not a 1-in-1-out node
            if Out(v) > 0
                for each outgoing edge (v, w) from v
                    NonBranchingPath ← the path consisting of the single edge (v, w)
                    while w is a 1-in-1-out node
                        extend NonBranchingPath by the outgoing edge (w, u) from w
                        w ← u
                    add NonBranchingPath to the set Paths
    for each isolated cycle Cycle in Graph
        add Cycle to Paths
    return Paths
```

Detours

A short history of DNA sequencing technologies

In 1988, Radoje Drmanac, Andrey Mirzabekov, and Edwin Southern simultaneously and independently proposed the futuristic and at the time completely implausible method of **DNA arrays** for DNA sequencing. None of these three biologists knew of the work of Euler, Hamilton, and de Bruijn; none could have possibly imagined that the implications of his own experimental research would eventually bring him face-to-face with these mathematical giants.

A decade earlier, Frederick Sanger had sequenced the tiny 5,386 nucleotide-long genome of the φX174 virus. By the late 1980s, biologists were routinely sequencing viruses containing hundreds of thousands of nucleotides, but the idea of sequencing bacterial (let alone human) genomes remained preposterous, both experimentally and computationally. Indeed, generating a single read in the late 1980s cost more than a dollar, pricing mammalian genome sequences in the billions. DNA arrays were therefore invented with the goal of cheaply generating a genome's k-mer composition, albeit with a smaller read length k than the original DNA sequencing technology. For example, whereas Sanger's expensive sequencing technique generated 500 nucleotide-long reads in 1988, the DNA array inventors initially aimed at producing reads of length only 10.

DNA arrays work as follows. We first synthesize all 4^k possible DNA k-mers and attach them to a DNA array, which is a grid on which each k-mer is assigned a unique location. Next, we fluorescently label a single-stranded DNA fragment (with unknown sequence) and apply a solution containing this labeled DNA to the DNA array. The k-mers in a DNA fragment will hybridize (bind) to their reverse complementary k-mers on the array. All we need to do is use spectroscopy to analyze which sites on the array emit the fluorescence; the reverse complements of k-mers corresponding to these sites must therefore belong to the (unknown) DNA fragment. Thus, the set of fluorescent k-mers on the array reveals the composition of a DNA fragment (Figure 3.42).

At first, few believed that DNA arrays would work, because both the biochemical problem of synthesizing millions of short DNA fragments and the algorithmic problem of sequence reconstruction appeared too complicated. In 1988, *Science* magazine wrote that given the amount of work required to synthesize a DNA array, "using DNA arrays for sequencing would simply be substituting one horrendous task for another". It turned out that *Science* was only half right. In the mid-1990s, a number of companies perfected technologies for designing large DNA arrays, but DNA arrays ultimately failed to realize the dream that motivated their inventors because the fidelity of DNA hybridization with the array was too low and because the value of k was too small.

AAA	AGA	CAA	CGA	GAA	GGA	TAA	TGA
AAC	AGC	CAC	CGC	GAC	GGC	TAC	TGC
AAG	AGG	CAG	CGG	GAG	GGG	TAG	TGG
AAT	AGT	CAT	CGT	GAT	GGT	TAT	TGT
ACA	ATA	CCA	CTA	GCA	GTA	TCA	TTA
ACC	ATC	CCC	CTC	GCC	GTC	TCC	TTC
ACG	ATG	CCG	CTG	GCG	GTG	TCG	TTG
ACT	ATT	CCT	CTT	GCT	GTT	TCT	TTT

FIGURE 3.42 A toy DNA array containing all possible 3-mers. Taking the reverse complements of the fluorescently labeled 3-mers yields the collection {ACC, ACG, CAC, CCG, CGC, CGT, GCA, GTT, TAC, TTA}, which is the 3-mer composition of the twelve nucleotide-long string CGCACGTTACCG. Note that this DNA array provides no information regarding 3-mer multiplicities.

Yet the failure of DNA arrays was a spectacular one; while the original goal (DNA sequencing) dangled out of reach, two new unexpected applications of DNA arrays emerged. Today, arrays are used to measure gene expression as well as to analyze genetic variations. These unexpected applications transformed DNA arrays into a multi-billion dollar industry that included Hyseq, founded by Radoje Drmanac, one of the original inventors of DNA arrays.

After founding Hyseq, Drmanac did not abandon his dream of inventing an alternative DNA sequencing technology. In 2005, he founded Complete Genomics, one of the first Next Generation Sequencing (NGS) companies. Complete Genomics, Illumina, Life Technologies, and other NGS companies subsequently developed the technology to cheaply generate almost all k-mers from a genome, thus at last enabling the method of Eulerian assembly. While these technologies are quite different from the DNA array technology proposed in 1988, one can still recognize the intellectual legacy of DNA arrays in NGS approaches, a testament to the fact that good ideas never die, even if they fail at first.

Similarly to DNA arrays, NGS technologies initially generated millions of rather short error-prone reads (of length about 20 nucleotides) when the NGS revolution began in 2005. However, within a span of just a few years, NGS companies were able to increase read length and improve accuracy by an order of magnitude. Moreover, Pacific

Biosciences and Oxford Nanopore Technologies already generate error-ridden reads containing thousands of nucleotides. Perhaps your own start-up will find a way to generate a single read spanning the entire genome, thus making this chapter a footnote in the history of genome sequencing. Whatever the future brings, recent developments in NGS have already revolutionized genomics, and biologists are preparing to assemble the genomes of all mammalian species on Earth ... all while relying on a simple idea that Leonhard Euler conceived in 1735.

Repeats in the human genome

A **transposon** is a DNA fragment that can change its position within the genome, often resulting in a duplication (repeat). A transposon that inserts itself into a gene will most likely disable that gene. Diseases that are caused by transposons include hemophilia, porphyria, Duchenne muscular dystrophy, and many others. Transposons make up a large fraction of the human genome and are divided into two classes according to their mechanism of transposition, which can be described as either **retrotransposons** or **DNA transposons**.

Retrotransposons are copied in two stages: first they are transcribed from DNA to RNA, and the RNA produced is then **reverse transcribed** to DNA by a **reverse transcriptase**. This copied DNA fragment is then inserted at a new position into the genome. DNA transposons do not involve an RNA intermediate but instead are catalyzed by **transposases**. The transposase cuts out the DNA transposon, which is then inserted into a new site in the genome, resulting in a repeat.

The first transposons were discovered in maize by Barbara McClintock, for which she was awarded a Nobel Prize in 1983. About 85% of the maize genome and 50% of the human genome consist of transposons. The most common transposon in humans is the Alu sequence, which is approximately 300 bases long and accounts for approximately a million repeats (with mutations) in the human genome. The **Mariner sequence** is another transposon in the human genome that has about 14,000 repeats, making up 2.6 million base pairs. Mariner-like transposons exist in many species and can even be transmitted from one species to another. The Mariner transposon was used to develop the **Sleeping Beauty transposon system**, a synthetic DNA transposon constructed for the purposes of introducing new traits in animals and for gene therapy.

Graphs

The use of the word "graph" in this book is different from its use in high school mathematics; we do not mean a chart of data. You can think of a graph as a diagram showing cities connected by roads.

The first panel in Figure 3.43 shows a 4×4 chessboard with the four corner squares removed. A knight can move two steps in any of four directions (left, right, up, and down) followed by one step in a perpendicular direction. For example, a knight at square 1 can move to square 7 (two down and one left), square 9 (two down and one right), or square 6 (two right and one down).

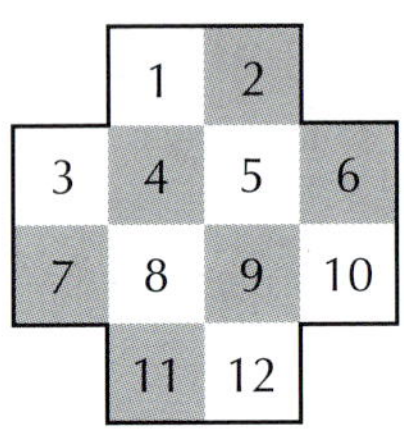

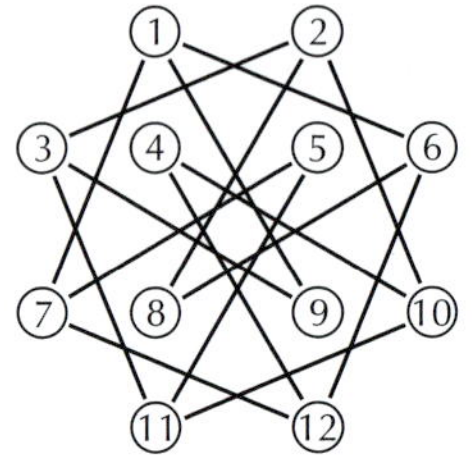

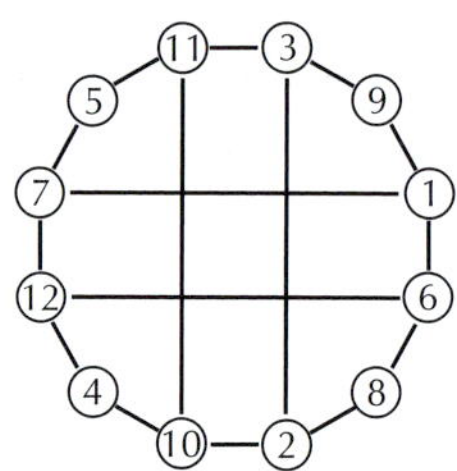

FIGURE 3.43 (Left) A hypothetical chessboard. (Middle) The Knight Graph represents each square by a node and connects two nodes with an edge if a knight can travel between the respective squares in a single move. (Right) An equivalent representation of the Knight Graph.

STOP and Think: Can a knight travel around this board, pass through each square exactly once, and return to the same square it started on?

The second panel in Figure 3.43 represents each of the chessboard's twelve squares as a node. Two nodes are connected by an edge if a knight can move between them in a single step. For example, node 1 is connected to nodes 6, 7, and 9. Connecting nodes in this manner produces a "Knight Graph" consisting of twelve nodes and sixteen edges.

We can describe a graph by its set of nodes and edges, where every edge is written as the pair of nodes that it connects. The graph in the second panel of Figure 3.43 is described by the node set

$$1, 2, 3, 4, 5, 6, 7, 8, 9, 10, 11, 12\,,$$

and the following edge set:

$1 — 6 \quad 1 — 7 \quad 1 — 9 \quad 2 — 3 \quad 2 — 8 \quad 2 — 10 \quad 3 — 9 \quad 3 — 11$
$4 — 10 \quad 4 — 12 \quad 5 — 7 \quad 5 — 11 \quad 6 — 8 \quad 6 — 12 \quad 7 — 12 \quad 10 — 11$

A **path** in a graph is a sequence of edges, where each successive edge begins at the node where the previous edge ends. For example, the path $8 \rightarrow 6 \rightarrow 1 \rightarrow 9$ in Figure 3.43 starts at node 8, ends at node 9, and consists of 3 edges. Paths that start and end at the same node are referred to as **cycles**. The cycle $3 \rightarrow 2 \rightarrow 10 \rightarrow 11 \rightarrow 3$ starts and ends at node 3 and consists of 4 edges.

The way a graph is drawn is irrelevant; two graphs with the same node and edge sets are equivalent, even if the particular pictures that represent the graph are different. The only thing that is important is which nodes are connected and which are not. Therefore, the graph in the second panel of Figure 3.43 is identical to the graph in the third panel. This graph reveals a cycle that visits every node in the Knight Graph once and describes a sequence of knight moves that visit every square exactly once.

Exercise Break: How many knight's tours exist for the chessboard in Figure 3.43?

The number of edges incident to a given node v is called the **degree** of v. For example, node 1 in the Knight Graph has degree 3, while node 5 has degree 2. The sum of degrees of all twelve nodes is, in this case, 32 (eight nodes have degree 3 and four nodes have degree 2), which is twice the number of edges in the graph.

STOP and Think: Can you connect seven phones in such a way that each is connected to exactly three others?

Many bioinformatics problems analyze **directed graphs**, in which every edge is directed from one node to another, as shown by the arrows in Figure 3.44. You can think of a directed graph as a diagram showing cities connected by one-way roads. Every node in a directed graph is characterized by indegree (the number of incoming edges) and outdegree (the number of outgoing edges).

STOP and Think: Prove that for every directed graph, the sum of indegrees of all nodes is equal to the sum of outdegrees of all nodes.

An undirected graph is **connected** if every two nodes have a path connecting them. Disconnected graphs can be partitioned into disjoint connected components.

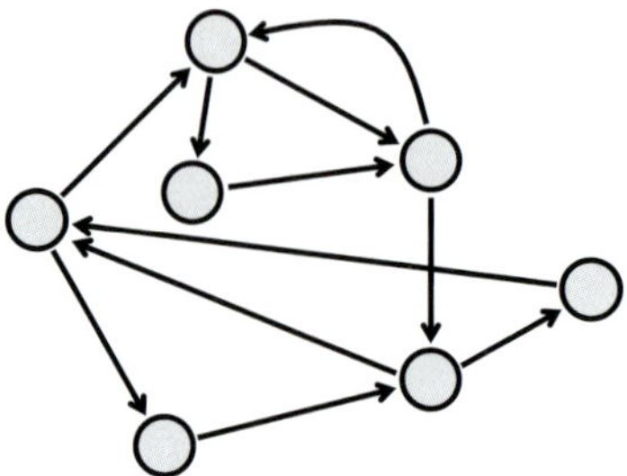

FIGURE 3.44 A directed graph.

The icosian game

We make an historical detour to Dublin, with the creation in 1857 of the **icosian game** by the Irish mathematician William Hamilton. This "game", which was a commercial flop, consisted of a wooden board with twenty pegholes and some lines connecting the holes, as well as twenty numbered pegs (Figure 3.45 (left)). The objective was to place the numbered pegs in the holes in such a way that peg 1 would be connected by a line on the board to peg 2, which would in turn be connected by a line to peg 3, and so on, until finally peg 20 would be connected by a line back to peg 1. In other words, if we follow the lines on the board from peg to peg in ascending order, we reach every peg exactly once and then arrive back at our starting peg.

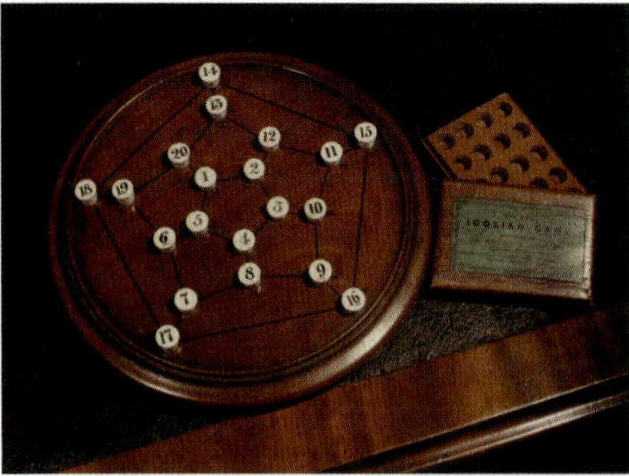

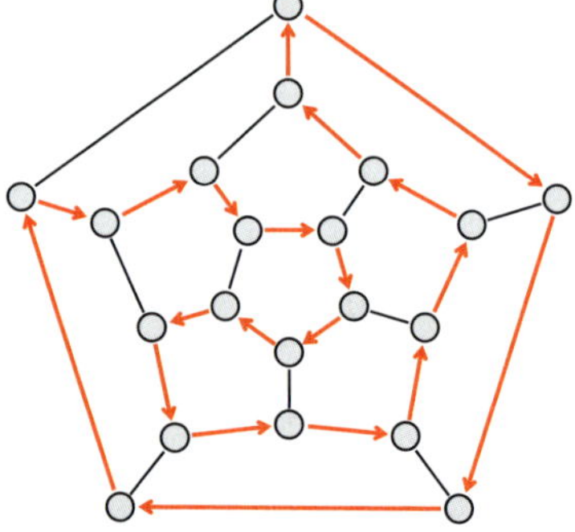

FIGURE 3.45 (Left) Hamilton's icosian game, with a winning placement of the pegs shown. (Right) The winning placement of the pegs can be represented as a Hamiltonian cycle in a graph. Each node in this graph represents a peghole on the board, and an edge connects two pegholes that are connected by a line segment on the board.

We can model the icosian game using a graph if we represent each peghole by a node and then transform lines that connect pegholes into edges that connect the corresponding nodes. This graph does have a Hamiltonian cycle solving the icosian game; one of them is shown in Figure 3.45 (right). Although a brute force approach to the Hamiltonian Cycle Problem works for these small graphs, it quickly becomes infeasible for large graphs.

Tractable and intractable problems

Inspired by Euler's Theorem, you probably are wondering whether there exists such a simple result leading to a fast algorithm for the Hamiltonian Cycle Problem. The key challenge is that while we are guided by Euler's Theorem in solving the Eulerian Cycle Problem, an analogous simple condition for the Hamiltonian Cycle Problem remains unknown. Of course, you could always explore all walks through the graph and report back if you find a Hamiltonian cycle. The problem with this brute force method is that for a graph on just a thousand nodes, there may be more walks through the graph than there are atoms in the universe!

For years, the Hamiltonian Cycle Problem eluded all attempts to solve it by some of the world's most brilliant researchers. After years of fruitless effort, computer scientists began to wonder whether this problem is **intractable**, i.e., that their failure to find a polynomial-time algorithm was not attributable to a lack of insight, but rather because such an algorithm for solving the Hamiltonian Cycle Problem simply does not exist. In the 1970s, computer scientists discovered thousands more algorithmic problems with the same fate as the Hamiltonian Cycle Problem. While these problems may appear to be simple, no one has been able to find fast algorithms for solving them. A large subset of these problems, including the Hamiltonian Cycle Problem, are now collectively known as ***NP*-complete**. A formal definition of *NP*-completeness is beyond the scope of this text.

All the *NP*-complete problems are *equivalent* to each other: an instance of any *NP*-complete problem can be transformed into an instance of any other *NP*-complete problem in polynomial time. Thus, if you find a fast algorithm for one *NP*-complete problem, you will automatically be able to use this algorithm to design a fast algorithm for any other *NP*-complete problem! The problem of efficiently solving *NP*-complete problems, or finally proving that they are intractable, is so fundamental that it was named one of seven "Millennium Problems" by the Clay Mathematics Institute in 2000. Find an efficient algorithm for any *NP*-complete problem, or show that one of these problems is intractable, and the Clay Institute will award you a prize of one million dollars.

Think twice before you embark on solving an *NP*-complete problem. So far, only one of the seven Millennium Problems has been solved; in 2003, Grigori Perelman proved the Poincaré Conjecture. A true mathematician, Perelman would later turn down the million-dollar prize, believing that the purity and beauty of mathematics is above all worldly compensation.

NP-complete problems fall within a larger class of difficult computational problems. A problem *A* is ***NP*-hard** if there is some *NP*-complete problem that can be reduced to *A* in polynomial time. Because *NP*-complete problems can be reduced to each other in polynomial time, all *NP*-complete problems are *NP*-hard. However, not every *NP*-hard problem is *NP*-complete (meaning that the former are "at least as difficult" to solve as the latter). One example of an *NP*-hard problem that is not *NP*-complete is the **Traveling Salesperson Problem**, in which we are given a graph with weighted edges and we must produce a Hamiltonian cycle of minimum total weight.

From Euler to Hamilton to de Bruijn

Euler's presented his solution of the Bridges of Königsberg Problem to the Imperial Russian Academy of Sciences in St. Petersburg in 1735. Figure 3.46 shows Euler's drawing of the Seven Bridges of Königsberg.

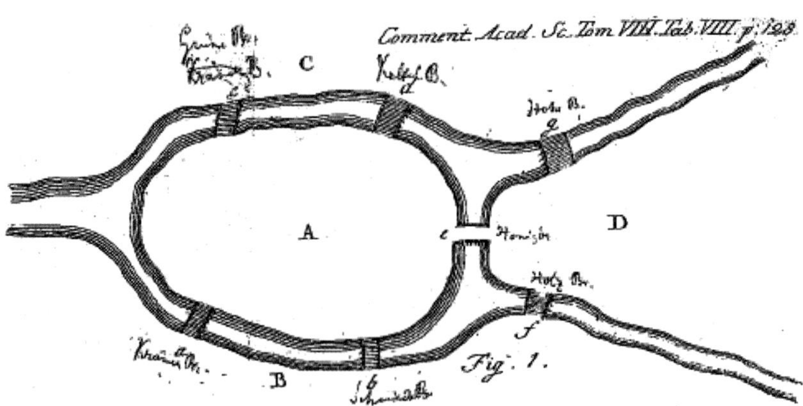

FIGURE 3.46 Euler's illustration of Königsberg, showing each of the four parts of the city labeled A, B, C, and D, along with the seven bridges crossing the different arms of the Pregel river.

Euler was the most prolific mathematician of all time; besides graph theory, he first introduced the notation *f(x)* to represent a function, *i* for the square root of -1, and π for the circular constant. Working very hard throughout his entire life, he became blind in his right eye in 1735. He kept working. In 1766, he lost the use of his left eye and commented: "Now I will have fewer distractions." He kept working. Even after going completely blind, he published hundreds of papers.

In this chapter, we have met three mathematicians of three different centuries, Euler, Hamilton, and de Bruijn, spread out across Europe (Figure 3.47). We might be inclined to feel a sense of adventure at their work and how it converged to this singular point in modern biology. Yet the first biologists who worked on DNA sequencing had no idea of how graph theory could be applied to this subject. What's more, the first paper applying the three mathematicians' ideas to genome assembly was published lifetimes after the deaths of Euler and Hamilton, when de Bruijn was in his seventies. So perhaps we might think of these three men not as adventurers, but instead as lonely wanderers. As is so often the mathematician's curse, each of the three passionately pursued abstract questions while having no idea where the answers might one day lead without him.

FIGURE 3.47 Leonhard Euler (left), William Hamilton (center), and Nicolaas de Bruijn (right).

The seven bridges of Kaliningrad

Königsberg was largely destroyed during World War II; its ruins were captured by the Soviet army. The city was renamed Kaliningrad in 1946 in honor of Soviet revolutionary Mikhail Kalinin.

Since the 18th Century, much has changed in the layout of Königsberg, and it just so happens that the bridge graph drawn today for the city of Kaliningrad still does not contain an Eulerian cycle. However, this graph does contain an Eulerian path, which means that Kaliningrad residents can walk crossing every bridge exactly once, but cannot do so and return to where they started. Thus, the citizens of Kaliningrad finally achieved at least a small part of the goal set by the citizens of Königsberg (although they have to take a taxi home). Yet strolling around Kaliningrad is not as pleasant as it would have been in 1735, since the beautiful old Königsberg was ravaged by the combination of Allied bombing in 1944 and dreadful Soviet architecture in the years following World War II.

Pitfalls of assembling double-stranded DNA

To assemble real genomes, bioinformaticians must handle reads from both DNA strands without knowing in advance which strand each read comes from. To address this challenge, they first add the reverse complement of each read to the collection of reads, effectively doubling the number of reads. In an ideal world, the de Bruijn graph formed from all these reads would consist of two (topologically identical but differently labeled) connected components, one for each DNA strand (see Figure 3.48).

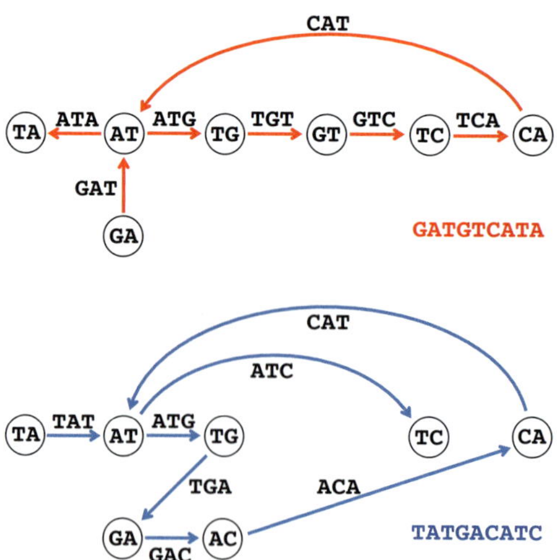

FIGURE 3.48 Reconstructing genomes **GATGTCATA** and **TATGACATC** from their corresponding graphs DEBRUIJN3(**GATGTCATA**) (top) and DEBRUIJN3(**TATGACATC**) (bottom) can be done easily, as there is only one way to traverse each graph.

In reality, these two components will be glued together, since there are many reverse complementary *k*-mers within a single strand in genomes. Indeed, in addition to direct repeats (like ATG in **GATGTATGA**), genomes have many **inverted repeats**, in which one substring is the reverse complement of another (like ATG/CAT in **GATGTCATA**). As a result, while the single strand **GATGTCATA** has no repeated 3-mers, making its assembly trivial, the reverse complementary strands **GATGTCATA** and **TATGACATC** have repeated 3-mers. Figure 3.49 shows the de Bruijn graph for the reverse complementary strings **GATGTCATA** and **TATGACATC**, which requires gluing nodes from two different strings, thus complicating assembly.

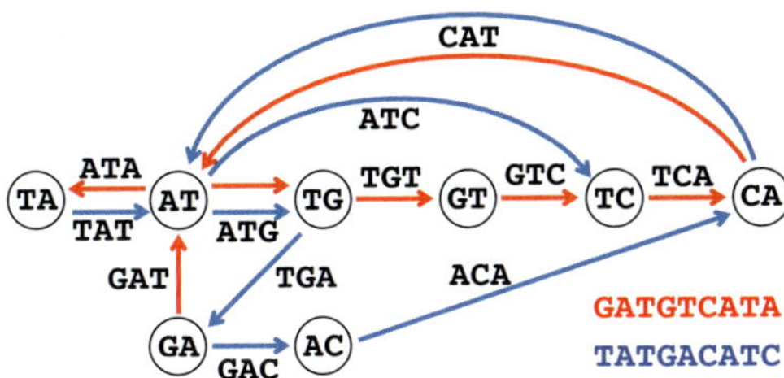

FIGURE 3.49 The de Bruijn graph formed by combining 3-mers from **GATGTCATA** and its reverse complement **TATGACATC**. Reconstructing the original genome is now a nontrivial problem.

The BEST Theorem

Given an adjacency matrix $A(G)$ of a directed Eulerian graph G, we define the matrix $A^*(G)$ by replacing the i-th diagonal entry of $-A(G)$ by INDEGREE(i) for each node i in G (Figure 3.50).

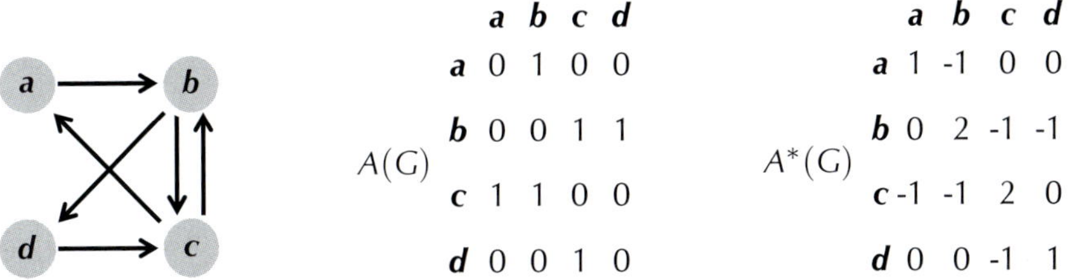

FIGURE 3.50 (Left) A graph G with two Eulerian cycles. (Middle) The adjacency matrix $A(G)$ of G. (Right) The matrix $A^*(G)$. Each i-cofactor of $A^*(G)$ is equal to 2. Thus, the BEST theorem computes the number of Eulerian cycles as $2 \cdot 0! \cdot 1! \cdot 1! \cdot 0! = 2$.

The ***i*-cofactor** of a matrix M is the determinant of the matrix obtained from M by deleting its i-th row and i-th column. It can be shown that for a given Eulerian graph G, all i-cofactors of $A^*(G)$ have the same value, which we denote as $c(G)$.

The following theorem provides a formula computing the number of Eulerian cycles in a graph. Its name is an acronym of its discoverers: de **B**ruijn, van Aardenne-**E**hrenfest, **S**mith, and **T**utte.

BEST Theorem: *The number of Eulerian cycles in an Eulerian graph is equal to*

$$c(G) \cdot \prod_{\text{all nodes } v \text{ in graph } G} (\text{INDEGREE}(v) - 1)!$$

The proof of the BEST Theorem, which is beyond the scope of this text, provides us with an alternative way to construct all Eulerian cycles in an Eulerian directed graph.

Bibliography Notes

After Euler's work on the Königsberg Bridge Problem (Euler, 1758), graph theory was forgotten for over a hundred years but was revived in the second half of the 19th Century. Graph theory flourished in the 20th Century, when it became an important area of mathematics with many practical applications. The de Bruijn graph was introduced independently by Nicolaas de Bruijn (de Bruijn, 1946) and I. J. Good (Good, 1946).

DNA sequencing methods were invented independently and simultaneously in 1977 by groups led by Frederick Sanger (Sanger, Nicklen, and Coulson, 1977) and Walter Gilbert (Maxam and Gilbert, 1977). A year previously, Walter Flyers and colleagues had sequenced a shorter virus called MS2, but Sanger's approach scaled for larger genomes. DNA arrays were proposed simultaneously and independently in 1988 by Radoje Drmanac (Drmanac et al., 1989), Andrey Mirzabekov (Lysov et al., 1988) and Edwin Southern (Southern, 1988). The Eulerian approach to DNA arrays was described in 1989 (Pevzner, 1989).

The Eulerian approach to DNA sequencing was described by Idury and Waterman, 1995 and further developed by Pevzner, Tang, and Waterman, 2001. To address the challenge of assembly from short reads produced by next generation sequencing technologies, a number of assembly tools that are based on de Bruijn graphs have been developed (Zerbino and Birney, 2008, Butler et al., 2008). Paired de Bruijn graphs were introduced by Medvedev et al., 2011.

The Sleeping Beauty transposon system was developed by Ivics et al., 1997.

HOW DO WE SEQUENCE ANTIBIOTICS?
Brute Force Algorithms
147 · 119 · 128
TWO MEN AND ONE MONKEY ON THE MOST FANTASTIC JOURNEY OF THEIR LIVES...

The Discovery of Antibiotics

In August 1928, before leaving for vacation, Scottish microbiologist Alexander Fleming stacked his cultures of infection-causing *Staphylococcus* bacteria on a laboratory bench. When he returned to work a few weeks later, Fleming noticed that one culture had been contaminated with *Penicillium* fungus, and that the colony of *Staphylococcus* surrounding it had been destroyed! Fleming named the bacteria-killing substance **penicillin**, and he suggested that it could be used to treat bacterial infections in humans.

When Fleming published his discovery in 1929, his article had little immediate impact. Subsequent experiments struggled to isolate the **antibiotic** agent (i.e., the compound that actually killed bacteria) from the fungus. As a result, Fleming eventually concluded that penicillin could not be practically applied to treat bacterial infections and abandoned his antibiotics research.

Searching for new drugs to treat wounded soldiers, the American and British governments intensified their search for antibiotics after the start of World War II; however, the challenge of mass-producing antibiotics remained. In March 1942, half of the total supply of penicillin owned by pharmaceutical giant Merck was used to treat a single infected patient.

Also in 1942, Russian biologists Georgy Gause and Maria Brazhnikova noticed that the *Bacillus brevis* bacterium killed the pathogenic bacterium *Staphylococcus aureus*. In contrast to Fleming's efforts with penicillin, they successfully isolated the antibiotic compound from *Bacillus brevis* and named it *Gramicidin Soviet*. Within a year, this antibiotic was distributed to Soviet military hospitals.

Meanwhile, American scientists were scouring various food markets for rotten groceries and finally found a moldy cantaloupe in Illinois with a high concentration of penicillin. This mundane discovery allowed the United States to produce 2 million doses of penicillin in time for the Allied invasion of Normandy in 1944, thus saving thousands of wounded soldiers' lives.

Gause continued his research into *Gramicidin Soviet* after World War II but failed to elucidate its chemical structure. Taking the torch from Gause, English biochemist Richard Synge studied *Gramicidin Soviet* and a wide array of other antibiotics produced by *Bacillus brevis*. A few years after World War II ended, he demonstrated that they represent short amino acid sequences (i.e., mini-proteins) called **peptides**. Gause received the Stalin Prize in 1946, and Synge won the Nobel Prize in 1952. The former award proved more valuable as it protected Gause from execution during the period of **Lysenkoism**, the Soviet campaign against "bourgeois" geneticists that intensified in the postwar era. See DETOUR: Gause and Lysenkoism.

The mass-production of antibiotics initiated an arms race between pharmaceutical companies and pathogenic bacteria. The former worked to develop new antibiotic drugs, while the latter acquired resistance against these drugs. Although modern medicine won every battle for six decades, the last ten years have witnessed an alarming rise in antibiotic-resistant bacterial infections that cannot be treated even by the most powerful antibiotics. In particular, the *Staphylococcus aureus* bacterium that Gause had studied in 1942 mutated into a resistant strain known as **Methicillin-resistant *Staphylococcus aureus* (MRSA)**. MRSA is now the leading cause of death from infections in hospitals; its death rate has even passed that of AIDS in the United States.

With the rise of MRSA at hand, developing new antibiotics represents a central challenge to modern medicine. A difficult problem in antibiotics research is that of **sequencing** newly discovered antibiotics, or determining the order of amino acids making up the antibiotic peptide.

How Do Bacteria Make Antibiotics?

How peptides are encoded by the genome

Let's begin by considering **Tyrocidine B1**, one of many antibiotics produced by *Bacillus brevis*. Tyrocidine B1 is defined by the 10 amino acid-long sequence shown below (using both the one-letter and three-letter notations for amino acids). Our goal in this section is to figure out how *Bacillus brevis* could have made this antibiotic.

Three-letter notation	Val-	Lys-	Leu-	Phe-	Pro-	Trp-	Phe-	Asn-	Gln-	Tyr
One-letter notation	V	K	L	F	P	W	F	N	Q	Y

The **Central Dogma of Molecular Biology** states that "DNA makes RNA makes protein." According to the Central Dogma, a gene from a genome is first **transcribed** into a strand of RNA composed of four **ribonucleotides**: adenine, cytosine, guanine, and uracil. A strand of RNA can be represented as an **RNA string**, formed over the four-letter alphabet {A, C, G, U}. You can think of the genome as a large cookbook, in which case the gene and its RNA transcript form a recipe in this cookbook. Then, the RNA transcript is **translated** into an amino acid sequence of a protein.

Much like replication, the chemical machinery underlying transcription and translation is fascinating, but from a computational perspective, both processes are straightforward. Transcription simply transforms a DNA string into an RNA string by replacing all occurrences of T with U. The resulting strand of RNA is translated into an amino acid sequence as follows. During translation, the RNA strand is partitioned into non-overlapping 3-mers called **codons**. Then, each codon is converted into one of 20 amino acids via the **genetic code**; the resulting sequence can be represented as an **amino acid string** over a 20-letter alphabet. As illustrated in Figure 4.1, each of the 64 RNA codons encodes its own amino acid (some codons encode the same amino acid), with the exception of three **stop codons** that do not translate into amino acids and serve to halt translation (see **DETOUR: Discovery of Codons**). For example, the DNA string **TATACGAAA** transcribes into the RNA string **UAUACGAAA**, which in turn translates into the amino acid string **YTK**.

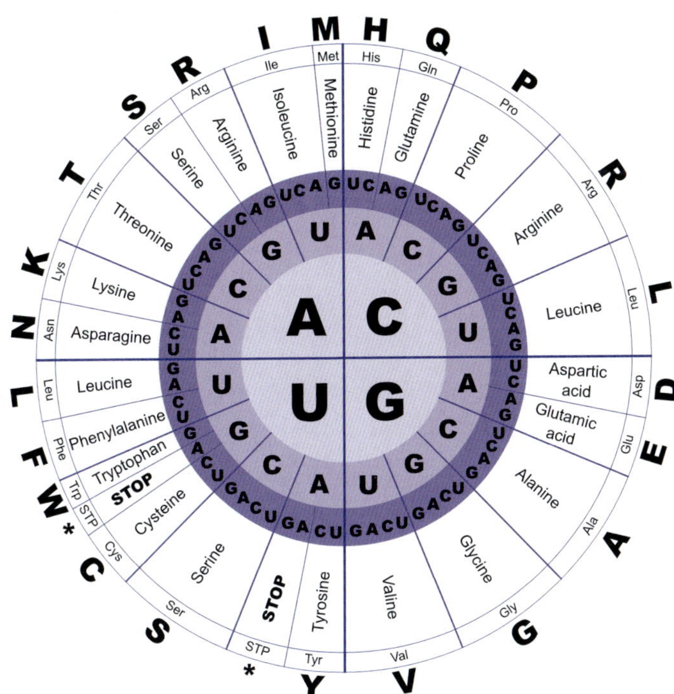

FIGURE 4.1 The genetic code describes the translation of an RNA 3-mer (codon) into one of 20 amino acids. The first three circles, moving from the inside out, represent the first, second, and third nucleotides of a codon. The fourth, fifth, and sixth circles define the translated amino acid in three ways: the amino acid's full name, its 3-letter abbreviation, and its single-letter abbreviation. Three of the 64 total RNA codons are stop codons, which halt translation.

We will represent the genetic code as an array GENETICCODE containing 64 elements, as shown in Figure 4.2. The following problem asks you to find the translation of an RNA string into an amino acid string.

Protein Translation Problem:
Translate an RNA string into an amino acid string.

Input: An RNA string *Pattern* and the array GENETICCODE.
Output: The translation of *Pattern* into an amino acid string *Peptide*.

Exercise Break: How many DNA strings of length 30 transcribe and translate into Tyrocidine B1?

0	AAA	K	16	CAA	Q	32	GAA	E	48	UAA	⋆
1	AAC	N	17	CAC	H	33	GAC	D	49	UAC	Y
2	AAG	K	18	CAG	Q	34	GAG	E	50	UAG	⋆
3	AAU	N	19	CAU	H	35	GAU	D	51	UAU	Y
4	ACA	T	20	CCA	P	36	GCA	A	52	UCA	S
5	ACC	T	21	CCC	P	37	GCC	A	53	UCC	S
6	ACG	T	22	CCG	P	38	GCG	A	54	UCG	S
7	ACU	T	23	CCU	P	39	GCU	A	55	UCU	S
8	AGA	R	24	CGA	R	40	GGA	G	56	UGA	⋆
9	AGC	S	25	CGC	R	41	GGC	G	57	UGC	C
10	AGG	R	26	CGG	R	42	GGG	G	58	UGG	W
11	AGU	S	27	CGU	R	43	GGU	G	59	UGU	C
12	AUA	I	28	CUA	L	44	GUA	V	60	UUA	L
13	AUC	I	29	CUC	L	45	GUC	V	61	UUC	F
14	AUG	M	30	CUG	L	46	GUG	V	62	UUG	L
15	AUU	I	31	CUU	L	47	GUU	V	63	UUU	F

FIGURE 4.2 The array GENETICCODE contains 64 elements, each of which is an amino acid or a stop codon (represented by ⋆).

Where is Tyrocidine encoded in the Bacillus brevis genome?

Thousands of different DNA 30-mers could code for Tyrocidine B1, and we would like to know which one appears in the *Bacillus brevis* genome. There are three different ways to divide a DNA string into codons for translation, one starting at each of the first three positions of the string. These different ways of dividing a DNA string into codons are called **reading frames**. Since DNA is double-stranded, a genome has six reading frames (three on each strand), as shown in Figure 4.3.

```
                           ------------->
                             GluThrPheSerLeuValSTPSerIle
Translated peptides         STPAsnPhePheLeuGlyLeuIleAsn
                           ValLysLeuPheProTrpPheAsnGlnTyr

Transcribed RNA            GUGAAACUUUUUCCUUGGUUUAAUCAAUAU

                        5' GTGAAACTTTTTCCTTGGTTTAATCAATAT 3'
DNA                     3' CACTTTGAAAAAGGAACCAAATTAGTTATA 5'

Transcribed RNA            CACUUUGAAAAAGGAACCAAAUUAGUUAUA

                           HisPheLysLysArgProLysIleLeuIle
Translated peptides         SerValLysGluLysThrSTPAspIle
                             PheSerLysGlyGlnAsnLeuSTPTyr
                                             <-------------
```

FIGURE 4.3 Six different reading frames give six different ways for the same fragment of DNA to be transcribed and translated (three from each strand). The top three amino acid strings are read from left to right, whereas the bottom three strings are read from right to left. The highlighted amino acid string spells out the sequence of Tyrocidine B1. Stop codons are represented by STP.

We say that a DNA string *Pattern* **encodes** an amino acid string *Peptide* if the RNA string transcribed from either *Pattern* or its reverse complement $\overline{Pattern}$ translates into *Peptide*. For example, the DNA string GAAACT is transcribed into GAAACU and translated into ET. The reverse complement of this DNA string, AGTTTC, is transcribed into AGUUUC and translated into SF. Thus, GAAACT encodes both ET and SF.

Peptide Encoding Problem:
Find substrings of a genome encoding a given amino acid sequence.

Input: A DNA string *Text* and an amino acid string *Peptide*.
Output: All substrings of *Text* encoding *Peptide* (if any such substrings exist).

STOP and Think: Solve the Peptide Encoding Problem for *Bacillus brevis* and Tyrocidine B1. Which starting positions in *Bacillus brevis* encode this peptide?

By solving the Peptide Encoding Problem for Tyrocidine B1, we should find a 30-mer in the *Bacillus brevis* genome encoding Tyrocidine B1, and yet no such 30-mer exists!

STOP and Think: How could a bacterium produce a peptide that is not encoded by the bacterium's genome?

From linear to cyclic peptides

Neither Gause nor Synge was aware of it, but tyrocidines and gramicidins are actually **cyclic peptides**. The cyclic representation for Tyrocidine B1 is shown in Figure 4.4 (left). Thus, Tyrocidine B1 has ten different linear representations, and we should run the Peptide Encoding Problem on every one of these sequences to find potential 30-mers coding for Tyrocidine B1. Yet when we solve the Peptide Encoding Problem for each of the ten strings in Figure 4.4 (right), we find no 30-mer in the *Bacillus brevis* genome encoding Tyrocidine B1!

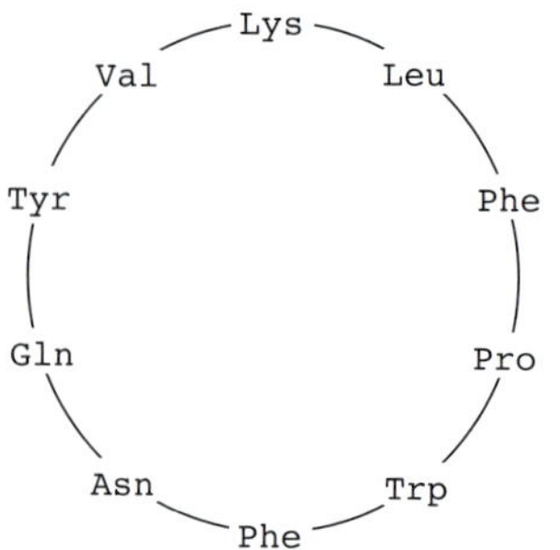


```
1  Val-Lys-Leu-Phe-Pro-Trp-Phe-Asn-Gln-Tyr
2  Lys-Leu-Phe-Pro-Trp-Phe-Asn-Gln-Tyr-Val
3  Leu-Phe-Pro-Trp-Phe-Asn-Gln-Tyr-Val-Lys
4  Phe-Pro-Trp-Phe-Asn-Gln-Tyr-Val-Lys-Leu
5  Pro-Trp-Phe-Asn-Gln-Tyr-Val-Lys-Leu-Phe
6  Trp-Phe-Asn-Gln-Tyr-Val-Lys-Leu-Phe-Pro
7  Phe-Asn-Gln-Tyr-Val-Lys-Leu-Phe-Pro-Trp
8  Asn-Gln-Tyr-Val-Lys-Leu-Phe-Pro-Trp-Phe
9  Gln-Tyr-Val-Lys-Leu-Phe-Pro-Trp-Phe-Asn
10 Tyr-Val-Lys-Leu-Phe-Pro-Trp-Phe-Asn-Gln
```

FIGURE 4.4 Tyrocidine B1 is a cyclic peptide (left), and so it has ten different linear representations (right).

Dodging the Central Dogma of Molecular Biology

Hopefully, you are perplexed, because the Central Dogma of Molecular Biology implies that *all* peptides must be encoded by the genome. Nobel laureate Edward Tatum was just as confused, and in 1963, he devised an ingenious experiment. Protein translation is carried out by a molecular machine called a **ribosome**, and so Tatum reasoned that if he inhibited the ribosome, all protein production in *Bacillus brevis* should grind to a halt. To his amazement, all proteins did indeed shut down — except for tyrocidines and gramicidins! His experiment led Tatum to hypothesize that some yet unknown *non-ribosomal* mechanism must assemble these peptides.

In 1969, Fritz Lipmann (another Nobel laureate) demonstrated that tyrocidines and gramicidins are **non-ribosomal peptides (NRPs)**, synthesized not by the ribosome, but by a giant protein called **NRP synthetase**. This enzyme pieces together antibiotic peptides without any reliance on RNA or the genetic code! We now know that every NRP synthetase assembles peptides by growing them one amino acid at a time, as shown in Figure 4.5.

The reason why many NRPs have pharmaceutical applications is that they have been optimized by eons of evolution as "molecular bullets" that bacteria and fungi use to kill their enemies. If these enemies happen to be pathogens, researchers are eager to borrow these bullets as antibacterial drugs. However, NRPs are not limited to antibiotics: many of them represent anti-tumor agents and immunosuppressors, while others are used by bacteria to communicate with other cells (see DETOUR: Quorum Sensing). PAGE 219

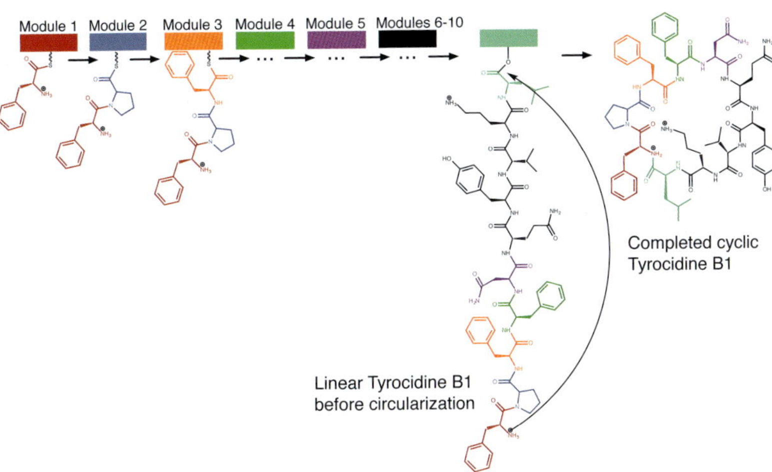

FIGURE 4.5 NRP synthetase is a giant multi-module protein that assembles a cyclic peptide one amino acid at a time. Each of ten different modules (shown by different colors) adds a single amino acid to the peptide, which in the figure is one of many tyrocidines produced by *Bacillus brevis*. In a final step, the peptide is circularized.

Sequencing Antibiotics by Shattering Them into Pieces

Introduction to mass spectrometry

Since NRPs do not adhere to the Central Dogma, we cannot infer them from the genome, which brings us back to where we started. What makes sequencing these peptides even more difficult is that many NRPs (including tyrocidines and gramicidins) are cyclic. Thus, the standard tools for sequencing linear peptides, which we will describe later, are not applicable to NRP analysis.

The workhorse of peptide sequencing is the **mass spectrometer**, an expensive molecular scale that shatters molecules into pieces and then weighs the resulting fragments. The mass spectrometer measures the mass of a molecule in **daltons (Da)**; 1 Da is approximately equal to the mass of a single nuclear particle (i.e., a proton or neutron).

We will approximate the mass of a molecule by simply adding the number of protons and neutrons found in the molecule's constituent atoms, which yields the molecule's **integer mass**. For example, the amino acid glycine, which has chemical formula C_2H_3ON, has an integer mass of 57, since $2 \cdot 12 + 3 \cdot 1 + 1 \cdot 16 + 1 \cdot 14 = 57$. Yet 1 Da is not exactly equal to the mass of a proton/neutron, and we may need to account for different naturally occurring isotopes of each atom when weighing a molecule (see DETOUR: Molecular Mass). As a result, amino acids typically have non-integer masses (e.g., glycine has total integer mass equal to approximately 57.02 Da); for simplicity, however, we will work with the integer mass table given in Figure 4.6. Tyrocidine B1, which is represented by `VKLFPWFNQY`, has total mass 1322 Da (99 + 128 + 113 + 147 + 97 + 186 + 147 + 114 + 128 + 163 = 1322).

G	A	S	P	V	T	C	I	L	N	D	K	Q	E	M	H	F	R	Y	W
57	71	87	97	99	101	103	113	113	114	115	128	128	129	131	137	147	156	163	186

FIGURE 4.6 Integer masses of amino acids. The integer mass of each amino acid is computed by adding the number of protons and neutrons in the amino acid molecule.

The mass spectrometer can break each molecule of Tyrocidine B1 into two linear fragments, and it analyzes samples that may contain billions of identical copies of the peptide, with each copy breaking in its own way. One copy may break into `LFP` and `WFNQYVK` (with respective masses 357 and 965), whereas another may break into `PWFN` and `QYVKLF`. Our goal is to use the masses of these and other fragments to sequence the peptide. The collection of all the fragment masses generated by the mass spectrometer is called an **experimental spectrum**.

STOP and Think: How can we use the experimental spectrum to sequence a peptide?

The Cyclopeptide Sequencing Problem

For now, we will assume for simplicity that the mass spectrometer breaks the copies of a cyclic peptide at every possible two bonds, so that the resulting experimental spectrum contains the masses of all possible linear fragments of the peptide, which are called **subpeptides**. For example, the cyclic peptide `NQEL` has 12 subpeptides: `N`, `Q`, `E`, `L`, `NQ`, `QE`, `EL`, `LN`, `NQE`, `QEL`, `ELN`, and `LNQ`. We will also assume that subpeptides may occur more than once if an amino acid occurs multiple times in the peptide; for example, `ELEL` also has 12 subpeptides: `E`, `L`, `E`, `L`, `EL`, `LE`, `EL`, `LE`, `ELE`, `LEL`, `ELE`, and `LEL`.

Exercise Break: How many subpeptides does a cyclic peptide of length n have?

The **theoretical spectrum** of a cyclic peptide *Peptide* is the collection of all of the masses of its subpeptides, in addition to the mass 0 and the mass of the entire peptide, with masses ordered from smallest to largest. We will denote the theoretical spectrum of *Peptide* by CYCLOSPECTRUM(*Peptide*) and assume that this theoretical spectrum can contain duplicate elements, as is the case for `NQEL` (shown below), where **`NQ`** and **`EL`** have the same mass.

	L	N	Q	E	LN	**NQ**	**EL**	QE	LNQ	ELN	QEL	NQE	NQEL
0	113	114	128	129	227	**242**	**242**	257	355	356	370	371	484

Generating Theoretical Spectrum Problem:
Generate the theoretical spectrum of a cyclic peptide.

Input: An amino acid string *Peptide*.
Output: CYCLOSPECTRUM(*Peptide*).

Charging Station (Generating the Theoretical Spectrum of a Peptide): One way to solve the Generating Theoretical Spectrum Problem is to construct a list containing all subpeptides of *Peptide*, then find the mass of each subpeptide by adding the integer masses of its constituent amino acids. This approach will work, but check out this Charging Station to see a more elegant algorithm.

Generating the theoretical spectrum of a known peptide is easy, but our aim is to solve the reverse problem of reconstructing an *unknown* peptide from its *experimental* spectrum. We will start by assuming that a biologist is lucky enough to generate an **ideal spectrum**, which is one coinciding with the peptide's theoretical spectrum.

STOP and Think: Consider the theoretical spectrum for Tyrocidine B1 shown in Figure 4.7. If an experiment produced this spectrum, how would you reconstruct the amino acid sequence of Tyrocidine B1?

0	97	99	113	114	128	128	147	147	163	186	227
241	242	244	260	261	262	283	291	333	340	357	388
389	390	390	405	430	430	447	485	487	503	504	518
543	544	552	575	577	584	631	632	650	651	671	672
690	691	738	745	747	770	778	779	804	818	819	835
837	875	892	892	917	932	932	933	934	965	982	989
1031	1039	1060	1061	1062	1078	1080	1081	1095	1136	1159	1175
1175	1194	1194	1208	1209	1223	1225	1322				

FIGURE 4.7 The theoretical spectrum for Tyrocidine B1 (VKLFPWFNQY), whose integer representation is 99-128-113-147-97-186-147-114-128-163.

Cyclopeptide Sequencing Problem:
Given an ideal spectrum, find a cyclic peptide whose theoretical spectrum matches the experimental spectrum.

> **Input**: A collection of (possibly repeated) integers *Spectrum* corresponding to an ideal spectrum.
> **Output**: An amino acid string *Peptide* such that CYCLOSPECTRUM(*Peptide*) = *Spectrum* (if such a string exists).

From now on, we will sometimes work directly with amino acid masses, taking the liberty to represent a peptide by a sequence of integers denoting the peptide's constituent amino acid masses. For example, we represent NQEL as 114-128-129-113 and Tyrocidine B1 (VKLFPWFNQY) as 99-128-113-147-97-186-147-114-128-163. We have therefore replaced an alphabet of 20 amino acids with an alphabet of only 18 integers because, recalling Figure 4.6, two amino acid pairs (I/L and K/Q) have the same integer mass.

Note that in the general case that we are not restricted to the amino acid alphabet, the Cyclopeptide Sequencing Problem could have multiple solutions. For example, "peptides" 1-1-3-3 and 1-2-1-4 have the same theoretical spectrum $\{1,1,2,3,3,4,4,5,5,6,7,7\}$.

STOP and Think: Can you find two peptides (in the alphabet of 18 amino acid masses) with identical theoretical spectra?

A Brute Force Algorithm for Cyclopeptide Sequencing

We first encountered brute force algorithms when looking for motifs in Chapter 2. In this chapter, we will discuss how to speed up brute force algorithms for peptide sequencing to make them practical.

Let's design a straightforward brute force algorithm for the Cyclopeptide Sequencing Problem. We denote the total mass of an amino acid string *Peptide* as MASS(*Peptide*). In mass spectrometry experiments, whereas the peptide that generated *Spectrum* is unknown, the peptide's mass is typically known and is denoted PARENTMASS(*Spectrum*).

For the sake of simplicity, we will also assume that for all experimental spectra, PARENTMASS(*Spectrum*) is equal to the largest mass in *Spectrum*. The following brute force cyclopeptide sequencing algorithm generates all possible peptides whose mass is equal to PARENTMASS(*Spectrum*) and then checks which of these peptides has theoretical spectra matching *Spectrum*.

```
BFCyclopeptideSequencing(Spectrum)
    for every Peptide with Mass(Peptide) equal to ParentMass(Spectrum)
        if Spectrum = Cyclospectrum(Peptide)
            output Peptide
```

There should be no question that **BFCYCLOPEPTIDESEQUENCING** will solve the Cyclopeptide Sequencing Problem. However, we should be concerned about its running time: how many peptides are there having mass equal to PARENTMASS(*Spectrum*)?

Counting Peptides with Given Mass Problem:
Compute the number of peptides of given mass.

Input: An integer m.
Output: The number of linear peptides having integer mass m.

If you have difficulty solving this problem or getting the runtime down, please return to it after learning more about dynamic programming algorithms in Chapter 5. It turns out that there are *trillions* of peptides having the same integer mass (1322) as Tyrocidine B1 (Figure 4.8). Therefore, **BFCYCLOPEPTIDESEQUENCING** is completely impractical, and we will not even bother asking you to implement it.

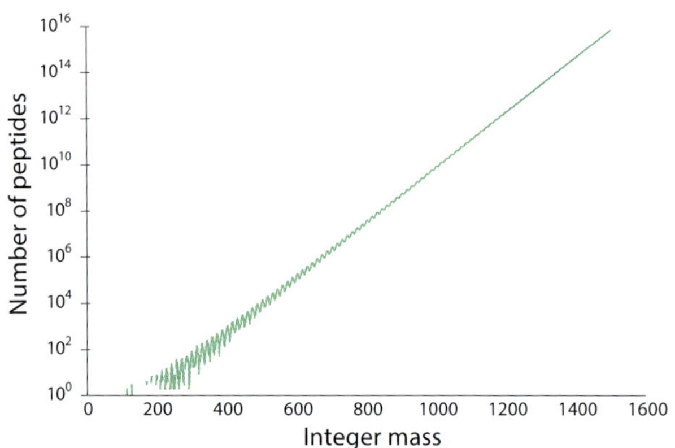

FIGURE 4.8 The number of peptides with given integer mass grows exponentially.

STOP and Think: Figure 4.8 suggests that for large m, the number of peptides with given integer mass m can be approximated as $k \cdot C^m$, where k and C are constants. Can you find these constants?

A Branch-and-Bound Algorithm for Cyclopeptide Sequencing

Just because the algorithm from the previous section failed miserably does not mean that all brute force approaches are doomed. Can we design a faster brute force algorithm based on a different idea?

Instead of checking all *cyclic* peptides with a given mass, our new approach to solving the Cyclopeptide Sequencing Problem will "grow" candidate *linear* peptides whose theoretical spectra are "consistent" with the experimental spectrum.

STOP and Think: What should it mean for a linear peptide to be consistent with an experimental spectrum? Would you classify `VKF` as consistent with the spectrum shown in Figure 4.7? What about `VKY`?

Given an experimental spectrum *Spectrum*, we will form a collection *Peptides* of candidate linear peptides initially consisting of the **empty peptide**, which is just an empty string (denoted `""`) having mass 0. At the next step, we will expand *Peptides* to contain all linear peptides of length 1. We continue this process, creating 18 new peptides of length $k+1$ for each amino acid string *Peptide* of length k in *Peptides* by appending every possible amino acid mass to the end of *Peptide*.

To prevent the number of candidate peptides from increasing exponentially, every time we expand *Peptides*, we trim it by keeping only those linear peptides that remain consistent with the experimental spectrum. We then check if any of these new linear peptides has mass equal to MASS(*Spectrum*). If so, we circularize this peptide and check whether it provides a solution to the Cyclopeptide Sequencing Problem.

More generally, brute force algorithms that enumerate all candidate solutions but discard large subsets of hopeless candidates by using various consistency conditions are known as **branch-and-bound algorithms**. Each such algorithm consists of a **branching step** to increase the number of candidate solutions, followed by a **bounding step** to remove hopeless candidates. In our branch-and-bound algorithm for the Cyclopeptide Sequencing Problem, the branching step will extend each candidate peptide of length k into 18 peptides of length $k+1$, and the bounding step will remove inconsistent peptides from consideration.

Note that the spectrum of a *linear* peptide does not contain as many masses as the spectrum of a *cyclic* peptide with the same amino acid sequence. For instance, the theoretical spectrum of the cyclic peptide `NQEL` contains 14 masses (corresponding to `""`, `N`, `Q`, `E`, `L`, `LN`, `NQ`, `QW`, `EL`, `ELN`, `LNQ`, `NQE`, `QEL`, and `NQEL`). However, the theoretical spectrum of the *linear* peptide `NQEL`, shown in Figure 4.9, does not contain masses corresponding to `LN`, `LNQ`, or `ELN`, since these subpeptides "wrap around" the end of the linear peptide.

0	113	114	128	129	242	242	257	370	371	484
""	L	N	Q	E	NQ	EL	QE	QEL	NQE	NQEL

FIGURE 4.9 The theoretical spectrum for the linear peptide `NQEL`.

Exercise Break: How many subpeptides does a linear peptide of length n have?

Given an experimental spectrum *Spectrum* of a cyclic peptide, a linear peptide is **consistent** with *Spectrum* if every mass in its theoretical spectrum is contained in *Spectrum*. If a mass appears more than once in the theoretical spectrum of the linear peptide, then it must appear *at least* that many times in *Spectrum* in order for the linear peptide to be consistent with *Spectrum*. For example, a linear peptide can still be consistent with the theoretical spectrum of `NQEL` if the peptide's spectrum contains 242 twice. But a peptide cannot be consistent with the theoretical spectrum of `NQEL` if its spectrum contains 113 twice.

The key to our new algorithm is that every linear subpeptide of a cyclic peptide *Peptide* is consistent with CYCLOSPECTRUM(*Peptide*). Thus, to solve the Cyclopeptide Sequencing Problem for *Spectrum*, we can safely ban all peptides that are *inconsistent* with *Spectrum* from the growing set *Peptides*, which powers the bounding step that we described above. For example, the linear peptide `VKF` (with spectrum $\{0, 99, 128, 147, 227, 275, 374\}$) will be banned because it is inconsistent with Tyrocidine B1's spectrum in Figure 4.7. But the linear peptide `VKY` will not be banned because every mass in its theoretical spectrum ($\{0, 99, 128, 163, 227, 291, 390\}$) is present in Figure 4.7.

What about the branching step? Given the current collection of linear peptides *Peptides*, define EXPAND(*Peptides*) as a new collection containing all possible extensions of peptides in *Peptides* by a single amino acid mass. We can now provide the pseudocode for the branch-and-bound algorithm, called **CYCLOPEPTIDESEQUENCING**.

```
CYCLOPEPTIDESEQUENCING(Spectrum)
    Peptides ← a set containing only the empty peptide
    while Peptides is nonempty
        Peptides ← EXPAND(Peptides)
        for each peptide Peptide in Peptides
            if MASS(Peptide) = PARENTMASS(Spectrum)
                if CYCLOSPECTRUM(Peptide) = Spectrum
                    output Peptide
                remove Peptide from Peptides
            else if Peptide is not consistent with Spectrum
                remove Peptide from Peptides
```

Charging Station (How Fast is CYCLOPEPTIDESEQUENCING?): After the failure of **BFCYCLOPEPTIDESEQUENCING**, you may be hesitant to implement **CYCLOPEPTIDESEQUENCING**. The potential problem with this algorithm is that it may generate incorrect k-mers at intermediate stages (i.e., k-mers that are not subpeptides of a correct solution). In practice, however, this is not a concern. Check out this Charging Station to see an example.

It is hard to imagine a worst-case scenario in which **CYCLOPEPTIDESEQUENCING** takes a long time to run, but no one has been able to guarantee that this algorithm will not generate many incorrect *k*-mers at intermediate stages. **BFCYCLOPEPTIDESEQUENCING** is exponential, and although in practice, **CYCLOPEPTIDESEQUENCING** is much faster, this algorithm has not been proven to be polynomial. Thus, from the perspective of an introductory algorithms course focusing on theoretical computer science, the practical **CYCLOPEPTIDESEQUENCING** is just as inefficient as **BFCYCLOPEPTIDESEQUENCING**, since neither algorithm's running time can be bounded by a polynomial. See the Open Problems section to learn more about the algorithmic challenges related to this problem.

Mass Spectrometry Meets Golf

From theoretical to real spectra

Although **CYCLOPEPTIDESEQUENCING** successfully reconstructed Tyrocidine B1, this algorithm only works in the case of an ideal spectrum, i.e., when the experimental spectrum of a peptide coincides exactly with its theoretical spectrum. This inflexibility of **CYCLOPEPTIDESEQUENCING** presents a practical barrier, since mass spectrometers generate "noisy" spectra that are far from ideal — they are characterized by having both **false masses** and **missing masses**. A false mass is present in the experimental spectrum but absent from the theoretical spectrum; a missing mass is present in the theoretical spectrum but absent from the experimental spectrum (Figure 4.10).

Theoretical:	0		113	114	128	**129**	227	**242**	**242**	257		355	356	370	371	484
Experimental:	0	**99**	113	114	128		227			257	**299**	355	356	370	371	484

FIGURE 4.10 Theoretical and simulated experimental spectra for NQEL. Masses that are missing from the experimental spectrum are shown in blue, and false masses in the experimental spectrum are shown in green.

What is particularly worrisome about the example in Figure 4.10 is that the mass of the amino acid E (**129**) is missing, and the mass of the amino acid V (**99**) is false; as a result, the first step of **CYCLOPEPTIDESEQUENCING** would establish {V, L, N, Q} as the amino acid composition of our candidate peptides, which is incorrect. In fact, *any* false or missing mass will cause **CYCLOPEPTIDESEQUENCING** to throw out the correct peptide, because its theoretical spectrum differs from the experimental spectrum.

STOP and Think: How would you reformulate the Cyclopeptide Sequencing Problem to handle experimental spectra with errors?

Adapting cyclopeptide sequencing for spectra with errors

To generalize the Cyclopeptide Sequencing Problem to handle noisy spectra, we need to relax the requirement that a candidate peptide's theoretical spectrum must match the experimental spectrum *exactly*, and instead incorporate a **scoring function** that will select the peptide whose theoretical spectrum matches the given experimental spectrum *the most closely*. Given a cyclic peptide *Peptide* and a spectrum *Spectrum*, we define SCORE(*Peptide*, *Spectrum*) as the number of masses shared between CYCLOSPECTRUM(*Peptide*) and *Spectrum*. Recalling our example above, if

$$Spectrum = \{0, 99, 113, 114, 128, 227, 257, 299, 355, 356, 370, 371, 484\},$$

then SCORE(NQEL, *Spectrum*) = 11.

The scoring function should take into account the **multiplicities** of shared masses, i.e., how many times they occur in each spectrum. For example, suppose that *Spectrum* is the theoretical spectrum of NQEL (Figure 4.7); for this spectrum, mass 242 has multiplicity 2. If 242 has multiplicity 1 in the theoretical spectrum of *Peptide*, then 242 contributes 1 to SCORE(*Peptide*, *Spectrum*). If 242 has larger multiplicity in the theoretical spectrum of *Peptide*, then 242 contributes 2 to SCORE(*Peptide*, *Spectrum*).

Cyclopeptide Scoring Problem:
Compute the score of a cyclic peptide against a spectrum.

Input: An amino acid string *Peptide* and a collection of integers *Spectrum*.
Output: The score of *Peptide* against *Spectrum*, SCORE(*Peptide*, *Spectrum*).

We now can redefine the Cyclopeptide Sequencing Problem for noisy spectra.

Cyclopeptide Sequencing Problem (for spectra with errors):
Find a cyclic peptide having maximum score against an experimental spectrum.

Input: A collection of integers *Spectrum*.
Output: A cyclic peptide *Peptide* maximizing SCORE(*Peptide*, *Spectrum*) over all peptides *Peptide* with mass equal to PARENTMASS(*Spectrum*).

Our goal is to adapt the **CyclopeptideSequencing** algorithm to find a peptide with maximum score. Remember that this algorithm had a stringent bounding step in which all candidate linear peptides having inconsistent spectra were thrown out. For example, we saw that the linear peptide `VKF` is inconsistent with the theoretical spectrum in Figure 4.7. However, we perhaps should not ban `VKF` in the case of experimental spectra, since they can have missing masses. Thus, we need to revise the bounding step to include more candidate linear peptides, while still ensuring that the number of peptides that we consider does not grow out of control.

STOP and Think: How can we limit the growth of the list of candidate linear peptides in the case of experimental spectra?

To limit the number of candidate linear peptides under consideration, we will replace the list *Peptides* with the list *Leaderboard*, which holds the *N* highest scoring candidate peptides for further extension. At each step, we will expand all candidate peptides found in *Leaderboard*, then eliminate those peptides whose newly calculated scores are not high enough to keep them on the *Leaderboard*. This idea is similar to the notion of a "cut" in a golf tournament; after the cut, only the top *N* golfers are allowed to play in the next round, since they are the only players who have a reasonable chance of winning.

To be fair, a cut should include anyone who is tied with the *N*th-place competitor. Thus, *Leaderboard* should be trimmed down to the "*N* highest-scoring linear peptides including ties", which may include more than *N* peptides. Given a list of peptides *Leaderboard*, a spectrum *Spectrum*, and an integer *N*, define Trim(*Leaderboard*, *Spectrum*, *N*) as the collection of the top *N* highest-scoring linear peptides in *Leaderboard* (including ties) with respect to *Spectrum*.

STOP and Think: Our peptide scoring function currently assumes that peptides are circular, but peptides should technically be scored as linear peptides until the final step. How would you score a linear peptide against a spectrum?

Note that Score(*Peptide*, *Spectrum*) currently only scores *Peptide* against *Spectrum* if *Peptide* is cyclic. However, to generalize this scoring function when *Peptide* is linear, we simply exclude those subpeptides of *Peptide* that wrap around the end of the string, resulting in a function LinearScore(*Peptide*, *Spectrum*). For example, if *Spectrum* is the experimental spectrum of `NQEL` from Figure 4.10, then you can verify that LinearScore(`NQEL`, *Spectrum*) = 8. This brings us to the pseudocode for **LeaderboardCyclopeptideSequencing**.

```
LEADERBOARDCYCLOPEPTIDESEQUENCING(Spectrum, N)
    Leaderboard ← set containing only the empty peptide
    LeaderPeptide ← empty peptide
    while Leaderboard is non-empty
        Leaderboard ← EXPAND(Leaderboard)
        for each Peptide in Leaderboard
            if MASS(Peptide) = PARENTMASS(Spectrum)
                if SCORE(Peptide, Spectrum) > SCORE(LeaderPeptide, Spectrum)
                    LeaderPeptide ← Peptide
            else if MASS(Peptide) > PARENTMASS(Spectrum)
                remove Peptide from Leaderboard
        Leaderboard ← TRIM(Leaderboard, Spectrum, N)
    output LeaderPeptide
```

Charging Station (Trimming the Peptide Leaderboard): The tricky aspect of implementing **LEADERBOARDCYCLOPEPTIDESEQUENCING** is making sure that the TRIM function is implemented correctly. Check out this Charging Station for some help with trimming *Leaderboard*.

We point out that because a linear peptide giving rise to the highest-scoring cyclic peptide may be trimmed early on, **LEADERBOARDCYCLOPEPTIDESEQUENCING** is a heuristic, not guaranteed to correctly solve the Cyclopeptide Sequencing Problem. When we develop a heuristic, we must ask: *how accurate is it?* Consider the simulated spectrum $Spectrum_{10}$ of Tyrocidine B1 shown in Figure 4.11 (top), with approximately 10% **missing**/**false** masses. Applying **LEADERBOARDCYCLOPEPTIDESEQUENCING** to this spectrum (with $N = 1000$) results in the correct cyclic peptide VKLFPWFNQY, which has a score of 86.

So far, **LEADERBOARDCYCLOPEPTIDESEQUENCING** has worked well, but as the number of errors increases, so does the likelihood that this algorithm will return an incorrect peptide. Let's see how this algorithm performs on a noisier simulated spectrum; in Figure 4.11 (bottom), we show $Spectrum_{25}$ for Tyrocidine B1, which has 25% **missing**/**false** masses.

When run on $Spectrum_{25}$, **LEADERBOARDCYCLOPEPTIDESEQUENCING** (with $N = 1000$) identifies VKLFP**AD**FNQY (score: 83) as a highest-scoring cyclic peptide instead of the correct peptide VKLFP**W**FNQY (score: 82). These two peptides are similar, owing to the fact that the combined mass of **A** (71) and **D** (115) is equal to the mass of **W** (186).

0	97	99	**113**	114	**128**	128	147	147	163	186	227
241	242	244	260	261	262	283	291	333	340	357	**385**
388	389	390	390	405	430	430	447	485	487	503	504
518	543	544	552	575	577	584	**631**	632	650	651	671
672	690	691	738	745	747	770	778	779	804	818	819
820	835	837	875	892	**892**	917	932	932	933	934	965
982	989	**1030**	**1031**	1039	1060	1061	1062	1078	1080	1081	1095
1136	1159	1175	1175	1194	1194	1208	1209	1223	1225	1322	

0	97	99	113	114	**115**	128	128	147	147	163	186
227	241	242	**244**	244	**256**	260	261	262	283	291	**309**
330	333	340	**347**	**357**	**385**	388	389	390	390	405	**430**
430	**435**	447	485	487	503	504	518	**543**	544	552	575
577	584	**599**	**608**	631	632	650	651	**653**	**671**	672	690
691	**717**	738	745	**747**	770	**778**	779	804	818	819	**827**
835	837	875	892	892	917	932	932	933	934	965	982
989	**1031**	1039	1060	**1061**	1062	1078	1080	1081	1095	1136	1159
1175	1175	1194	1194	1208	1209	1223	**1225**	1322			

FIGURE 4.11 (Top) A simulated experimental spectrum $Spectrum_{10}$ of Tyrocidine B1. This spectrum has approximately 10% missing (blue) and false (green) masses. Note that the blue masses are not actually in the spectrum, but we show them so that it is clear which masses are missing. (Bottom) A simulated experimental spectrum $Spectrum_{25}$ of Tyrocidine B1, with approximately 25% missing and false masses.

STOP and Think: How could we have eliminated the incorrect peptide `VKLFPADFNQY` from consideration for $Spectrum_{25}$?

Notice that although the correct and incorrect peptides are similar, their amino acid compositions differ. If we could figure out the amino acid composition of Tyrocidine B1 from its spectrum alone and run **LEADERBOARDCYCLOPEPTIDESEQUENCING** on this smaller alphabet (rather than on the alphabet of all amino acids), then we could eliminate the incorrect peptide `VKLFPADFNQY` from consideration.

From 20 to More than 100 Amino Acids

Until now, we have assumed that just 20 amino acids form the building blocks of proteins; these building blocks are called **proteinogenic amino acids**. There are actually two additional proteinogenic amino acids, called **selenocysteine** and **pyrrolysine**,

which are incorporated into proteins by special biosynthetic mechanisms (see **DETOUR: Selenocysteine and Pyrrolysine**). Yet in addition to the 22 proteinogenic amino acids, NRPs contain **non-proteinogenic amino acids**, which expand the number of possible building blocks for antibiotic peptides from 20 to over 100.

Enlarging the amino acid alphabet spells trouble for our current approach to cyclopeptide sequencing. Indeed, the correct peptide now must "compete" with many more incorrect ones for a place on the leaderboard, increasing the chance that the correct peptide will be cut along the way.

For example, although Tyrocidine B1 contains only proteinogenic amino acids, its close relative, Tyrocidine B (`Val`-**`Orn`**-`Leu-Phe-Pro-Trp-Phe-Asn-Gln-Tyr`), contains a non-proteinogenic amino acid called **ornithine** (**`Orn`**). Because so many non-proteinogenic amino acids exist, bioinformaticians often assume that any integer between 57 and 200 may represent the mass of an amino acid; the "lightest" amino acid, `Gly`, has mass 57 Da, and most amino acids have masses smaller than 200 Da.

STOP and Think: Apply LEADERBOARDCYCLOPEPTIDESEQUENCING on the extended amino acid alphabet (containing the 144 integers between 57 and 200) to $Spectrum_{10}$, and identify the highest-scoring peptides.

When we apply LEADERBOARDCYCLOPEPTIDESEQUENCING for the extended alphabet to $Spectrum_{10}$, one of the highest-scoring peptides is `VKLFPWFNQ`**`XZ`**, where `X` has mass 98 and `Z` has mass 65. Apparently, non-standard amino acids successfully competed with standard amino acids for the limited number of positions on the leaderboard, resulting in `VKLFPWFNQ`**`XZ`** winning over the correct peptide `VKLFPWFNQ`**`Y`**. Since LEADERBOARDCYCLOPEPTIDESEQUENCING fails to identify the correct peptide even with only 10% false and missing masses, our stated aim from the previous section is now even more important. We must determine the amino acid composition of a peptide from its spectrum so that we may run LEADERBOARDCYCLOPEPTIDESEQUENCING on this smaller alphabet of amino acids.

STOP and Think: How can we determine which amino acids are present in an unknown peptide using only an experimental spectrum?

The Spectral Convolution Saves the Day

One way to determine the amino acid composition of a peptide from its experimental spectrum would be to take the smallest masses present in the spectrum (between 57 and 200 Da). However, even if only a single amino acid mass is missing, then this approach will fail to reconstruct the peptide's amino acid composition.

Let's take a different approach. Say that our experimental spectrum contains the masses of subpeptides NQ**E** and NQ. If we subtract these two masses, then we will obtain the mass **E** for free, even if it was not present in the experimental spectrum! If the underlying peptide is NQEL, then we can also find the mass of **E** by subtracting the masses of Q**E** and Q or NQ**E**L and LNQ.

Following this example, we define the **convolution** of a spectrum by taking all positive differences of masses in the spectrum. Figure 4.12 shows the convolutions of the theoretical (top) and simulated (bottom) spectra of NQEL from Figure 4.10.

As predicted, some of the values in Figure 4.12 appear more frequently than others. For example, **113** (the mass of **L**) has multiplicity 8. Six of the eight occurrences of **113** correspond to subpeptide pairs differing in an **L**: **L** and ""; **L**N and N; E**L** and E; **L**NQ and NQ; QE**L** and QE; NQE**L** and NQE. Interestingly, **129** (the mass of **E**) pops up three times in the convolution of the simulated spectrum, even though **129** was missing from the spectrum itself.

We now should feel confident about using the most frequently appearing integers in the convolution as a guess for the amino acid composition of an unknown peptide. In our simulated spectrum for NQEL, the most frequent elements of the convolution in the range from 57 to 200 are (multiplicities in parentheses):

$$\mathbf{113}\ (4),\ \mathbf{114}\ (4),\ \mathbf{128}\ (4),\ \mathbf{99}\ (3),\ \mathbf{129}\ (3)\,.$$

Note that these most frequent elements capture all four amino acids in NQEL.

Spectral Convolution Problem:
Generate the convolution of a spectrum.

> **Input**: A collection of integers *Spectrum*.
> **Output**: The list of elements in the convolution of *Spectrum* in decreasing order of their multiplicities.

	""	L	N	Q	E	LN	NQ	EL	QE	LNQ	ELN	QEL	NQE
	0	113	114	128	129	227	242	242	257	355	356	370	371
0													
113	**113**												
114	**114**	1											
128	**128**	15	14										
129	**129**	16	15	1									
227	227	**114**	**113**	99	98								
242	242	**129**	**128**	**114**	**113**	15							
242	242	**129**	**128**	**114**	**113**	15							
257	257	144	143	**129**	**128**	30	15	15					
355	355	242	241	227	226	**128**	**113**	**113**	98				
356	356	243	242	228	227	**129**	**114**	**114**	99	1			
370	370	257	256	242	241	143	**128**	**128**	**113**	15	14		
371	371	258	257	243	242	144	**129**	**129**	**114**	16	15	1	
484	484	371	370	356	355	257	242	242	227	**129**	**128**	**114**	**113**

	""	false	L	N	Q	LN	QE	false	LNQ	ELN	QEL	NQE
	0	99	113	114	128	227	257	299	355	356	370	371
0												
99	**99**											
113	**113**	14										
114	**114**	15	1									
128	**128**	29	15	14								
227	227	**128**	**114**	**113**	**99**							
257	257	158	144	143	**129**	30						
299	299	200	186	185	171	72	42					
355	355	256	242	241	227	**128**	98	56				
356	356	257	243	242	228	**129**	**99**	57	1			
370	370	271	257	256	242	143	**113**	71	15	14		
371	371	272	258	257	243	144	**114**	72	16	15	1	
484	484	385	371	370	356	257	227	185	**129**	**128**	**114**	**113**

FIGURE 4.12 (Top) Spectral convolution for the theoretical spectrum of NQEL. The most frequent elements in the convolution between 57 and 200 are (multiplicities in parentheses): **113** (8), **114** (8), **128** (8), **129** (8). (Bottom) Spectral convolution for the simulated spectrum of NQEL. The most frequent elements in the convolution between 57 and 200 are (multiplicities in parentheses): **113** (4), **114** (4), **128** (4), **99** (3), **129** (3).

Recall that **LEADERBOARDCYCLOPEPTIDESEQUENCING** failed to reconstruct Tyrocidine B1 from $Spectrum_{10}$ when using the extended alphabet of amino acids. The ten most frequent elements of its spectral convolution in the range from 57 to 200 are (multiplicities in parentheses):

147 (35)	128 (31)	97 (28)	113 (28)	114 (26)
186 (23)	57 (21)	163 (21)	99 (18)	145 (18)

Every mass in this list except for 57 and 145 captures an amino acid in Tyrocidine B1!

We now have the outline for a new cyclopeptide sequencing algorithm. Given an experimental spectrum, we first compute the convolution of an experimental spectrum. We then select the M most frequent elements between 57 and 200 in the convolution to form an extended alphabet of amino acid masses. In order to be fair, we should include the top M elements of the convolution "with ties". Finally, we run **LEADERBOARDCYCLOPEPTIDESEQUENCING**, where amino acid masses are restricted to this alphabet. We call this algorithm **CONVOLUTIONCYCLOPEPTIDESEQUENCING**.

STOP and Think: Run **CONVOLUTIONCYCLOPEPTIDESEQUENCING** on the simulated spectra $Spectrum_{10}$ and $Spectrum_{25}$ with $N = 1000$ and $M = 20$. Identify the highest scoring peptides.

CONVOLUTIONCYCLOPEPTIDESEQUENCING (with $N = 1000$ and $M = 20$) now correctly reconstructs Tyrocidine B1 from $Spectrum_{10}$. The true test of this algorithm is whether it will work on a noisier spectrum. Recall that our previous algorithm failed on $Spectrum_{25}$; in contrast, **CONVOLUTIONCYCLOPEPTIDESEQUENCING** (with $N = 1000$ and $M = 20$) now correctly identifies Tyrocidine B1 from this spectrum!

Epilogue: From Simulated to Real Spectra

In this chapter, we have sheltered you from the gruesome realities of mass spectrometry by providing simulated spectra that are relatively easy to sequence (even those with false and missing masses). We committed a sin of omission by loosely describing the mass spectrometer as a "scale" and assuming that this complex machine simply weighs tiny peptide fragments one at a time. In truth, the mass spectrometer first converts subpeptides into ions (i.e., charged particles). Ionization of particles helps the mass spectrometer sort the ions by using an electromagnetic field; ions are separated not by their mass, but rather according to their **mass/charge ratio**. If fragment ion `NQY` (integer mass: $114 + 128 + 163 = 405$) has charge +1, then it contains one additional proton, resulting in a total integer mass of 406 and a mass/charge ratio of $406/1 = 406$. To be more precise, the *monoisotopic* mass of NQY is approximately $114.043 + 128.058 + 163.063 = 405.164$, and the mass of a proton is 1.007 Da, which makes the mass charge/ratio more closely equal to $(405.164 + 1.007)/1 = 406.171$ (see **DETOUR: Molecular Mass**).

PAGE 219

The mass spectrometer outputs a collection of **peaks**, which are shown in Figure 4.13 for a real Tyrocidine B1 spectrum. Each peak's x-coordinate represents an ion's mass/charge ratio, and its height represents the **intensity** (i.e., relative abundance) of ions having that mass/charge ratio. For example, in the experimental spectrum of Tyrocidine B1 shown in Figure 4.13, you will find a small (almost invisible) peak with a mass/charge ratio of 406.30, which corresponds to the fragment ion NQY having mass/charge ratio 406.171, with an error of approximately 0.13 Da.

As you can imagine, we must navigate a few practical barriers in order to analyze real spectra. First, the charge of each peak is unknown, often forcing researchers to try all possible charges from 1 to some parameter *maxCharge*, where the particular choice of *maxCharge* depends on the fragmentation technology used. This procedure generates *maxCharge* masses for each peak, so that the larger the value of *maxCharge*, the more false masses in the spectrum.

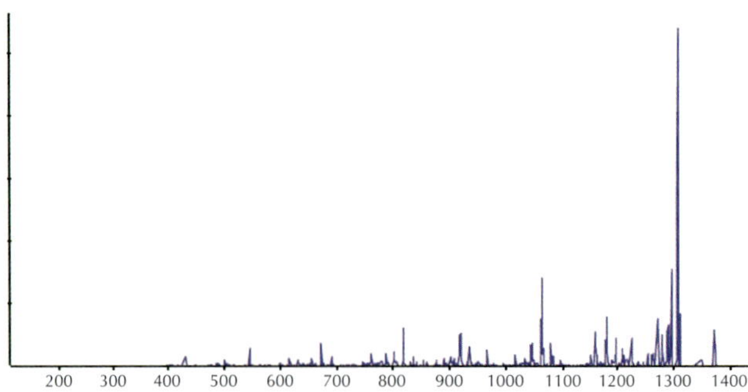

FIGURE 4.13 A real spectrum for Tyrocidine B1. A peak's x-coordinate represents its mass/charge ratio, and its height represents the intensity of ions having that mass/charge ratio.

Second, the spectrum in Figure 4.13 has nearly 1,000 peaks, most of which are **false peaks**, meaning that their mass/charge ratio does not correspond to any subpeptide's mass/charge ratio (for any charge value). Fortunately, false peaks typically have low intensities, necessitating a pre-processing step that removes low-intensity peaks before applying an algorithm. Figure 4.14 shows the list of the 95 mass/charge ratios for peaks that survived this preprocessing step for the Tyrocidine B1 spectrum in Figure 4.13. Their intensities may nevertheless vary by 2-3 orders of magnitude; for example, the intensity of the peak having mass/charge ratio 372.2 is 300 times smaller than the intensity of the peak with mass/charge ratio 1306.5.

372.2	397.2	402.0	**406.3**	415.1	**431.2**	**448.3**	449.3	452.2
471.3	**486.3**	**488.2**	500.5	**505.3**	516.1	536.1	**544.2**	**545.3**
562.5	571.3	599.2	614.4	615.4	616.4	618.2	**632.0**	655.5
656.3	**672.5**	**673.3**	677.3	**691.4**	**692.4**	712.1	722.3	**746.5**
760.4	761.6	762.5	**771.6**	788.4	802.3	803.3	818.5	**819.4**
831.4	**836.3**	853.3	875.5	**876.5**	901.5	915.9	916.5	917.8
918.4	**933.4**	**934.7**	**935.5**	949.4	**966.2**	995.4	1015.6	1027.5
1029.5	1031.5	1044.5	1046.5	**1061.5**	**1063.4**	**1079.2**	1083.7	
1088.4	1093.5	**1096.5**	1098.4	1158.5	1159.5	**1176.6**	1177.7	
1178.6	1192.7	**1195.4**	1207.5	**1210.4**	**1224.6**	1252.5	1270.5	
1271.5	1278.6	1279.6	1295.6	1305.6	1306.5	1307.5	1309.6	

FIGURE 4.14 The 95 mass/charge ratios having the highest intensity in Figure 4.13. The values shown in bold correspond to masses of subpeptides of Tyrocidine B1, if we allow a mass discrepancy of up to 0.3 Da (see Figure 4.15).

Mass	Subpeptide	Mass	Subpeptide	Mass	Subpeptide
406.2	NQY	431.2	FPW	448.2	WFN
486.2	KLFP	488.2	VKLF	505.2	NQYV
544.2	LFPW	545.2	PWFN	632.3	QYVKL
672.3	KLFPW	673.3	PWFNQ	691.3	LFPWF
692.3	FPWFN	746.3	NQYVKL	771.3	VKLFPW
819.4	KLFPWF	836.4	PWFNQY	876.4	QYVKLFP
918.4	VKLFPWF	933.4	LFPWFNQ	934.4	YVKLFPW
935.4	PWFNQYV	966.4	WFNQYVK	1061.5	KLFPWFNQ
1063.5	PWFNQYVK	1079.5	WFNQYVKL	1096.5	LFPWFNQY
1176.5	NQYVKLFPW	1195.6	LFPWFNQYV	1210.6	FPWFNQYVK
1224.6	KLFPWFNQY				

FIGURE 4.15 Values from Figure 4.14 corresponding to masses of subpeptides of Tyrocidine B1, with *maxCharge* = 1 and an allowable discrepancy of up to 0.3 Da. These 31 masses represent less than a third of the 95 subpeptides in the theoretical spectrum of Tyrocidine B1.

Only 31 of these 95 mass/charge ratios (shown in bold in Figure 4.14) can be matched to subpeptides of Tyrocidine B1, as illustrated in Figure 4.15.

You can now see that sequencing Tyrocidine B1 from a real spectrum, for which two-thirds of all masses are false, presents a much more difficult problem than sequencing this peptide from the simulated $Spectrum_{25}$. In the following challenge problem, you will need to further develop the methods we studied in this chapter to analyze a real spectrum.

Challenge Problem: Tyrocidine B1 is just one of many known NRPs produced by *Bacillus brevis*. A single bacterial species may produce dozens of different antibiotics, and even after 70 years of research, there are likely undiscovered antibiotics produced by *Bacillus brevis*. Try to sequence the tyrocidine corresponding to the real experimental spectrum below. Since the fragmentation technology used for generating this spectrum tends to produce ions with charge +1, you can safely assume that all charges are +1.

371.5	375.4	390.4	392.2	409.0	420.2	427.2	443.3	446.4	461.3
471.4	477.4	491.3	505.3	506.4	519.2	536.1	546.5	553.3	562.3
588.2	600.3	616.2	617.4	618.3	633.4	634.4	636.2	651.5	652.4
702.5	703.4	712.5	718.3	721.0	730.3	749.4	762.6	763.4	764.4
779.6	780.4	781.4	782.4	797.3	862.4	876.4	877.4	878.6	879.4
893.4	894.4	895.4	896.5	927.4	944.4	975.5	976.5	977.4	979.4
1005.5	1007.5	1022.5	1023.7	1024.5	1039.5	1040.3	1042.5	1043.4	1057.5
1119.6	1120.6	1137.6	1138.6	1139.5	1156.5	1157.6	1168.6	1171.6	1185.4
1220.6	1222.5	1223.6	1239.6	1240.6	1250.5	1256.5	1266.5	1267.5	1268.6

Hint: since the peptide from which this spectrum was generated is in the tyrocidine family, this peptide should be similar to Tyrocidine B1.

Open Problems

The Beltway and Turnpike Problems

In the case of the alphabet of arbitrary integers, the Cyclopeptide Sequencing Problem corresponds to a computer science problem known as the **Beltway Problem**. The Beltway Problem asks you to find a set of points on a circle such that the distances between all pairs of points (where distance is measured around the circle) match a given collection of integers.

The Beltway Problem's analogue in the case when the points lie along a line segment instead of on a circle is called the **Turnpike Problem**. The terms "beltway" and "turnpike" arise from an analogy with exits on circular and linear roads, respectively. In the case of n points on a circle and line, the inputs for the Beltway and Turnpike Problems consist of $n(n-1)+2$ and $\frac{n(n-1)}{2}+2$ distances, respectively (these formulas include the distance 0 as well as the length of the entire segment).

Various attempts to design polynomial algorithms for the Beltway and Turnpike Problems (or to prove that they are intractable) have failed. However, there is a **pseudo-polynomial algorithm** for the Turnpike Problem (see **DETOUR: Pseudo-polynomial Algorithm for the Turnpike Problem**). In contrast to a truly polynomial algorithm, which can be bounded by a polynomial in the length of the input, a pseudo-polynomial algorithm for the Turnpike Problem is polynomial in the total length of the line segment. For example, if n points are separated by huge distances, say on the order of 2^{100}, then a polynomial algorithm would still be fast, whereas a pseudo-polynomial algorithm would be prohibitively slow. Note that although the distances themselves will be huge, each distance can be stored using only about 100 bits, implying that the length of the input is small even for such large distances.

PAGE 221

Pseudo-polynomial algorithms are useful in practice because practical instances of the problems typically do not include huge distances. Interestingly, although a pseudo-polynomial algorithm exists for the Turnpike Problem, such an algorithm for the seemingly similar Beltway Problem remains undiscovered. Can you develop such an algorithm?

Sequencing cyclic peptides in primates

Bacteria and fungi do not have a monopoly on producing cyclic peptides; animals and plants make them too (albeit through a completely different mechanism). The first cyclic peptide found in animals (called **θ-defensin**) was discovered in 1999 in macaques. θ-defensin prevents viruses from entering cells and has strong anti-HIV activity. Yet the question of how primates make θ-defensin remains a mystery.

Needless to say, there is no 54-mer in the macaque genome encoding the 18 amino acid-long θ-defensin. Instead, this cyclic peptide is formed by concatenating two 9 amino acid-long peptides excised from two different proteins called RTD1a and RTD1b, as shown in Figure 4.16. It remains unclear which enzymes do this elaborate cutting and pasting.

Interestingly, macaques and baboons produce θ-defensin, whereas humans and chimpanzees do not. This discrepancy makes us wonder whether a mutation occurred

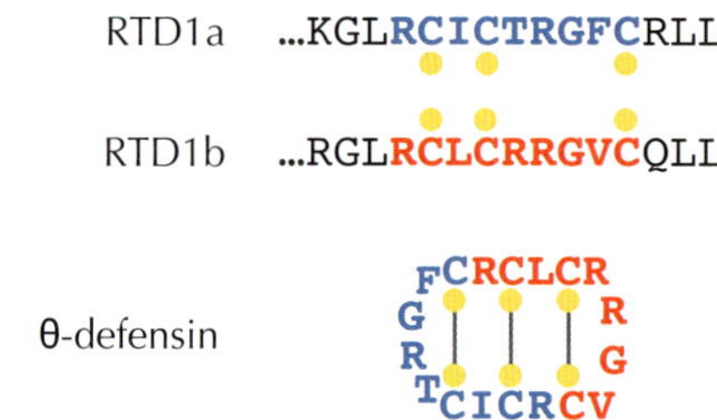

FIGURE 4.16 The 18 amino acid-long θ-defensin peptide is formed by cutting two 9 amino acid-long peptides **RCICTRGFC** and **RCLCRRGVC** from the RTD1a and RTD1b proteins, concatenating them, and then circularizing the resulting peptide (along with introducing three **disulfide bridges** that form bonds across the peptide).

in the human-chimpanzee ancestor that resulted in the loss of this very useful peptide. Interestingly, genes very similar to RTD1a and RTD1b do exist in humans, but a codon in one of these genes mutated into a stop codon, thus shortening the encoded protein. Since this stop codon is located before the 9 amino acid-long peptide contributing to θ-defensin, humans do not produce this peptide and thus cannot produce θ-defensin.

In a remarkable experiment, Alexander Cole demonstrated that humans could get θ-defensin back! Certain drugs can force the ribosome to ignore stop codons and continue translating RNA, even after encountering a stop codon. The researchers demonstrated that after treatment with such a drug, human cells began producing the human version of θ-defensin. The surprising conclusion of this experiment is that although humans and chimpanzees lost θ-defensin millions of years ago, we still possess the mysterious enzymes required to cut and paste its constituent peptides.

Some biologists believe that since the enzymes making θ-defensin still work in humans, they must be needed for something else. If these enzymes did not provide some selective advantage, then over time, mutations would cause their genes to become **pseudogenes**, or non-functional remnants of previously working genes. The most natural explanation for why these enzymes are still functional is that humans produce still undiscovered cyclic peptides, and that the enzymes needed for θ-defensin are also used to "cut-and-paste" other (still undiscovered) cyclic peptides. The hypothesis that we may possess undiscovered cyclic peptides is not as improbable as you might think because biologists still lack robust algorithms for cyclopeptide discovery from the billions of spectra generated in hundreds of labs analyzing the human proteome.

By default, researchers assume that all spectra ever acquired in human proteome studies originated from linear peptides. Could they be wrong? Can you devise a fast cyclopeptide sequencing algorithm that can analyze all these spectra and hopefully discover human cyclopeptides?

Charging Stations

Generating the theoretical spectrum of a peptide

Given an amino acid string *Peptide*, we will begin by assuming that it represents a *linear* peptide. Our approach to generating its theoretical spectrum is based on the assumption that the mass of any subpeptide is equal to the difference between the masses of two prefixes of *Peptide*. We can compute an array PREFIXMASS storing the masses of each prefix of *Peptide* in increasing order, e.g., for *Peptide* = NQEL, PREFIXMASS = $(0, 114, 242, 371, 484)$. Then, the mass of the subpeptide of *Peptide* beginning at position $i+1$ and ending at position j can be computed as PREFIXMASS(j) − PREFIXMASS(i). For example, when *Peptide* = NQEL,

$$\text{MASS}(\text{QE}) = \text{PREFIXMASS}(3) - \text{PREFIXMASS}(1) = 371 - 114 = 257\,.$$

The following pseudocode implements this idea. It also represents the alphabet of 20 amino acids and their integer masses as a pair of 20-element arrays AMINOACID and AMINOACIDMASS, corresponding to the top and bottom rows of Figure 4.6, respectively.

```
LINEARSPECTRUM(Peptide, AMINOACID, AMINOACIDMASS)
    PREFIXMASS(0) ← 0
    for i ← 1 to |Peptide|
        for j ← 1 to 20
            if AMINOACID(j) = i-th amino acid in Peptide
                PREFIXMASS(i) ← PREFIXMASS(i − 1) + AMINOACIDMASS(j)
    LinearSpectrum ← a list consisting of the single integer 0
    for i ← 0 to |Peptide| − 1
        for j ← i + 1 to |Peptide|
            add PREFIXMASS(j) − PREFIXMASS(i) to LinearSpectrum
    return the sorted list LinearSpectrum
```

If the amino acid string *Peptide* represents a cyclic peptide instead, then the masses in its theoretical spectrum can be divided into those found by **LINEARSPECTRUM** and those corresponding to subpeptides wrapping around the end of the linearized version of *Peptide*. Furthermore, each such subpeptide has mass equal to the difference between MASS(*Peptide*) and a subpeptide mass identified by **LINEARSPECTRUM**. For example, when *Peptide* = **NQEL**,

$$\text{MASS}(\textbf{LN}) = \text{MASS}(\textbf{NQEL}) - \text{MASS}(\textbf{QE}) = \textbf{484} - 257 = 227\,.$$

Thus, we can generate a cyclic spectrum by making only a small modification to the pseudocode of **LINEARSPECTRUM**.

```
CYCLICSPECTRUM(Peptide, AMINOACID, AMINOACIDMASS)
    PREFIXMASS(0) ← 0
    for i ← 1 to Peptide
        for j ← 1 to 20
            if AMINOACID(j) = i-th amino acid in Peptide
                PREFIXMASS(i) ← PREFIXMASS(i − 1) + AMINOACIDMASS(j)
    peptideMass ← PREFIXMASS(|Peptide|)
    CyclicSpectrum ←  a list consisting of the single integer 0
    for i ← 0 to |Peptide| − 1
        for j ← i + 1 to |Peptide|
            add PREFIXMASS(j) − PREFIXMASS(i) to CyclicSpectrum
            if i > 0 and j < |Peptide|
                add peptideMass - (PREFIXMASS(j) − PREFIXMASS(i)) to CyclicSpectrum
    return sorted list CyclicSpectrum
```

How fast is **CYCLOPEPTIDESEQUENCING**?

Let's run **CYCLOPEPTIDESEQUENCING** on the following *Spectrum*:

0	97	97	99	101	103	196	198	198	200	202
295	297	299	299	301	394	396	398	400	400	497

CYCLOPEPTIDESEQUENCING first expands the set *Peptides* into the set of all 1-mers consistent with *Spectrum*:

97	99	101	103
P	V	T	C

The algorithm next appends each of the 18 amino acid masses to each of the 1-mers above. The resulting set *Peptides* containing $4 \cdot 18 = 72$ peptides of length 2 is then trimmed to keep only the 10 peptides that are consistent with *Spectrum*:

97-99 PV	97-101 PT	97-103 PC	99-97 VP	99-101 VT
99-103 VC	101-97 TP	101-99 TV	103-97 CP	103-99 CV

After expansion and trimming in the next iteration, the set *Peptides* contains 15 consistent 3-mers:

97-99-103 PVC	**97-99-101** **PVT**	97-101-97 PTP	**97-101-99** **PTV**	97-103-99 PCV
99-97-103 **VPC**	99-97-101 VPT	**99-101-97** **VTP**	99-103-97 VCP	101-97-99 TPV
101-97-103 TPC	**101-99-97** **TVP**	103-97-101 CPT	**103-97-99** **CPV**	103-99-97 CVP

With one more iteration, the set *Peptides* contains ten consistent 4-mers. Observe that the six 3-mers highlighted in red above failed to expand into any 4-mers below, and so we now know that **CYCLOPEPTIDESEQUENCING** may generate some incorrect k-mers at intermediate iterations.

97-99-103-97 PVCP	97-101-97-99 PTPV	97-101-97-103 PTPC	97-103-99-97 PCVP
99-97-101-97 VPTP	99-103-97-101 VCPT	101-97-99-103 TPVC	101-97-103-99 TPCV
	103-97-101-97 CPTP	103-99-97-101 CVPT	

In the final iteration, we generate ten consistent 5-mers:

97-99-103-97-101 PVCPT	97-101-97-99-103 PTPVC	97-101-97-103-99 PTPCV
97-103-99-97-101 PCVPT	99-97-101-97-103 VPTPC	99-103-97-101-97 VCPTP
101-97-99-103-97 TPVCP	101-97-103-99-97 TPCVP	103-97-101-97-99 CPTPV
	103-99-97-101-97 CVPTP	

All these linear peptides correspond to the same cyclic peptide PVCPT, thus solving the Cyclopeptide Sequencing Problem. You can verify that **CYCLOPEPTIDESEQUENCING** also quickly reconstructs Tyrocidine B1 from the spectrum in Figure 4.7.

Trimming the peptide leaderboard

Note: This Charging Station uses some notation from **Charging Station: Generating the Theoretical Spectrum of a Peptide**.

To implement the TRIM function in **LEADERBOARDCYCLOPEPTIDESEQUENCING**, we first will generate the theoretical spectra of all linear peptides from *Leaderboard*. Then, we will compute the scores of each theoretical spectrum against an experimental spectrum *Spectrum*. This requires implementing LINEARSCORE(*Peptide*, *Spectrum*).

Figure 4.17 shows a leaderboard of ten linear peptides represented as a list *Leaderboard* along with a ten-element array LINEARSCORES containing their scores.

Leaderboard	PVT	PTP	PTV	PCP	VPC	VTP	VCP	TPV	TPC	TVP
LINEARSCORES	6	2	4	6	5	2	5	4	4	3

FIGURE 4.17 A collection of peptides *Leaderboard* (top) along with an array LINEARSCORES (bottom) holding the score of each peptide.

The **TRIM** algorithm, shown below, sorts all peptides in *Leaderboard* according to their scores, resulting in a sorted *Leaderboard* (Figure 4.18). **TRIM** then retains the top N scoring peptides including ties (e.g., for $N = 5$, the seven top-scoring peptides shown in blue will be retained), and removes all other peptides from *Leaderboard*.

```
TRIM(Leaderboard, Spectrum, N, AMINOACID, AMINOACIDMASS)
    for j ← 1 to |Leaderboard|
        Peptide ← j-th peptide in Leaderboard
        LINEARSCORES(j) ← LINEARSCORE(Peptide, Spectrum)
    sort Leaderboard according to the decreasing order of scores in LINEARSCORES
    sort LINEARSCORES in decreasing order
    for j ← N + 1 to |Leaderboard|
        if LINEARSCORES(j) < LINEARSCORES(N)
            remove all peptides starting from the j-th peptide from Leaderboard
            return Leaderboard
    return Leaderboard
```

Leaderboard	PVT	PCP	VPC	VCP	PTV	TPV	TPC	TVP	PTP	VTP
LINEARSCORES	6	6	5	5	4	4	4	3	2	2

FIGURE 4.18 Arrays *Leaderboard* and LINEARSCORES from Figure 4.17 sorted according to score. The seven highest-scoring peptides, which will be retained after applying **TRIM** with $N = 5$, are shown in blue; remaining peptides are shown in red and will be removed from the leaderboard.

Detours

Gause and Lysenkoism

The term **Lysenkoism** refers to the politicization of genetics in the Soviet Union that began in the late 1920s and lasted for three decades until the death of Stalin. Lysenkoism was built on theories of inheritance by acquired characteristics, which ran counter to Mendelian laws.

In 1928, Trofim Lysenko, the son of Ukrainian peasants, claimed to have found a way to vastly increase the crop yield of wheat. During Stalin's rule, Soviet propaganda focused on inspirational stories of working class citizens, and it portrayed Lysenko as a genius, even though he had manufactured his experimental data. Empowered by his sudden hero status, Lysenko denounced genetics and started promoting his own "scientific" views. He called geneticists "fly lovers and people haters" and claimed that they were trying to undermine the onward march of Soviet agriculture.

Gause found himself among the few Soviet biologists who were not afraid of publicly denouncing Lysenko. By 1935, Lysenko announced that by opposing his theories, geneticists were directly opposing the teachings of Marxism. Stalin, who was in the audience, was the first to applaud, calling out, "Bravo, Comrade Lysenko!" This event gave Lysenko free reign to slander any geneticists who spoke out against him; many of Lysenkoism's opponents were imprisoned or even executed.

After World War II, Lysenko did not forget Gause's criticism: Lysenko's supporters demanded that Gause be expelled from the Russian Academy of Sciences. Lysenkoists made various attempts to invite Gause to denounce genetics and accept their pseudoscience. Gause was probably the only Soviet biologist at that time who could simply ignore such "invitations", the only other contemporary opponents of Lysenkoism being top Soviet nuclear physicists. However, Stalin left Gause and the physicists alone; in Stalin's mind, the development of antibiotics and the atomic bomb were too important. In 1949, when the director of the Russian secret police (Lavrentiy Beria) told Stalin of the dissident scientists, Stalin responded, "Make sure that our scientists have everything needed to do their job", adding, "there will always be time to execute them [later]".

Discovery of codons

In 1961, Sydney Brenner and Francis Crick established the rule of "one codon, one amino acid" during protein translation. They observed that deleting a single nucleotide or two consecutive nucleotides in a gene dramatically altered the protein product. Paradoxically, deleting *three* consecutive nucleotides resulted in only minor changes in the protein. For example, the phrase

THE · SLY · FOX · AND · THE · SHY · DOG

turns into gibberish after deleting one letter:

THE · SYF · OXA · NDT · HES · HYD · OG

or after deleting two letters:

THE · SFO · XAN · DTH · ESH · YDO · G

but it makes sense after deleting three letters:

THE · FOX · AND · THE · SHY · DOG

In 1964, Charles Yanofsky demonstrated that a gene and the protein that it produces are **collinear**, meaning that the first codon codes for the first amino acid in the protein, the second codon codes for the second amino acid, etc. For the next thirteen years, biologists believed that a protein was encoded by a long string of *contiguous* nucleotide triplets. However, the discovery of split genes in 1977 proved otherwise and necessitated the computational problem of predicting the locations of genes using only the genomic sequence (see DETOUR: Split Genes).

Quorum sensing

The traditional view that bacteria act as loners and have few interactions with the rest of their colony has been challenged by the discovery of a communication method called **quorum sensing**. This finding has shown that bacteria are capable of coordinated activity when migrating to a better nutrient supply or adopting a **biofilm** formation for defense within hostile environments. The "language" used in quorum sensing is often based on the exchange of peptides (as well as other molecules) called **bacterial pheromones**. The nature of communications between bacteria can be amicable or adversarial.

When a single bacterium releases pheromones into its environment, their concentration is often too low to be detected; however, once the population density increases, pheromone concentrations reach a threshold level that allows the bacteria to activate certain genes in response.

For example, *Burkholderia cepacia* is a pathogen affecting individuals with **cystic fibrosis**. Most patients colonized with *B. cepacia* are coinfected with *Pseudomonas aeruginosa*. The correlation of the two strains in these patients led biologists to hypothesize that interspecies communication with *P. aeruginosa* may help *B. cepacia* enhance its own pathogenicity. Indeed, the addition of *P. aeruginosa* to clones of *B. cepacia* results in a significant increase in the synthesis of proteases (i.e., enzymes needed to break down proteins), suggesting the presence of quorum sensing — *B. cepacia* may profit from pheromones made by a different species in order to improve its own chances of survival.

Molecular mass

The **dalton** (abbreviated **Da**) is the unit used for measuring atomic masses on a molecular scale. One dalton is equivalent to one twelfth of the mass of carbon-12 and has a value of approximately $1.66 \cdot 10^{-27}$ kg. The **monoisotopic mass** of a molecule is equal to the sum of the masses of the atoms in that molecule, using the mass of the most abundant **isotope** for each element. Figure 4.19 provides the elemental composition and monoisotopic masses of all 20 standard amino acids.

Amino acid	3-letter code	Chemical formula	Mass (Da)
Alanine	Ala	C_3H_5NO	71.03711
Cysteine	Cys	C_3H_5NOS	103.00919
Aspartic acid	Asp	$C_4H_5NO_3$	115.02694
Glutamic acid	Glu	$C_5H_7NO_3$	129.04259
Phenylalanine	Phe	C_9H_9NO	147.06841
Glycine	Gly	C_2H_3NO	57.02146
Histidine	His	$C_6H_7N_3O$	137.05891
Isoleucine	Ile	$C_6H_{11}NO$	113.08406
Lysine	Lys	$C_6H_{12}N_2O$	128.09496
Leucine	Leu	$C_6H_{11}NO$	113.08406
Methionine	Met	C_5H_9NOS	131.04049
Asparagine	Asn	$C_4H_6N_2O_2$	114.04293
Proline	Pro	C_5H_7NO	97.05276
Glutamine	Gln	$C_5H_8N_2O_2$	128.05858
Arginine	Arg	$C_6H_{12}N_4O$	156.10111
Serine	Ser	$C_3H_5NO_2$	87.03203
Threonine	Thr	$C_4H_7NO_2$	101.04768
Valine	Val	C_5H_9NO	99.06841
Tryptophan	Trp	$C_{11}H_{10}N_2O$	186.07931
Tyrosine	Tyr	$C_9H_9NO_2$	163.06333

FIGURE 4.19 Elemental composition and monoisotopic masses of amino acids.

Selenocysteine and pyrrolysine

Selenocysteine is a proteinogenic amino acid that exists in all kingdoms of life as a building block of a special class of proteins called selenoproteins. Unlike other amino acids, selenocysteine is not directly encoded in the genetic code. Instead, it is encoded in a special way by a `UGA` codon, which is normally a stop codon, through a mechanism known as **translational recoding**.

Pyrrolysine is a proteinogenic amino acid that exists in some archaea and methane-producing bacteria. In organisms incorporating pyrrolysine, this amino acid is encoded by `UAG`, which also normally acts as a stop codon.

Pseudo-polynomial algorithm for the Turnpike Problem

If $A = (a_1 = 0, a_2, \ldots, a_n)$ is a set of n points on a line segment in increasing order ($a_1 < a_2 < \cdots < a_n$), then ΔA denotes the collection of all pairwise differences between points in A. For example, if $A = (0, 2, 4, 7)$, then

$$\Delta A = (-7, -5, -4, -3, -2, -2, 0, 0, 0, 0, 2, 2, 3, 4, 5, 7)\,.$$

The turnpike problem asks us to reconstruct A from ΔA.

Turnpike Problem:
Given all pairwise distances between points on a line segment, reconstruct the positions of those points.

Input: A collection of integers L.
Output: A set of integers A such that $\Delta A = L$.

We will now outline an approach to solving the Turnpike Problem that is polynomial in the length of the line segment. Given a collection of integers $A = (a_1, \ldots, a_n)$, the **generating function** of A is the polynomial

$$A(x) = \sum_{i=1}^{n} x^{a_i}\,.$$

For example, if $A = (0, 2, 4, 7)$, then

$$A(x) = x^0 + x^2 + x^4 + x^7$$
$$\Delta A(x) = x^{-7} + x^{-5} + x^{-4} + x^{-3} + 2x^{-2} + 4x^0 + 2x^2 + x^3 + x^4 + x^5 + x^7$$

You can verify that the above generating function for $\Delta A(x)$ is equal to $A(x) \cdot A(x^{-1})$. Thus, the Turnpike Problem reduces to a problem about polynomial factorization. Just as an integer can be broken down into its prime factors, a polynomial with integer coefficients can be factored into "prime" polynomials having integer coefficients. If we can factor $\Delta A(x)$ and determine which prime factors contribute to $A(x)$ and which prime factors contribute to $A(x^{-1})$, then we will know $A(x)$ and therefore A. In 1982, Rosenblatt and Seymour described such a method to represent $\Delta A(x)$ as $A(x) \cdot A(x^{-1})$. Since a polynomial can be factored in time polynomial in its maximum exponent, $\Delta A(x)$ can be factored in time polynomial in the total length of the line segment, which yields the desired pseudo-polynomial algorithm for the Turnpike Problem.

STOP and Think: Can the generating function approach be modified to address the case when there are errors in the pairwise differences?

Split genes

In 1977, Phillip Sharp and Richard Roberts independently discovered **split genes**, which are genes formed by discontiguous intervals of DNA.

Sharp hybridized RNA encoding an adenovirus protein called **hexon** against a single-strand of adenovirus DNA. If the hexon gene were contiguous, then he expected to see a one-to-one hybridization of RNA bases with DNA bases.

Yet to Sharp's surprise, when he viewed the RNA-DNA hybridization under an electron microscope, he saw three loop structures, rather than the continuous duplex segment suggested by the contiguous gene model (Figure 4.20). This observation implied that the hexon mRNA must be built from four non-contiguous fragments of the adenovirus genome. These four segments, called **exons**, are separated by three fragments (the loops in Figure 4.20) called **introns**, to form a split gene. Split genes are analogous to a magazine article printed on pages 12, 17, 40, and 95, with many pages of advertising appearing in-between.

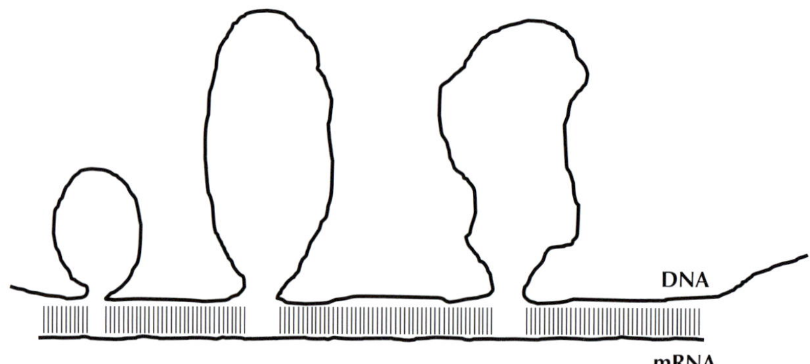

FIGURE 4.20 A rendering of Sharp's electron microscopy experiment that led to the discovery of split genes. When hexon RNA is hybridized against the DNA that generated it, three distinct loops are formed. Because the loops are present in the DNA and are not present in RNA, these loops (called introns) must be removed during the process of RNA formation.

The discovery of split genes caused an interesting quandary: *What happens to the introns?* In other words, the RNA that is transcribed from a split gene (called **precursor**

mRNA (pre-mRNA) should be longer than the RNA that is used as a template for protein synthesis (called **messenger RNA (mRNA)**. Some biological process must remove the introns in pre-mRNA and concatenate the exons into a single mRNA string. This process of converting pre-mRNA into mRNA is known as **splicing**, and it is carried out by a molecular machine called the **spliceosome**.

The discovery of split genes led to many new avenues of research. Biologists still debate what purpose introns serve; some introns are viewed as "junk DNA", while others contain important regulatory elements. Furthermore, the partition of a gene into exons often varies from species to species. For example, a gene in the chicken genome may have a different number of exons than the related gene in the human genome.

Bibliography Notes

The pseudo-polynomial algorithm for the Turnpike Problem was proposed by Rosenblatt and Seymour, 1982. The first cyclopeptide sequencing algorithm was proposed by Ng et al., 2009. Tang et al., 1999 discovered θ-defensin. Venkataraman et al., 2009 showed that humans cells can be tricked into producing θ-defensin.

How Do We Compare DNA Sequences?

Dynamic Programming

Cracking the Non-Ribosomal Code

The RNA Tie Club

Following Watson & Crick's publication of DNA's double helix structure in 1953, physicist George Gamow founded the "RNA Tie Club" for renowned scientists. A necktie embroidered with a double helix signified membership in this club, which was restricted to twenty regular members (one for each amino acid) as well as four honorary members (one for each nucleotide). Gamow wanted the RNA Tie Club to serve more than a social function; by convening top scientific minds, he hoped to decode the message hidden within DNA by determining how RNA is converted into amino acids. Indeed, Sydney Brenner and Francis Crick struck first one year later by discovering that amino acids are translated from codons (i.e., triplets of nucleotides).

The RNA Tie Club would eventually boast eight Nobel laureates, but scientists from outside of the club would decipher the genetic code. In 1961, Marshall Nirenberg synthesized RNA strands consisting only of uracil (UUUUUUUUUUUU...), added ribosomes and amino acids, and produced a peptide consisting only of phenylalanine (PhePhePhePhe...). Nirenberg thus concluded that the RNA codon UUU codes for the amino acid phenylalanine. Following Nirenberg's success, Har Gobind Khorana synthesized the RNA strand **UCUCUCUCUCUC**... and demonstrated that it translates into **SerLeuSerLeu**... Following these insights, the rest of the *ribosomal* genetic code was rapidly elucidated.

Nearly four decades later, Mohamed Marahiel set out to solve the much more challenging puzzle of cracking the **non-ribosomal code**. You will recall from Chapter 4 that bacteria and fungi produce antibiotics and other non-ribosomal peptides (NRPs) without any reliance on the ribosome and the genetic code. Instead, these organisms manufacture NRPs by employing a giant protein called NRP synthetase:

$$\text{DNA} \longrightarrow \text{RNA} \longrightarrow \text{NRP synthetase} \longrightarrow \text{NRP}$$

The NRP synthetase that encodes the 10 amino acid-long antibiotic Tyrocidine B1 (which we worked with in Chapter 4) includes 10 segments called **adenylation domains (A-domains)**; each A-domain is about 500 amino acids long and is responsible for adding a single amino acid to Tyrocidine B1.

A generation earlier, the RNA Tie Club had asked, "How does RNA encode an amino acid?" Now Marahiel set out to answer the far more challenging question, "How does each A-domain encode an amino acid?"

From protein comparison to the non-ribosomal code

Fortunately, Marahiel already knew the amino acid sequences of some A-domains, along with the amino acids that they add to the growing peptide. Below are three of these A-domains (taken from three different bacteria), which code for aspartic acid (`Asp`), ornithine (`Orn`), and valine (`Val`), respectively. In the interest of space, we will show you only short fragments taken from the three A-domains.

```
YAFDLGYTCMFPVLLGGGELHIVQKETYTAPDEIAHYIKEHGITYIKLTPSLFHTIVNTASFAFDANFESLRLIVLGGEKIIPIDVIAFRKMYGHTEFINHYGPTEATIGA
AFDVSAGDFARALLTGGQLIVCPNEVKMDPASLYAIIKKYDITIFEATPALVIPLMEYIYEQKLDISQLQILIVGSDSCSMEDFKTLVSRFGSTIRIVNSYGVTEACIDS
IAFDASSWEIYAPLLNGGTVVCIDYYTTIDIKALEAVFKQHHIRGAMLPPALLKQCLVSAPTMISSLEILFAAGDRLSSQDAILARRAVGSGVYNAYGPTENTVLS
```

STOP and Think: You now have a portion of the data that Marahiel had in 1999 when he discovered the non-ribosomal code. What would you do to infer the non-ribosomal code?

Marahiel conjectured that since A-domains have similar function (i.e., adding an amino acid to the growing peptide), different A-domains should have similar parts. A-domains should also have differing parts to incorporate different amino acids. However, only three conserved columns (shown in red below) are common to the three sequences and have likely arisen by pure chance:

```
YAFDLGYTCMFPVLLGGGELHIVQKETYTAPDEIAHYIKEHGITYIKLTPSLFHTIVNTASFAFDANFESLRLIVLGGEKIIPIDVIAFRKMYGHTEFINHYGPTEATIGA
AFDVSAGDFARALLTGGQLIVCPNEVKMDPASLYAIIKKYDITIFEATPALVIPLMEYIYEQKLDISQLQILIVGSDSCSMEDFKTLVSRFGSTIRIVNSYGVTEACIDS
IAFDASSWEIYAPLLNGGTVVCIDYYTTIDIKALEAVFKQHHIRGAMLPPALLKQCLVSAPTMISSLEILFAAGDRLSSQDAILARRAVGSGVYNAYGPTENTVLS
```

STOP and Think: How else are the three sequences similar?

If we slide the second sequence only one amino acid to the right, adding a **space symbol** ("-") to the beginning of the sequence, then we find 11 conserved columns!

```
YAFDLGYTCMFPVLLGGGELHIVQKETYTAPDEIAHYIKEHGITYIKLTPSLFHTIVNTASFAFDANFESLRLIVLGGEKIIPIDVIAFRKMYGHTEFINHYGPTEATIGA
-AFDVSAGDFARALLTGGQLIVCPNEVKMDPASLYAIIKKYDITIFEATPALVIPLMEYIYEQKLDISQLQILIVGSDSCSMEDFKTLVSRFGSTIRIVNSYGVTEACIDS
IAFDASSWEIYAPLLNGGTVVCIDYYTTIDIKALEAVFKQHHIRGAMLPPALLKQCLVSAPTMISSLEILFAAGDRLSSQDAILARRAVGSGVYNAYGPTENTVLS
```

Adding a few more space symbols reveals 14 conserved columns:

```
YAFDLGYTCMFPVLLGGGELHIVQKETYTAPDEIAHYIKEHGITYIKLTPSLFHTIVNTASFAFDANFESLRLIVLGGEKIIPIDVIAFRKMYGHTEFINHYGPTEATIGA
-AFDVSAGDFARALLTGGQLIVCPNEVKMDPASLYAIIKKYDITIFEATPALVIPLMEYI-YEQKLDISQLQILIVGSDSCSMEDFKTLVSRFGSTIRIVNSYGVTEACIDS
IAFDASSWEIYAPLLNGGTVVCIDYYTTIDIKALEAVFKQHHIRGAMLPPALLKQCLVSA----PTMISSLEILFAAGDRLSSQDAILARRAVGSGVYNAYGPTENTVLS
```

and even more sliding reveals 19 conserved columns:

```
YAFDLGYTCMFPVLLGGGELHIVQKETYTAPDEIAHYIKEHGITYIKLTPSLFHTIVNTASFAFDANFESLRLIVLGGEKIIPIDVIAFRKMYGHTE-FINHYGPTEATIGA
-AFDVSAGDFARALLTGGQLIVCPNEVKMDPASLYAIIKKYDITIFEATPALVIPLMEYI-YEQKLDISQLQILIVGSDSCSMEDFKTLVSRFGSTIRIVNSYGVTEACIDS
IAFDASSWEIYAPLLNGGTVVCIDYYTTIDIKALEAVFKQHHIRGAMLPPALLKQCLVSA----PTMISSLEILFAAGDRLSSQDAILARRAVGSGV-Y-NAYGPTENTVLS
```

It turns out that the red columns represent the **conserved core** shared by many A-domains. Now that Marahiel knew how to correctly *align* the A-domains, he hypothesized that some of the remaining variable columns should code for `Asp`, `Orn`, and `Val`. He discovered that the non-ribosomal code is defined by 8 amino acid-long non-ribosomal **signatures**, which are shown as purple columns below.

```
YAFDLGYTCMFPVLLGGGELHIVQKETYTAPDEIAHYIKEHGITYIKLTPSLFHTIVNTASFAFDANFESLRLIVLGGEKIIPIDVIAFRKMYGHTE-FINHYGPTEATIGA
-AFDVSAGDFARALLTGGQLIVCPNEVKMDPASLYAIIKKYDITIFEATPALVIPLMEYI-YEQKLDISQLQILIVGSDSCSMEDFKTLVSRFGSTIRIVNSYGVTEACIDS
IAFDASSWEIYAPLLNGGTVVCIDYYTTIDIKALEAVFKQHHIRGAMLPPALLKQCLVSA----PTMISSLEILFAAGDRLSSQDAILARRAVGSGV-Y-NAYGPTENTVLS
```

The purple columns define the signatures **`LTKVGHIG`**, **`VGEIGSID`**, and **`AWMFAAVL`**, coding for `Asp`, `Orn`, and `Val`, respectively:

`LTKVGHIG` → `Asp`
`VGEIGSID` → `Orn`
`AWMFAAVL` → `Val`

It is important to note that without first constructing the conserved core, Marahiel would not have been able to infer the non-ribosomal code, since the 24 amino acids in the signatures above do not line up in the original alignment:

```
YAFDLGYTCMFPVLLGGGELHIVQKETYTAPDEIAHYIKEHGITYIKLTPSLFHTIVNTASFAFDANFESLRLIVLGGEKIIPIDVIAFRKMYGHTEFINHYGPTEATIGA
AFDVSAGDFARALLTGGQLIVCPNEVKMDPASLYAIIKKYDITIFEATPALVIPLMEYIYEQKLDISQLQILIVGSDSCSMEDFKTLVSRFGSTIRIVNSYGVTEACIDS
IAFDASSWEIYAPLLNGGTVVCIDYYTTIDIKALEAVFKQHHIRGAMLPPALLKQCLVSAPTMISSLEILFAAGDRLSSQDAILARRAVGSGVYNAYGPTENTVLS
```

Even after identifying the conserved core, you may be wondering whether Marahiel had a crystal ball; why did he choose these particular 8 purple columns? Why should signatures have 8 amino acids and not 5, or better yet 3? See **DETOUR: Fireflies and the Non-Ribosomal Code** for a better appreciation of the complexities underlying Marahiel's work. Suffice it to say that fifteen years after Marahiel's initial discovery, the non-ribosomal code is still not fully understood. PAGE 286

What do oncogenes and growth factors have in common?

Marahiel's cracking of the non-ribosomal code is just one of many biological problems that have benefited from sequence comparison. Another striking example of the power of sequence comparison was established in 1983 when Russell Doolittle compared the newly sequenced **platelet derived growth factor (PDGF)** gene with all other genes known at the time. Doolittle stunned cancer biologists when he showed that PDGF was very similar to the sequence of a gene known as **v-sis**. The two genes' similarity was puzzling because their functions differ greatly; the PDGF gene encodes a protein stimulating cell growth, whereas v-sis is an **oncogene**, or a gene in viruses that causes a cancer-like transformation of infected human cells. Following Doolittle's discovery, scientists hypothesized that some forms of cancer might be caused by a good gene doing the right thing at the wrong time. The link between PDGF and v-sis established a new paradigm; searching all new sequences against sequence databases is now the first order of business in genomics.

However, the question remains: what is the best way to compare sequences algorithmically? Returning to the A-domain example, the insertion of spaces to reveal the conserved core probably looked like a magic trick to you. It is completely unclear what algorithm we have used to decide where to insert the space symbols, or how we should quantify the "best" alignment of the three sequences.

Introduction to Sequence Alignment

Sequence alignment as a game

To simplify matters, we will compare only two sequences at a time, returning to multiple sequence comparison at the end of the chapter. The Hamming distance, which counts mismatches in two strings, rigidly assumes that we align the *i*-th symbol of one sequence against the *i*-th symbol of the other. However, since biological sequences are subject to insertions and deletions, it is often the case that the *i*-th symbol of one sequence corresponds to a symbol at a completely different position in the other sequence. The goal, then, is to find the most appropriate correspondence of symbols.

For example, ATGCATGC and TGCATGCA have no matching positions, and so their Hamming distance is equal to 8:

```
ATGCATGC
TGCATGCA
```

Yet these strings have seven matching positions if we align them differently:

```
ATGCATGC-
-TGCATGCA
```

Strings ATGCTTA and TGCATTAA have more subtle similarities:

```
ATGC-TTA-
-TGCATTAA
```

These examples lead us to postulate a notion of a good alignment as one that matches as many symbols as possible. You can think about maximizing the number of matched symbols in two strings as a single-person game (Figure 5.1). At each turn, you have two choices. You can remove the first symbol from each sequence, in which case you earn a point if the symbols match; alternatively, you can remove the first symbol from either of the two sequences, in which case you earn no points but may set yourself up to earn more points in later moves. Your goal is to maximize the number of points.

STOP and Think: Figure 5.1 shows just one of many possible ways to play the alignment game for the strings `ATGCATGC` and `TGCATGCA`. Can you find an even better way to play this game for the strings?

Sequence alignment and the longest common subsequence

We now define an **alignment** of sequences v and w as a two-row matrix such that the first row contains the symbols of v (in order), the second row contains the symbols of w (in order), and space symbols (called **gap symbols**) may be interspersed throughout both strings, as long as two space symbols are not aligned against each other. Here is the alignment of `ATGTTATA` and `ATCGTCC` from Figure 5.1.

```
A T - G T T A T A
A T C G T - C - C
```

An alignment presents one possible scenario by which v could have evolved into w. Columns containing the same letter in both rows are called **matches** and represent conserved nucleotides, whereas columns containing different letters are called **mismatches** and represent single-nucleotide substitutions. Columns containing a space symbol are called **indels**: a column containing a space symbol in the top row of the alignment is called an **insertion**, as it implies the insertion of a symbol when transforming v into w; a column containing a space symbol in the bottom row of the alignment is called a **deletion**, as it indicates the deletion of a symbol when transforming v into w. The alignment above has four matches, two mismatches, one insertion, and two deletions.

The matches in an alignment of two strings define a **common subsequence** of the two strings, or a sequence of symbols appearing in the same order (although not necessarily consecutively) in both strings. For example, the alignment in Figure 5.1 indicates that **ATGT** is a common subsequence of **ATGT**TATA and **AT**C**GT**CC. An alignment of two strings maximizing the number of matches therefore corresponds to a **longest**

Growing alignment	Remaining symbols	Score
	A T G T T A T A A T C G T C C	
A A	T G T T A T A T C G T C C	+1
A T A T	G T T A T A C G T C C	+1
A T - A T C	G T T A T A G T C C	
A T - G A T C G	T T A T A T C C	+1
A T - G T A T C G T	T A T A C C	+1
A T - G T T A T C G T -	A T A C C	
A T - G T T A A T C G T - C	T A C	
A T - G T T A T A T C G T - C -	A C	
A T - G T T A T A A T C G T - C - C		

FIGURE 5.1 One way of playing the alignment game for the strings ATGTTATA and ATCGTCC, with score 4. At each step, we choose to remove either one or both symbols from the left of the two sequences in the "remaining symbols" column. If we remove both symbols, then we align them in the "growing alignment". If we remove only one symbol, then we align this symbol with a space symbol in the growing alignment. Matched symbols are shown in red (and receive score 1). Mismatched symbols are shown in purple; symbols aligned against space symbols are shown in blue or green depending on which sequence they were removed from.

common subsequence (LCS) of these strings. Note that two strings may have more than one longest common subsequence.

Longest Common Subsequence Problem:
Find a longest common subsequence of two strings.

> **Input**: Two strings.
> **Output**: A longest common subsequence of these strings.

Exercise Break: Find all longest common subsequences of the strings ACTGCA and CATCGC. How many such subsequences did you find?

If we limit our attention to the two A-domains coding for Asp and Orn from the introduction, then in addition to the 19 matches that we have already found, we can find 10 more matches (shown in blue below), yielding a common subsequence of length 29.

```
YAFDLGYTCMFPVLLGGGELHIVQKETYTAPDEIAHYIKEHGITYIKLTPSLFHTIVNTASFAFDANFESLRLIVLGGEKIIPIDVIAFRKMYGHTE-FINHYGPTEATIGA
-AFDVSAGDFARALLTGGQLIVCPNEVKMDPASLYAIIKKYDITIFEATPALVIPLMEYI-YEQKLDISQLQILIVGSDSCSMEDFKTLVSRFGSTIRIVNSYGVTEACIDS
```

STOP and Think: What is the longest common subsequence of these strings?

None of the algorithmic approaches we have studied thus far will help us solve the Longest Common Subsequence Problem, and so before asking you to solve this problem, we will change course to describe a different problem that may seem completely unrelated to sequence alignment.

The Manhattan Tourist Problem

What is the best sightseeing strategy?

Imagine you are a tourist in Midtown Manhattan, and you want to see as many sights as possible on your way from the corner of 59th Street and 8th Avenue to the corner of 42nd Street and 3rd Avenue (Figure 5.2 (left)). However, you are short on time, and at each intersection, you can only move south (↓) or east (→). You can choose from many different paths through the map, but no path will visit all the sights. The challenge of finding a legal path through the city that visits the most sights is called the **Manhattan Tourist Problem**.

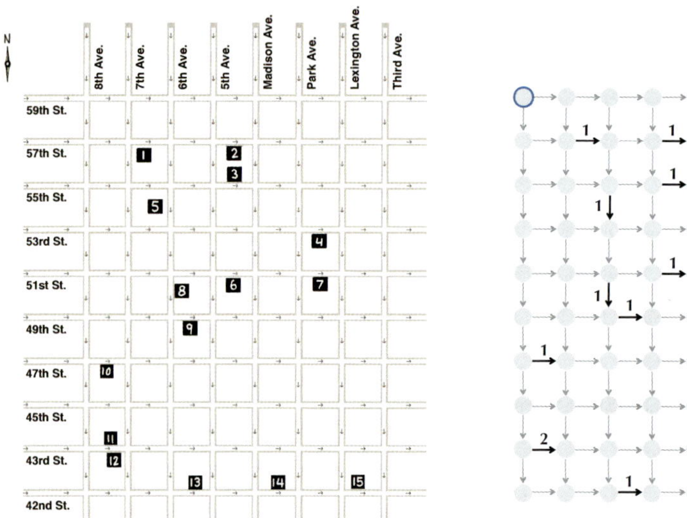

FIGURE 5.2 (Left) A simplification of Midtown Manhattan. You start at the intersection of 59th Street and 8th Avenue in the northwest corner and end at the intersection of 42nd Street and 3rd Avenue in the southeast corner, traveling only south (↓) or east (→) between intersections. The attractions shown are: Carnegie Hall (1), Tiffany & Co. (2), the Sony Building (3), the Museum of Modern Art (4), the Four Seasons Hotel (5), St. Patrick's Cathedral (6), the General Electric Building (7), Radio City Music Hall (8), Rockefeller Center (9), the Paramount Building (10), the New York Times Building (11), Times Square (12), the General Society of Mechanics and Tradesmen (13), Grand Central Terminal (14), and the Chrysler Building (15). (Right) The directed graph *ManhattanGraph* in which every edge is weighted by the number of attractions along that city block (edge weights equal to 0 are not shown).

We will represent the map of Manhattan as a directed graph *ManhattanGraph* in which we model each intersection as a node and each city block between two intersections as a directed edge indicating the legal direction of travel (↓ or →), as shown in Figure 5.2 (right). We then assign each directed edge a **weight** equal to the number of attractions along the corresponding block. The starting (blue) node is called the **source node**, and the ending (red) node is called the **sink node**. Adding the weights along a path from the source to the sink yields the number of attractions along that path. Therefore, to solve the Manhattan Tourist Problem, we need to find a **maximum-weight path** connecting the source to the sink (also called a **longest path**) in *ManhattanGraph*.

We can model any rectangular grid of streets using a similar directed graph; Figure 5.3 (left) shows the graph for a hypothetical city with even more attractions. In contrast to the Cartesian plane, we orient the axes of this grid down and to the right. Therefore, the blue source node is assigned the coordinates $(0,0)$, and the red sink node is assigned the coordinates (n,m). This implies the following generalization of our original problem.

Manhattan Tourist Problem:
Find a longest path in a rectangular city.

> **Input**: A weighted $n \times m$ rectangular grid with $n+1$ rows and $m+1$ columns.
> **Output**: A longest path from source $(0,0)$ to sink (n,m) in the grid.

Exercise Break: How many different paths are there from source to sink in an $n \times m$ rectangular grid?

Applying a brute force approach to the Manhattan Tourist Problem is impractical because the number of paths is huge. A sensible greedy approach would choose between the two possible directions at each node (→ or ↓) according to how many attractions you would see if you moved only one block south versus moving one block east. For example, in Figure 5.3 (right), we start off by moving east from $(0,0)$ rather than south because the horizontal edge has three attractions, while the vertical edge has only one. Unfortunately, this greedy strategy may miss the longest path in the long run. Figure 5.3 (right).

STOP and Think: Find a longer path than the one in Figure 5.3 (right).

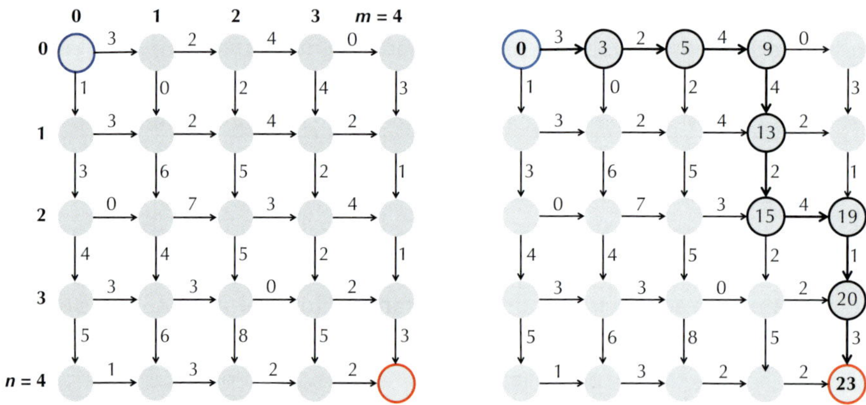

FIGURE 5.3 (Left) An $n \times m$ city grid represented as a graph with weighted edges for $n = m = 4$. The bottom left node is indexed as (4, 0), and the upper right node is indexed as (0, 4). (Right) A path through the graph found by the greedy algorithm is not the longest path.

Sightseeing in an arbitrary directed graph

In reality, the streets in Midtown Manhattan do not form a perfect rectangular grid because Broadway Avenue cuts diagonally across the grid, but the network of streets can still be represented by a directed graph. In fact, the Manhattan Tourist Problem is just a special case of the more general problem of finding the longest path in an arbitrary directed graph, such as the ones in Figure 5.4.

Longest Path in a Directed Graph Problem:
Find a longest path between two nodes in an edge-weighted directed graph.

Input: An edge-weighted directed graph with source and sink nodes.
Output: A longest path from source to sink in the directed graph.

STOP and Think: What is the length of a longest path between the source and sink in the directed graph shown in Figure 5.4 (right)?

If a directed graph contained a directed cycle (e.g., the four central edges of weight 1 in Figure 5.4 (right)), then a tourist could traverse this cycle indefinitely, revisiting

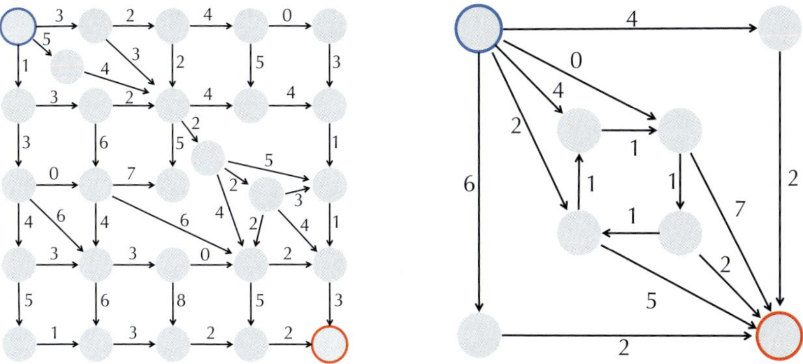

FIGURE 5.4 Directed graphs corresponding to hypothetical irregular city grids.

the same attractions over and over again and creating a path of huge length. For this reason, the graphs that we will consider in this chapter do not contain directed cycles; such graphs are called **directed acyclic graphs (DAGs)**.

Longest Path in a DAG Problem:
Find a longest path between two nodes in an edge-weighted DAG.

Input: An edge-weighted DAG with source and sink nodes.
Output: A longest path from source to sink in the DAG.

STOP and Think: Do you see any similarities between the Longest Path in a DAG Problem and the Longest Common Subsequence Problem?

Sequence Alignment is the Manhattan Tourist Problem in Disguise

In Figure 5.5, we add two arrays of integers to an alignment of `ATGTTATA` and `ATCGTCC`. The array [0 1 2 2 3 4 5 6 7 8] holds the number of symbols of `ATGTTATA` used up to a given column in the alignment. Similarly, the array [0 1 2 3 4 5 5 6 6 7] holds the number of symbols of `ATCGTCC` used up to a given column. In Figure 5.5, we have added a third array, [$\searrow$ $\searrow$ $\rightarrow$ $\searrow$ $\searrow$ $\downarrow$ $\searrow$ $\downarrow$ $\searrow$], recording whether each column represents a **match**/**mismatch** ($\searrow$/$\searrow$), an **insertion** ($\rightarrow$), or a **deletion** ($\downarrow$).

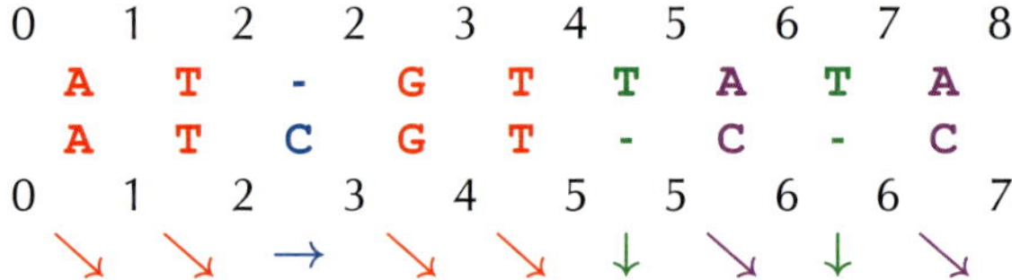

FIGURE 5.5 An alignment of ATGTTATA and ATCGTCC. The array in the first row counts the number of symbols of ATGTTATA used up to a given position. The array in the fourth row counts the number of symbols of ATCGTCC used up to a given position. And the array in the last row records whether each column of the alignment represents a **match/mismatch** (↘/↘), **insertion** (→), or **deletion** (↓).

This third array corresponds to a path from source to sink in an 8×7 rectangular grid, shown in Figure 5.6 (left). The i-th node of this path is made up of the i-th element of [0 1 2 2 3 4 5 6 7 8] and the i-th element of [0 1 2 3 4 5 5 6 6 7]:

$$(0,0)\searrow(1,1)\searrow(2,2)\rightarrow(2,3)\searrow(3,4)\searrow(4,5)\downarrow(5,5)\searrow(6,6)\downarrow(7,6)\searrow(8,7)$$

Note that in addition to horizontal and vertical edges, we have added diagonal edges connecting (i,j) to $(i+1,j+1)$ in Figure 5.6.

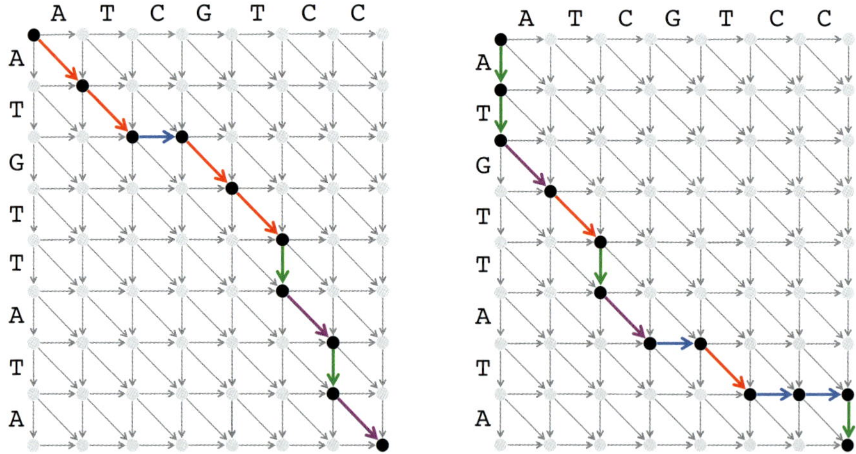

FIGURE 5.6 Every alignment corresponds to a path in the alignment graph from source to sink, and vice-versa. (Left) The path (0, 0) ↘ (1, 1) ↘ (2, 2) → (2, 3) ↘ (3, 4) ↘ (4, 5) ↓ (5, 5) ↘ (6, 6) ↓ (7, 6) ↘ (8, 7) is highlighted above and corresponds to the alignment of ATGTTATA and ATCGTCC in Figure 5.5. (Right) Another path in the alignment graph.

We call the DAG in Figure 5.6 the **alignment graph** of strings v and w, denoted ALIGNMENTGRAPH(v, w), and we call a path from source to sink in this DAG an **alignment path**. Every alignment of v and w can be viewed as a set of instructions to construct a unique alignment path in ALIGNMENTGRAPH(v, w), where each **match**/**mismatch**, **insertion**, and **deletion** corresponds to an edge ↘/↘, →, and ↓, respectively. Thus, each alignment of strings v and w corresponds to a path in ALIGNMENTGRAPH(v, w), and vice-versa.

Exercise Break: Construct the alignment of ATGTTATA and ATCGTCC corresponding to the alignment path in Figure 5.6 (right).

STOP and Think: Can you use the alignment graph to find a longest common subsequence of two strings?

Recall that finding a longest common subsequence of two strings is equivalent to finding an alignment of these strings maximizing the number of matches. In Figure 5.7, we highlight all diagonal edges of ALIGNMENTGRAPH(ATGTTATA, ATCGTCC) corresponding to matches. If we assign a weight of 1 to all these edges and 0 to all other edges, then the Longest Common Subsequence Problem is equivalent to finding a longest path in this weighted DAG!

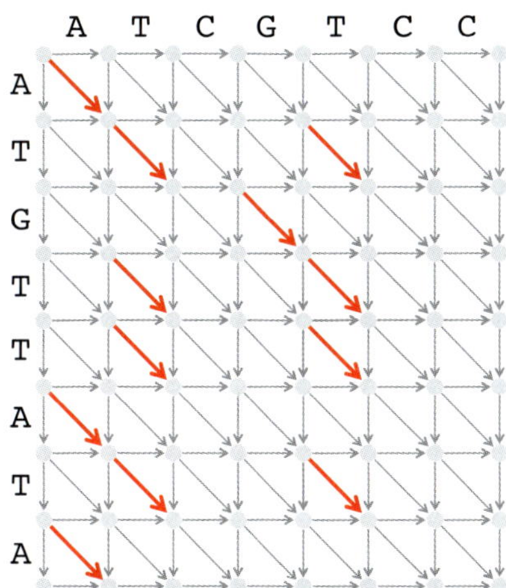

FIGURE 5.7 ALIGNMENTGRAPH(ATGTTATA, ATCGTCC) with all edges of weight 1 colored red (all other edges have weight 0). These edges correspond to potential matched symbols in an alignment of the two strings.

Thus, we need to design an algorithm for the Longest Path in a DAG Problem, but to do so, we need to know more about **dynamic programming**, a powerful algorithmic paradigm that is used for solving thousands of problems from various scientific fields. If you are already familiar with dynamic programming, then you may want to skip the next section.

An Introduction to Dynamic Programming: The Change Problem

Changing money greedily

Imagine that you bought this textbook in a bookstore for $69.24, which you paid for with $70 in cash. You are due 76 cents in change, and the cashier now must make a decision whether to give you a fistful of 76 1-cent coins or just four coins ($25 + 25 + 25 + 1 = 76$). Making change in this example is easy, but it casts light on a more general problem: how can a cashier make change using the fewest number of coins?

Different currencies have different possible coin values, or **denominations**. In the United States, the coin denominations are $(100, 50, 25, 10, 5, 1)$; in the Roman Republic, they were $(120, 40, 30, 24, 20, 10, 5, 4, 1)$. The heuristic used by cashiers all over the world to make change, which we call **GREEDYCHANGE**, iteratively selects the largest coin denomination possible.

```
GreedyChange(money)
    Change ← empty collection of coins
    while money > 0
        coin ← largest denomination that is less than or equal to money
        add a coin with denomination coin to the collection of coins Change
        money ← money − coin
    return Change
```

STOP and Think: Does **GREEDYCHANGE** always return the minimum possible number of coins?

Say we want to change 48 units of currency (denarii) in ancient Rome. **GREEDYCHANGE** returns five coins ($48 = 40 + 5 + 1 + 1 + 1$), and yet we can make change using only two coins ($48 = 24 + 24$). Thus, **GREEDYCHANGE** is suboptimal for some denominations!

STOP and Think: During the reign of Augustus, Roman coin denominations were changed to $(1600, 800, 400, 200, 100, 50, 25, 2, 1)$. Why did these denominations make Roman cashiers' lives easier? More generally, find a condition on coin denominations that dictates when **GREEDYCHANGE** will make change with the fewest number of coins.

Since **GREEDYCHANGE** is incorrect, we need to devise a different approach. We can represent coins from d arbitrary denominations by an array of integers

$$\text{COINS} = (coin_1, \ldots, coin_d)\,,$$

where the values $coin_i$ are given in decreasing order. We say that an array of d positive integers $(change_1, \ldots, change_d)$ with the **number of coins** $change_1 + \cdots + change_d$ **changes** an integer *money* (for the denominations COINS) if

$$coin_1 \cdot change_1 + \cdots + coin_d \cdot change_d = money\,.$$

For example, for the Roman denominations COINS $= (120, 40, 30, 24, 20, 10, 5, 4, 1)$, both $(0, 1, 0, 0, 0, 0, 0, 1, 0, 3)$ and $(0, 0, 0, 2, 0, 0, 0, 0, 0)$ change $money = 48$.

We will consider the problem of finding the minimum number of coins needed to make change, instead of actually producing these coins. Let MINNUMCOINS(*money*) denote the minimum number of coins needed to change *money* for a given collection of denominations (e.g., for the Roman denominations, MINNUMCOINS(48) = 2).

Change Problem:
Find the minimum number of coins needed to make change.

Input: An integer *money* and an array COINS of d positive integers.
Output: The minimum number of coins with denominations COINS that changes *money*.

Changing money recursively

Since the greedy solution used by Roman cashiers to solve the Change Problem is incorrect, we will consider a different approach. Suppose you need to change 76 denarii, and you only have coins of the three smallest denominations: COINS $= (5, 4, 1)$. A minimal collection of coins totaling 76 denarii must be one of the following:

- a minimal collection of coins totaling 75 denarii, plus one 1-denarius coin;
- a minimal collection of coins totaling 72 denarii, plus one 4-denarius coin;
- a minimal collection of coins totaling 71 denarii, plus one 5-denarius coin.

For the general denominations COINS = ($coin_1, \ldots, coin_d$), MINNUMCOINS($money$) is equal to the minimum of d numbers:

$$\text{MinNumCoins}(money) = \min \begin{cases} \text{MinNumCoins}(money - coin_1) + 1 \\ \vdots \\ \text{MinNumCoins}(money - coin_d) + 1 \end{cases}$$

We have just produced a **recurrence relation**, or an equation for MINNUMCOINS($money$) in terms of MINNUMCOINS(m) for smaller values m. The above recurrence relation motivates the following recursive algorithm, which solves the Change Problem by computing MINNUMCOINS(m) for smaller and smaller values of m. In this algorithm, |COINS| refers to the number of denominations in COINS. See **DETOUR: The Towers of Hanoi** if you did not encounter recursive algorithms in Chapter 1.

PAGE 61

```
RecursiveChange(money, Coins)
    if money = 0
        return 0
    minNumCoins ← ∞
    for i ← 1 to |Coins|
        if money ≥ coin_i
            numCoins ← RecursiveChange(money − coin_i, Coins)
            if numCoins + 1 < minNumCoins
                minNumCoins ← numCoins + 1
    return minNumCoins
```

STOP

STOP and Think: Implement RECURSIVECHANGE and run it on $money = 76$ with COINS $= (5, 4, 1)$. What happens?

RECURSIVECHANGE may appear efficient, but it is completely impractical because it recalculates the optimal coin combination for a given value of $money$ over and over again. For example, when $money = 76$ and COINS $= (5, 4, 1)$, MINNUMCOINS(70) gets computed six times, five of which are shown in Figure 5.8. This may not seem like a problem, but MINNUMCOINS(30) will be computed billions of times!

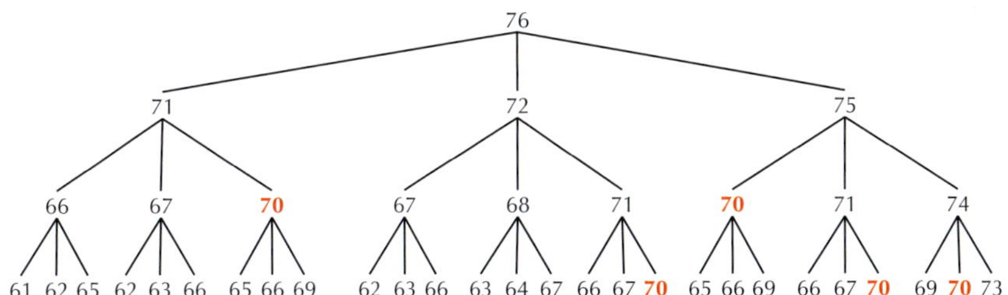

FIGURE 5.8 A tree illustrating the computation of MINNUMCOINS(76) for the denominations COINS $= (5, 4, 1)$. The edges of this tree represent recursive calls of **RECURSIVECHANGE** for different input values, with five of the six computations of MINNUMCOINS(70) highlighted in red. When MINNUMCOINS(70) is computed for the sixth time (corresponding to the path $76 \rightarrow 75 \rightarrow 74 \rightarrow 73 \rightarrow 72 \rightarrow 71 \rightarrow 70$), **RECURSIVECHANGE** has already been called hundreds of times.

Changing money using dynamic programming

To avoid the many recursive calls needed to compute MINNUMCOINS(*money*), we will use a dynamic programming strategy. Wouldn't it be nice to know all the values of MINNUMCOINS($money - coin_i$) by the time we compute MINNUMCOINS(*money*)? Instead of making time-consuming calls to **RECURSIVECHANGE**($money - coin_i$, COINS), we could simply look up the values of MINNUMCOINS($money - coin_i$) in an array and thus compute MINNUMCOINS(*money*) using just |COINS| comparisons.

The key to dynamic programming is to take a step that may seem counterintuitive. Instead of computing MINNUMCOINS(m) for every value of m from 76 *downward* toward $m = 1$ via recursive calls, we will invert our thinking and compute MINNUMCOINS(m) from $m = 1$ *upward* toward 76, storing all these values in an array so that we only need to compute MINNUMCOINS(m) once for each value of m. MINNUMCOINS(m) is still computed via the same recurrence relation:

$$\text{MinNumCoins}(m) = \min \begin{cases} \text{MinNumCoins}(m-5)+1 \\ \text{MinNumCoins}(m-4)+1 \\ \text{MinNumCoins}(m-1)+1 \end{cases}$$

For example, assuming that we have already computed MINNUMCOINS(m) for $m < 6$,

$$\text{MinNumCoins}(6) = \min \begin{cases} \text{MinNumCoins}(1) + 1 = 2 \\ \text{MinNumCoins}(2) + 1 = 3 \\ \text{MinNumCoins}(5) + 1 = 2 \end{cases}$$
$$= 2.$$

Following the same reasoning,

$$\text{MinNumCoins}(7) = \min \begin{cases} \text{MinNumCoins}(2) + 1 = 3 \\ \text{MinNumCoins}(3) + 1 = 4 \\ \text{MinNumCoins}(6) + 1 = 3 \end{cases}$$
$$= 3.$$

Continuing these calculations results in Figure 5.9.

m	0	1	2	3	4	5	6	7	8	9	10	11	12
MinNumCoins(*m*)	0	1	2	3	1	1	2	3	2	2	2	3	3

FIGURE 5.9 MinNumCoins(m) for values of m between 1 and 12.

Exercise Break: Use dynamic programming to fill in the next ten values of MinNumCoins(m) in Figure 5.9.

Notice that MinNumCoins(2) is used in the computation of both MinNumCoins(6) and MinNumCoins(7), but instead of draining computational resources by having to compute this value both times, we simply consult the pre-computed value in the array. The following dynamic programming algorithm calculates MinNumCoins(*money*) with runtime $\mathcal{O}(money \cdot |\text{Coins}|)$.

```
DPChange(money, Coins)
    MinNumCoins(0) ← 0
    for m ← 1 to money
        MinNumCoins(m) ← ∞
        for i ← 1 to |Coins|
            if m ≥ coin_i
                if MinNumCoins(m − coin_i) + 1 < MinNumCoins(m)
                    MinNumCoins(m) ← MinNumCoins(m − coin_i) + 1
    return MinNumCoins(money)
```

STOP and Think: If $money = 10^9$, **DPCHANGE** requires a huge array of size 10^9. Modify the **DPCHANGE** algorithm so that the array size required does not exceed the value of the largest coin denomination.

STOP and Think: Recall that our original goal was to make change, not just compute MINNUMCOINS(*money*). Modify **DPCHANGE** so that it not only computes the minimum number of coins but also returns these coins.

The Manhattan Tourist Problem Revisited

You should now be ready to implement an algorithm solving the Manhattan Tourist Problem. The following pseudocode computes the length of the longest path to node (i, j) in a rectangular grid and is based on the observation that the only way to reach node (i, j) in the Manhattan Tourist Problem is either by moving south ($\downarrow$) from $(i-1, j)$ or east ($\rightarrow$) from $(i, j-1)$.

```
SOUTHOREAST(i, j)
    if i = 0 and j = 0
        return 0
    x ← −∞, y ← −∞
    if i > 0
        x ← SOUTHOREAST(i − 1, j)+ weight of vertical edge into (i, j)
    if j > 0
        y ← SOUTHOREAST(i, j − 1)+ weight of horizontal edge into (i, j)
    return max{x, y}
```

STOP and Think: How many times is **SOUTHOREAST**$(3, 2)$ called in the computation of **SOUTHOREAST**$(9, 7)$?

Similarly to **RECURSIVECHANGE**, **SOUTHOREAST** suffers from a huge number of recursive calls, and we need to reframe this algorithm using dynamic programming. Remember how **DPCHANGE** worked from small instances *upward*? To find the length of a longest path from source $(0, 0)$ to sink (n, m), we will first find the lengths of longest paths from the source to *all* nodes (i, j) in the grid, expanding slowly *outward* from the source.

At first glance, you may think that we have created additional work for ourselves by solving $n \times m$ different problems instead of a single problem. Yet **SOUTHOREAST** also solves all these smaller problems, just as **RECURSIVECHANGE** and **DPCHANGE** both computed MINNUMCOINS(m) for all values of $m < money$. The trick behind dynamic programming is to solve each of the smaller problems once rather than billions of times.

We will henceforth denote the length of a longest path from $(0,0)$ to (i,j) as $s_{i,j}$. Computing $s_{0,j}$ (for $0 \leq j \leq m$) is easy, since we can only reach $(0,j)$ by moving right ($\rightarrow$) and do not have any flexibility in our choice of path. Thus, $s_{0,j}$ is the sum of the weights of the first j horizontal edges leading out from the source. Similarly, $s_{i,0}$ is the sum of the weights of the first i vertical edges from the source (Figure 5.10).

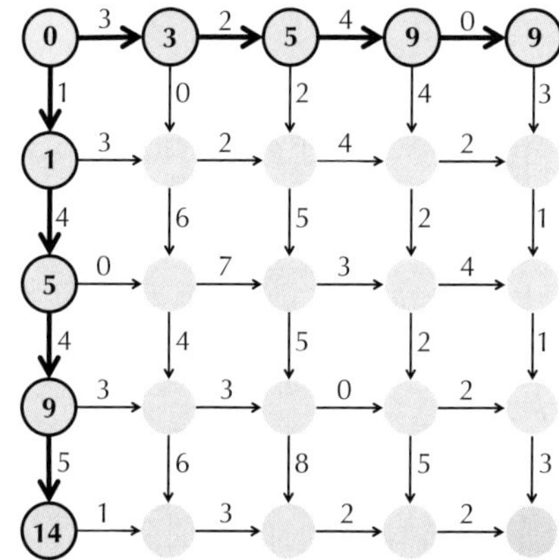

FIGURE 5.10 Computing $s_{i,0}$ and $s_{0,j}$ is easy because there is only one path from the source to $(i,0)$ and only one path from the source to $(0,j)$.

For $i > 0$ and $j > 0$, the only way to reach node (i,j) is by moving down from node $(i-1,j)$ or by moving right from node $(i,j-1)$. Thus, $s_{i,j}$ can be computed as the maximum of two values:

$$s_{i,j} = \max \begin{cases} s_{i-1,j} + \text{ weight of vertical edge from } (i-1,j) \text{ to } (i,j) \\ s_{i,j-1} + \text{ weight of horizontal edge from } (i,j-1) \text{ to } (i,j) \end{cases}$$

Now that we have computed $s_{0,1}$ and $s_{1,0}$, we can compute $s_{1,1}$. You can arrive at $(1,1)$ by traveling down from $(0,1)$ or right from $(1,0)$. Therefore, $s_{1,1}$ is the maximum of two values:

$$s_{1,1} = \max \begin{cases} s_{0,1} + \text{ weight of vertical edge from } (0,1) \text{ to } (1,1) & = 3+0=3 \\ s_{1,0} + \text{ weight of horizontal edge from } (1,0) \text{ to } (1,1) & = 1+3=4 \end{cases}$$

Since our goal is to find the longest path from $(0,0)$ to $(1,1)$, we conclude that $s_{1,1} = 4$. Because we chose the horizontal edge from $(1,0)$ to $(1,1)$, the longest path through $(1,1)$ must use this edge, which we highlight in Figure 5.11 (top left). Similar logic allows us to compute the rest of the values in column 1; for each $s_{i,1}$, we highlight the edge that we chose leading into $(i,1)$, as shown in Figure 5.11 (top right).

Continuing column-by-column (Figure 5.11 (bottom left)), we can compute every score $s_{i,j}$ in a single sweep of the graph, eventually calculating $s_{4,4} = 34$.

For each node (i,j), we will highlight the edge leading into (i,j) that we used to compute $s_{i,j}$. However, note that we have a tie when we compute $s_{3,3}$:

$$s_{3,3} = \max \begin{cases} s_{2,3} + \text{ weight of vertical edge from } (2,3) \text{ to } (3,3) & = 20+2=22 \\ s_{3,2} + \text{ weight of horizontal edge from } (3,2) \text{ to } (3,3) & = 22+0=22 \end{cases}$$

To reach $(3,3)$, we could have used *either* the horizontal or vertical incoming edge, and so we highlight both of these edges in the completed graph in Figure 5.11 (bottom left).

STOP and Think: Thus far, we have only discussed how to find the length of a longest path. How could you use the highlighted edges in Figure 5.11 (bottom left) to reconstruct a longest path?

We now have the outline of a dynamic programming algorithm for finding the length of a longest path in the Manhattan Tourist Problem, called **MANHATTANTOURIST**. In the pseudocode below, $down_{i,j}$ and $right_{i,j}$ are the respective weights of the vertical and horizontal edges entering node (i,j). We denote the matrices holding $(down_{i,j})$ and $(right_{i,j})$ as *Down* and *Right*, respectively.

Exercise Break: Modify **MANHATTANTOURIST** in order to find the length of the longest path from source to sink in the graph shown in Figure 5.11 (bottom right).

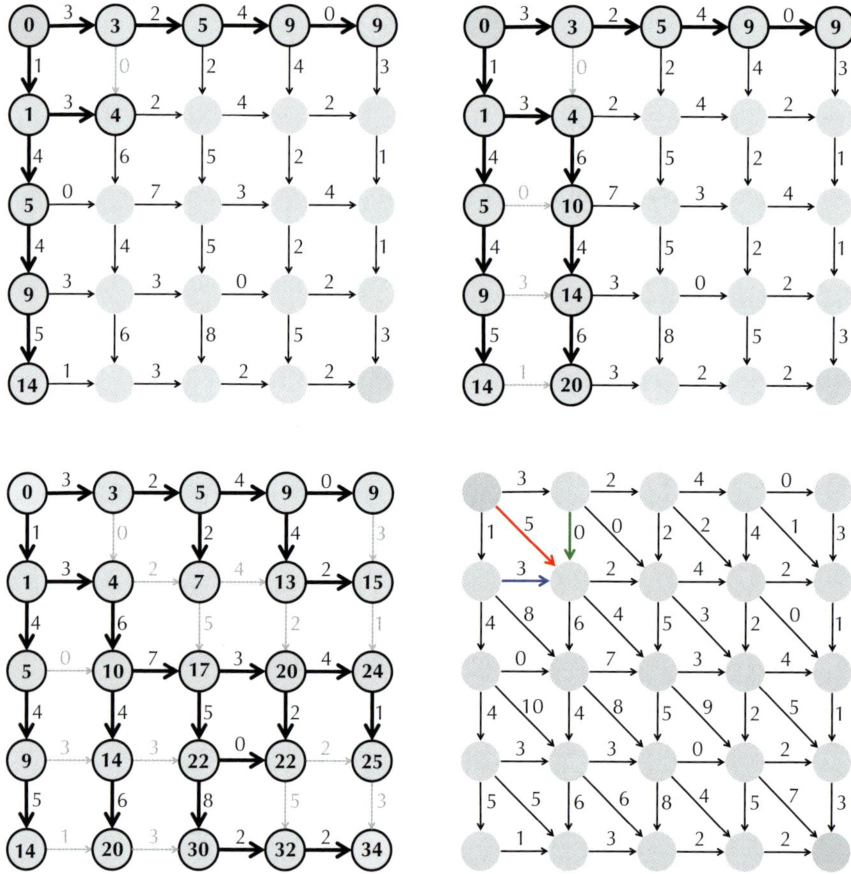

FIGURE 5.11 (Top left) Computing $s_{1,1}$ uses the horizontal edge from (1, 0), which is highlighted. (Top right) Computing all values $s_{i,1}$ in column 1. (Bottom left) The graph displaying all scores $s_{i,j}$. (Bottom right) A graph with diagonal edges constructed for an imaginary city. Node (1, 1) has three predecessors ((0, 0), (0, 1), and (1, 0)) that are used in the computation of $s_{1,1}$.

```
MANHATTANTOURIST(n, m, Down, Right)
    s_{0,0} ← 0
    for i ← 1 to n
        s_{i,0} ← s_{i−1,0} + down_{i,0}
    for j ← 1 to m
        s_{0,j} ← s_{0,j−1} + right_{0,j}
    for i ← 1 to n
        for j ← 1 to m
            s_{i,j} ← max{s_{i−1,j} + down_{i,j}, s_{i,j−1} + right_{i,j}}
    return s_{n,m}
```

From Manhattan to an Arbitrary Directed Acyclic Graph

Sequence alignment as building a Manhattan-like graph

After seeing how dynamic programming solved the Manhattan Tourist Problem, you should be prepared to adapt **MANHATTANTOURIST** for alignment graphs with diagonal edges. Recall Figure 5.7, in which we modeled the Longest Common Subsequence Problem as finding the longest path in an alignment graph "city" whose "attractions"' (matches) all lie on diagonal edges with weight 1.

You can probably work out the recurrence relation for the alignment graph on your own, but imagine for a second that you have not already learned that an LCS can be represented by a longest path in the alignment graph. As DETOUR: Finding a Longest Common Subsequence without Building a City explains, we don't need to build a Manhattan-like city to compute the length of an LCS. However, the arguments required to do so are tedious. More importantly, various alignment applications are much more complex than the Longest Common Subsequence Problem and require building a DAG with appropriately chosen edge weights in order to model the specifics of a biological problem. Rather than treating each subsequent alignment application as a frightening new challenge, we would like to equip you with a generic dynamic programming algorithm that will find a longest path in any DAG. Moreover, many bioinformatics problems have nothing to do with alignment, yet they can also be solved as applications of the Longest Path in a DAG Problem.

Dynamic programming in an arbitrary DAG

Given a node b in a DAG, let s_b denote the length of a longest path from the source to b. We call node a a **predecessor** of node b if there is an edge connecting a to b in the DAG; note that the indegree of a node is equal to the number of its predecessors. The **score** s_b of node b with indegree k is computed as a maximum of k terms:

$$s_b = \max_{\text{all predecessors } a \text{ of node } b} \{s_a + \text{weight of edge from } a \text{ to } b\}.$$

For example, in the graph shown in Figure 5.11 (bottom right), node $(1,1)$ has three predecessors. You can arrive at $(1,1)$ by traveling right from $(1,0)$, down from $(0,1)$, or diagonally from $(0,0)$, Assuming that we have already computed $s_{0,0}$, $s_{0,1}$, and $s_{1,0}$, we can therefore compute $s_{1,1}$ as the maximum of three values:

$$s_{1,1} = \max \begin{cases} s_{0,1} + \text{weight of edge } \downarrow \text{ connecting } (0,1) \text{ to } (1,1) = 3+0 = 3 \\ s_{1,0} + \text{weight of edge } \rightarrow \text{ connecting } (1,0) \text{ to } (1,1) = 1+3 = 4 \\ s_{0,0} + \text{weight of edge } \searrow \text{ connecting } (0,0) \text{ to } (1,1) = 0+5 = 5 \end{cases}$$

To compute scores for any node (i,j) of this graph, we use the following recurrence:

$$s_{i,j} = \max \begin{cases} s_{i-1,j} & + \text{ weight of edge } \downarrow \text{ between } (i-1,j) \text{ and } (i,j) \\ s_{i,j-1} & + \text{ weight of edge } \rightarrow \text{ between } (i,j-1) \text{ and } (i,j) \\ s_{i-1,j-1} & + \text{ weight of edge } \searrow \text{ between } (i-1,j-1) \text{ and } (i,j) \end{cases}$$

An analogous argument can be applied to the alignment graph to compute the length of an LCS between sequences v and w. Since in this case all edges have weight 0 except for diagonal edges of weight 1 that represent matches ($v_i = w_j$), we obtain the following recurrence for computing the length of an LCS:

$$s_{i,j} = \max \begin{cases} s_{i-1,j} & +0 \\ s_{i,j-1} & +0 \\ s_{i-1,j-1} & +1, \text{ if } v_i = w_j \end{cases}$$

STOP and Think: The above recurrence does not incorporate mismatch edges. Why is this not a problem?

A similar approach can be developed to find the longest path in any DAG, as suggested by the next exercise.

Exercise Break: What is the length of a longest path between the blue and red nodes in the DAG shown in Figure 5.12?

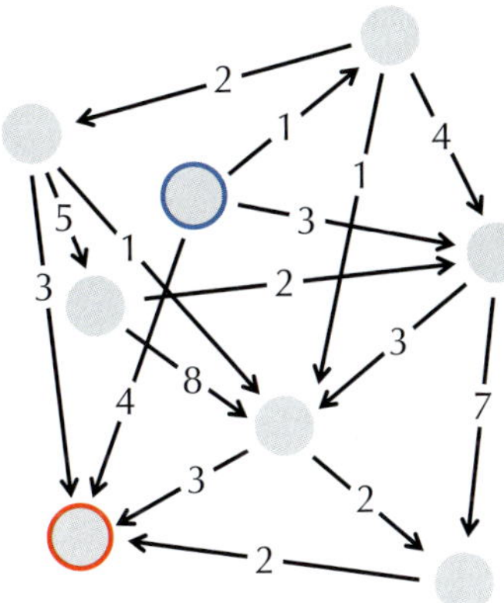

FIGURE 5.12 A weighted DAG without the obvious order inherent in the graphs previously encountered in this chapter.

Topological orderings

Do not worry if you struggled to solve the last exercise. The hitch to using dynamic programming in order to find the length of a longest path in a DAG is that we must decide on the *order* in which to visit nodes when computing the values s_b according to the recurrence

$$s_b = \max_{\text{all predecessors } a \text{ of node } b} \{s_a + \text{weight of edge from } a \text{ to } b\}.$$

This ordering of nodes is important, since by the time we reach node b, the values s_a for all its predecessors must have already been computed. We have managed to hide this issue for rectangular grids because the order in which we have computed the $s_{i,j}$ ensured that we would never consider a node before visiting all of its predecessors.

To illustrate the importance of visiting nodes in the correct order, consider the DAG in Figure 5.13, which corresponds to a "Dressing Challenge Problem". How would you order the nodes of this graph so that you don't put on your boots before your tights?

FIGURE 5.13 The DAG associated with the Dressing Challenge Problem. If a directed edge connects two items of clothing, then the first item must be put on before the second item.

To solve the Dressing Challenge Problem, we need to arrange the nodes in the DAG in Figure 5.13 along a line so that every directed edge connects a node to a node on its right (Figure 5.14). To get dressed without any mishaps, you can simply visit nodes from left to right.

To find a longest path in an arbitrary DAG, we first need to order the nodes of the DAG so that every node falls after all its predecessors. Formally, an ordering of nodes $(a_1, \ldots, a_k)$ in a DAG is called a **topological ordering** if every edge (a_i, a_j) of the DAG connects a node with a smaller index to a node with a larger index, i.e., $i < j$.

Exercise Break: Construct a topological ordering of the DAG in Figure 5.12.

Exercise Break: How many topological orderings does the Dressing Challenge DAG have?

MANHATTANTOURIST is able to find a longest path in a rectangular grid because its pseudocode implicitly orders nodes according to the "row-by-row" topological ordering shown in Figure 5.15 (left). The "column-by-column" ordering (Figure 5.15 (right)) gives another topological ordering of a rectangular grid.

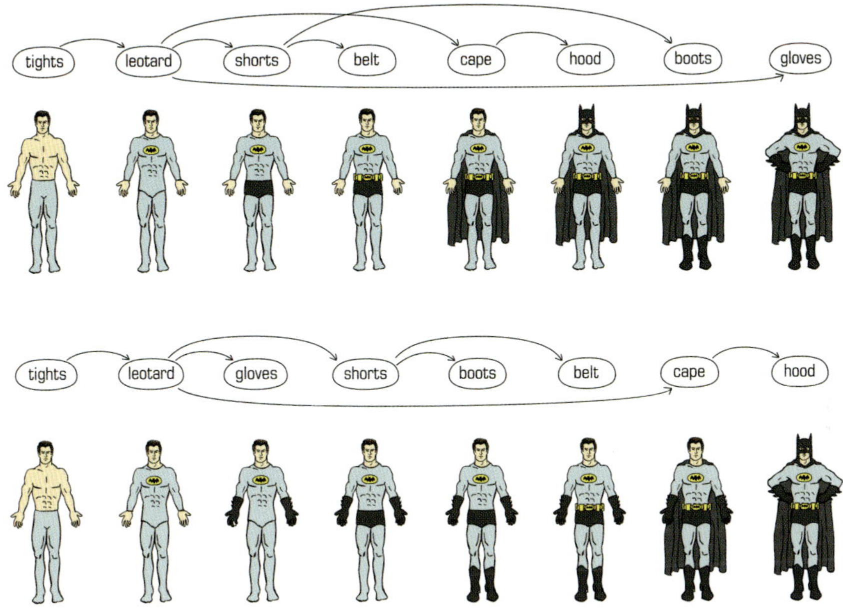

FIGURE 5.14 Two different topological orderings of the Dressing Challenge DAG from Figure 5.13.

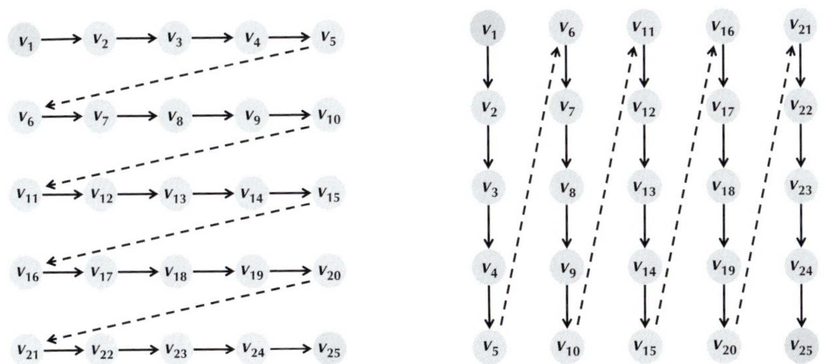

FIGURE 5.15 The row-by-column (left) and column-by-column (right) topological orderings of a rectangular grid.

STOP and Think: Rewrite the MANHATTANTOURIST pseudocode based on the topological ordering shown in Figure 5.16.

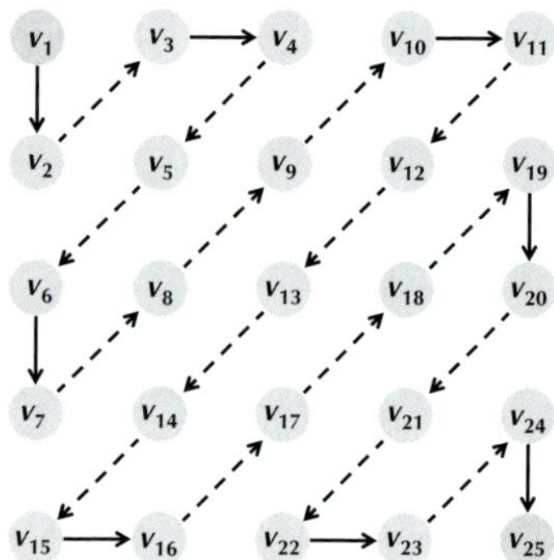

FIGURE 5.16 Another topological ordering of the rectangular grid from Figure 5.15.

PAGE 287

It can be proven that any DAG has a topological ordering, and that this topological ordering can be constructed in time proportional to the number of edges in the graph (see **DETOUR: Constructing a Topological Ordering**). Once we have a topological ordering, we can compute the length of the longest path from source to sink by visiting the nodes of the DAG in the order dictated by the topological ordering, which is achieved by the following algorithm. For simplicity, we assume that the source node is the only node with indegree 0 in *Graph*.

```
LONGESTPATH(Graph, source, sink)
    for each node b in Graph
        s_b ← −∞
    s_source ← 0
    topologically order Graph
    for each node b in Graph (following the topological order)
        s_b ← max_{all predecessors a of node b}{s_a + weight of edge from a to b}
    return s_sink
```

Since every edge participates in only a single recurrence, the runtime of **LONGESTPATH** is proportional to the number of edges in the DAG *Graph*.

We can now efficiently compute the *length* of a longest path in an arbitrary DAG, but we do not yet know how to convert **LONGESTPATH** into an algorithm that will *construct* this longest path. In the next section, we will use the Longest Common Subsequence Problem to explain how to construct a longest path in a DAG.

Backtracking in the Alignment Graph

In Figure 5.11 (bottom left), we highlight each edge chosen by **MANHATTANTOURIST**. To form a longest path, we simply need to find a path from source to sink formed by highlighted edges (more than one such path may exist). However, if we were to walk from source to sink along the highlighted edges, we might reach a dead end, such as the node $(1, 2)$. In contrast, every path from the sink will bring us back to the source if we **backtrack** in the direction opposite to each highlighted edge (Figure 5.17).

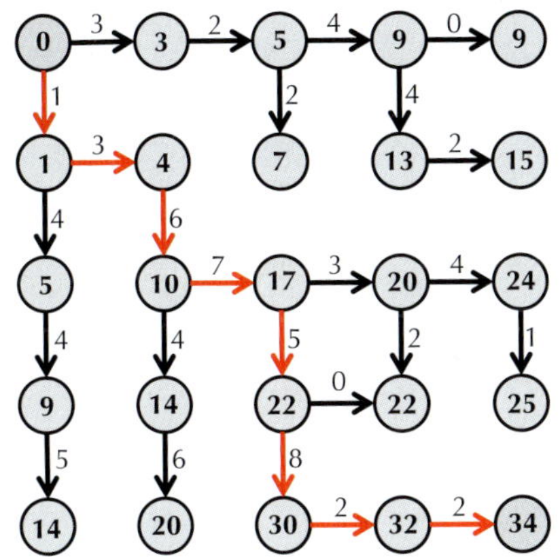

FIGURE 5.17 Following edges from the sink backwards to the source produces the red path in the DAG above, which highlights a longest path from the source to sink.

We can use this backtracking idea to construct an LCS of strings v and w. We know that if we assign a weight of 1 to the edges in ALIGNMENTGRAPH(v, w) corresponding to matches and assign a weight of 0 to all other edges, then $s_{|v|,|w|}$ gives the length of an LCS. The following algorithm maintains a record of which edge in ALIGNMENTGRAPH(v, w) was used to compute each value $s_{i,j}$ by utilizing **backtracking pointers**, which take one of the three values ↓, →, or ↘. Backtracking pointers are stored in a matrix *Backtrack*.

STOP and Think: How does changing the order of the three "if" statements in the **LCSBACKTRACK** pseudocode affect the computation of *Backtrack*?

```
LCSBACKTRACK(v, w)
    for i ← 0 to |v|
        s_{i,0} ← 0
    for j ← 0 to |w|
        s_{0,j} ← 0
    for i ← 1 to |v|
        for j ← 1 to |w|
                            ⎧ s_{i−1,j}
            s_{i,j} ← max   ⎨ s_{i,j−1}
                            ⎩ s_{i−1,j−1} + 1, if v_i = w_j
            if s_{i,j} = s_{i−1,j}
                Backtrack_{i,j} ← "↓"
            else if s_{i,j} = s_{i,j−1}
                Backtrack_{i,j} ← "→"
            else if s_{i,j} = s_{i−1,j−1} + 1 and v_i = w_j
                Backtrack_{i,j} ← "↘"
    return Backtrack
```

We now need to find a path from the source to the sink formed by the highlighted edges. The algorithm below solves the Longest Common Subsequence Problem by using the information in *Backtrack*.**OUTPUTLCS**(*Backtrack*, v, i, j) outputs an LCS between the i-prefix of v and the j-prefix of w. The initial invocation that outputs an LCS of v and w is **OUTPUTLCS**(*Backtrack*, $v, |v|, |w|$).

```
OUTPUTLCS(Backtrack, v, i, j)
    if i = 0 or j = 0
        return
    if Backtrack_{i,j} = ↓
        OUTPUTLCS(Backtrack, v, i − 1, j)
    else if Backtrack_{i,j} = →
        OUTPUTLCS(Backtrack, v, i, j − 1)
    else
        OUTPUTLCS(Backtrack, v, i − 1, j − 1)
        output v_i
```

STOP and Think: **OUTPUTLCS** is a recursive algorithm, but it is efficient. What makes it different from the inefficient recursive algorithms for making change and finding a longest path in a DAG?

The backtracking method can be generalized to construct a longest path in any DAG. Whenever we compute s_b as

$$s_b = \max_{\text{all predecessors } a \text{ of node } b} \{s_a + \text{weight of edge from } a \text{ to } b\},$$

we simply need to store a predecessor of b that was used in the computation of s_b so that we can backtrack later on. You are now ready to use backtracking to find the longest path in an arbitrary DAG.

Exercise Break: Currently, **OUTPUTLCS** finds a single LCS of two strings. Modify **OUTPUTLCS** (and **LCSBACKTRACK**) to find every LCS of two strings.

Scoring Alignments

What is wrong with the LCS scoring model?

Recall Marahiel's alignment of the A-domains coding for `Asp` and `Orn`, which had 19 + 10 matches:

```
YAFDLGYTCMFPVLLGGGELHIVQKETYTAPDEIAHYIKEHGITYIKLTPSLFHTIVNTASFAFDANFESLRLIVLGGEKIIPIDVIAFRKMYGHTE-FINHYGPTEATIGA
-AFDVSAGDFARALLTGGQLIVCPNEVKMDPASLYAIIKKYDITIFEATPALVIPLMEYI-YEQKLDISQLQILIVGSDSCSMEDFKTLVSRFGSTIRIVNSYGVTEACIDS
```

It is not difficult to construct an alignment having even more matches at the expense of introducing more indels. Yet the more indels we add, the less biologically relevant the alignment becomes, as it diverges further and further from the biologically correct alignment found by Marahiel. Below is the alignment with the maximum number of matches, representing an LCS of length 19 + 8 + 19 = 46 (the green symbols represent new matches). This alignment is so long that we cannot fit it on a single line.

```
YAFDL--G-YTCMFP--VLL-GGGELHIV---Q-K-E--T-YTAPDEIAHYIK--EHGITYI---KLTPSL-FHT
-AFDVSAGD----FARA-LLTGG-QL-IVCPNEVKMDPASLY-A---I---IKKYD--IT-IFEA--TPALV---

IVNTASFAFDANFE-----S-LR-LIVLGG-----EKIIPIDVIAFRK-M---YGHTEFI---NHYGPTEATIGA
IPLMEYIY-----EQKLDISQLQILIV-GSDSCSME-----D---F-KTLVSRFGST--IRIVNSYGVTEACIDS
```

STOP and Think: If Marahiel had constructed this alignment, would he have been able to infer the eight amino acid-long signatures of the non-ribosomal code?

Below, we highlight the purple amino acids representing the non-ribosomal signatures. Although these signatures are grouped in eight conserved columns in Marahiel's alignment from the beginning of the chapter, only five of these columns have "survived" in the LCS alignment above, making it impossible to infer the non-ribosomal signatures:

```
YAFDL--G-YTCMFP--VLL-GGGELHIV---Q-K-E--T-YTAPDEIAHYIK--EHGITYI---KLTPSL-FHT
-AFDVSAGD----FARA-LLTGG-QL-IVCPNEVKMDPASLY-A---I---IKKYD--IT-IFEA--TPALV---

IVNTASFAFDANFE-----S-LR-LIVLGG-----EKIIPIDVIAFRK-M---YGHTEFI---NHYGPTEATIGA
IPLMEYIY-----EQKLDISQLQILIV-GSDSCSME-----D---F-KTLVSRFGST--IRIVNSYGVTEACIDS
```

The frivolous matches hiding the real evolutionary scenario have appeared because nothing stopped us from introducing an excessive number of indels when building an LCS. Recalling our original alignment game in which we rewarded matched symbols, we need some way of *penalizing* indels and mismatches. First, let's handle indels. Say that in addition to assigning matches a premium of +1, we decide to assess each indel a penalty of -4. The top-scoring alignment of the A-domains gets closer to the biologically correct alignment, with six of the columns corresponding to correctly aligned signatures.

```
YAFDLGYTCMFP-VLL-GGGELHIV-QKETYTAPDEI-AHYIKEHGITYI-KLTPSLFHTIVNTASFAFDANFE
-AFDVS-AGDFARALLTGG-QL-IVCPNEVKMDPASLYA-IIKKYDIT-IFEATPAL--VIPLME-YIYEQKLD

-S-LR-LIVLGGEKIIPIDVIAFRKM---YGHTE-FINHYGPTEATIGA
ISQLQILIV-GSDSC-SME--DFKTLVSRFGSTIRIVNSYGVTEACIDS
```

Scoring matrices

To generalize the alignment scoring model, we still award +1 for matches, but we also penalize mismatches by some positive constant μ (the **mismatch penalty**) and indels by some positive constant σ (the **indel penalty**). As a result, the **score** of an alignment is equal to

$$\#\,\text{matches} - \mu \cdot \#\,\text{mismatches} - \sigma \cdot \#\,\text{indels}\,.$$

For example, with the parameters $\mu = 1$ and $\sigma = 2$, the alignment below will be assigned a score of -4:

A	T	-	G	T	T	A	T	A
A	T	C	G	T	-	C	-	C
+1	+1	-2	+1	+1	-2	-1	-2	-1

Biologists have further refined this cost function to allow for the fact that some mutations may be more likely than others, which calls for mismatches and indel penalties that differ depending on the specific symbols involved. We will extend the k-letter alphabet to include the space symbol and then construct a $(k+1) \times (k+1)$ **scoring matrix** *Score* holding the score of aligning every pair of symbols. The scoring matrix for comparing DNA sequences ($k = 4$) when all mismatches are penalized by μ and all indels are penalized by σ is shown below.

	A	C	G	T	–
A	**+1**	-μ	-μ	-μ	-σ
C	-μ	**+1**	-μ	-μ	-σ
G	-μ	-μ	**+1**	-μ	-σ
T	-μ	-μ	-μ	**+1**	-σ
–	-σ	-σ	-σ	-σ	

Although scoring matrices for DNA sequence comparison are usually defined only by the parameters μ and σ, scoring matrices for protein sequence comparison weight different mutations differently and become quite involved (see **DETOUR: PAM Scoring Matrices**). PAGE 288

From Global to Local Alignment

Global alignment

You should now be ready to modify the alignment graph to solve a generalized form of the alignment problem that takes a scoring matrix as input.

Global Alignment Problem:
Find a highest-scoring alignment of two strings as defined by a scoring matrix.

Input: Two strings and a scoring matrix *Score*.
Output: An alignment of the strings whose alignment score (as defined by *Score*) is maximized among all possible alignments of the strings.

To solve the Global Alignment Problem, we still must find a longest path in the alignment graph after updating the edge weights to reflect the values in the scoring matrix

(Figure 5.18). Recalling that deletions correspond to vertical edges (↓), insertions correspond to horizontal edges (→), and matches/mismatches correspond to diagonal edges (↘/↘), we obtain the following recurrence for $s_{i,j}$, the length of a longest path from $(0,0)$ to (i,j):

$$s_{i,j} = \max \begin{cases} s_{i-1,j} & + \; Score(v_i, -) \\ s_{i,j-1} & + \; Score(-, w_j) \\ s_{i-1,j-1} & + \; Score(v_i, w_j). \end{cases}$$

When the match reward is +1, the mismatch penalty is μ, and the indel penalty is σ, the alignment recurrence can be written as follows:

$$s_{i,j} = \max \begin{cases} s_{i-1,j} & - \sigma \\ s_{i,j-1} & - \sigma \\ s_{i-1,j-1} + 1, & \text{if } v_i = w_j \\ s_{i-1,j-1} - \mu, & \text{if } v_i \neq w_j. \end{cases}$$

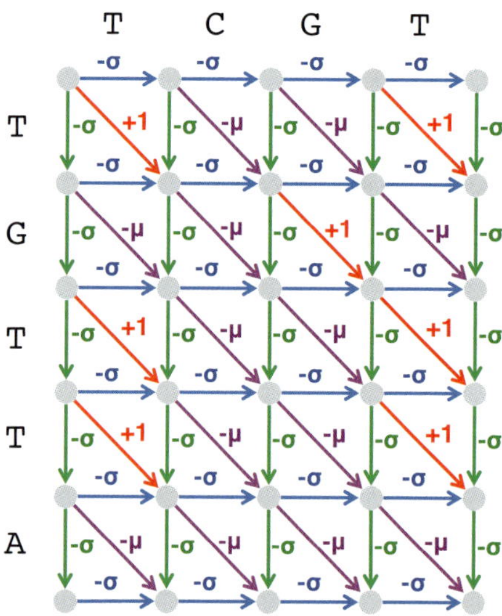

FIGURE 5.18 ALIGNMENTGRAPH(TGTTA, TCGT), with each edge colored according to whether it represents a match, mismatch, insertion, or deletion.

Limitations of global alignment

Analysis of **homeobox genes** offers an example of a problem for which global alignment may fail to reveal biologically relevant similarities. These genes regulate embryonic development and are present in a large variety of species, from flies to humans. Homeobox genes are long, and they differ greatly between species, but an approximately 60 amino acid-long region in each gene, called the **homeodomain**, is highly conserved. For instance, consider the mouse and human homeodomains below.

Mouse

```
...ARRSRTHFTKFQTDILIEAFEKNRFPGIVTREKLAQQTGIPESRIHIWFQNRRARHPDPG...
...ARQKQTFITWTQKNRLVQAFERNPFPDTATRKKLAEQTGLQESRIQMWFQKQRSLYLKKS...
```

Human

The immediate question is how to find this conserved segment within the much longer genes and ignore the flanking areas, which exhibit little similarity. Global alignment seeks similarities between two strings across their entire length; however, when searching for homeodomains, we are looking for smaller, *local* regions of similarity and do not need to align the entire strings. For example, the global alignment below has 22 matches, 18 indels, and 2 mismatches, resulting in the score $22 - 18 - 2 = 2$ (if $\sigma = \mu = 1$):

```
GCC-C-AGTC-TATGT-CAGGGGGCACG--A-GCATGCACA-
GCCGCC-GTCGT-T-TTCAG----CA-GTTATGT-T-CAGAT
```

However, these sequences can be aligned differently (with 17 matches and 32 indels) based on a highly conserved interval represented by the substrings CAGTCTATGTCAG and CAGTTATGTTCAG:

```
---G----C-----C--CAGTCTATG-TCAGGGGGCACGAGCATGCACA
GCCGCCGTCGTTTTCAGCAGT-TATGTTCAG-----A------T-----
```

This alignment has fewer matches and a lower score of $17 - 32 = -15$, even though the conserved region of the alignment contributes a score of $12 - 2 = 10$, which is hardly an accident.

Figure 5.19 shows the two alignment paths corresponding to these two different alignments. The upper path, corresponding to the second alignment above, loses out because it contains many heavily penalized indels on either side of the diagonal corresponding to the conserved interval. As a result, global alignment outputs the biologically irrelevant lower path.

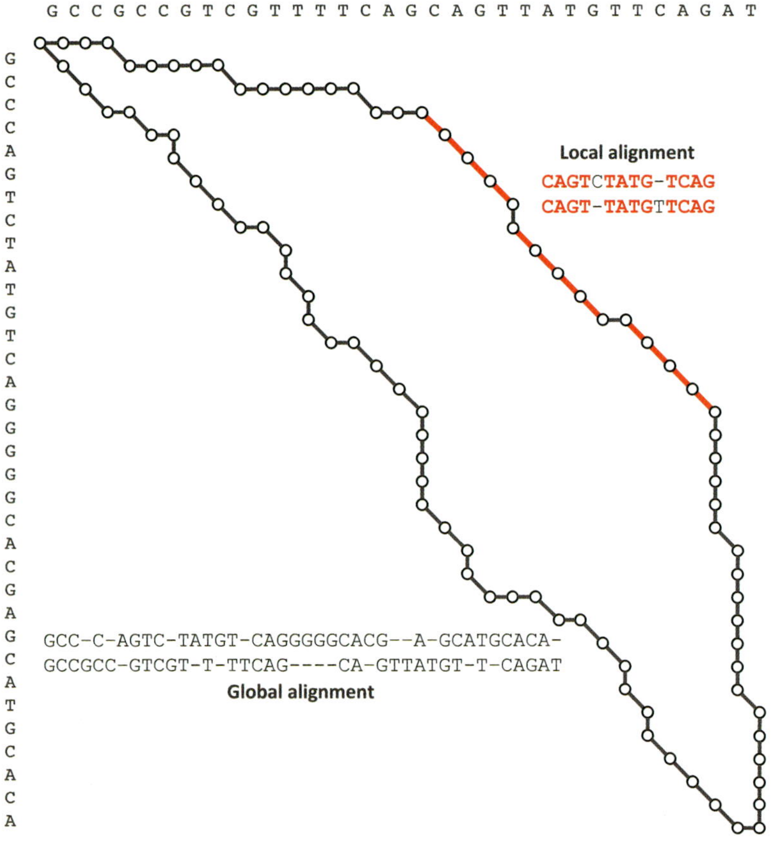

FIGURE 5.19 Global and local alignments of two DNA strings that share a highly conserved interval. The relevant alignment that captures this interval (upper path) loses to an irrelevant alignment (lower path), since the former incurs heavy indel penalties.

When biologically significant similarities are present in some parts of sequences v and w and absent from others, biologists attempt to ignore global alignment and instead align *substrings* of v and w, which yields a **local alignment** of the two strings. The problem of finding substrings that maximize the global alignment score over all substrings of v and w is called the **Local Alignment Problem**.

Local Alignment Problem:
Find the highest-scoring local alignment between two strings.

> **Input**: Strings v and w as well as a scoring matrix *Score*.
> **Output**: Substrings of v and w whose global alignment score (as defined by *Score*) is maximized among all substrings of v and w.

The straightforward but inefficient way to solve the Local Alignment Problem is to find the longest path connecting every pair of nodes in the alignment graph (rather than just those connecting the source and sink, as in the Global Alignment Problem), and then to select the path having maximum weight over all these longest paths.

STOP and Think: What is the runtime of this approach?

STOP

Free taxi rides in the alignment graph

For a faster local alignment approach, imagine a "free taxi ride" from the source $(0,0)$ to the node representing the start node of the conserved (red) interval in Figure 5.19. Imagine also a free taxi ride from the end node of the conserved interval to the sink. If such rides were available (Figure 5.20), then you could reach the starting node of the conserved interval for free, instead of incurring heavy penalties as in global alignment. Then, you could travel along the conserved interval to its end node, accumulating many positive match scores while encountering few mismatch and deletion penalties. Finally, you could take another free ride from the end node of the conserved interval to the sink. The resulting score of this ride is equal to the alignment score of only the conserved intervals, as desired.

The only problem with the "free rides" idea is that we don't know in advance where the local alignment (i.e., the conserved red interval in Figure 5.19) starts and ends! Yet we can provide free rides from the source to *all* nodes and from *all* nodes to the sink.

Connecting the source $(0,0)$ to every other node by adding a zero-weight edge and connecting every node to the sink (n,m) by a zero-weight edge will result in a DAG perfectly suited for solving the Local Alignment Problem (Figure 5.21). Because of free taxi rides, we no longer need to construct a longest path between every pair of nodes in the graph — the longest path from source to sink yields an optimal local alignment!

The total number of edges in the graph in Figure 5.21 is $\mathcal{O}(|v| \cdot |w|)$, which is still small. Since the runtime of finding a longest path in a DAG is defined by the number

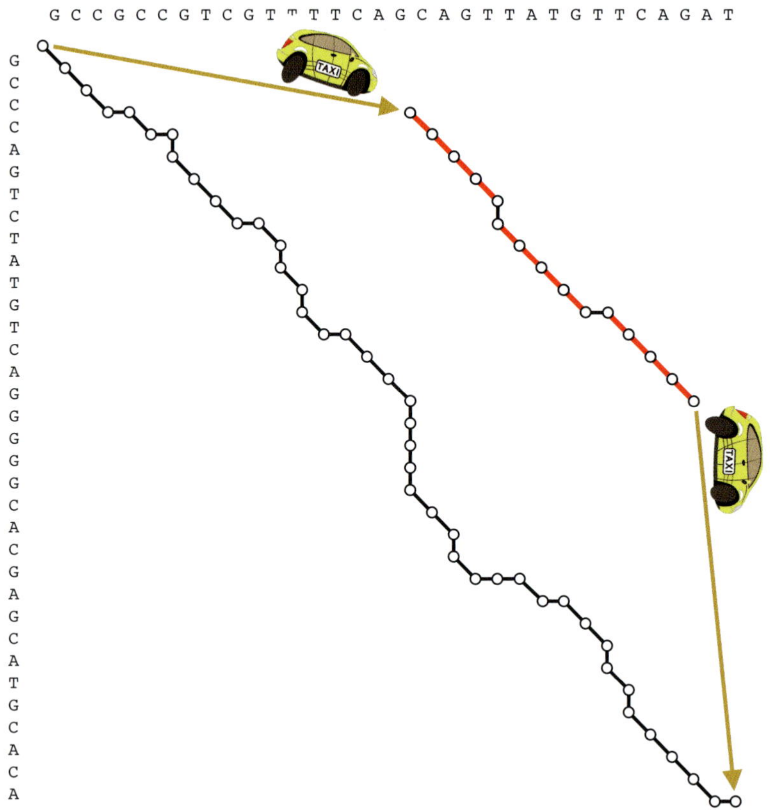

FIGURE 5.20 Modifying Figure 5.19 by adding "free taxi ride" edges (having weight 0) connecting the source to the start node of the conserved interval and connecting the end node of the conserved interval to the sink. These new edges allow us to score only the local alignment containing the conserved interval.

of edges in the graph, the resulting local alignment algorithm will be fast. As for computing the values $s_{i,j}$, adding zero-weight edges from $(0,0)$ to every node has made the source node $(0,0)$ a predecessor of every node (i,j). Therefore, there are now four edges entering (i,j), which adds only one new term to the longest path recurrence relation:

$$s_{i,j} = \max \begin{cases} 0 \\ s_{i-1,j} \quad + Score(v_i, -) \\ s_{i,j-1} \quad + Score(-, w_j) \\ s_{i-1,j-1} + Score(v_i, w_j) \end{cases}$$

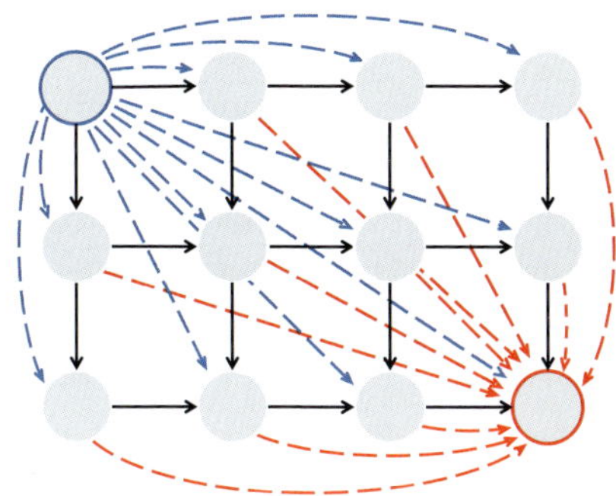

FIGURE 5.21 The local alignment algorithm introduces zero-weight edges (shown by blue dashed lines) connecting the source $(0, 0)$ to every other node in the alignment graph, as well as zero-weight edges (shown by red dashed lines) connecting every node to the sink node.

The recurrence above incorporates free rides from *source* $= (0, 0)$, but it does not incorporate free rides into *sink* $= (n, m)$. Since *sink* has every other node as a predecessor, $s_{n,m}$ is equal to the largest value of $s_{i,j}$ over the entire alignment graph,

$$s_{n,m} = \max_{0 \leq i \leq n, 0 \leq j \leq n} s_{i,j}.$$

STOP and Think: After computing all values $s_{i,j}$, how can you find where the path corresponding to the best local alignment starts and ends in the alignment graph?

You might still be wondering why we are allowed to free taxi rides through the alignment graph. The point is that you are in charge of designing whatever Manhattan-like DAG you like, as long as it adequately models the specific alignment problem at hand. Transformations like free taxi rides will become a common theme in this chapter. Various alignment problems can be solved by constructing an appropriate DAG with as few edges as possible (to minimize runtime), assigning edge weights to model the requirements of the problem, and then finding a longest path in this DAG.

The Changing Faces of Sequence Alignment

In this section, we will describe three sequence comparison problems and let you apply what you have already learned to solve them. Hint: the idea is to frame each problem as an instance of the Longest Path in a DAG Problem.

Edit distance

In 1966, Vladimir Levenshtein introduced the notion of the **edit distance** between two strings as the minimum number of **edit operations** needed to transform one string into another. Here, an edit operation is the insertion, deletion, or substitution of a single symbol. For example, TGCATACT can be transformed into ATCCGAT with six edit operations, implying that the edit distance between these strings is at most 6.

```
 TGCATACT
        ↓     delete last nucleotide
 TGCATAC
       ↓      delete last nucleotide
 TGCATA
      ↓       delete last nucleotide
 TGCAT
↓             insert A at the front
ATGCAT
  ↓           substitute G for C in the 3rd position
ATCCAT
    ↓         insert G after the 4th position
ATCCGAT
```

STOP and Think: Can you transform TGCATACT into ATCCGAT using fewer operations?

In fact, the edit distance between TGCATACT and ATCCGAT is 5:

```
 TGCATACT
↓             insert A at the front
ATGCATACT
     ↓        delete the 6th nucleotide
ATGCAACT
    ↓         substitute A for G in the 5th position
ATGCGACT
      ↓       delete the 7th nucleotide
ATGCGAT
  ↓           substitute G for C in the 3rd position
ATCCGAT
```

Levenshtein introduced edit distance but did not describe an algorithm for computing it, which we leave to you.

Edit Distance Problem:
Find the edit distance between two strings.

Input: Two strings.
Output: The edit distance between these strings.

Fitting alignment

Say that we wish to compare the approximately 20,000 amino acid-long NRP synthetase from *Bacillus brevis* with the approximately 600 amino acid-long A-domain from *Streptomyces roseosporus*, the bacterium that produces the powerful antibiotic Daptomycin. We hope to find a region within the longer protein sequence v that has high similarity with all of the shorter sequence w. Global alignment will not work because it tries to align all of v to all of w; local alignment will not work because it tries to align substrings of both v and w. Thus, we have a distinct alignment application called the **Fitting Alignment Problem**.

"Fitting" w to v requires finding a substring v' of v that maximizes the global alignment score between v' and w among all substrings of v. For example, the best global, local, and fitting alignments of v = CGTAGGCTTAAGGTTA and w = ATAGATA are shown in Figure 5.22 (with mismatch and indel penalties equal to 1).

Global	Local	Fitting
CGTAGGCTTAAGGTTA	CGTAGGCTTAAGGTTA	CGTAGGCTTAAGGTTA
A-TAG----A---T-A	ATAGATA	ATAGA--TA

FIGURE 5.22 Optimal global (left), local (middle), and fitting (right) alignments for the strings CGTAGGCTTAAGGTTA and ATAGATA, with match reward 1 and mismatch/indel penalties equal to 1. Black symbols do not contribute to the local and fitting alignments and are only shown to illustrate flanking regions. (Left) The optimal global alignment includes one mismatch and nine deletions. (Middle) The shared substring TAG corresponds to an optimal local alignment. (Right) A fitting alignment is most reasonable in this case, since it allows for the fact that all of ATAGATA aligns well with a region of CGTAGGCTTAAGGTTA.

Note in Figure 5.22 that the optimal local alignment (with score 3) is not a valid fitting alignment. On the other hand, the score of the optimal global alignment (**6** − **9** − **1** = −4) is smaller than that of the best fitting alignment (**5** − **2** − **2** = +1).

Fitting Alignment Problem:
Construct a highest-scoring fitting alignment between two strings.

Input: Strings v and w as well as a scoring matrix *Score*.
Output: A highest-scoring fitting alignment of v and w as defined by *Score*.

Overlap alignment

In Chapter 3, we discussed how to use overlapping reads to assemble a genome, a problem that was complicated by errors in reads. Aligning the ends of the hypothetical reads shown below offers a way to find overlaps between error-prone reads.

```
ATGCATGCCGG
     T-CC-GAAAC
```

An **overlap alignment** of strings $v = v_1 \ldots v_n$ and $w = w_1 \ldots w_m$ is a global alignment of a suffix of v with a prefix of w. An optimal overlap alignment of strings v and w maximizes the global alignment score between an i-suffix of v and a j-prefix of w (i.e., between $v_i \ldots v_n$ and $w_1 \ldots w_j$) among all possible i and j.

Overlap Alignment Problem:
Construct a highest-scoring overlap alignment between two strings.

Input: Two strings and a scoring matrix *Score*.
Output: A highest-scoring overlap alignment between the two strings as defined by *Score*.

Penalizing Insertions and Deletions in Sequence Alignment

Affine gap penalties

We have seen that introducing mismatch and indel penalties can produce more biologically adequate global alignments. However, even with this more robust scoring model, the A-domain alignment that we previously constructed (with indel penalty $\sigma = 4$) still reveals only six of the eight conserved purple columns corresponding to the non-ribosomal signatures:

```
YAFDLGYTCMFP-VLL-GGGELHIV-QKETYTAPDEI-AHYIKEHGITYI-KLTPSLFHTIVNTASFAFDANFE
-AFDVS-AGDFARALLTGG-QL-IVCPNEVKMDPASLYA-IIKKYDIT-IFEATPAL--VIPLME-YIYEQKLD

-S-LR-LIVLGGEKIIPIDVIAFRKM---YGHTE-FINHYGPTEATIGA
ISQLQILIV-GSDSC-SME--DFKTLVSRFGSTIRIVNSYGVTEACIDS
```

STOP and Think: Would increasing the indel penalty from $\sigma = 4$ to $\sigma = 10$ reveal the biologically correct alignment?

STOP

In our previously defined **linear scoring model**, if σ is the penalty for the insertion or deletion of a single symbol, then $\sigma \cdot k$ is the penalty for the insertion or deletion of an interval of k symbols. This cost model unfortunately results in inadequate scoring for biological sequences. Mutations are often caused by errors in DNA replication that insert or delete an entire interval of k nucleotides as a single event instead of as k independent insertions or deletions. Thus, penalizing such an indel by $\sigma \cdot k$ represents an excessive penalty. For example, the alignment on the right is more adequate than the alignment on the left, but they would currently receive the same score.

```
GATCCAG     GATCCAG
GA-C-AG     GA--CAG
```

A **gap** is a contiguous sequence of space symbols in a row of an alignment. One way to score gaps more appropriately is to define an **affine penalty** for a gap of length k as $\sigma + \epsilon \cdot (k-1)$, where σ is the **gap opening penalty**, assessed to the first symbol in the gap, and ϵ is the **gap extension penalty**, assessed to each additional symbol in the gap. We select ϵ to be smaller than σ so that the affine penalty for a gap of length k is smaller than the penalty for k independent single-nucleotide indels ($\sigma \cdot k$). For example, with affine gap penalties, the alignment on the left above is penalized by 2σ, whereas the alignment on the right is only penalized by $\sigma + \epsilon$.

Alignment with Affine Gap Penalties Problem:
Construct a highest-scoring global alignment of two strings (with affine gap penalties).

> **Input**: Two strings, a scoring matrix *Score*, and numbers σ and ϵ.
> **Output**: A highest scoring global alignment between these strings, as defined by *Score* and by the gap opening and extension penalties σ and ϵ.

STOP and Think: How would you modify the alignment graph to solve this problem?

Figure 5.23 illustrates how affine gap penalties can be modeled in the alignment graph by introducing a new "long" edge for each gap. Since we do not know in advance where gaps should be located, we need to add edges accounting for every possible gap. Thus, affine gap penalties can be accommodated by adding all possible vertical and horizontal edges in the alignment graph to represent all possible gaps. Specifically, we add edges connecting (i, j) to both $(i + k, j)$ and $(i, j + k)$ with weights $\sigma + \epsilon \cdot (k - 1)$ for all possible gap sizes k, as illustrated in Figure 5.24. For two sequences of length n, the number of edges in the resulting alignment graph modeling affine gap penalties increases from $\mathcal{O}(n^2)$ to $\mathcal{O}(n^3)$.

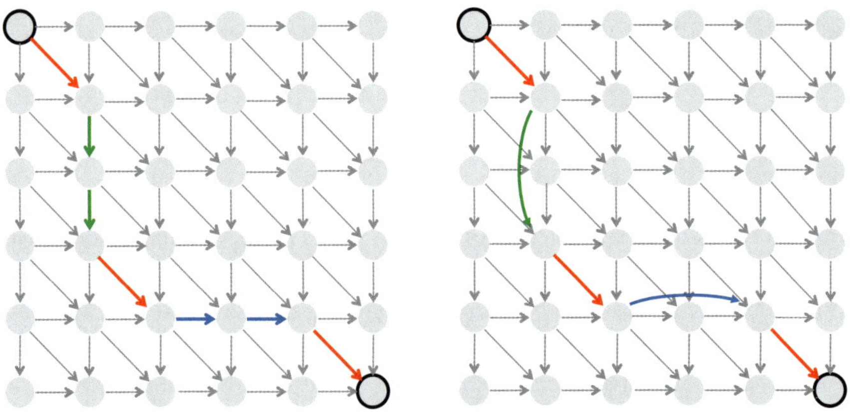

FIGURE 5.23 Representing gaps in the alignment graph on the left as "long" insertion and deletion edges in the alignment graph on the right. For a gap of length k, the weight of the corresponding long edge is equal to $\sigma + \varepsilon \cdot (k - 1)$.

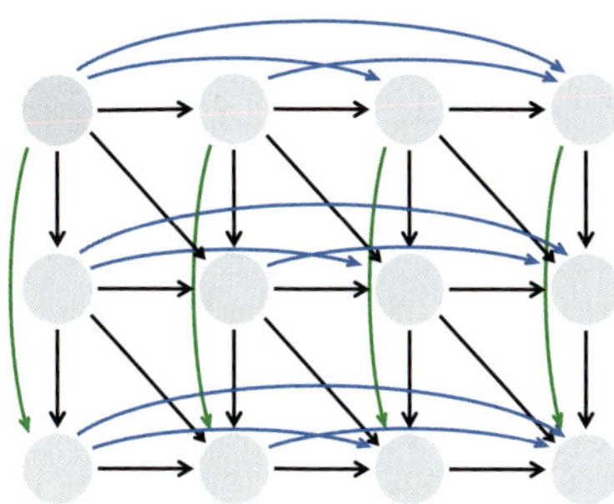

FIGURE 5.24 Adding edges corresponding to indels for all possible gap sizes adds a large number of edges to the alignment graph.

STOP and Think: Can you design a DAG with just $\mathcal{O}(n^2)$ edges to solve the Alignment with Affine Gap Penalties Problem?

Building Manhattan on three levels

The trick to decreasing the number of edges in the DAG for the Alignment with Affine Gap Penalties Problem is to *increase* the number of nodes. To this end, we will build an alignment graph on three levels; for each node (i, j), we will construct three different nodes: $(i, j)_{\text{lower}}$, $(i, j)_{\text{middle}}$, and $(i, j)_{\text{upper}}$. The middle level will contain diagonal edges of weight $Score(v_i, w_j)$ representing matches and mismatches. The lower level will have only vertical edges with weight $-\epsilon$ to represent gap extensions in v, and the upper level will have only horizontal edges with weights $-\epsilon$ to represent gap extensions in w (Figure 5.25).

Life in such a three-level city would be difficult because there is currently no way to move between different levels. To address this issue, we add edges responsible for opening and closing a gap. To model gap opening, we connect each node $(i, j)_{\text{middle}}$ to both $(i+1, j)_{\text{lower}}$ and $(i, j+1)_{\text{upper}}$; we then weight these edges with $-\sigma$. Closing gaps does not carry a penalty, and so we introduce zero-weight edges connecting nodes $(i, j)_{\text{lower}}$ and $(i, j)_{\text{upper}}$ with the corresponding node $(i, j)_{\text{middle}}$. As a result, a gap of length k starts and ends at the middle level and is charged $-\sigma$ for the first symbol, $-\epsilon$ for each subsequent symbol, and 0 to close the gap, producing a total penalty of $\sigma + \epsilon \cdot (k-1)$, as desired. Figure 5.26 illustrates how the path in Figure 5.23 traverses the three-level alignment graph.

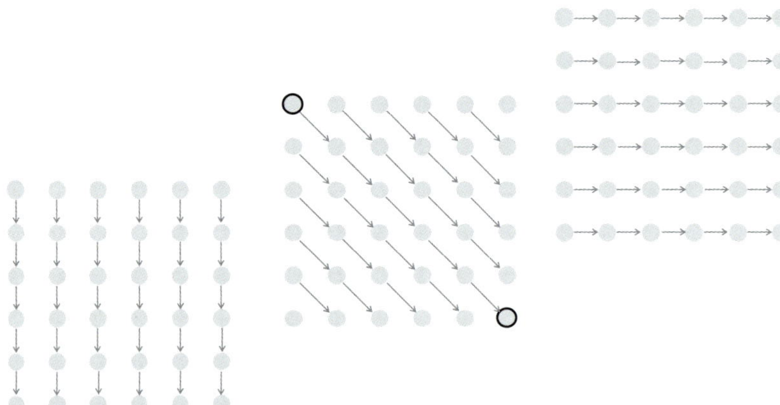

FIGURE 5.25 Building a three-level graph for alignment with affine gap penalties. The lower level corresponds to gap extensions in v, the middle level corresponds to matches and mismatches, and the upper level corresponds to gap extensions in w.

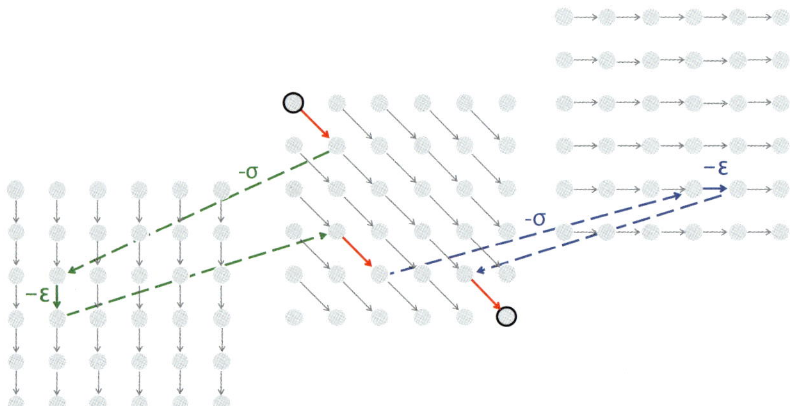

FIGURE 5.26 Every path from source to sink in the standard alignment graph shown in Figure 5.23 corresponds to a path from source to sink in the three-level graph of the same length (and vice-versa). Every node in the middle level has one outgoing (blue) edge to the upper level and one outgoing (green) edge to the lower level, both represented by dashed edges and having weight equal to the gap opening penalty. Every node in the middle level also has one incoming blue edge from the upper level and one incoming green edge from the lower level, both represented by dashed edges and having zero weight (these edges close a gap).

Exercise Break: Prove that the number of edges in the graph described in Figure 5.26 is at most $7 \cdot n \cdot m$ for sequences of length n and m.

The DAG in Figure 5.26 may be complicated, but it uses only $\mathcal{O}(n \cdot m)$ edges for sequences of length n and m, and a longest path in this graph still constructs an optimal alignment with affine gap penalties. The three-level alignment graph translates into the system of three recurrence relations shown below. Here, $lower_{i,j}$, $middle_{i,j}$, and $upper_{i,j}$ are the lengths of the longest paths from the source node to $(i,j)_{\text{lower}}$, $(i,j)_{\text{middle}}$, and $(i,j)_{\text{upper}}$, respectively.

$$lower_{i,j} = \max \begin{cases} lower_{i-1,j} & - \epsilon \\ middle_{i-1,j} & - \sigma \end{cases}$$

$$middle_{i,j} = \max \begin{cases} lower_{i,j} \\ middle_{i-1,j-1} + Score(v_i, w_j) \\ upper_{i,j} \end{cases}$$

$$upper_{i,j} = \max \begin{cases} upper_{i,j-1} & - \epsilon \\ middle_{i,j-1} & - \sigma \end{cases}$$

The variable $lower_{i,j}$ computes the score of an optimal alignment between the i-prefix of v and the j-prefix of w ending with a deletion (i.e., a vertical edge), whereas the variable $upper_{i,j}$ computes the score of an optimal alignment of these prefixes ending with an insertion (i.e., a horizontal edge), and the variable $middle_{i,j}$ computes the score of an optimal alignment ending with a match or mismatch. The first term in the recurrences for $lower_{i,j}$ and $upper_{i,j}$ corresponds to extending the gap, whereas the second term corresponds to initiating the gap.

STOP and Think: Compute an optimal alignment with affine gap penalties for the A-domains considered in the beginning of this section. How does varying the gap opening and extension penalties affect the quality of the alignment?

Exercise Break: Design an algorithm for computing optimal local (rather than global) alignment with affine gap penalties.

Space-Efficient Sequence Alignment

Computing alignment score using linear memory

To introduce fitting alignments, we used the example of aligning a 20,000 amino acid-long NRP synthetase from *Bacillus brevis* against a 600 amino acid-long A-domain from

Streptomyces roseosporus. However, you may not be able to construct this alignment on your computer because the memory required to store the dynamic programming matrix is substantial.

The runtime of the dynamic programming algorithm for aligning two strings of lengths n and m is proportional to the number of edges in their alignment graph, which is $\mathcal{O}(n \cdot m)$. The memory required by this algorithm is also $\mathcal{O}(n \cdot m)$, since we need to store the backtracking references. We will now demonstrate how to construct an alignment in just $\mathcal{O}(n)$ space at the expense of doubling the runtime, meaning that the runtime is still $\mathcal{O}(n \cdot m)$. In this section, for the sake of simplicity, we will consider the case of global alignment with non-affine gap penalties, i.e., when each symbol in a gap contributes the same score.

A **divide-and-conquer algorithm** often works when a solution to a large problem can be constructed from solutions of smaller problem instances. Such a strategy proceeds in two phases. The **divide** phase splits a problem instance into smaller instances and solves them; the **conquer** phase stitches the smaller solutions into a solution to the original problem (see **DETOUR: Divide-and-Conquer Algorithms**). PAGE 291

Before we proceed with the divide-and-conquer algorithm for linear-space alignment, note that if we only wish to compute the *score* of an alignment rather than the alignment itself, then the space required can easily be reduced to just twice the number of nodes in a single column of the alignment graph, or $\mathcal{O}(n)$. This reduction derives from the observation that the only values needed to compute the alignment scores in column j are the alignment scores in column $j - 1$. Therefore, the alignment scores in the columns before column $j - 1$ can be discarded when computing the alignment scores for column j, as illustrated in Figure 5.27. Unfortunately, finding the longest path requires us to store an entire matrix of backtracking pointers, which causes the quadratic space requirement. The idea of the space reduction technique is that we don't need to store any backtracking pointers if we are willing to spend a little more time. In fact, we will show how to construct an alignment without storing any backtracking pointers.

The Middle Node Problem

Given strings $v = v_1 \ldots v_n$ and $w = w_1 \ldots w_m$, define *middle* = $\lfloor m/2 \rfloor$. The **middle column** of ALIGNMENTGRAPH(v, w) is the column containing all nodes $(i, middle)$ for $0 \leq i \leq n$. A longest path from source to sink in the alignment graph must cross the middle column somewhere, and our first task is to figure out where using only $\mathcal{O}(n)$ memory. We refer to a node on this longest path belonging to the middle column as a

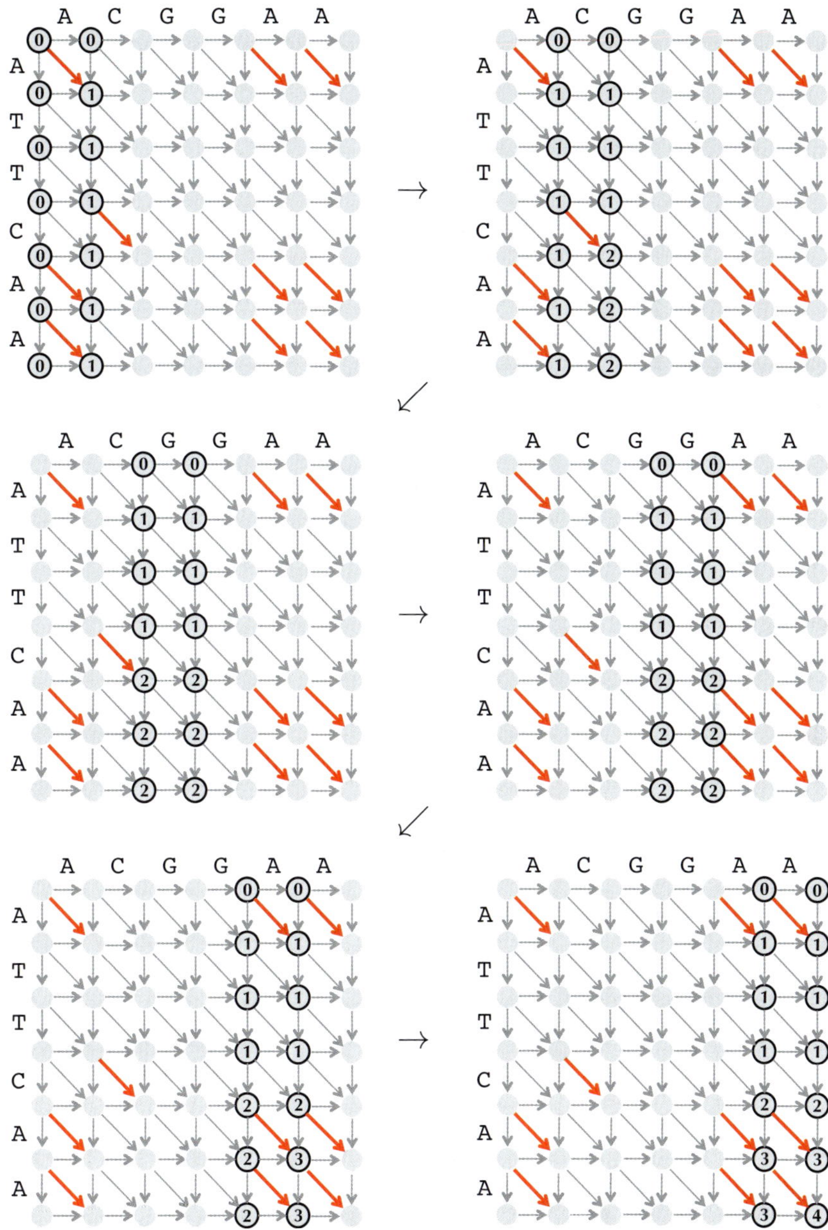

FIGURE 5.27 Computing an LCS alignment score by storing scores in just two columns of the alignment graph. As in Figure 5.7, all red edges have weight 1, and all other edges have weight 0.

middle node. (Note that different longest paths may have different middle nodes, and a given longest path may have more than one middle node.) In Figure 5.28 (top left), $middle = 3$, and the alignment path crosses the middle column at the (unique) middle node $(4, 3)$.

Exercise Break: Find all middle nodes for all longest paths when finding a longest common subsequence of ACCA and CAAC.

The key observation is that we can find a longest path's middle node without having to construct this path in the alignment graph. We will classify a path from the source to the sink as an ***i*-path** if it passes through the middle column at row i. For example, the highlighted path in Figure 5.28 is a 4-path because it passes through the middle column at row 4. For each i between 0 and n, we would like to find the length of a longest i-path (denoted LENGTH(i)) because the largest value of LENGTH(i) over all i will reveal a middle node.

Let FROMSOURCE(i) denote the length of the longest path from the source ending at $(i, middle)$ and TOSINK(i) denote the length of the longest path from $(i, middle)$ to the sink. Certainly,

$$\text{LENGTH}(i) = \text{FROMSOURCE}(i) + \text{TOSINK}(i),$$

and so we need to compute FROMSOURCE(i) and TOSINK(i) for each i.

STOP and Think: Can you compute FROMSOURCE(i) and TOSINK(i) in linear space? How much time would it take?

The value of FROMSOURCE(i) is simply $s_{i,middle}$, which we already know can be computed in linear space. Thus, the values of FROMSOURCE(i) for all i are stored as $s_{i,middle}$ in the middle column of the alignment graph, and computing them requires sweeping through half of the alignment graph from column 0 to the middle column. Since we need to explore roughly half the edges of the alignment graph to compute FROMSOURCE(i), we say that the runtime needed for computing all values FROMSOURCE(i) is proportional to half of the "area" of the alignment graph, or $n \cdot m/2$ (Figure 5.28 (top right)).

Computing TOSINK(i) is equivalent to finding the longest path from the sink to $(i, middle)$ if all the edge directions are reversed. Instead of reversing the edges, we can reverse the *strings* $v = v_1 \ldots v_n$ and $w = w_1 \ldots w_m$ and find $s_{n-i,m-middle}$ in the alignment graph for $v_n \ldots v_1$ and $w_m \ldots w_1$. Computing TOSINK(i) is therefore similar

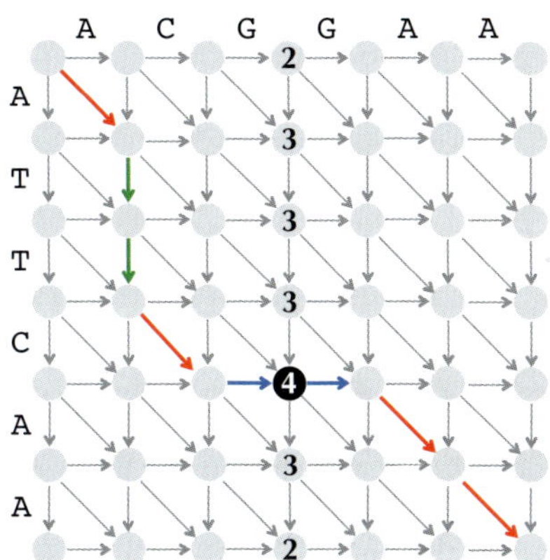

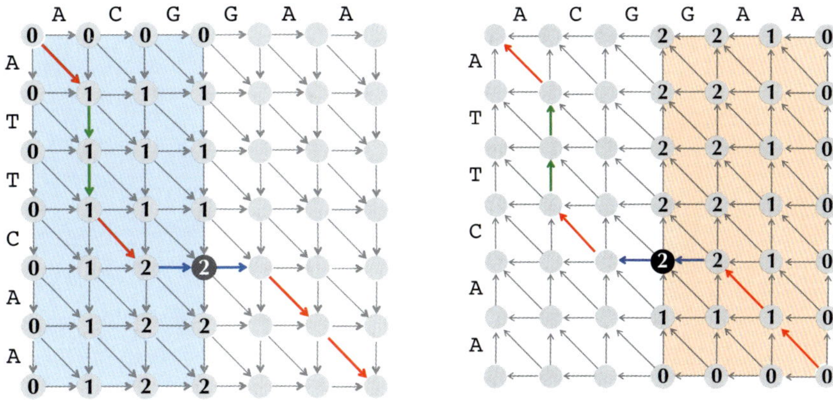

FIGURE 5.28 (Top) The alignment graph of ATTCAA and ACGGAA, with the optimal LCS alignment path highlighted. The number inside each node (i, *middle*) in the middle column is equal to LENGTH(i). The node maximizing LENGTH(i) is a middle node (multiple middle nodes may exist); the middle node in this graph is colored black. (Bottom left) Computing FROMSOURCE(i) for all i can be done in $\mathcal{O}(n)$ space and $\mathcal{O}(n \cdot m/2)$ time. (Bottom right) Computing TOSINK(i) for all i can also be done in $\mathcal{O}(n)$ space and $\mathcal{O}(n \cdot m/2)$ time; this requires reversing the direction of all edges and treating the sink as the source. The middle column in the upper matrix is equal to the sum of the middle columns in the two matrices on the bottom.

to computing FROMSOURCE(i) and can also be done in $\mathcal{O}(n)$ space and runtime proportional to $n \cdot m/2$, or half of the area of the alignment graph (Figure 5.28 (bottom)). In total, we can compute all values LENGTH(i) = FROMSOURCE(i) + TOSINK(i) in linear space with runtime proportional to $n \cdot m/2 + n \cdot m/2 = n \cdot m$, which is the total area of the alignment graph.

It looks like we have wasted a lot of time just to find a single node on the alignment path! You may think that this approach is doomed to fail because we have already spent $\mathcal{O}(n \cdot m)$ time (the entire area of the alignment graph) to gain very little information.

A surprisingly fast and memory-efficient alignment algorithm

Once we have found a middle node, we automatically know two rectangles through which a longest path must travel on either side of this node. As shown in Figure 5.29, one of these rectangles consists of all nodes above and to the left of the middle node, whereas the other rectangle consists of all nodes below and to the right of the middle node. Thus, the area of the two highlighted rectangles is half the total area of the alignment graph.

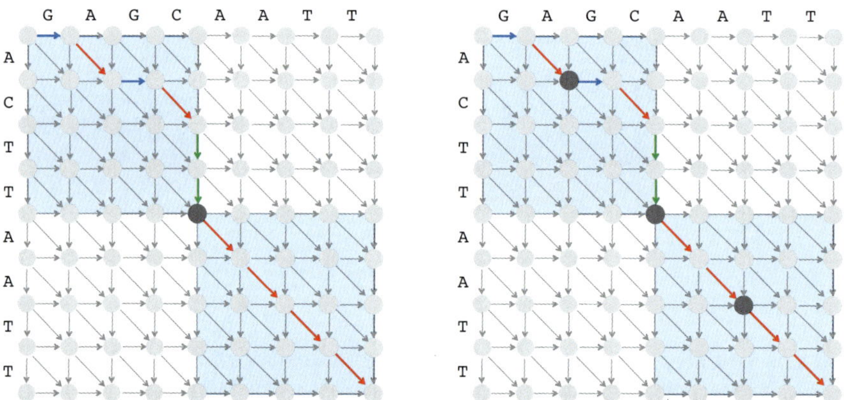

FIGURE 5.29 (Left) A middle node (shown in black) defines two highlighted rectangles and illustrates that an optimal path passing through this middle node must travel within these rectangles. We can therefore eliminate the remaining parts of the alignment graph from consideration for an optimal alignment path. (Right) Finding middle nodes (shown as two more black circles) within previously identified rectangles.

We can now *divide* the problem of finding the longest path from $(0,0)$ to (n,m) into two subproblems: finding a longest path from (0,0) to the middle node; and finding a longest path from the middle node to (n, m). The *conquer* step finds the two middle nodes within the smaller rectangles, which can be done in time proportional to the sum of the areas of these rectangles, or $n \cdot m/2$ (Figure 5.29 (right)). Note that we have now reconstructed three nodes of an optimal path. In the next iteration, we will divide and conquer to find four middle nodes in time equal to the sum of the areas of the even smaller blue rectangles, which have total area $n \cdot m/4$ (Figure 5.30). We have now reconstructed nearly all nodes of an optimal alignment path!

STOP and Think: How much time would it take to find all nodes on an optimal alignment path?

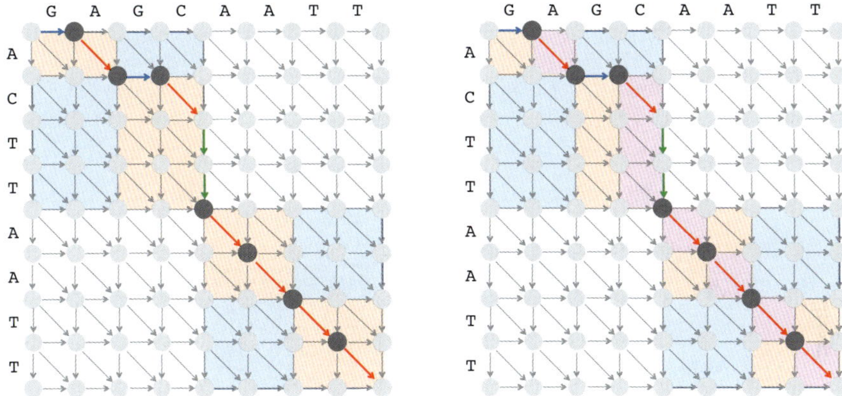

FIGURE 5.30 Finding middle nodes (highlighted as black circles) within previously identified blue rectangles.

In general, at each new step before the final step, we double the number of middle nodes found while halving the runtime required to find middle nodes. Proceeding in this way, we will find middle nodes of all rectangles (and thus construct the entire alignment!) in time equal to

$$n \cdot m + \frac{n \cdot m}{2} + \frac{n \cdot m}{4} + \cdots < 2 \cdot n \cdot m = \mathcal{O}(n \cdot m).$$

Thus, we have arrived at a quadric-time alignment algorithm that requires only linear space.

The Middle Edge Problem

Instead of asking you to implement a divide-and conquer algorithm based on finding a middle node, we will use a more elegant approach based on finding a **middle edge**, or an edge in an optimal alignment path starting at the middle node (more than one middle edge may exist for a given middle node). Once we find the middle edge, we again know two rectangles through which the longest path must travel on either side of the middle edge. But now these two rectangles take up even less than half of the area of the alignment graph (Figure 5.31), which is an advantage of selecting the middle edge instead of the middle node.

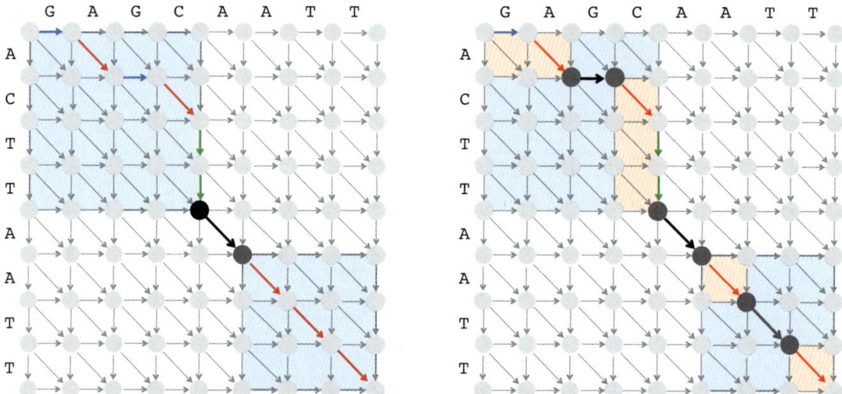

FIGURE 5.31 (Left) A middle edge (shown in bold) starts at the middle node (shown as a black circle). The optimal path travels inside the first highlighted rectangle, passes the middle edge, and travels inside the second highlighted rectangle afterwards. We can eliminate the remaining parts of the alignment graph, which takes up over half of the area formed by the graph, from further consideration. (Right) Finding middle edges (shown in bold) within previously identified rectangles.

Middle Edge in Linear Space Problem:
Find a middle edge in the alignment graph in linear space.

Input: Two strings and a scoring matrix *Score*.
Output: A middle edge in the alignment graph of these strings (where the edge lengths are defined by *Score*).

The pseudocode below for **LINEARSPACEALIGNMENT** describes how to recursively find a longest path in the alignment graph constructed for a substring $v_{top+1} \ldots v_{bottom}$ of v and a substring $w_{left+1} \ldots w_{right}$ of w. **LINEARSPACEALIGNMENT** calls the function MIDDLENODE($top, bottom, left, right$), which returns the coordinate i of the middle node (i, j) defined by the substrings $v_{top+1} \ldots v_{bottom}$ and $w_{left+1} \ldots w_{right}$. The algorithm also calls MIDDLEEDGE($v, w, top, bottom, left, right$), which returns "$\rightarrow$" , "$\downarrow$", or "$\searrow$" depending on whether the middle edge is horizontal, vertical, or diagonal. The alignment of strings v and w is constructed by calling **LINEARSPACEALIGNMENT**$(0, n, 0, m)$. The case $left = right$ describes the alignment of an empty string against the string $v_{top+1} \ldots v_{bottom}$, which is trivially computed as the score of a gap formed by $bottom - top$ vertical edges.

```
LINEARSPACEALIGNMENT(v, w, top, bottom, left, right)
    if left = right
        output path formed by bottom − top vertical edges
    if top = bottom
        output path formed by right − left horizontal edges
    middle ← ⌊(left + right) / 2⌋
    midEdge ← MIDDLEEDGE(v, w, top, bottom, left, right)
    midNode ← vertical coordinate of the initial node of midEdge
    LINEARSPACEALIGNMENT(v, w, top, midNode, left, middle)
    output midEdge
    if midEdge = "→" or midEdge = "↘"
        middle ← middle + 1
    if midEdge = "↓" or midEdge = "↘"
        midNode ← midNode + 1
    LINEARSPACEALIGNMENT(v, w, midNode, bottom, middle, right)
```

One straightforward way to implement **LINEARSPACEALIGNMENT** is to output each edge as a symbol `V` (vertical), `H` (horizontal), or `D` (diagonal). The resulting string constructed from the 3-symbol alphabet {`V`, `D`, `H`} defines a path through the alignment graph and allows us to reconstruct the alignment.

Exercise Break: Design a space-efficient algorithm for local sequence alignment.

Epilogue: Multiple Sequence Alignment

Building a three-dimensional Manhattan

Amino acid sequences of proteins performing the same function are likely to be somewhat similar, but these similarities may be elusive in the case of distant species. You now possess an arsenal of algorithms for aligning pairs of sequences, but if sequence similarity is weak, pairwise alignment may not identify biologically related sequences. However, simultaneous comparison of many sequences often allows us to find similarities that pairwise sequence comparison fails to reveal. Bioinformaticians sometimes say that while pairwise alignment whispers, multiple alignment shouts.

We are now ready to use pairwise sequence analysis to build up our intuition for comparison of multiple sequences. In our three-way alignment of A-domains from the introduction, we found 19 conserved columns:

```
YAFDLGYTCMFPVLLGGGELHIVQKETYTAPDEIAHYIKEHGITYIKLTPSLFHTIVNTA
-AFDVSAGDFARALLTGGQLIVCPNEVKMDPASLYAIIKKYDITIFEATPALVIPLMEYI
IAFDASSWEIYAPLLNGGTVVCIDYYTTIDIKALEAVFKQHHIRGAMLPPALLKQCLVSA

SFAFDANFESLRLIVLGGEKIIPIDVIAFRKMYGHTE-FINHYGPTEATIGA
-YEQKLDISQLQILIVGSDSCSMEDFKTLVSRFGSTIRIVNSYGVTEACIDS
----PTMISSLEILFAAGDRLSSQDAILARRAVGSGV-Y-NAYGPTENTVLS
```

However, similarities between A-domains are not limited to these 19 columns, as we can find $\mathbf{10} + \mathbf{9} + \mathbf{12} = 31$ semi-conservative columns, each of which has two matching amino acids:

```
YAFDLGYTCMFPVLLGGGELHIVQKETYTAPDEIAHYIKEHGITYIKLTPSLFHTIVNTA
-AFDVSAGDFARALLTGGQLIVCPNEVKMDPASLYAIIKKYDITIFEATPALVIPLMEYI
IAFDASSWEIYAPLLNGGTVVCIDYYTTIDIKALEAVFKQHHIRGAMLPPALLKQCLVSA

SFAFDANFESLRLIVLGGEKIIPIDVIAFRKMYGHTE-FINHYGPTEATIGA
-YEQKLDISQLQILIVGSDSCSMEDFKTLVSRFGSTIRIVNSYGVTEACIDS
----PTMISSLEILFAAGDRLSSQDAILARRAVGSGV-Y-NAYGPTENTVLS
```

A **multiple alignment** of t strings $v^1, \ldots, v^t$, also called a ***t*-way alignment**, is specified by a matrix having t rows, where the i-th row contains the symbols of v^i in order, interspersed with space symbols. We also assume that no column in a multiple alignment contains only space symbols. In the 3-way alignment below, we have highlighted the most popular symbol in each column using upper case letters:

```
A T - G T T a T A
A g C G a T C - A
A T C G T - C T c
```

```
0 1 2 2 3 4 5 6 7 8
0 1 2 3 4 5 6 7 7 8
0 1 2 3 4 5 5 6 7 8
```

The multiple alignment matrix is a generalization of the pairwise alignment matrix to more than two sequences. The three arrays shown below this alignment record the respective number of symbols in **ATGTTATA**, **AGCGATCA**, and **ATCGTCTC** encountered up to a given position (compare to Figure 5.5). Together, these three arrays correspond to a path in a three-dimensional grid:

$$(\mathbf{0}, \mathbf{0}, \mathbf{0}) \to (\mathbf{1}, \mathbf{1}, \mathbf{1}) \to (\mathbf{2}, \mathbf{2}, \mathbf{2}) \to (\mathbf{2}, \mathbf{3}, \mathbf{3}) \to (\mathbf{3}, \mathbf{4}, \mathbf{4}) \to (\mathbf{4}, \mathbf{5}, \mathbf{5}) \to (\mathbf{5}, \mathbf{6}, \mathbf{5}) \to (\mathbf{6}, \mathbf{7}, \mathbf{6}) \to (\mathbf{7}, \mathbf{7}, \mathbf{7}) \to (\mathbf{8}, \mathbf{8}, \mathbf{8})$$

As the alignment graph for two sequences is a grid of squares, the alignment graph for three sequences is a grid of cubes. Every node in the 3-way alignment graph has up to seven incoming edges, as shown in Figure 5.32.

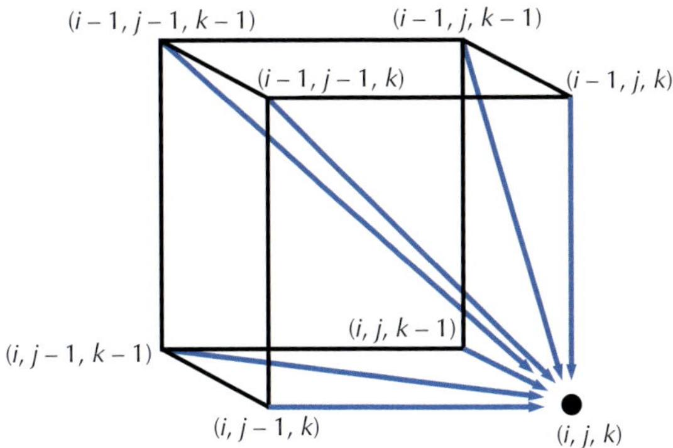

FIGURE 5.32 One cube making up the alignment graph for three sequences. Each node in the alignment graph for three sequences has up to seven incoming edges.

The score of a multiple alignment is defined as the sum of scores of the alignment columns (or, equivalently, weights of edges in the alignment path), with an optimal alignment being one that maximizes this score. In the case of an amino acid alphabet, we can use a very general scoring method that is defined by a t-dimensional matrix containing 21^t entries that describes the scores of all possible combinations of t symbols (representing 20 amino acids and the space symbol). See **DETOUR: Scoring Multiple Alignments**. Intuitively, we should reward more conserved columns with higher scores. For example, in the Multiple Longest Common Subsequence Problem, the score of a column is equal to 1 if all of the column's symbols are identical, and 0 if even one symbol disagrees.

PAGE 293

Multiple Alignment Problem:
Find the highest-scoring alignment between multiple strings under a given scoring matrix.

Input: A collection of t strings and a t-dimensional matrix *Score*.
Output: A multiple alignment of these strings whose score (as defined by the matrix *Score*) is maximized among all possible alignments of these strings.

A straightforward dynamic programming algorithm applied to the t-dimensional alignment graph solves the Multiple Alignment Problem for t strings. For three sequences v, w, and u, we define $s_{i,j,k}$ as the length of the longest path from the source $(0,0,0)$ to node (i,j,k) in the alignment graph. The recurrence for $s_{i,j,k}$ in the three-dimensional case is similar to the recurrence for pairwise alignment:

$$s_{i,j,k} = \max \begin{cases} s_{i-1,j,k} & + \, Score(v_i, -, -) \\ s_{i,j-1,k} & + \, Score(-, w_j, -) \\ s_{i,j,k-1} & + \, Score(-, -, u_k) \\ s_{i-1,j-1,k} & + \, Score(v_i, w_j, -) \\ s_{i-1,j,k-1} & + \, Score(v_i, -, u_k) \\ s_{i,j-1,k-1} & + \, Score(-, w_j, u_k) \\ s_{i-1,j-1,k-1} & + \, Score(v_i, w_j, u_k) \end{cases}$$

In the case of t sequences of length n, the alignment graph consists of approximately n^t nodes, and each node has up to $2^t - 1$ incoming edges, yielding a total runtime of $\mathcal{O}(n^t \cdot 2^t)$. As t grows, the dynamic programming algorithm becomes impractical. Many heuristics for suboptimal multiple alignments have been proposed to address this runtime bottleneck.

A greedy multiple alignment algorithm

Note that the multiple alignment

```
AT-GTTaTA
AgCGaTC-A
ATCGT-CTc
```

induces three pairwise alignments:

```
AT-GTTaTA    AT-GTTaTA    AgCGaTC-A
AgCGaTC-A    ATCGT-CTc    ATCGT-CTc
```

But can we work in the opposite direction, combining optimal pairwise alignments into a multiple alignment?

STOP and Think:

1. Does an optimal multiple alignment induce optimal pairwise alignments?

2. Try combining the pairwise alignments below into a multiple alignment of the strings CCCCTTTT, TTTTGGGG, and GGGGCCCC.

```
CCCCTTTT----    ----CCCCTTTT    TTTTGGGG----
----TTTTGGGG    GGGGCCCC----    ----GGGGCCCC
```

Unfortunately, we cannot always combine optimal pairwise alignments into a multiple alignment because some pairwise alignments may be incompatible. Indeed, the first pairwise alignment in the above question implies that CCCC occurs before TTTT in the multiple alignment constructed from these three pairwise alignments. The third pairwise alignment implies that TTTT occurs before GGGG in the multiple alignment. But the second pairwise alignment implies that GGGG occurs before CCCC in the multiple alignment. Thus, CCCC must occur before TTTT, which must occur before GGGG, which must occur before CCCC, a contradiction.

To avoid incompatibility, some multiple alignment algorithms attempt to greedily construct a multiple alignment from pairwise alignments that are not necessarily optimal. The greedy heuristic starts by selecting the two strings having the highest scoring pairwise alignment (among all possible pairs of strings) and then uses this pairwise alignment as a building block for iteratively adding one string at a time to the growing multiple alignment. We align the two closest strings at the first step because they often provide the best chance of building a reliable multiple alignment. For the same reason, we then select the string having maximum score against the current alignment at each stage. But what does it mean to align a string against an alignment of other strings?

An alignment of nucleotide sequences with k columns can be represented as a $4 \times k$ profile matrix like the one in Figure 5.33, which holds the nucleotide frequencies from each column (amino acid alignments are represented by $20 \times k$ profile matrices). The greedy multiple alignment heuristic adds a string to the current alignment by constructing a pairwise alignment between the string and the profile of the current alignment. As a result, the problem of constructing a multiple alignment of t sequences is reduced to constructing $t - 1$ pairwise alignments.

Alignment	T	C	G	G	G	-	g	T	T	T	t	t
	c	C	-	-	t	G	A	c	T	T	a	C
	a	C	G	-	G	G	A	T	T	T	t	C
	T	t	G	G	G	-	A	c	T	T	t	t
	a	-	-	-	G	-	-	-	T	-	C	-
	T	t	G	G	G	G	A	c	T	T	C	C
	T	C	G	-	-	G	A	T	T	c	a	t
	-	-	-	G	G	G	A	T	T	c	C	-
	T	a	G	G	G	G	A	a	c	-	-	C
	T	C	G	G	G	t	A	T	a	a	C	C
Profile A:	.2	.1	0	0	0	0	**.8**	.1	.1	.1	.2	0
C:	.1	**.5**	0	0	0	0	0	.3	.1	.2	**.4**	**.5**
G:	0	0	**.7**	**.6**	**.8**	**.6**	.1	0	0	0	0	0
T:	**.6**	.2	0	0	.1	.1	0	**.5**	**.8**	**.6**	.2	.3

FIGURE 5.33 A profile matrix for a multiple alignment of ten sequences. Each column of the profile matrix sums to 1 minus the frequency of the space symbol. The most popular nucleotides in each column are shown as uppercase colored letters.

STOP and Think: Design an algorithm for aligning a string against a profile matrix. How would you score the columns in such an alignment?

Although greedy multiple alignment algorithms work well for similar sequences, their performance deteriorates for dissimilar sequences because greedy methods may be misled by a spurious pairwise alignment. If the first two sequences picked for building a multiple alignment are aligned in a way that is incompatible with the optimal multiple alignment, then the error in this initial pairwise alignment will propagate all the way through to the final multiple alignment.

After learning how to align multiple sequences, you are now ready to solve a challenge problem that is comparable to the task that Marahiel and his collaborators faced in 1999.

Challenge Problem: In 1999, Marahiel derived the non-ribosomal code for only 14 of the 20 proteinogenic amino acids (`Ala`, `Asn`, `Asp`, `Cys`, `Gln`, `Glu`, `Ile`, `Leu`, `Phe`, `Pro`, `Ser`, `Thr`, `Tyr`, `Val`) because he lacked A-domain sequences for the remaining 6 amino acids. With the availability of many more A-domains, you now have the chance to fill in the gaps in Marahiel's original paper. Construct the multiple alignment of 397 A-domains, reveal the conservative columns in this alignment, and make your best guess about the signatures encoding all 20 proteinogenic amino acids based on Marahiel's alignment.

Detours

Fireflies and the non-ribosomal code

When Marahiel started his groundbreaking work on decoding the non-ribosomal code in the late 1990s, 160 A-domains had already been sequenced, and the amino acids that they encode had been experimentally identified. However, it was still not clear exactly how each A-domain encoded a specific amino acid.

We showed an alignment of three A-domain intervals in the main chapter, but Marahiel actually aligned all of the 160 identified A-domains in order to reveal the conservative core. Yet it still remained unclear which columns in this alignment defined the non-ribosomal signatures.

Help came from an unusual ally: *Photinus pyralis*, the most common species of firefly. Fireflies produce an enzyme called **luciferase** that helps them emit light to attract mates at night. Different species have different glow patterns, and females respond to males from the same species by identifying the color, duration, and intensity of their flash.

What do fireflies have to do with the non-ribosomal code? Firefly luciferase belongs to a class of enzymes called **adenylate-forming enzymes**, which share similarities with adenylation domains. For this reason, when Peter Brick published the three-dimensional structure of firefly luciferase in 1996, Marahiel was quick to take notice. In 1997, he and Brick joined forces and used firefly luciferase as a scaffold to reconstruct the first three-dimensional structure of an A-domain, which coded for phenylalanine (`Phe`); it is worth noting that this A-domain belonged to an NRP synthetase encoding *Gramicidin Soviet*, the first mass-produced antibiotic.

Marahiel and Brick actually constructed the three-dimensional structure of a larger complex containing both the A-domain and `Phe`. This three-dimensional structure provided information about amino acid residues in the A-domain located close to `Phe`, a hypothetical **active pocket** of the A-domain. Marahiel further demonstrated experimentally and computationally that amino acids in this active pocket define the non-ribosomal code, thus producing the 8 purple columns in the 3-way alignment that we showed at the beginning of the chapter.

Figure 5.34 shows the partial non-ribosomal code deduced by Marahiel. Although he was able to deduce some signatures, the non-ribosomal code is very **redundant**, meaning that multiple mutated variants of a signature all code for the same amino acid. For some amino acids, this redundancy is pronounced, e.g., Marahiel identified three very different signatures `AWMFAAVL`, `AFWIGGTF`, and `FESTAAVY` coding for `Val`.

Amino acid	Signature	Amino acid	Signature
Ala	LLFGIAVL	Leu	AFMLGMVF
Asn	LTKLGEVG	Orn	MENLGLIN
Asp	LTKVGHIG	Orn	VGEIGSID
Cys	HESDVGIT	Phe	AWTIAAVC
Cys	LYNLSLIW	Pro	VQLIAHVV
Gln	AQDLGVVD	Ser	VWHLSLID
Glu	AWHFGGVD	Thr	FWNIGMVH
Glu	AKDLGVVD	Tyr	GTITAEVA
Ile	GFFLGVVY	Tyr	ALVTGAVV
Ile	AFFYGITF	Tyr	ASTVAAVC
Leu	AWFLGNVV	Val	AFWIGGTF
Leu	AWLYGAVM	Val	FESTAAVY
Leu	GAYTGEVV	Val	AWMFAAVL

FIGURE 5.34 Marahiel's partial non-ribosomal code. Some proteinogenic amino acids are missing from this table because they were not present in Marahiel's dataset.

Finding a longest common subsequence without building a city

Define the ***i*-prefix** of a string as the substring formed by its first i letters. Given strings v and w, let $LCS_{i,j}$ denote an LCS between the i-prefix of v and the j-prefix of w, and let $s_{i,j}$ be the length of $LCS_{i,j}$. By definition, $s_{i,0} = s_{0,j} = 0$ for all values of i and j. Next, $LCS_{i,j}$ could contain both v_i and w_j, in which case these symbols match, and $LCS_{i,j}$ extends an LCS of the shorter prefixes $v_1 \ldots v_{i-1}$ and $w_1 \ldots w_{j-1}$. Otherwise, either v_i or w_j is not present in $LCS_{i,j}$. If v_i is not present in $LCS_{i,j}$, then this LCS is therefore also an LCS of $v_1 \ldots v_{i-1}$ and $w_1 \ldots w_j$. A similar argument applies to the case when w_j is not present in $LCS_{i,j}$. Therefore, $s_{i,j}$ satisfies the following recurrence, the same one we derived in the main text using a Manhattan-like grid:

$$s_{i,j} = \max \begin{cases} s_{i-1,j} \\ s_{i,j-1} \\ s_{i-1,j-1} + 1, \text{ if } v_i = w_j \end{cases}$$

Constructing a topological ordering

The first applications of topological ordering resulted from large management projects in an attempt to schedule a sequence of tasks based on their dependencies (such as the Dressing Challenge). In these projects, tasks are represented by nodes, and an edge connects node a to node b if task a must be completed before task b can be started.

STOP and Think: Prove that every DAG has a node with no incoming edges and a node with no outgoing edges.

The following algorithm for constructing a topological ordering is based on the observation that every DAG has at least one node with no incoming edges. We will label one of these nodes as v_1 and then remove this node from the graph along with all its outgoing edges. The resulting graph is also a DAG, which in turn must have a node with no incoming edges; we label this node v_2, and again remove it from the graph along with its outgoing edges. The resulting algorithm proceeds until all nodes have been removed, producing a topological order $v_1, \ldots, v_n$. This algorithm runs in time proportional to the number of edges in the input DAG.

```
TopologicalOrdering(Graph)
    List ← empty list
    Candidates ← set of all nodes in Graph with no incoming edges
    while Candidates is non-empty
        select an arbitrary node a from Candidates
        add a to the end of List and remove it from Candidates
        for each outgoing edge from a to another node b
            remove edge (a, b) from Graph
            if b has no incoming edges
                add b to Candidates
    if Graph has edges that have not been removed
        return "the input graph is not a DAG"
    else
        return List
```

PAM scoring matrices

Mutations of a gene's nucleotide sequence often change the amino acid sequence of the translated protein. Some of these mutations impair the protein's ability to function, making them rare events in molecular evolution. Asn, Asp, Glu, and Ser are the most "mutable" amino acids, whereas Cys and Trp are the least mutable. Knowledge of the likelihood of each possible mutation allows biologists to construct amino acid scoring matrices for biologically sound sequence alignments in which different substitutions are penalized differently. The (i, j)-th entry of the amino acid scoring matrix *Score* usually reflects how often the i-th amino acid substitutes the j-th amino acid in alignments of related protein sequences. As a result, optimal alignments of amino acid sequences may have very few matches but still represent biologically adequate alignments.

How do biologists know which mutations are more likely than others? If we know a large set of pairwise alignments of *related sequences* (e.g., sharing at least 90% of amino acids), then computing $Score(i, j)$ is based on counting how many times the corresponding amino acids are aligned. However, we need to know the scoring matrix in advance in order to build this set of starter alignments — a catch-22!

Fortunately, the correct alignment of very similar sequences is so obvious that it can be constructed even with a primitive scoring scheme that does not account for varying mutation propensities (such as +1 for matches and -1 for mismatches and indels), thus resolving the conundrum. After constructing these obvious alignments, we can use them to compute a new scoring matrix that we can use iteratively to form less and less obvious alignments.

This simplified description hides some details. For example, the probability of `Ser` mutating into `Phe` in species that diverged 1 million years ago is smaller than the probability of the same mutation in species that diverged 100 million years ago. This observation implies that scoring matrices for protein comparison should depend on the similarity of the organisms and the speed of evolution of the proteins of interest. In practice, the proteins that biologists use to create an initial alignment are extremely similar, having 99% of their amino acids conserved (e.g., most proteins shared by humans and chimpanzees). Sequences that are 99% similar are said to be 1 **PAM unit** diverged ("PAM" stands for "point accepted mutation"). You can think of a PAM unit as the amount of time in which an "average" protein mutates 1% of its amino acids.

The **PAM$_1$ scoring matrix** is defined as follows from many pairwise alignments of 99% similar proteins. Given a set of pairwise alignments, let $M(i, j)$ be the number of times that the i-th and j-th amino acids appear in the same column, divided by the total number of times that the i-th amino acid appears in all sequences. Let $f(j)$ be the frequency of the j-th amino acid in the sequences, or the number of times it appears across all sequences divided by the combined lengths of the two sequences. The (i, j)-th entry of the PAM$_1$ matrix is defined as

$$\log\left(\frac{M(i,j)}{f(j)}\right).$$

For a larger number of PAM units n, the PAM$_n$ matrix is computed based on the observation that the matrix M^n (the result of multiplying M by itself n times) holds the empirical probabilities that one amino acid mutates to another during n PAM units. The (i, j)-th entry of the PAM_n scoring matrix is thus given by

$$\log\left(\frac{M^n(i,j)}{f(j)}\right).$$

The PAM_{250} scoring matrix is shown in Figure 5.35.

This approach assumes that the frequencies of the amino acids $f(j)$ remain constant over time, and that the mutational processes in an interval of 1 PAM unit operate consistently over long periods. For large n, the resulting PAM matrices often allow us to find related proteins, even when the alignment has few matches.

	A	C	D	E	F	G	H	I	K	L	M	N	P	Q	R	S	T	V	W	Y	-
A	2	-2	0	0	-3	1	-1	-1	-1	-2	-1	0	1	0	-2	1	1	0	-6	-3	-8
C	-2	12	-5	-5	-4	-3	-3	-2	-5	-6	-5	-4	-3	-5	-4	0	-2	-2	-8	0	-8
D	0	-5	4	3	-6	1	1	-2	0	-4	-3	2	-1	2	-1	0	0	-2	-7	-4	-8
E	0	-5	3	4	-5	0	1	-2	0	-3	-2	1	-1	2	-1	0	0	-2	-7	-4	-8
F	-3	-4	-6	-5	9	-5	-2	1	-5	2	0	-3	-5	-5	-4	-3	-3	-1	0	7	-8
G	1	-3	1	0	-5	5	-2	-3	-2	-4	-3	0	0	-1	-3	1	0	-1	-7	-5	-8
H	-1	-3	1	1	-2	-2	6	-2	0	-2	-2	2	0	3	2	-1	-1	-2	-3	0	-8
I	-1	-2	-2	-2	1	-3	-2	5	-2	2	2	-2	-2	-2	-2	-1	0	4	-5	-1	-8
K	-1	-5	0	0	-5	-2	0	-2	5	-3	0	1	-1	1	3	0	0	-2	-3	-4	-8
L	-2	-6	-4	-3	2	-4	-2	2	-3	6	4	-3	-3	-2	-3	-3	-2	2	-2	-1	-8
M	-1	-5	-3	-2	0	-3	-2	2	0	4	6	-2	-2	-1	0	-2	-1	2	-4	-2	-8
N	0	-4	2	1	-3	0	2	-2	1	-3	-2	2	0	1	0	1	0	-2	-4	-2	-8
P	1	-3	-1	-1	-5	0	0	-2	-1	-3	-2	0	6	0	0	1	0	-1	-6	-5	-8
Q	0	-5	2	2	-5	-1	3	-2	1	-2	-1	1	0	4	1	-1	-1	-2	-5	-4	-8
R	-2	-4	-1	-1	-4	-3	2	-2	3	-3	0	0	0	1	6	0	-1	-2	2	-4	-8
S	1	0	0	0	-3	1	-1	-1	0	-3	-2	1	1	-1	0	2	1	-1	-2	-3	-8
T	1	-2	0	0	-3	0	-1	0	0	-2	-1	0	0	-1	-1	1	3	0	-5	-3	-8
V	0	-2	-2	-2	-1	-1	-2	4	-2	2	2	-2	-1	-2	-2	-1	0	4	-6	-2	-8
W	-6	-8	-7	-7	0	-7	-3	-5	-3	-2	-4	-4	-6	-5	2	-2	-5	-6	17	0	-8
Y	-3	0	-4	-4	7	-5	0	-1	-4	-1	-2	-2	-5	-4	-4	-3	-3	-2	0	10	-8
-	-8	-8	-8	-8	-8	-8	-8	-8	-8	-8	-8	-8	-8	-8	-8	-8	-8	-8	-8	-8	

FIGURE 5.35 The PAM_{250} scoring matrix for protein alignment with indel penalty 8.

Divide-and-conquer algorithms

We will use the problem of sorting a list of integers as an example of a divide-and-conquer algorithm. We begin from the problem of **merging**, in which we want to combine two sorted lists $List_1$ and $List_2$ into a single sorted list (Figure 5.36). The **MERGE** algorithm combines two sorted lists into a single sorted list in $\mathcal{O}(|List_1| + |List_2|)$ time by iteratively choosing the smallest remaining element in $List_1$ and $List_2$ and moving it to the growing sorted list.

$List_1$	2 5 7 8	2 5 7 8	2 5 7 8	2 5 7 8	2 5 7 8	2 5 7 8
$List_2$	3 4 6	3 4 6	3 4 6	3 4 6	3 4 6	3 4 6
SortedList	2	3	4	5	6	7 8

FIGURE 5.36 Merging the sorted lists $(2, 5, 7, 8)$ and $(3, 4, 6)$ results in the sorted list $(2, 3, 4, 5, 6, 7, 8)$.

```
MERGE(List1, List2)
    SortedList ← empty list
    while both List1 and List2 are non-empty
        if the smallest element in List1 is smaller than the smallest element in List2
            move the smallest element from List1 to the end of SortedList
        else
            move the smallest element from List2 to the end of SortedList
    move any remaining elements from either List1 or List2 to the end of SortedList
    return SortedList
```

MERGE would be useful for sorting an arbitrary list if we knew how to divide an arbitrary (unsorted) list into two already sorted half-sized lists. However, it may seem that we are back to where we started, except now we have to sort two smaller lists instead of one big one. Yet sorting two smaller lists is a preferable computational problem. To see why, let's consider the **MERGESORT** algorithm, which **divides** an unsorted list into two parts and then recursively **conquers** each smaller sorting problem before merging the sorted lists.

```
MERGESORT(List)
    if List consists of a single element
        return List
    FirstHalf ← first half of List
    SecondHalf ← second half of List
    SortedFirstHalf ← MERGESORT(FirstHalf)
    SortedSecondHalf ← MERGESORT(SecondHalf)
    SortedList ← MERGE(SortedFirstHalf, SortedSecondHalf)
    return SortedList
```

STOP and Think: What is the runtime of **MERGESORT**?

Figure 5.37 shows the **recursion tree** of **MERGESORT**, consisting of $\log_2 n$ levels, where n is the size of the original unsorted list. At the bottom level, we must merge two sorted lists of approximately $n/2$ elements each, requiring $\mathcal{O}(n/2 + n/2) = \mathcal{O}(n)$ time.

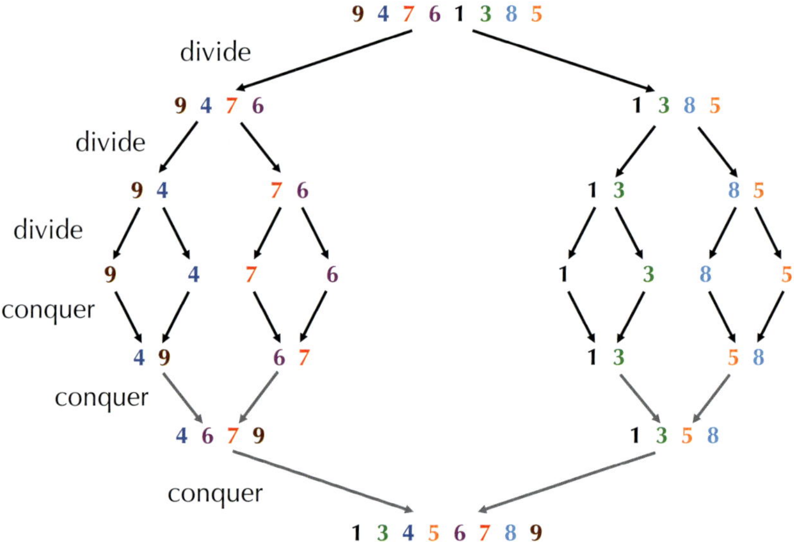

FIGURE 5.37 The recursion tree for sorting an 8-element list with **MERGESORT**. The *divide* (upper) steps consist of $\log_2 8 = 3$ levels, where the input list is split into smaller and smaller sublists. The *conquer* (lower) steps consist of the same number of levels, as the sorted sublists are merged back together.

At the next highest level, we must merge four lists of $n/4$ elements, each requiring $\mathcal{O}(n/4+n/4+n/4+n/4) = \mathcal{O}(n)$ time. This pattern can be generalized: the i-th level contains 2^i lists, each having approximately $n/2^i$ elements, and requires $\mathcal{O}(n)$ time to merge. Since there are $\log_2 n$ levels in the recursion tree, **MERGESORT** therefore requires $\mathcal{O}(n \cdot \log_2 n)$ runtime overall, which offers a speedup over naive $\mathcal{O}(n^2)$ sorting algorithms.

Scoring multiple alignments

The choice of scoring function can drastically affect the quality of a multiple alignment. In the main chapter, we described a way to score t-way alignments by using a t-dimensional scoring matrix. Below, we describe more practical approaches to scoring alignments.

The columns of a t-way alignment describe a path in a t-dimensional alignment graph whose edge weights are defined by the scoring function. Using the statistically motivated entropy score, the score of a multiple alignment is defined as the sum of the entropies of its columns. Recall from Chapter 2 that the entropy of a column is equal to $-\sum p_x \cdot \log_2 p_x$, where the sum is taken over all symbols x present in the column, and p_x is the frequency of symbol x in the column.

In Chapter 2, we saw that more highly conserved columns will have lower entropy scores. Because we wish to maximize the alignment score, we use the negative of entropy in order to ensure that more highly conserved columns receive higher scores. Finding a longest path in the t-dimensional alignment graph therefore corresponds to finding a multiple alignment with minimal entropy.

Another popular scoring approach is the **Sum-of-Pairs score** (**SP-score**). A multiple alignment *Alignment* of t sequences induces a pairwise alignment between the i-th and j-th sequences, having score $s(Alignment, i, j)$. The SP-score for a multiple alignment simply adds the scores of each induced pairwise alignment:

$$\text{SP-SCORE}(Alignment) = \sum_{1 \le i < j \le t} s(Alignment, i, j).$$

Exercise Break: Compute the entropy score and SP-score of Marahiel's 3-way alignment, reproduced below.

```
YAFDLGYTCMFPVLLGGGELHIVQKETYTAPDEIAHYIKEHGITYIKLTPSLFHTIVNTA
-AFDVSAGDFARALLTGGQLIVCPNEVKMDPASLYAIIKKYDITIFEATPALVIPLMEYI
IAFDASSWEIYAPLLNGGTVVCIDYYTTIDIKALEAVFKQHHIRGAMLPPALLKQCLVSA

SFAFDANFESLRLIVLGGEKIIPIDVIAFRKMYGHTE-FINHYGPTEATIGA
-YEQKLDISQLQILIVGSDSCSMEDFKTLVSRFGSTIRIVNSYGVTEACIDS
----PTMISSLEILFAAGDRLSSQDAILARRAVGSGV-Y-NAYGPTENTVLS
```

Bibliography Notes

Edit distance was introduced by Levenshtein, 1966. The local alignment algorithm described in the text was proposed by Smith and Waterman, 1981. When Doolittle et al., 1983 revealed the similarities between oncogenes and PDGF, they did not know about the Smith-Waterman algorithm. The 3-D structures of firefly luciferase and the A-domain from *Bacilus brevis* were published by Conti, Franks, and Brick, 1996 as well as Conti et al., 1997. The non-ribosomal code was first presented by Stachelhaus, Mootz, and Marahiel, 1999.

ARE THERE FRAGILE REGIONS IN THE HUMAN GENOME?

Combinatorial Algorithms

Of Mice and Men

> *"I have further been told," said the cat, "that you can also transform yourself into the smallest of animals, for example, a rat or a mouse. But I can scarcely believe that. I must admit to you that I think it would be quite impossible."*
>
> *"Impossible!" cried the ogre. "You shall see!"*
>
> *He immediately changed himself into a mouse and began to run about the floor. As soon as the cat saw this, he fell upon him and ate him up.*

How different are the human and mouse genomes?

When Charles Perrault described the transformation of an ogre into a mouse in "Puss in Boots", he could hardly have anticipated that three centuries later, research would show that the human and mouse genomes are surprisingly similar. Nearly every human gene has a mouse counterpart, although mice greatly outperform us when it comes to the olfactory genes responsible for smell. We are essentially mice without tails — we even have the genes needed to make a tail, but these genes have been "silenced" during our evolution.

We started with a fairy tale question: "How can an ogre transform into a mouse?" Since we share most of the same genes with mice, we now ask a question about mammalian evolution: "What evolutionary forces have transformed the genome of the human-mouse ancestor into the present-day human and mouse genomes?"

If a precocious child had grown out of reading fairy tales and wanted to learn about how the human and mouse genomes differ, then here is what we would tell her. You can cut the 23 human chromosomes into 280 pieces, shuffle these DNA fragments, and then glue the pieces together in a new order to form the 20 mouse chromosomes. The truth, however, is that evolution has not employed a single dramatic cut-and-paste operation; instead, it applies smaller changes known as **genome rearrangements**, which will be our focus in this chapter.

Unfortunately, our bioinformatics time machine won't take us more than a few centuries into the past. If it did, we could travel 75 million years back in time, watching humans slowly change into a small, furry animal that lived with dinosaurs. Then, we could travel back to the present, watching how this animal evolved into the mouse. In this chapter, we hope to understand the genome rearrangements that have separated the human and mouse genomes without having to revamp our time machine.

Synteny blocks

To simplify genome comparison, we will first focus on the X chromosome, which is one of the two sex-determining chromosomes in mammals and has retained nearly all its genes throughout mammalian evolution (see **DETOUR: Why is the Gene Content of Mammalian X Chromosomes So Conserved?**). We can therefore view the X chromosome as a "mini-genome" when comparing mice to humans, since this chromosome's genes have not jumped around onto different chromosomes (and vice-versa).

PAGE 346

It turns out not only that most human genes have mouse counterparts, but also that hundreds of similar genes often line up one after another in the same order in the two species genomes. Each of the eleven colored segments in Figure 6.1 represents such a procession of similar genes and is called a **synteny block**. Later, we will explain how to construct synteny blocks and what the left and right *directions* of the blocks signify.

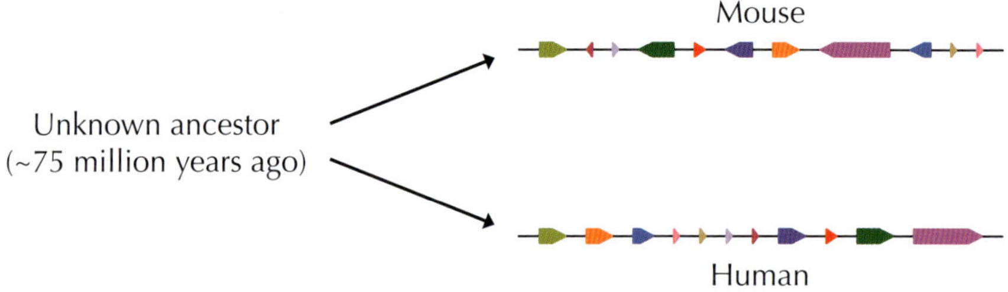

FIGURE 6.1 Mouse and human X chromosomes represented as eleven colored, directed segments (synteny blocks). The length of each synteny block is proportional to its length in the genome.

Synteny blocks simplify the comparison of the mouse and human X chromosomes from about 150 million base pairs to only eleven units. This simplification is analogous to comparing two similar photographs. If we compare the images one pixel at a time, we may be overwhelmed by the scale of the problem; instead, we need to zoom out in order to notice higher-level patterns. It is no accident that biologists use the term "resolution" to discuss the level at which genomes are analyzed.

Reversals

You have probably been wondering how the genome changes when it undergoes a genome rearrangement. Genome rearrangements were discovered 90 years ago when Alfred Sturtevant was studying fruit fly mutants with scarlet- and peach-colored eyes as well as abnormally shaped deltoid wings. Sturtevant analyzed the genes coding for these traits, called **scarlet**, **peach**, and **delta**, and he was amazed to find that the arrangement of these genes in *Drosophila melanogaster* (**scarlet**, **peach**, **delta**) differed from their arrangement in *Drosophila simulans* (**scarlet**, **delta**, **peach**). He immediately conjectured that the chromosomal segment containing **peach** and **delta** must have been flipped around (see **DETOUR: Discovery of Genome Rearrangements**). Sturtevant had witnessed the most common form of genome rearrangement, called a **reversal**, which flips around an interval of a chromosome and inverts the directions of any synteny blocks within the interval.

PAGE 346

Figure 6.2 shows a series of seven reversals transforming the mouse X chromosome into the human X chromosome. If this scenario is correct, then the X chromosome of the human-mouse ancestor must be represented by one of the intermediate synteny block orderings. Unfortunately, this series of seven reversals offers only one of 1,070 different seven-step scenarios transforming the mouse X chromosome into the human X chromosome. We have no clue which scenario is correct, or even whether the correct scenario had exactly seven reversals.

STOP and Think: Can you convert the mouse X chromosome into the human X chromosome using only six reversals?

Regardless of how many reversals separate the human and mouse X chromosomes, reversals must be rare genomic events. Indeed, genome rearrangements typically cause the death or sterility of the mutated organism, thus preventing it from passing the rearrangement on to the next generation. However, a tiny fraction of genome rearrangements may have a positive effect on survival and propagate through a species as the result of natural selection. When a population becomes isolated from the rest of its species for long enough, rearrangements can even create a new species.

Rearrangement hotspots

Geology provides a thought-provoking analogy for thinking about genome evolution. You might like to think of genome rearrangements as "genomic earthquakes" that dramatically change the chromosomal architecture of an organism. Genome rearrangements contrast with much more frequent point mutations, which work slowly and are analogous to "genomic erosion".

You can visualize a reversal as breaking the genome on both sides of a chromosomal interval, flipping the interval, and then gluing the resulting segments in a new order.

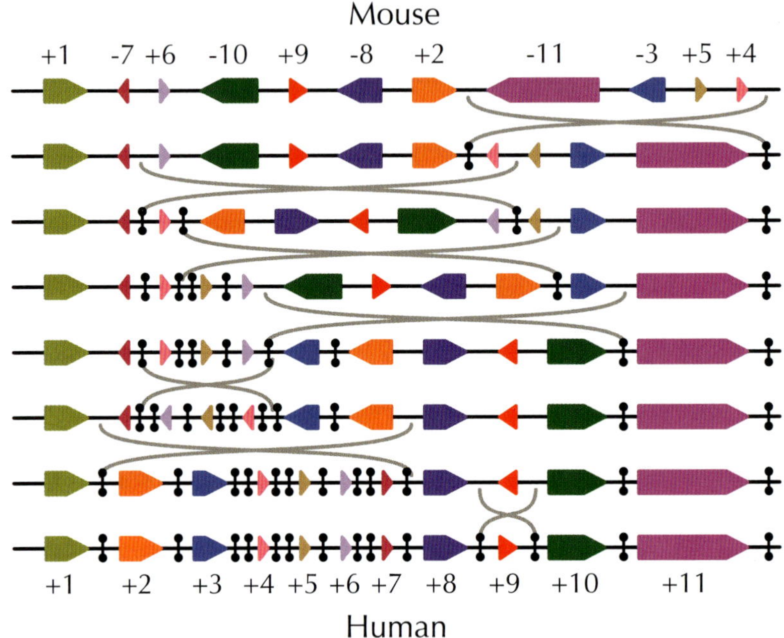

FIGURE 6.2 Transforming the mouse X chromosome into the human X chromosome with seven reversals. Each synteny block is uniquely colored and labeled with an integer between 1 and 11; the positive or negative sign of each integer indicates the synteny block's direction (pointing right or left, respectively). Two short vertical segments delineate the endpoints of the inverted interval in each reversal. Suppose that this evolutionary scenario is correct and that, say, the fifth synteny block arrangement from the top presents the true ancestral arrangement. Then the first four reversals happened on the evolutionary path from mice to the human-mouse common ancestor (traveling backward in time), and the final three reversals happened on the evolutionary path from the common ancestor to humans (traveling forward in time). In this chapter, we are not trying to reconstruct the ancestral genome and thus are not concerned with whether a certain reversal travels backward or forward in time.

Keeping in mind that earthquakes occur more frequently along fault lines, we wonder if a similar principle holds for reversals — are they occurring over and over again in the same genomic regions? A fundamental question in chromosome evolution studies is whether the **breakage points** of reversals (i.e., the ends of the inverted intervals) occur along "fault lines" called **rearrangement hotspots**. If such hotspots exist in the human genome, we want to locate them and determine how they might relate to genetic disorders, which are often attributable to rearrangements.

Of course, we should rigorously define what we mean by a "rearrangement hotspot". Re-examining the seven-reversal scenario changing the mouse X chromosome into the human X chromosome in Figure 6.2, we record the endpoints of each reversal using vertical segments. Regions affected by multiple reversals are indicated by multiple vertical segments in the human X chromosome. For example, the region adjacent to the pointed side of block 3 in Figure 6.2 is used as an endpoint of both the fourth and fifth reversals. As a result, we have placed two vertical lines between blocks 3 and 4 in the human X chromosome. However, just because we showed two breakage points in this region does not imply that this region is a rearrangement hotspot, since the reversals in Figure 6.2 represent just one possible evolutionary scenario. Because the true rearrangement scenario is unknown, it is not immediately clear how we could determine whether rearrangement hotspots exist.

The Random Breakage Model of Chromosome Evolution

In 1973, Susumu Ohno proposed the **Random Breakage Model** of chromosome evolution. This hypothesis states that the breakage points of rearrangements are selected randomly, implying that rearrangement hotspots in mammalian genomes do not exist. Yet Ohno's model lacked supporting evidence when it was introduced. After all, how could we possibly determine whether rearrangement hotspots exist without knowing the exact sequence of rearrangements separating two species?

STOP and Think: Consider the following questions.

1. Say that a series of random reversals result in one huge synteny block covering 90% of the genome in addition to 99 tiny synteny blocks covering the remaining 10% of the genome. Should we be surprised?
2. What if random reversals result in 100 synteny blocks of roughly the same length? Should we be surprised?

The idea that we wish to impress on you in the preceding questions is that we can test the Random Breakage Model by analyzing the distribution of synteny block lengths. For example, the lengths of the human-mouse synteny blocks on the X chromosome vary widely, with the largest block (block 11 in Figure 6.2) taking up nearly 25% of the entire length of the X chromosome. Is this variation in synteny block length consistent with the Random Breakage Model?

In 1984, Joseph Nadeau and Benjamin Taylor asked what the expected lengths of synteny blocks should be after N reversals occurring at random locations in the genome. If we rule out the unlikely event that two random reversals cut the chromosome in exactly the same position, then N random reversals cut the chromosome in $2N$ locations and produce $2N + 1$ synteny blocks. Figure 6.3 (top) depicts the result of a computational experiment in which 320 random reversals are applied to a simulated chromosome consisting of 25,000 genes, producing $2 \cdot 320 + 1 = 641$ synteny blocks.

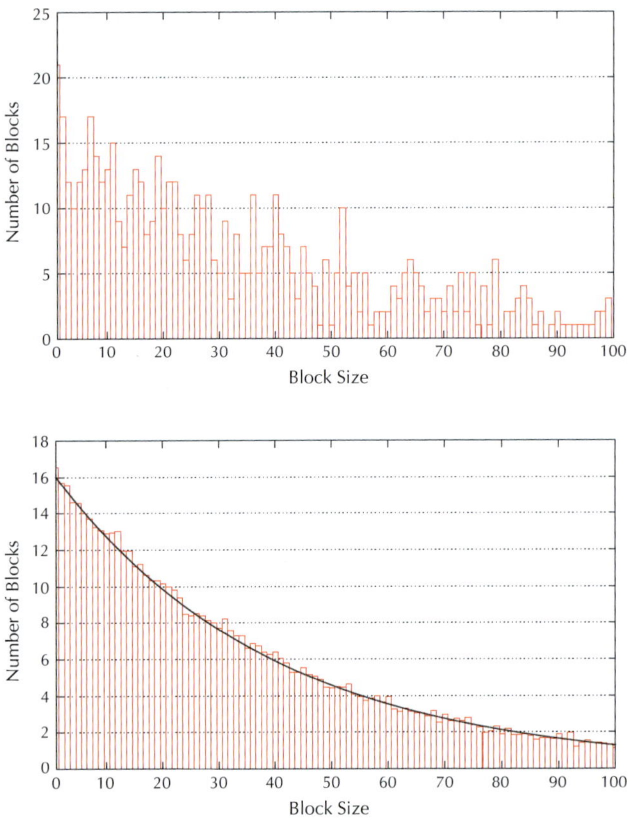

FIGURE 6.3 (Top) A histogram showing the number of blocks of each size for a simulated genome with 25,000 genes (an approximation for the number of genes in a mammalian genome) after 320 randomly chosen reversals. Blocks having more than 100 genes are not shown. (Bottom) A histogram of synteny block lengths averaged over 100 simulations, fitted by the exponential distribution.

STOP and Think: If N random reversals generate $2N + 1$ synteny blocks, then why did the seven reversals shown in Figure 6.2 generate only eleven synteny blocks?

The average synteny block size is $25{,}000/641 \approx 34$ genes, but this does not mean that all synteny blocks should have approximately 34 genes. If we select random locations for breakage points, then some blocks may have only a few genes, whereas other blocks may contain over a hundred. Figure 6.3 (bottom) averages the results of 100 such simulations and illustrates that the distribution of synteny block lengths can be approximated by a curve corresponding to an **exponential distribution** (see DETOUR: The Exponential Distribution). The exponential distribution predicts that there will be about seven blocks having 34 genes and one or two much larger blocks having 100 genes.

PAGE 347

?

What happens when we look at the histogram for the real human and mouse synteny blocks? When Nadeau and Taylor constructed this histogram for the limited genetic data available in 1984, they observed that the lengths of blocks fit the exponential distribution well. In the 1990s, more accurate synteny block data fit the exponential distribution even better (Figure 6.4). Case closed — even though we don't know the exact rearrangements causing our genome to evolve over the last 75 million years, these rearrangements must have followed the Random Breakage Model!

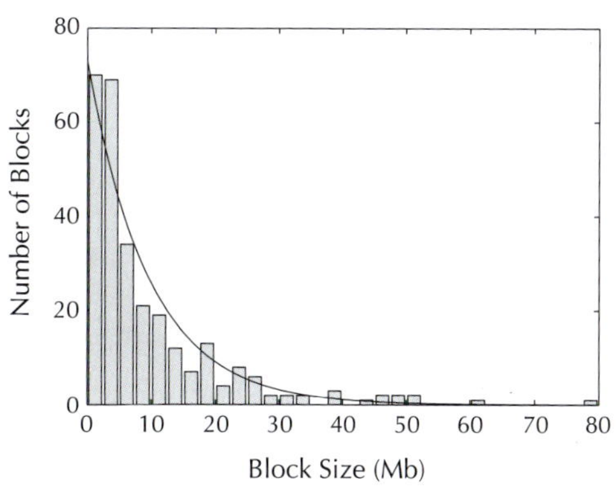

FIGURE 6.4 Histogram of human-mouse synteny block lengths (only synteny blocks longer than 1 million nucleotides are shown). The histogram is fitted by an exponential distribution.

STOP and Think: Do you agree with the logic behind this argument?

Sorting by Reversals

We now have evidence in favor of the Random Breakage Model, but this evidence is far from conclusive. To test this model, let's start building a mathematical model for rearrangement analysis. We will therefore return to a problem that we hinted at in the introduction, which is finding the minimum number of reversals that could transform the mouse X chromosome into the human X chromosome.

STOP and Think: From a biological perspective, why do you think we want to find the minimum possible number of reversals?

We ask for the minimum number of reversals in accordance with a principle called **Occam's razor**. When presented with some quandary, we should explain it using the simplest hypothesis that is consistent with what we already know. In this case, it seems most reasonable that evolution would take the "shortest path" between two species, i.e., the most **parsimonious** evolutionary scenario. Evolution may not always take the shortest path, but even when it does not, the number of steps in the true evolutionary scenario often comes close to the number of steps in the most parsimonious scenario. How, then, can we find the length of this shortest path?

Genome rearrangement studies typically ignore the lengths of synteny blocks and represent chromosomes by **signed permutations**. Each block is labeled by a number, which is assigned a positive/negative sign depending on the block's direction. The number of elements in a signed permutation is its **length**. As you can see from Figure 6.2, the human and mouse X chromosomes can be represented by the following signed permutations of length 11:

$$\begin{array}{rccccccccccc} \textbf{Mouse:} & (+1 & -7 & +6 & -10 & +9 & -8 & +2 & -11 & -3 & +5 & +4) \\ \textbf{Human:} & (+1 & +2 & +3 & +4 & +5 & +6 & +7 & +8 & +9 & +10 & +11) \end{array}$$

In the rest of the chapter, we will refer to signed permutations as **permutations** for short. Because we assume that each synteny block is unique, we do not allow repeated numbers in permutations (e.g., $(+1\ -2\ +3\ +2)$ is not a permutation).

Exercise Break: How many permutations of length n are there?

We can model reversals by inverting the elements within an interval of a permutation, then switching the signs of any elements within the inverted interval. For example, the cartoon in Figure 6.5 illustrates how a reversal changes the permutation $(+1\ +2\ +3\ +4$

+5 +6 +7 +8 +9 +10) into (+1 +2 +3 −8 −7 −6 −5 −4 +9 +10). This reversal can be viewed as first breaking the permutation between +3 and +4 as well as between +8 and +9,

$$(+1\ +2\ +3 \mid +4\ +5\ +6\ +7\ +8 \mid +9\ +10)\,.$$

It then inverts the middle segment,

$$(+1\ +2\ +3 \mid -8\ -7\ -6\ -5\ -4 \mid +9\ +10)\,,$$

and finally glues the three segments back together to form a new permutation,

$$(+1\ +2\ +3\ -8\ -7\ -6\ -5\ -4\ +9\ +10)\,.$$

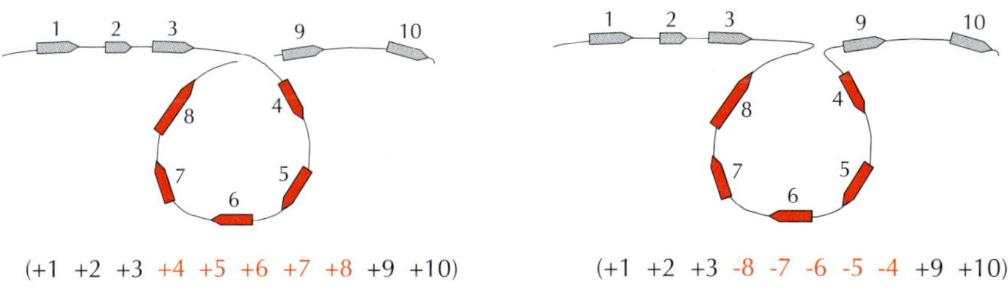

FIGURE 6.5 A cartoon illustrating how a reversal breaks a chromosome in two places and inverts the segment between the two breakage points. Note that the reversal changes the sign of each element within the permutation's inverted segment.

Exercise Break: How many different reversals can be applied to a permutation of length n?

We define the **reversal distance** between permutations P and Q, denoted $d_{\text{rev}}(P, Q)$, as the minimum number of reversals required to transform P into Q.

Reversal Distance Problem:
Calculate the reversal distance between two permutations.

Input: Two permutations of equal length.
Output: The reversal distance between these permutations.

We represented the human X chromosome by $(+1\ +2\ +3\ +4\ +5\ +6\ +7\ +8\ +9\ +10\ +11)$; such a permutation, in which blocks are ordered from smallest to largest with positive directions, is called the **identity permutation**. The reason why we used the identity permutation of length 11 to represent the human X chromosome is that when comparing two genomes, we can label the synteny blocks in one of the genomes however we like. The block labeling for which the human X chromosome is the identity permutation automatically induces the representation of the mouse chromosome as

$$(+1\ -7\ +6\ -10\ +9\ -8\ +2\ -11\ -3\ +5\ +4)\,.$$

Of course, as shown in Figure 6.6, we could have instead encoded the mouse X chromosome as the identity permutation, which would have induced the encoding of the human X chromosome as

$$(+1\ +7\ -9\ +11\ +10\ +3\ -2\ -6\ +5\ -4\ -8)\,.$$

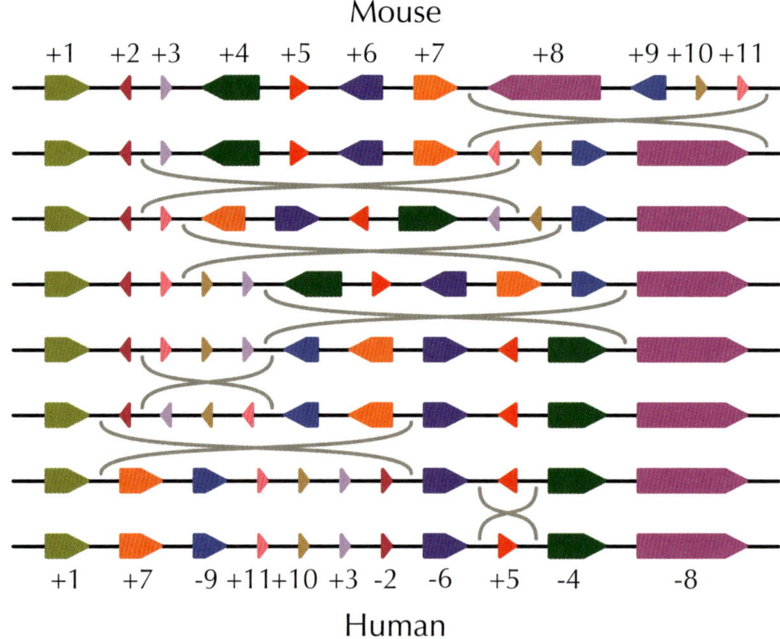

FIGURE 6.6 Encoding the mouse X chromosome as the identity permutation implies encoding the human X chromosome as $(+1\ +7\ -9\ +11\ +10\ +3\ -2\ -6\ +5\ -4\ -8)$.

Because we have the freedom to label synteny blocks however we like, we will consider an equivalent version of the Reversal Distance Problem in which permutation Q is the identity permutation $(+1\ +2\ \ldots +n)$. This computational problem is called **sorting by reversals**, and we denote the minimum number of reversals required to sort P into the identity permutation as $d_{\text{rev}}(P)$. The history of sorting by reversals is founded in a culinary application and involves two celebrities (see **DETOUR: Bill Gates and David X. Cohen Flip Pancakes**). PAGE 348

Sorting by Reversals Problem:
Compute the reversal distance between a permutation and the identity permutation.

Input: A permutation P.
Output: The reversal distance $d_{\text{rev}}(P)$.

Here is a sorting of $(+2\ {-4}\ {-3}\ {+5}\ {-8}\ {-7}\ {-6}\ {+1})$ using five reversals, with the inverted interval at each step shown in red:

$$
\begin{array}{l}
(+2\ \ \mathbf{-4}\ \ \mathbf{-3}\ \ +5\ \ -8\ \ -7\ \ -6\ \ +1)\\
(+2\ \ +3\ \ +4\ \ +5\ \ \mathbf{-8}\ \ \mathbf{-7}\ \ \mathbf{-6}\ \ +1)\\
(+2\ \ +3\ \ +4\ \ +5\ \ +6\ \ +7\ \ +8\ \ \mathbf{+1})\\
(\mathbf{+2}\ \ \mathbf{+3}\ \ \mathbf{+4}\ \ \mathbf{+5}\ \ \mathbf{+6}\ \ \mathbf{+7}\ \ \mathbf{+8}\ \ -1)\\
(\mathbf{-8}\ \ \mathbf{-7}\ \ \mathbf{-6}\ \ \mathbf{-5}\ \ \mathbf{-4}\ \ \mathbf{-3}\ \ \mathbf{-2}\ \ \mathbf{-1})\\
(+1\ \ +2\ \ +3\ \ +4\ \ +5\ \ +6\ \ +7\ \ +8)
\end{array}
$$

STOP and Think: Can you sort this permutation using fewer reversals?

Here is a faster sorting:

$$
\begin{array}{l}
(+2\ \ \mathbf{-4}\ \ \mathbf{-3}\ \ +5\ \ -8\ \ -7\ \ -6\ \ +1)\\
(\mathbf{+2}\ \ \mathbf{+3}\ \ \mathbf{+4}\ \ \mathbf{+5}\ \ -8\ \ -7\ \ -6\ \ +1)\\
(-5\ \ -4\ \ -3\ \ -2\ \ \mathbf{-8}\ \ \mathbf{-7}\ \ \mathbf{-6}\ \ \mathbf{+1})\\
(\mathbf{-5}\ \ \mathbf{-4}\ \ \mathbf{-3}\ \ \mathbf{-2}\ \ \mathbf{-1}\ \ +6\ \ +7\ \ +8)\\
(+1\ \ +2\ \ +3\ \ +4\ \ +5\ \ +6\ \ +7\ \ +8)
\end{array}
$$

STOP and Think: Consider the following questions.

1. Is it possible to sort this permutation even faster?
2. During sorting by reversals, the intermediate permutations in the example above are getting more and more "ordered". Can you come up with a quantitative measure of how ordered a permutation is?

A Greedy Heuristic for Sorting by Reversals

Let's see if we can design a greedy heuristic to approximate $d_{\text{rev}}(P)$. The simplest idea is to perform reversals that fix $+1$ in the first position, followed by reversals that fix $+2$ in the second position, and so on. For example, element 1 is already in the correct position and has the correct sign (+) in the mouse X chromosome, but element 2 is not in the correct position. We can keep element 1 fixed and move element 2 to the correct position by applying a single reversal.

$$(+1\ -7\ +6\ -10\ +9\ -8\ +2\ -11\ -3\ +5\ +4)$$
$$(+1\ -2\ +8\ -9\ +10\ -6\ +7\ -11\ -3\ +5\ +4)$$

One more reversal flips element 2 around so that it has the correct sign:

$$(+1\ -2\ +8\ -9\ +10\ -6\ +7\ -11\ -3\ +5\ +4)$$
$$(+1\ +2\ +8\ -9\ +10\ -6\ +7\ -11\ -3\ +5\ +4)$$

By iterating, we can successively move larger and larger elements to their correct positions in the identity permutation by following the reversals below. The inverted interval of each reversal is still shown in red, and elements that have been placed in the correct position are shown in blue.

$$(+1\ -7\ +6\ -10\ +9\ -8\ +2\ -11\ -3\ +5\ +4)$$
$$(+1\ -2\ +8\ -9\ +10\ -6\ +7\ -11\ -3\ +5\ +4)$$
$$(+1\ +2\ +8\ -9\ +10\ -6\ +7\ -11\ -3\ +5\ +4)$$
$$(+1\ +2\ +3\ +11\ -7\ +6\ -10\ +9\ -8\ +5\ +4)$$
$$(+1\ +2\ +3\ -4\ -5\ +8\ -9\ +10\ -6\ +7\ -11)$$
$$(+1\ +2\ +3\ +4\ -5\ +8\ -9\ +10\ -6\ +7\ -11)$$
$$(+1\ +2\ +3\ +4\ +5\ +8\ -9\ +10\ -6\ +7\ -11)$$
$$(+1\ +2\ +3\ +4\ +5\ +6\ -10\ +9\ -8\ +7\ -11)$$
$$(+1\ +2\ +3\ +4\ +5\ +6\ -7\ +8\ -9\ +10\ -11)$$
$$(+1\ +2\ +3\ +4\ +5\ +6\ +7\ +8\ -9\ +10\ -11)$$
$$(+1\ +2\ +3\ +4\ +5\ +6\ +7\ +8\ +9\ +10\ -11)$$
$$(+1\ +2\ +3\ +4\ +5\ +6\ +7\ +8\ +9\ +10\ +11)$$

This example motivates a greedy heuristic called **GREEDYSORTING**. We say that element k in permutation $P = (p_1 \dots p_n)$ is **sorted** if $p_k = +k$ and **unsorted** otherwise. We call P ***k*-sorted** if its first $k-1$ elements are sorted, but if element k is unsorted. For every k-sorted permutation P, there exists a single reversal, called the ***k*-sorting reversal**, that fixes the first $k-1$ elements of P and moves element k to the k-th position. In the case when $-k$ is already in the k-th position of P, the k-sorting reversal merely flips $-k$ around.

For example, in the sorting of the mouse X chromosome shown above, the 2-sorting reversal transforms $(+1\ -7\ +6\ -10\ +9\ -8\ +2\ -11\ -3\ +5\ +4)$ into $(+1\ -2\ +8\ -9\ +10\ -6\ +7\ -11\ -3\ +5\ +4)$. In this case, an additional 2-sorting reversal flipping -2 was needed to make element 2 sorted. The idea of **GREEDYSORTING**, then, is to apply k-sorting reversals for increasing values of k. Here, $|P|$ refers to the length of permutation P.

```
GREEDYSORTING(P)
    approxReversalDistance ← 0
    for k ← 1 to |P|
        if element k is not sorted
            apply the k-sorting reversal to P
            approxReversalDistance ← approxReversalDistance + 1
            if the k-th element of P is −k
                apply the k-sorting reversal to P
                approxReversalDistance ← approxReversalDistance + 1
    return approxReversalDistance
```

Exercise Break: How many reversals does **GREEDYSORTING** need to sort the permutation $(+n\ +(n-1)\ \dots +2\ +1)$?

In the case of the mouse X chromosome, **GREEDYSORTING** requires eleven reversals, but we already know that this permutation can be sorted with seven reversals, which causes us to wonder: how good of a heuristic is **GREEDYSORTING**?

Exercise Break: What is the largest number of reversals **GREEDYSORTING** could ever require to sort a permutation of length n?

Consider the permutation (−6 +1 +2 +3 +4 +5). You can verify that the greedy heuristic requires ten steps to sort this permutation, and yet it can be sorted using just two reversals!

(**−6 +1 +2 +3 +4 +5**)
(**−5 −4 −3 −2 −1** +6)
(+1 +2 +3 +4 +5 +6)

This example demonstrates that **GREEDYSORTING** provides a poor approximation for the reversal distance.

STOP and Think: Can you find a lower bound on $d_{\text{rev}}(P)$? For example, can you show that the mouse permutation (+1 −7 +6 −10 +9 −8 +2 −11 −3 +5 +4) cannot be sorted with fewer than seven reversals?

Breakpoints

What are breakpoints?

Consider the sorting by reversals shown in Figure 6.7. We would like to quantify how each subsequent permutation is moving closer to the identity as we apply subsequent reversals. For the first reversal, at the right endpoint of the inverted interval, it changes the consecutive elements (−11 +13) into the much more desirable (+12 +13). Less obvious is the work of the fourth reversal, which places −11 immediately left of −10 so that at the next step, the consecutive elements (−11 −10) can be part of an inverted interval, creating the desirable consecutive elements (+10 +11).

BREAKPOINTS(P)

|+3 +4 +5 |**−12** |**−8 −7 −6** |**+1 +2** |**+10** |**+9** |**−11** |+13 — 8

|**+3 +4 +5** |**+11** |**−9** |**−10** |**−2 −1** |+6 +7 +8 |+12 +13 — 7

+1 +2 |**+10** |**+9** |**−11** |**−5 −4 −3** |+6 +7 +8 |+12 +13 — 6

+1 +2 +3 +4 +5 |**+11** |**−9** |−10 |+6 +7 +8 |+12 +13 — 5

+1 +2 +3 +4 +5 |+9 |**−11 −10** |**+6 +7 +8** |+12 +13 — 4

+1 +2 +3 +4 +5 |**+9** |**−8 −7 −6** |+10 +11 +12 +13 — 3

+1 +2 +3 +4 +5 +6 +7 +8 |**−9** |+10 +11 +12 +13 — 2

+1 +2 +3 +4 +5 +6 +7 +8 +9 +10 +11 +12 +13 — 0

FIGURE 6.7 A sorting by reversals. The inverted interval of each reversal is shown in red, while breakpoints in each permutation are marked by vertical blue segments.

The intuition that we are trying to build is that consecutive elements like $(+12\ +13)$ are desirable because they appear in the same order as in the identity permutation. However, consecutive elements like $(-11\ -10)$ are also desirable, since these elements can be later inverted into the correct order, $(+10\ +11)$. The pairs $(+12\ +13)$ and $(-11\ -10)$ have something in common; the second element is equal to the first element plus 1. We therefore say that consecutive elements $(p_i\ p_{i+1})$ in permutation $P = (p_1 \ldots p_n)$ form an **adjacency** if $p_{i+1} - p_i$ is equal to 1. By definition, for any positive integer $k < n$, both $(k\ k+1)$ and $(-(k+1)\ -k)$ are adjacencies. If $p_{i+1} - p_i$ is not equal to 1, then we say that $(p_i\ p_{i+1})$ is a **breakpoint**.

We can think about a breakpoint intuitively as a pair of consecutive elements that are "out of order" compared to the identity permutation $(+1\ +2 \ldots +n)$. For example, the pair $(+5\ -12)$ is a breakpoint because +5 and -12 are not neighbors in the identity permutation. Similarly, $(-12\ -8)$, $(-6\ +1)$, $(+2\ +10)$, $(+9\ -11)$, and $(-11\ +13)$ are clearly out of order. But $(+10\ +9)$ is also a breakpoint (even though it is formed by consecutive integers) because its signs are out of order compared to the identity permutation.

STOP and Think: The permutation $(-5\ -4\ -3\ -2\ -1)$ is clearly not the identity permutation, but where are its breakpoints?

STOP

We will further represent the beginning and end of permutation P by adding **0** to the left of the first element and $\boldsymbol{n+1}$ to the right of the last element,

$$(\mathbf{0}\ p_1 \ldots p_n\ (\boldsymbol{n}+\mathbf{1}))\,.$$

As a result, there are $n+1$ pairs of consecutive elements:

$$(\mathbf{0}\ p_1),\ (p_1\ p_2),\ (p_2\ p_3),\ \ldots,\ (p_{n-1}\ \ p_n),\ (p_n\ \ (\boldsymbol{n}+\mathbf{1}))\,.$$

We use ADJACENCIES(P) and BREAKPOINTS(P) to denote the number of adjacencies and breakpoints of permutation P, respectively. Figure 6.7 illustrates how the number of breakpoints changes during sorting by reversals (note that 0 and $n+1$ are placeholders and cannot be affected by a reversal); the permutation at the beginning of this figure has eight breakpoints and six adjacencies.

Counting breakpoints

Because any pair of consecutive elements of a permutation form either a breakpoint or adjacency, we have the following identity for any permutation P of length n:

$$\text{ADJACENCIES}(P) + \text{BREAKPOINTS}(P) = n + 1.$$

STOP and Think: A permutation on n elements may have at most $n+1$ adjacencies. How many permutations on n elements have exactly $n+1$ adjacencies?

You can verify that the identity permutation $(+1\ +2\ \ldots +n)$ is the only permutation for which all consecutive elements are adjacencies, meaning that it has no breakpoints. Note also that the permutation $(-n\ -(n-1)\ \ldots -2\ -1)$ has adjacencies for every consecutive pair of elements except for the two breakpoints $(\mathbf{0}\ -n)$ and $(-1\ (\boldsymbol{n+1}))$.

Exercise Break: How many permutations of length n have exactly $n-1$ adjacencies?

Number of Breakpoints Problem:
Find the number of breakpoints in a permutation.

Input: A permutation.
Output: The number of breakpoints in this permutation.

STOP and Think: We defined a breakpoint between an arbitrary permutation and the identity permutation. Generalize the notion of a breakpoint between two arbitrary permutations, and design a linear-time algorithm for computing this number.

Sorting by reversals as breakpoint elimination

The reversals in Figure 6.7 reduce the number of breakpoints from 8 to 0. Note that the permutation becomes more and more "ordered" after every reversal as the number of breakpoints reduces at each step. You can therefore think of sorting by reversals as the process of breakpoint elimination — reducing the number of breakpoints in a permutation P from BREAKPOINTS(P) to 0.

STOP and Think: What is the maximum number of breakpoints that can be eliminated by a single reversal?

Consider the first reversal in Figure 6.7, which reduces the number of breakpoints from 8 to 7. On either side of the inverted interval, breakpoints and adjacencies certainly do not change; for example, the breakpoint $(0 +3)$ and the adjacency $(+13 +14)$ remain the same. Also note that every breakpoint within the inverted interval of a reversal remains a breakpoint after the reversal. In other words, if $(p_i\ p_{i+1})$ formed a breakpoint within the span of a reversal, i.e.,

$$p_{i+1} - p_i \neq 1,$$

then these consecutive elements will remain a breakpoint after the reversal changes them into $(-p_{i+1}\ -p_i)$:

$$-p_i - (-p_{i+1}) = p_{i+1} - p_i \neq 1\,.$$

For example, there are five breakpoints within the span of the following reversal on the permutation (0 +3 +4 +5 **−12 −8 −7 −6 +1 +2 +10 +9 −11** +13 +14 15):

(−12 −8) (−6 +1) (+2 +10) (+10 +9) (+9 −11)

After the reversal, these breakpoints become the following five breakpoints:

(+11 −9) (−9 −10) (−10 −2) (−1 +6) (+8 +12)

Since all breakpoints inside and outside the span of a reversal remain breakpoints after a reversal, the only breakpoints that could be eliminated by a reversal are the two breakpoints located on the boundaries of the inverted interval. The breakpoints on the boundaries of the first reversal in Figure 6.7 are (+5 −12) and (−11 +13); the reversal converts them into a breakpoint (+5 +11) and an adjacency (+12 +13), thus reducing the number of breakpoints by 1.

STOP and Think: Can the permutation (+3 +4 +5 −12 −8 −7 −6 +1 +2 +10 +9 −11 +13 +14), which has 8 breakpoints, be sorted with three reversals?

A reversal can eliminate at most two breakpoints, so two reversals can eliminate at most four breakpoints, three reversals can eliminate at most six breakpoints, and so on. This reasoning establishes the following theorem.

Breakpoint Theorem: $d_{\text{rev}}(P)$ *is greater than or equal to* BREAKPOINTS$(P)/2$.

It would be nice if we could *always* find a reversal that eliminates two breakpoints from a permutation, as this would imply a simple greedy algorithm for optimal sorting by reversals. Unfortunately, this is not the case. You can verify that there is no reversal that reduces the number of breakpoints in the permutation $P = (+2 +1)$, which has three breakpoints.

Exercise Break: How many permutations of length n have the property that no reversal applied to P decreases BREAKPOINTS(P)?

It turns out that every permutation of length n can be sorted using at most $n + 1$ reversals and that the permutation $(+n\ +(n-1)\ \ldots +1)$ requires $n + 1$ reversals to sort. Since this permutation has $n + 1$ breakpoints, there is a large gap between the lower bound of $(n + 1)/2$ provided by the Breakpoint Theorem and the reversal distance.

Exercise Break: Prove that there exists a shortest sequence of reversals sorting a permutation that never breaks a permutation at an adjacency.

You will soon see that the idea of breakpoints will help us return to our original aim of testing the Random Breakage Model. For now, we would like to move from permutations, which can only model single chromosomes, to a more general multichromosomal model. You may be surprised that we are moving to a seemingly more difficult model before resolving the unichromosomal case, which is already difficult. However, it turns out that our new multichromosomal model will be easier to analyze!

Rearrangements in Tumor Genomes

As we move toward a more robust model for genome comparison, we need to incorporate rearrangements that move genes from one chromosome to another. Indeed, with the notable exception of the X chromosome, the genes from a single human chromosome usually have their counterparts distributed over many mouse chromosomes (and vice-versa). We hope that there is a nagging voice in your head, wondering: *How can a genome rearrangement affect multiple chromosomes?*

Although multichromosomal rearrangements have occurred during species evolution over millions of years, we can witness them during a much narrower time frame in cancer cells, which exhibit many chromosomal aberrations. Some of these mutations have no direct effect on tumor development, but many types of tumors display recurrent rearrangements that trigger tumor growth by disrupting genes or altering gene regulation. By studying these rearrangements, we can identify genes that are important for tumor growth, leading to improved cancer diagnostics and therapeutics.

Figure 6.8 presents a rearrangement involving human chromosomes 9 and 22 in a rare form of cancer called **chronic myeloid leukemia (CML)**. In this type of rearrangement, called a **translocation**, two intervals of DNA are excised from the end of chromosomes 9 and 22 and then reattached on opposite chromosomes. One of the rearranged chromosomes is called the **Philadelphia chromosome**. This chromosome fuses together two genes called ABL and BCR that normally have nothing to do with each other. However, when joined on the Philadelphia chromosome, these two genes create a single **chimeric gene** coding for the **ABL-BCR fusion protein**, which has been implicated in the development of CML.

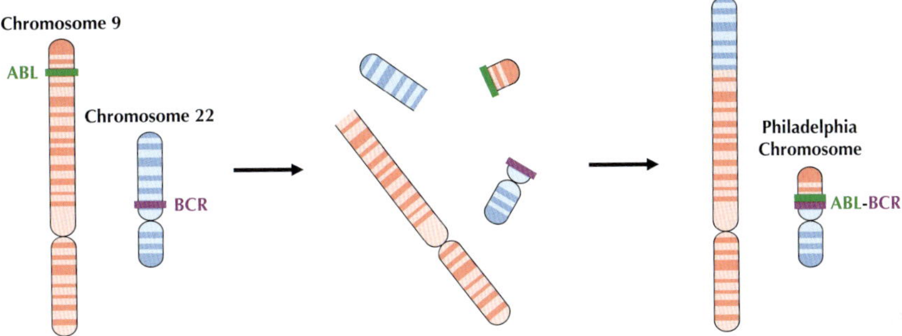

FIGURE 6.8 The Philadelphia chromosome is formed by a translocation affecting chromosomes 9 and 22. It fuses together the ABL gene and part of the BCR gene, forming a chimeric gene that can trigger CML.

Once scientists understood the root cause of CML, they started searching for a compound inhibiting ABL-BCR, which resulted in the introduction of a drug called **Gleevec** in 2001. Gleevec is a **targeted therapy** against CML that inhibits cancer cells but does not affect normal cells and has shown great clinical results. However, since it targets only the ABL-BCR fusion protein, Gleevec does not treat most other cancers. Nevertheless, the introduction of Gleevec has bolstered researchers' hopes that the search for specific rearrangements in other forms of cancer may produce additional targeted cancer therapies.

From Unichromosomal to Multichromosomal Genomes

Translocations, fusions, and fissions

To model translocations, we represent a multichromosomal genome with k chromosomes as a permutation that has been partitioned into k pieces. For example, the genome $(+1\ +2\ +3\ +4\ +5\ +6)(+7\ +8\ +9\ +10\ +11)$ is made up of the two chromosomes $(+1\ +2\ +3\ +4\ +5\ +6)$ and $(+7\ +8\ +9\ +10\ +11)$. A translocation exchanges segments of different chromosomes, e.g., a translocation of the two chromosomes

$$(+1\ +2\ +3\ +4\ +5\ +6)\ (+7\ +8\ +9\ +10\ +11)$$

may result in the two chromosomes

$$(+1\ +2\ +3\ +4\ +9\ +10\ +11)\ (+7\ +8\ +5\ +6)\,.$$

You can think about a translocation as first breaking each of the two chromosomes

$$(+1\ +2\ +3\ +4\ +5\ +6)\quad (+7\ +8\ +9\ +10\ +11)$$

into two parts,

$$(+1\ +2\ +3\ +4)\quad (+5\ +6)\qquad (+7\ +8)\quad (+9\ +10\ +11)\,,$$

and then gluing the resulting segments into two new chromosomes,

$$(+1\ +2\ +3\ +4\ +9\ +10\ +11)\quad (+7\ +8\ +5\ +6)\,.$$

Rearrangements in multichromosomal genomes are not limited to reversals and translocations. They also include chromosome **fusions**, which merge two chromosomes into a single chromosome, as well as **fissions**, which break a single chromosome into two chromosomes. For example, the two chromosomes

$$(+1\ +2\ +3\ +4\ +5\ +6)\quad (+7\ +8\ +9\ +10\ +11)$$

can be fused into the single chromosome

$$(+1\ +2\ +3\ +4\ +5\ +6\ +7\ +8\ +9\ +10\ +11)\,.$$

A subsequent fission of this chromosome could result in the two chromosomes

$$(+1\ +2\ +3\ +4)\quad (+5\ +6\ +7\ +8\ +9\ +10\ +11)\,.$$

Five million years ago, shortly after the human and chimpanzee lineages split, a fusion of two chromosomes (called 2A and 2B) in one of our ancestors created human chromosome 2 and reduced our chromosome count from 24 to 23.

STOP and Think: *A priori*, it could just as easily be the case that the human-chimpanzee ancestor had an intact chromosome 2, and that a fission split these two chromosomes into chimpanzee chromosomes 2A and 2B. How would you choose between the two scenarios? Hint: gorillas and orangutans, like chimpanzees, also have 24 chromosomes.

From a genome to a graph

We will henceforth assume that all chromosomes in a genome are circular. This assumption represents a slight distortion of biological reality, as mammalian chromosomes are linear. However, circularizing a linear chromosome by joining its endpoints will simplify the subsequent analysis without affecting our final conclusion.

We now have a multichromosomal genomic model, along with four types of rearrangements (reversals, translocations, fusions, and fissions) that can transform one genome into another. To model genomes with circular chromosomes, we will use a **genome graph**. First represent each synteny block by a directed gray edge indicating its direction, and then link gray edges corresponding to adjacent synteny blocks with a colored undirected edge. Figure 6.9 shows each circular chromosome as an **alternating cycle** of red and gray edges. In this model, the human genome can be represented using 280 human-mouse synteny blocks spread over 23 alternating cycles.

STOP and Think: Let P and Q be genomes consisting of linear chromosomes, and let P^* and Q^* be the circularized versions of these genomes. Can you convert a given series of reversals/translocations/fusions/fissions transforming P into Q into a series of rearrangements transforming P^* into Q^*? What about the reverse operation — can you convert a series of rearrangements transforming P^* into Q^* into a series of rearrangements transforming P into Q?

2-breaks

We now focus on one of the chromosomes in a multi-chromosomal genome and consider a reversal transforming the circular chromosome $P = (+a\ -b\ -c\ +d)$ into $Q = (+a\ -b\ -d\ +c)$. We can draw Q in a variety of ways, depending on how we choose to arrange its gray edges. Figure 6.10 shows two such equivalent representations.

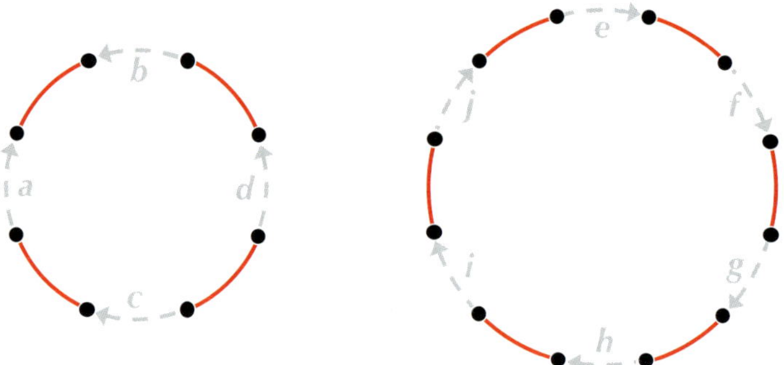

FIGURE 6.9 A genome with two circular chromosomes, (+a −b −c +d) and (+e +f +g +h +i +j). Gray directed edges represent synteny blocks, and red undirected edges connect adjacent synteny blocks. A circular chromosome with *n* elements can be written in 2*n* different ways; the chromosome on the left can be written as (+a −b −c +d), (−b −c +d +a), (−c +d +a −b), (+d +a −b −c), (−a −d +c +b) (−d +c +b −a), (+c +b −a −d), and (+b −a −d +c).

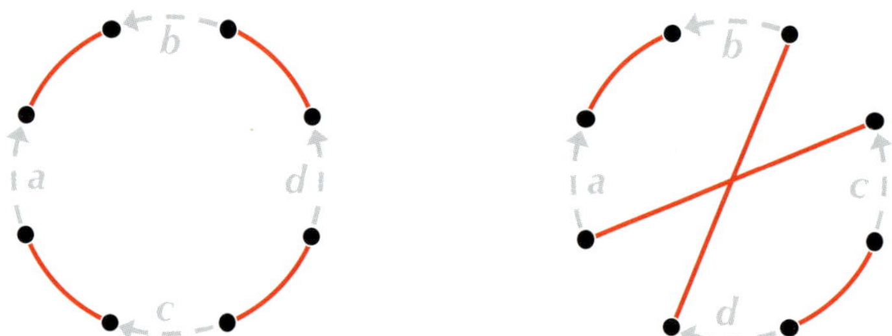

FIGURE 6.10 Two equivalent drawings of the circular chromosome Q = (+a −b −d +c).

Although the first drawing of *Q* in Figure 6.10 is its most natural representation, we will use the second representation because its gray edges are arranged around the circle in exactly the same order as they appear in the natural representation of $P = (+a\ -b\ -c\ +d)$. As illustrated in Figure 6.11, keeping the gray edges fixed allows us to visualize the effect of the reversal. As you can see, the reversal deletes ("breaks") two red edges in *P* (connecting *b* to *c* and *d* to *a*) and replaces them with two new red edges (connecting *b* to *d* and *c* to *a*).

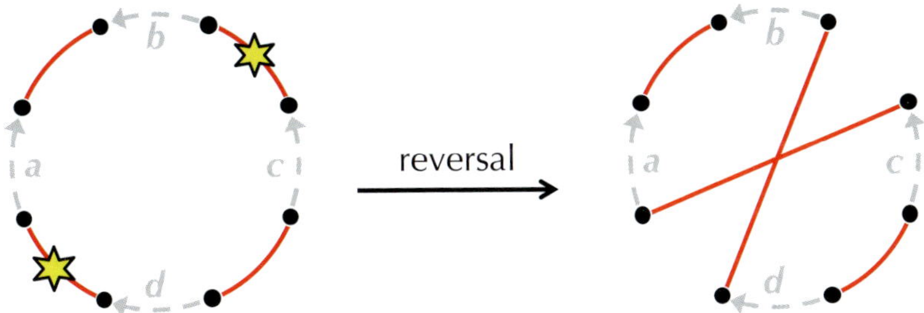

FIGURE 6.11 A reversal transforms $P = (+a\ -b\ -c\ +d)$ into $Q = (+a\ -b\ -d\ +c)$. We have arranged the gray edges of Q so that they have the same orientation and position as the gray edges in the natural representation of P. The reversal can be viewed as deleting the two red edges labeled by stars and replacing them with two new red edges on the same four nodes.

Figure 6.12 illustrates a fission of $P = (+a\ -b\ -c\ +d)$ into $Q = (+a\ -b)(-c\ +d)$; reversing this operation corresponds to a fusion of the two chromosomes of Q to yield P. Both the fusion and the fission operations, like the reversal, correspond to deleting two edges in one genome and replacing them with two new edges in the other genome.

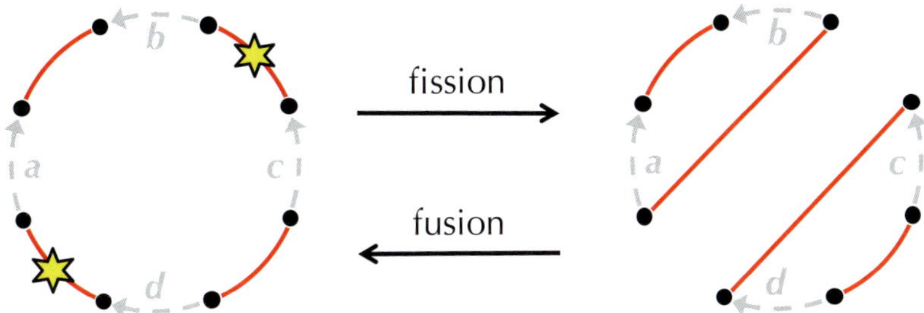

FIGURE 6.12 A fission of the single chromosome $P = (+a\ -b\ -c\ +d)$ into the genome $Q = (+a\ -b)(-c\ +d)$. The inverse operation is a fusion, transforming the two chromosomes of Q into a single chromosome by deleting two red edges of Q and replacing them with two other edges.

A translocation involving two *linear* chromosomes can also be mimicked by circularizing these chromosomes and then replacing two red edges with two different red edges, as shown in Figure 6.13. We have therefore found a common theme uniting the four different types of rearrangements. They all can be viewed as breaking two red edges of the genome graph and replacing them with two new red edges on the same four nodes. For this reason, we define the general operation on the genome graph in which two red edges are replaced with two new red edges on the same four nodes as a **2-break**.

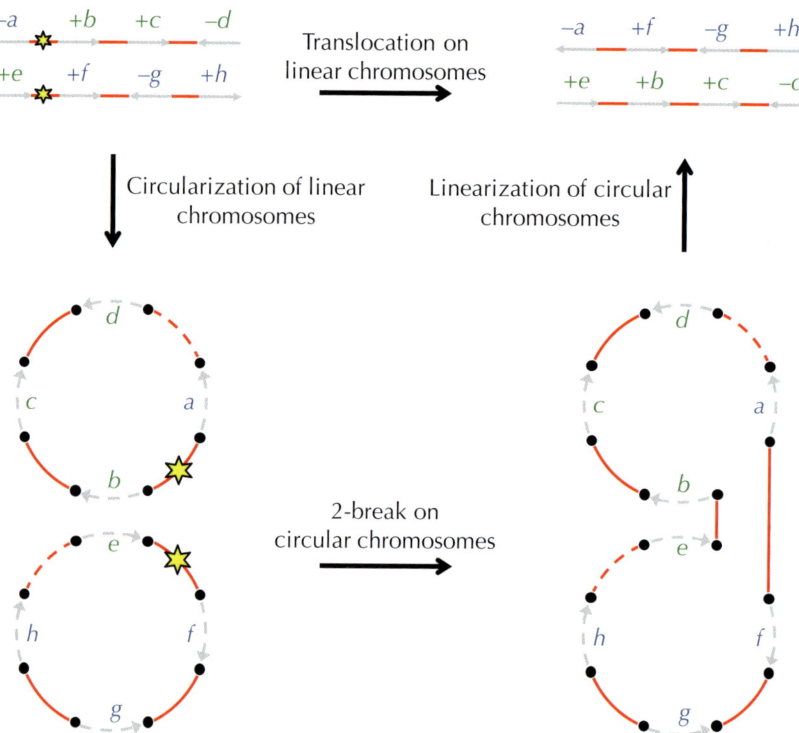

FIGURE 6.13 A translocation of linear chromosomes $(-a\ +b\ +c\ -d)$ and $(+e\ +f\ -g\ +h)$ transforms them into linear chromosomes $(-a\ +f\ -g\ +h)$ and $(+e\ +b\ +c\ -d)$. This translocation can also be accomplished by first circularizing the chromosomes (i.e., connecting the ends of each chromosome with a dashed red edge), then applying a 2-break to the new chromosomes, and finally converting the resulting circular chromosome into two linear chromosomes by removing the dashed red edges.

We would like to find a shortest sequence of 2-breaks transforming genome P into genome Q, and we refer to the number of operations in a shortest sequence of 2-breaks transforming P into Q as the **2-break distance** between P and Q, denoted $d(P, Q)$.

2-Break Distance Problem:
Find the 2-break distance between two genomes.

> **Input**: Two genomes with circular chromosomes on the same synteny blocks.
> **Output**: The 2-break distance between these genomes.

Breakpoint Graphs

To compute the 2-break distance, we will return to the notion of breakpoints to construct a graph for comparing two genomes. Consider the genomes $P = (+a\ -b\ -c\ +d)$ and $Q = (+a\ +c\ +b\ -d)$. Note that we have used red for the colored edges of P and blue for the colored edges of Q. As before, we rearrange the gray edges of Q so that they are arranged exactly as in P (Figure 6.14, middle). If we superimpose the genome graphs of P and Q, then we obtain the tri-colored **breakpoint graph** BREAKPOINTGRAPH(P, Q), shown in Figure 6.14.

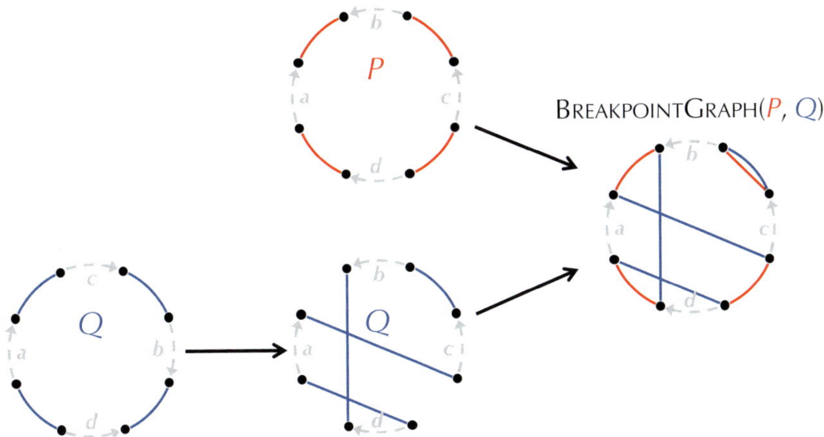

FIGURE 6.14 Constructing the breakpoint graph for the unichromosomal genomes $P = (+a\ -b\ -c\ +d)$ and $Q = (+a\ +c\ +b\ -d)$. After rearranging the gray edges of Q so that they are arranged the same as in P, the breakpoint graph BREAKPOINTGRAPH(P, Q) is formed by superimposing the graphs of P and Q. As shown on the right, there are two alternating red-blue cycles in this breakpoint graph.

Note that the red and gray edges in the breakpoint graph form P, and the blue and gray edges form Q. Moreover, the red and blue edges in the breakpoint graph form a collection of red-blue alternating cycles.

STOP and Think: Prove that the red and blue edges in any breakpoint graph form alternating cycles. Hint: How many red and blue edges meet at each node of the breakpoint graph?

We denote the number of red-blue alternating cycles in BREAKPOINTGRAPH(P, Q) as CYCLES(P, Q). For $P = (+a\ -b\ -c\ +d)$ and $Q = (+a\ +c\ +b\ -d)$, CYCLES(P, Q) = 2, as shown on the right in Figure 6.14. In what follows, we will be focusing on the red-blue alternating cycles in breakpoint graphs and often omit the gray edges.

Although Figure 6.14 illustrates the construction of the breakpoint graph for single-chromosomal genomes, the breakpoint graph can be constructed for genomes with multiple chromosomes in exactly the same way (Figure 6.15).

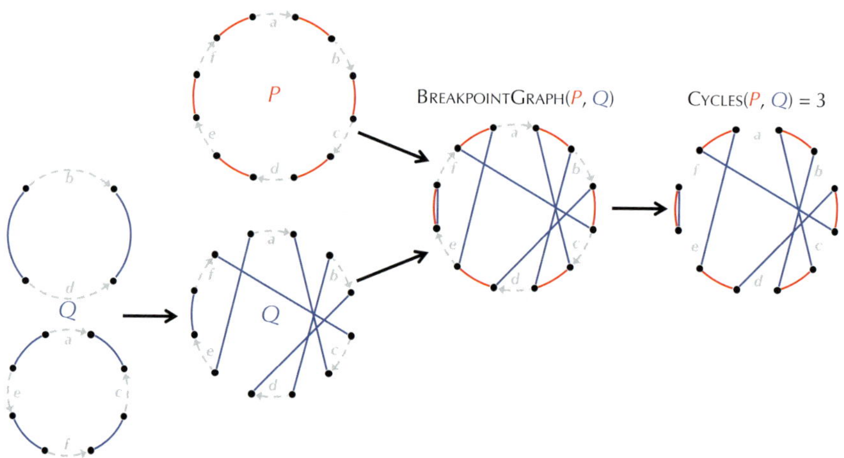

FIGURE 6.15 The construction of BREAKPOINTGRAPH(P, Q) for the unichromosomal genome $P = (+a\ +b\ +c\ +d\ +e\ +f)$ and the two-chromosome genome $Q = (+a\ -c\ -f\ -e)(+b\ -d)$. At the bottom, to illustrate the construction of the breakpoint graph, we first rearrange the gray edges of Q so that they are drawn in the same order along the cycle as in P.

STOP and Think: Given genome P, which genome Q maximizes CYCLES(P, Q)?

In the case that P and Q have the same number of synteny blocks, we denote the number of their synteny blocks as BLOCKS(P, Q). As shown in Figure 6.16, when P and Q are identical, their breakpoint graph consists of BLOCKS(P, Q) cycles of length 2, each containing one red and one blue edge. We refer to cycles of length 2 as **trivial cycles** and the breakpoint graph formed by identical genomes as the **trivial breakpoint graph**.

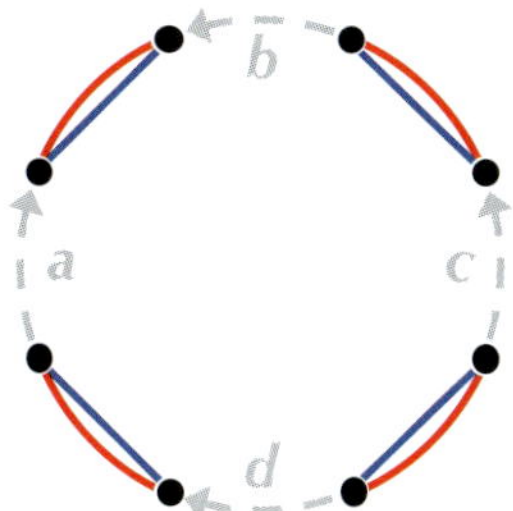

FIGURE 6.16 The trivial breakpoint graph BREAKPOINTGRAPH(P, P), formed by two copies of the genome $P = (+a\ -b\ -c\ +d)$. The breakpoint graph of *any* genome with itself consists only of length 2 alternating cycles.

Exercise Break: Prove that CYCLES(P, Q) is smaller than BLOCKS(P, Q) unless P is equal to Q.

You are likely wondering how the breakpoint graph is useful. We can view a 2-break transforming P into P' as an operation on BREAKPOINTGRAPH(P, Q) that yields BREAKPOINTGRAPH(P', Q) (Figure 6.17).

By extension, we can view a series of 2-breaks transforming P into Q as a series of 2-breaks transforming BREAKPOINTGRAPH(P, Q) into BREAKPOINTGRAPH(Q, Q), the trivial breakpoint graph. This means that solving the 2-Break Distance Problem for genomes P and Q is equivalent to finding a shortest series of 2-breaks transforming BREAKPOINTGRAPH(P, Q) into the trivial breakpoint graph. Figure 6.18 illustrates a transformation of a breakpoint graph with CYCLES(P, Q) $= 2$ into a trivial breakpoint graph with CYCLES(Q, Q) $= 4$ using two 2-breaks.

Since every transformation of P into Q transforms BREAKPOINTGRAPH(P, Q) into the trivial breakpoint graph BREAKPOINTGRAPH(Q, Q), any sorting by 2-breaks increases the number of red-blue cycles by

$$\text{CYCLES}(Q, Q) - \text{CYCLES}(P, Q)\,.$$

STOP and Think: How much can each 2-break contribute to this increase? In other words, if P' is obtained from P by a 2-break, how much bigger can CYCLES(P', Q) be than CYCLES(P, Q)?

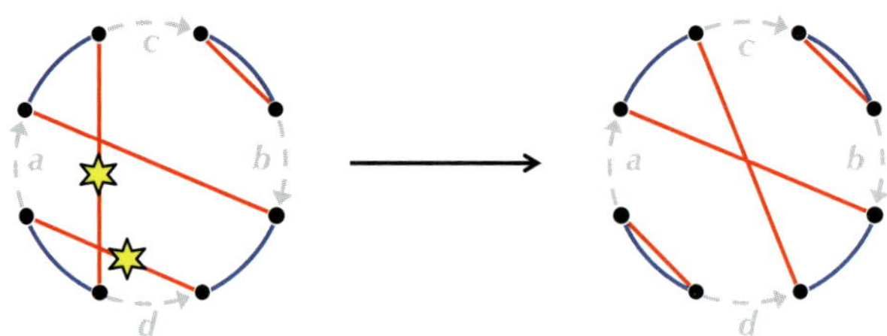

FIGURE 6.17 A 2-break (indicated by stars) transforming genome P into P' also transforms BREAKPOINTGRAPH(P, Q) into BREAKPOINTGRAPH(P', Q) for any genome Q. In this example, $P = (+a\ -b\ -c\ +d)$, $P' = (+a\ -b\ -c\ -d)$, and $Q = (+a\ +c\ +b\ -d)$.

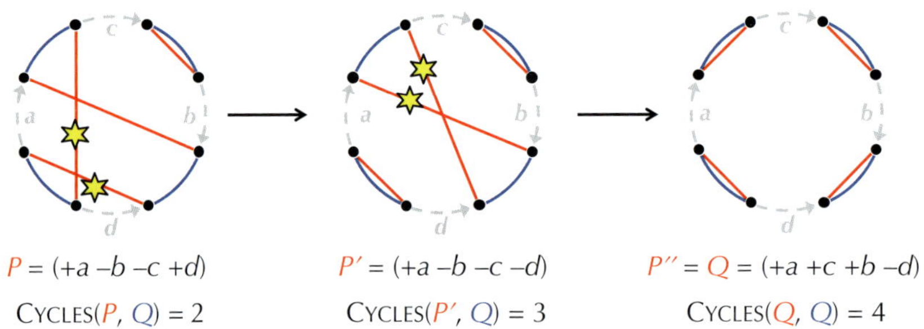

FIGURE 6.18 Every 2-break transformation of genome P into Q corresponds to a transformation of BREAKPOINTGRAPH(P, Q) into BREAKPOINTGRAPH(Q, Q). In the example shown, the number of red-blue cycles in the graph increases from CYCLES(P, Q) $= 2$ to CYCLES(Q, Q) $=$ BLOCKS(P, Q) $= 4$.

Computing the 2-Break Distance

The Breakpoint Theorem stated that a reversal applied to a chromosome P can reduce BREAKPOINTS(P) by at most 2. We now prove that a 2-break applied to a multichromosomal genome P can increase CYCLES(P, Q) by at most 1, i.e., for any 2-break transforming P into P', and for any genome Q, CYCLES(P', Q) cannot exceed CYCLES(P, Q) $+ 1$.

Cycle Theorem: *For genomes P and Q, any 2-break applied to P can increase* CYCLES(P, Q) *by at most 1.*

Proof. Figure 6.19 presents three cases that illustrate how a 2-break applied to P can affect the breakpoint graph. Each 2-break affects two red edges that either belong to the same cycle or to two different cycles in BREAKPOINTGRAPH(P, Q). In the former case, the 2-break either does not change CYCLES(P, Q), or it increases it by 1. In the latter case, it decreases CYCLES(P, Q) by 1. □

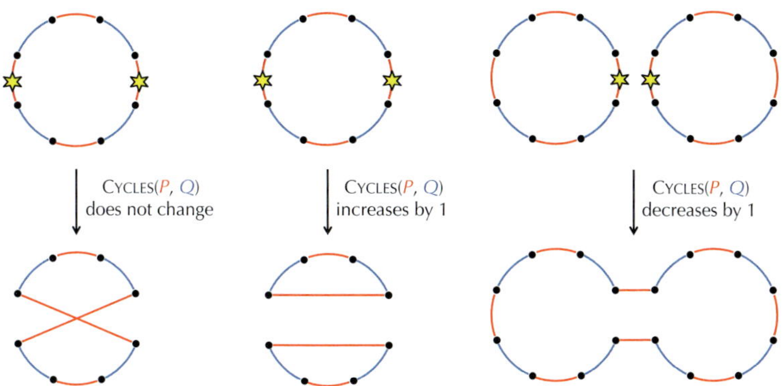

FIGURE 6.19 Three cases illustrating how a 2-break can affect the breakpoint graph; gray edges are not shown.

Although the preceding proof is short and intuitive, it is not a formal proof, but rather an invitation to examine Figure 6.19. If you are interested in a more rigorous mathematical argument, please read the next proof.

Proof. A 2-break adds two new red edges and thus forms at most **2** new cycles (containing two new red edges) in BREAKPOINTGRAPH(P, Q). At the same time, it deletes two red edges and thus deletes at least **1** old cycle (containing two old edges) from BREAKPOINTGRAPH(P, Q). Thus, the number of red-blue cycles in the breakpoint graph increases by at most $\mathbf{2} - \mathbf{1} = 1$, implying that CYCLES$(P, Q)$ increases by at most 1. □

Recall that there are permutations for which no reversal reduces the number of breakpoints, a fact that defeated our hopes for a greedy algorithm for sorting by reversals that reduces the number of breakpoints at each step. In the case of 2-breaks (on genomes with circular chromosomes), we now know that each 2-break can increase

CYCLES(P, Q) by at most 1. But is it *always* possible to find a 2-break that increases CYCLES(P, Q) by 1? As the following theorem illustrates, the answer is yes.

2-Break Distance Theorem: *The 2-break distance between genomes P and Q is equal to* BLOCKS(P, Q) − CYCLES(P, Q).

Proof. Recall that every sorting by 2-breaks must increase the number of alternating cycles by CYCLES(Q, Q) − CYCLES(P, Q), which equals BLOCKS(P, Q) − CYCLES(P, Q) because BLOCKS(P, Q) = CYCLES(Q, Q). The Cycle Theorem implies that each 2-break increases the number of cycles in the breakpoint graph by at most 1. This immediately implies in turn that $d(P, Q)$ is *at least* BLOCKS(P, Q) − CYCLES(P, Q). If P is not equal to Q, then there must be a **non-trivial cycle** in BREAKPOINTGRAPH(P, Q), i.e., a cycle with more than two edges. As shown in Figure 6.19 (middle), any non-trivial cycle in the breakpoint graph can be split into two cycles by a 2-break, implying that we can always find a 2-break increasing the number of red-blue cycles by 1. Therefore, $d(P, Q)$ is equal to BLOCKS(P, Q) − CYCLES(P, Q). □

Armed with this theorem, you should be ready to design an algorithm solving the 2-Break Distance Problem.

Charging Station (From Genomes to the Breakpoint Graph): You may be wondering how the graph representation that we have been using for breakpoint graphs could be transformed into an adjacency list. After all, we haven't even labeled the nodes of this graph! Check out this Charging Station to see how to implement the genome graph.

We now know how to compute the 2-break distance, but we would also like to reconstruct a collection of 2-breaks making up a shortest path between two genomes. This problem is called **2-break sorting**, which we leave to you as an exercise.

2-Break Sorting Problem:
Find a shortest transformation of one genome into another by 2-breaks.

Input: Two genomes with circular chromosomes on the same synteny blocks.
Output: The sequence of genomes resulting from applying a shortest sequence of 2-breaks transforming one genome into the other.

Charging Station (Solving the 2-Break Sorting Problem): The Breakpoint Theorem guarantees that there is always a 2-break reducing the number of red-blue cycles in the breakpoint graph by 1. However, it does not tell us how to find such a 2-break. Check out this Charging Station to see how this can be done.

Having proved the formula $d(P, Q) = \text{BLOCKS}(P, Q) - \text{CYCLES}(P, Q)$ for the 2-break distance between genomes with multiple circular chromosomes, we wonder whether we can find an analogous formula for the reversal distance between single linear chromosomes.

Exercise Break: Compute the 2-break distance between the circularized human and mouse X chromosomes. Can you transform a series of 2-breaks for these chromosomes into a series of reversals sorting the linear X chromosomes?

It turns out that a polynomial algorithm for sorting permutations by reversals does exist, yielding an exact formula for the reversal distance! Although this algorithm also relies on the notion of the breakpoint graph, it is unfortunately too complicated to present here (see **DETOUR: Sorting Linear Permutations by Reversals**).

The breakpoint graph constructed on the 280 human-mouse synteny blocks contains 35 alternating cycles, so that the 2-break distance between these genomes is $280 - 35 = 245$. Again, we don't know exactly how many 2-breaks happened in the last 75 million years, but we are certain that there were *at least* 245 steps. Remember this fact, since it will prove important in the next section.

Rearrangement Hotspots in the Human Genome

The Random Breakage Model meets the 2-Break Distance Theorem

You have probably anticipated from the beginning of the chapter that we would eventually argue against the Random Breakage Model. But it may still be unclear to you how the 2-break distance could possibly be used to do so.

Rearrangement Hotspots Theorem: *There are rearrangement hotspots in the human genome.*

Proof. Recall that if the Random Breakage Model is correct, then N reversals applied to a chromosome will produce approximately $2N$ breakpoints, since the probability is very low that two nearby locations in the genome will be used as the breakage point of more than one reversal. These $2N$ breakpoints form $2N + 1$ synteny blocks in the case of a linear chromosome and $2N$ synteny blocks in the case of a circular chromosome. Similarly, N random 2-breaks applied to circular chromosomes will produce $2N$ synteny blocks. Since there are 280 human-mouse synteny blocks, there must have been approximately $280/2 = 140$ 2-breaks on the evolutionary path between humans and mice. However, the 2-Break Distance Theorem tells us that there were at least 245 2-breaks on this evolutionary path.

STOP and Think: Is $245 \approx 140$?

Since 245 is much larger than 140, we have arrived at a contradiction, implying that one of our assumptions is incorrect! But the only assumption we made in this proof was "*If the Random Breakage Model is correct...*" Thus, this assumption must have been wrong. □

This argument, which is not a mathematical proof, is nevertheless logically solid. It offers an example of a **proof by contradiction**, in which we begin by assuming the statement that we intend to disprove and then demonstrate how this assumption cannot be true. As a result of the Rearrangement Hotspots Theorem, we conclude that there was breakpoint reuse on the human-mouse evolutionary path. This breakpoint reuse was extensive, as quantified by the large ratio between the actual 2-break distance and what the 2-break distance would have been under the Random Breakage Model ($245/140 = 1.75$).

Of course, our arguments need to be made statistically sound in order to ensure that the discrepancy between the Random Breakage Model's prediction and the 2-break distance is significant. After all, even though genomes are large, there is still a small chance that randomly chosen 2-breaks might occasionally break a genome more than once in a small interval. The necessary statistical analysis is beyond the scope of this book.

The Fragile Breakage Model

But wait — what about Nadeau and Taylor's argument in favor of the Random Breakage Model? We certainly cannot ignore that the lengths of the human-mouse synteny blocks resemble an exponential distribution.

STOP and Think: Can you find anything wrong with Nadeau and Taylor's logic?

The Nadeau and Taylor argument in favor of the Random Breakage Model exemplifies a classic logic fallacy. It is true that if breakage is random, then the histogram of synteny block lengths should follow the exponential distribution. But it is a completely different statement to conclude that just because synteny block lengths follow the exponential distribution, breakage must have been random. The distribution of synteny block lengths certainly provides support for the Random Breakage Model, but it does not prove that it is correct.

Nevertheless, any alternative hypothesis we put forth for the Random Breakage Model must account for the observation that the distribution of synteny block lengths for the human and mouse genomes is approximately exponential.

STOP and Think: Can you propose a different model of chromosome evolution that explains rearrangement hotspots and is consistent with the exponential distribution of synteny block lengths?

The contradiction of the Random Breakage Model led to an alternative **Fragile Breakage Model** of chromosome evolution, which was proposed in 2003. This model states that every mammalian genome is a mosaic of long solid regions, which are rarely affected by rearrangements, as well as short **fragile regions** that serve as putative rearrangement hotspots and that account only for a small fraction of the genome; for humans and mice, these fragile regions make up approximately 3% of the genome. We do not claim that every fragile region represents a rearrangement hotspot (i.e., was used as an endpoint of multiple rearrangements), but the Rearrangement Hotspots Theorem implies that many fragile regions must be rearrangement hotspots.

If we once again follow Occam's razor, then the most reasonable way to allow for exponentially distributed synteny block lengths is if the fragile regions themselves are distributed randomly in the genome; this is in contrast to the Random Breakage Model, which postulates that the breakage points of rearrangements are distributed randomly. Indeed, *randomly* selecting breakpoints within *randomly* distributed fragile regions is not unlike randomly selecting the endpoints of a rearrangement throughout the entire genome. Yet although we now have a model that fits our observations, many questions remain. For example, it is unclear where fragile regions corresponding to rearrangement hotspots are located, or what causes genomic fragility in the first place.

STOP and Think: Consider the following statement: "The exponential distribution of synteny block lengths and extensive breakpoint re-use imply that the Fragile Breakage Model must be true." Is this argument logically sound?

The point we are attempting to make by asking the preceding question is that we will never be able to prove a scientific theory like the Fragile Breakage Model in the same way that we have proved one of the mathematical theorems in this chapter. In fact, many biological theories are based on arguments that a mathematician would view as fallacious; the logical framework used in biology is quite different from that used in mathematics. To take an historical example, neither Darwin nor anyone else has ever proved that evolution by natural selection is the only — or even the most likely — explanation for how life on Earth evolved!

We have already given many reasons to biology professors to send us to Biology 101 boot camp, but now we will probably be rounded up and thrown into the Gulag alongside Intelligent Design proponents. However, the fact remains that not even Darwinism is unassailable; in the 20th Century, this theory was revised into Neo-Darwinism, and there is little doubt that it will continue to evolve.

Epilogue: Synteny Block Construction

Throughout our discussion of genome rearrangements, we assumed that we were given synteny blocks in advance. In this section, we will describe one way of constructing synteny blocks from genomic sequences.

Genomic dot-plots

Biologists sometimes visualize repeated k-mers within a string as a collection of points in the plane; a point with coordinates (x, y) represents identical k-mers occurring at positions x and y in the string. The top panels in Figure 6.20 present two of these **genomic dot plots**. Of course, since DNA is double-stranded, we should expand the notion of repeated k-mers to account for repeats occurring on the complementary strand. In the bottom left panel of Figure 6.20, blue points (x, y) indicate that the k-mers starting at positions x and y of the string are reverse complementary.

Finding shared k-mers

Recall that a synteny block is defined by many similar genes occurring in the same order in two genomes. Since similar genes often share the same k-mers (for an appropriately chosen value of k), let's first find the positions of all k-mers that are shared by the human and mouse X chromosomes. If we choose k to be sufficiently large (e.g., $k = 30$), then it is rather unlikely that shared k-mers represent spurious similarities. A more likely

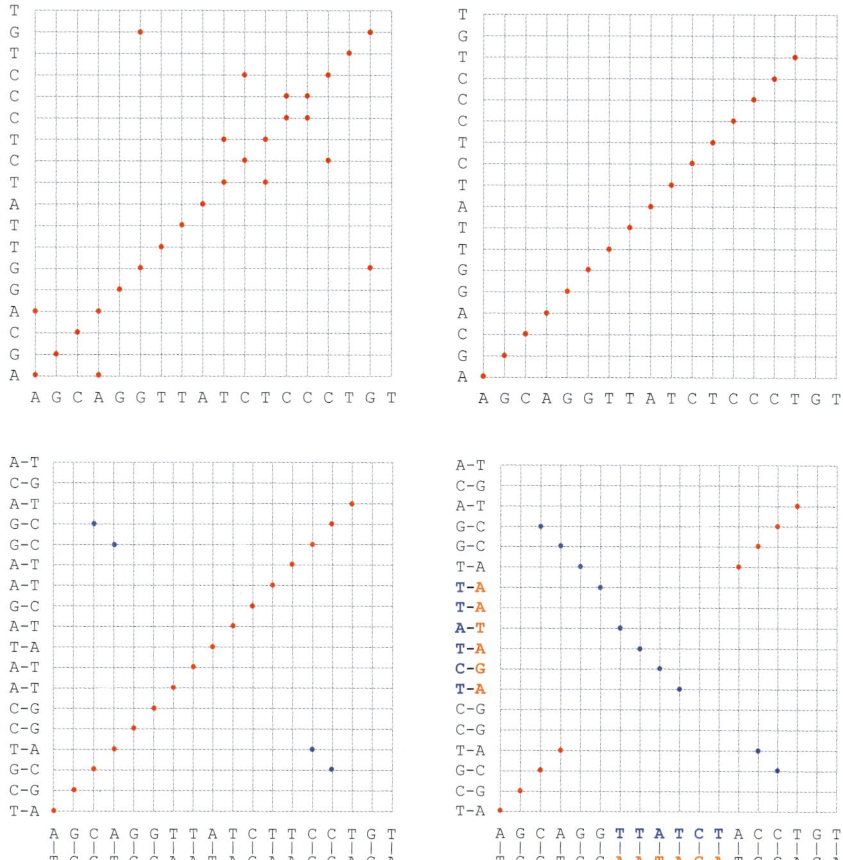

FIGURE 6.20 A visualization of repeated *k*-mers within the string AGCAGGTTATCTCCCTGT for $k = 2$ (top left) and $k = 3$ (top right). (Bottom left) We add blue points to the plot shown in the top right to indicate reverse complementary *k*-mers. For example, CCT and AGG are reverse complementary 3-mers in AGCAGGTTATCTTCCTGT, resulting in a blue dot at position $(14, 3)$ n the dot plot. (Bottom right): Genomic dot-plot showing shared 3-mers between AGCAGG**TTATCT**ACCTGT and AGCAGG**AGATAA**ACCTGT. The latter sequence resulted from the former sequence by a reversal of the segment **TTATCT**. Each point (x, y) corresponds to a *k*-mer shared by the two genomes. Red points indicate identical shared *k*-mers, whereas blue points indicate reverse complementary *k*-mers. Note that the dot-plot has six "spurious" blue points in the diagram: four in the upper left corner, and two in the bottom right corner. You will also notice that red dots can be connected into line segments with slope 1 and blue dots can be connected into line segments with slope -1. The resulting three synteny blocks (AGCAGG, **TTATCT**, and ACCTGT) correspond to three diagonals (each formed by four points) in the dot-plot.

explanation is that they come from related genes (or shared repeats) in the human and mouse genomes.

Formally, we say that a k-mer is **shared** by two genomes if either the k-mer or its reverse complement appears in each genome. Below are four pairs of 3-mers (shown in bold) that are shared by AAACTCATC and TTTCAAATC; note that the second pair of 3-mers are reverse complements of each other.

```
       0             0            4             6
       AAACTCATC     AAACTCATC    AAACTCATC     AAACTCATC
   TTTCAAATC         TTTCAAATC      TTTCAAATC   TTTCAAATC
       4             0              2           6
```

We can further generalize the genomic dot plot to analyze the shared k-mer content of two genomes. We color the point (x, y) red if the two genomes share a k-mer starting at respective positions x and y; we color (x, y) blue if the two genomes have reverse complementary k-mers at these starting positions. See Figure 6.20 (bottom right).

Exercise Break: Find all shared 2-mers of AAACTCATC and TTTCAAATC.

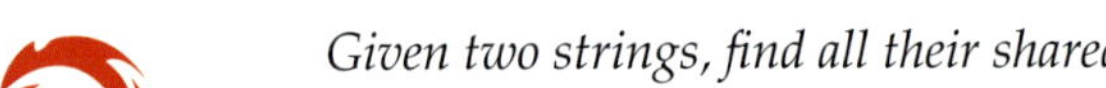

Shared k-mers Problem:
Given two strings, find all their shared k-mers.

Input: An integer k and two strings.
Output: All k-mers shared by these strings, in the form of ordered pairs (x, y) corresponding to starting positions of these k-mers in the respective strings.

Exercise Break: Answer the following questions regarding counting shared k-mers.

1. Compute the expected number of 30-mers shared by two random strings, each a billion nucleotides long.
2. How many shared 30-mers do the *E. coli* and *S. enterica* genomes share?

The *E. coli* and *S. enterica* genomes are both about five million nucleotides long. It can be shown that the expected number of shared 30-mers between two random 5 million nucleotide-long sequences is approximately $2 \cdot (5 \cdot 10^6)^2/4^{30} \approx 1/20{,}000$.

Yet solving the Shared *k*-mers Problem for *E. coli* and *S. enterica* yields over 200,000 pairs (x, y) corresponding to shared 30-mers. The surprisingly large number of shared 30-mers indicates that *E. coli* and *S. enterica* are close relatives that have retained many similar genes inherited from their common ancestor. However, these genes may be arranged in a different order in the two species: how can we infer synteny blocks from these genomes' shared *k*-mers? The genomic dot-plot plot for *E. coli* and *S. enterica* is shown in Figure 6.21.

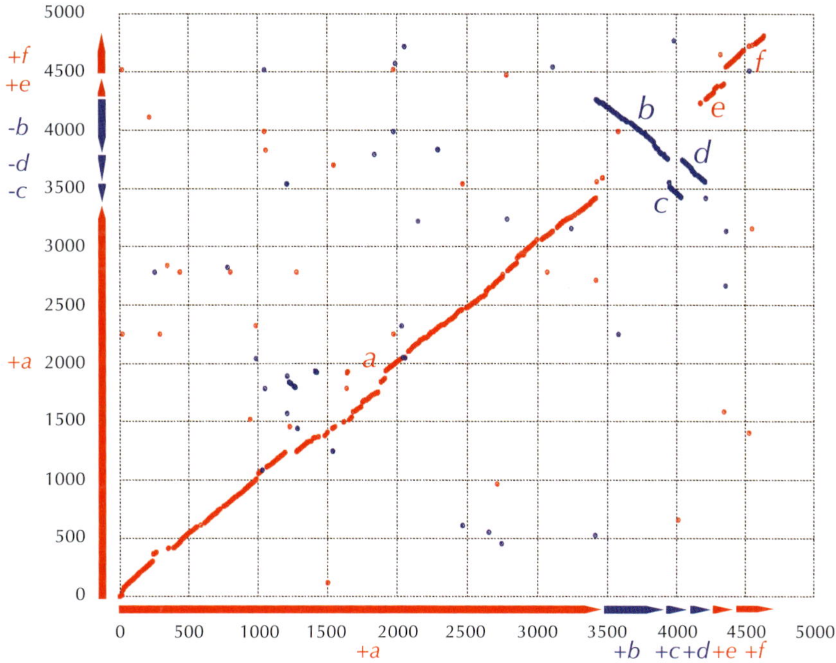

FIGURE 6.21 Genomic dot-plot of *E. coli* (horizontal axis) and *S. enterica* (vertical axis) for $k = 30$. Each point (x, y) corresponds to a *k*-mer shared by the two genomes. Red points indicate identical shared *k*-mers, whereas blue points indicate reverse complementary *k*-mers. Each axis is measured in kilobases (thousands of base pairs).

STOP and Think: Can you see the synteny blocks in the genomic dot-plot in Figure 6.21?

Constructing synteny blocks from shared k-mers

The genomic dot-plot in Figure 6.21 indicates six regions of similarity in the form of points that clump together into approximately diagonal segments. These segments are labeled by a, b, c, d, e, and f according to the order in which they appear in the *E. coli* genome. We ignore smaller diagonals such as the short blue diagonal starting around position 1.3 million in *E. coli* and around position 1.9 million in *S. enterica*. For example, while a corresponds to a long diagonal segment of slope 1 that covers approximately the first 3.5 million positions in both genomes, b corresponds to a shorter diagonal segment of slope -1 that starts shortly before position 3.5 million in *E. coli* and shortly after position 4 million in *S. enterica*. Although b appears small in Figure 6.21, don't be fooled by the scale of the figure; b is over 100,000 nucleotides long and contains nearly 100 genes.

The segments a, b, c, d, e, and f give us the synteny blocks that we have been looking for. If we project these blocks onto the x- and y-axes, then the ordering of blocks on each axis corresponds to the ordering of synteny blocks in the respective bacterium. The ordering of synteny blocks in *E. coli* (plotted on the x-axis) is $(+a\ +b\ +c\ +d\ +e\ +f)$, and the ordering in *S. enterica* (y-axis) is $(+a\ -c\ -d\ -b\ +e\ +f)$. Note that the blue letters in *S. enterica* are assigned a negative sign because these blocks were constructed from reverse complementary k-mers. Figure 6.21 also illustrates what the directions of blocks are — they respectively correspond to diagonals in the dot-plot with slope 1 (blocks with a "$+$" sign) and slope -1 (blocks with a "$-$" sign).

We have therefore represented the relationship between two bacterial genomes using just six synteny blocks. Of course, this simplification required us to throw out some points in the dot plot, corresponding to tiny regions of similarity that did not surpass a threshold length in order to be considered synteny blocks.

We are now ready to construct the eleven human-mouse synteny blocks originally presented in Figure 6.1 (page 298), but since the human and mouse X chromosomes are rather long, we will instead provide you with all positions (x, y) where they share significant similarities. Figure 6.22 (top left) presents the resulting genomic dot-plot for the human and mouse X chromosomes, where each point represents a long similar region rather than a shared k-mer. Our eyes immediately find eleven diagonals in this plot corresponding to the human-mouse X chromosome synteny blocks — problem solved! We state this problem as the Synteny Blocks Problem.

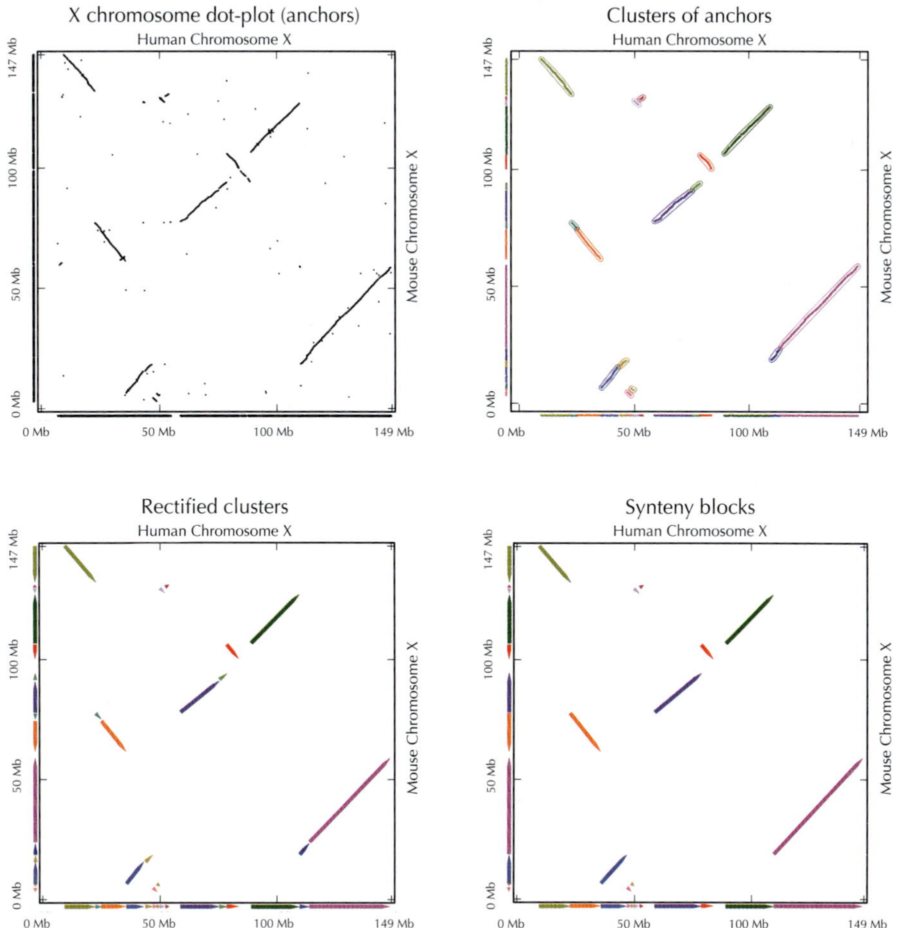

FIGURE 6.22 From local similarities to synteny blocks. (Top left) The genomic dot-plot for the human and mouse X chromosomes, representing all positions (x, y) where they share significant similarities. In contrast with Figure 6.21, we do not distinguish between red and blue dots. (Top right) Clusters (connected components) of points in the genomic dot-plot are formed by constructing the synteny graph. (Bottom left) Rectified clusters from the synteny graph transform each cluster into an exact diagonal of slope ±1. (Bottom right) Aggregated synteny blocks. Projection of the synteny blocks to the x-and y-axes results in the arrangements of synteny blocks in the respective human and mouse genomes (**+1 +2 +3 +4 +5 +6 +7 +8 +9 +10 +11**) and (**+1 -7 +6 -10 +9 -8 +2 -11 -3 +5 +4**), which correspond to the orderings of synteny blocks from Figure 6.2.

Synteny Blocks Problem:
Find diagonals in the genomic dot-plot.

> **Input**: A set of points *DotPlot* in the plane.
> **Output**: A set of diagonals in *DotPlot* representing synteny blocks.

Unfortunately, it remains unclear how to write a program to do what our eyes found to be so easy; we hope you have already noticed that the Synteny Blocks Problem is not a well-formulated computational problem. As we have mentioned, the diagonals in Figure 6.22 (top left) are not perfect. Moreover, there are many gaps within diagonals that cannot be seen by the human eye but will become apparent if we zoom into the genome plot. It is therefore unclear what method the human brain is using to transform the dots into the eleven diagonals in the genomic dot-plot.

STOP and Think: How can we translate the brain's ability to construct the diagonals that you see in Figure 6.22 (top left) into an algorithm that a computer can understand?

Synteny blocks as connected components in graphs

The reason why you can easily see the synteny blocks in a genomic dot-plot is that your brain is good at *clustering* nearby points in an image. To mimic this process with a computer, we therefore need a precise notion of clustering. Given a set of points *DotPlot* in the plane as well as a parameter *maxDistance*, we will construct the (undirected) **synteny graph** SYNTENYGRAPH(*DotPlot*, *maxDistance*) by connecting two points in *DotPlot* with an edge if the distance between them does not exceed *maxDistance*.

Every graph can be divided into disjoint connected subgraphs called **connected components**. The connected components in SYNTENYGRAPH(*DotPlot*, *maxDistance*) represent candidate synteny blocks between the two genomes. The larger the value of *maxDistance*, the fewer connected components, since previously isolated synteny blocks may be combined into a single block. This parameter offers a tradeoff, since we would like to have fewer connected components but do not want to combine spurious synteny blocks, as shown in Figure 6.23 (left).

The synteny graph for the human and mouse X chromosomes contains a huge number of small connected components (the exact number depends on our choice of the *maxDistance* parameter). However, we will ignore these small connected components, since they may represent spurious similarities. We thus introduce a parameter *minSize* representing the minimum number of points in a connected component that we will consider as forming a synteny block. Our goal is to return all connected components having at least *minSize* nodes.

The parameter *minSize* also offers a tradeoff. If *minSize* is too small, then short, spurious synteny blocks will complicate our analysis. If *minSize* is too large, then we may discard correct synteny blocks. The example in Figure 6.23 (right) demonstrates that even after picking an appropriate value of *maxDistance*, there may be no perfect value of *minSize*; in this case, the fact that one spurious synteny block has the same size as the true synteny blocks means that we are unable to exclude it from consideration. In practice, the hope is that correct synteny blocks will be large enough to prevent this issue.

```
SYNTENYBLOCKS(DotPlot, maxDistance, minSize)
    construct SYNTENYGRAPH(DotPlot, maxDistance)
    find the connected components in SYNTENYGRAPH(DotPlot, maxDistance)
    output connected components containing at least minSize nodes as candidate
            synteny blocks
```

As Figure 6.22 (top right) illustrates, **SYNTENYBLOCKS** has the tendency to partition a single diagonal (as perceived by the human eye) into multiple diagonals due to gaps that exceed the parameter *maxDistance*. However, this partitioning is not a problem, since the broken diagonals can be combined later into a single (aggregated) synteny block.

STOP and Think: We have defined synteny blocks as large connected components in SYNTENYGRAPH(*DotPlot*, *maxDistance*) but have not described how to determine where these synteny blocks are located in the original genomes. Using Figure 6.22 as a hint, design an algorithm for finding this information.

You should now be ready to solve the challenge problem and discover that the choice of parameters is one of the dark secrets of bioinformatics research.

Challenge Problem: Construct the synteny blocks for the human and mouse X chromosomes and compute the 2-break distance between the circularized human and mouse X chromosomes using the synteny blocks that you constructed. How does this distance change depending on the parameters *maxDistance* and *minSize*?

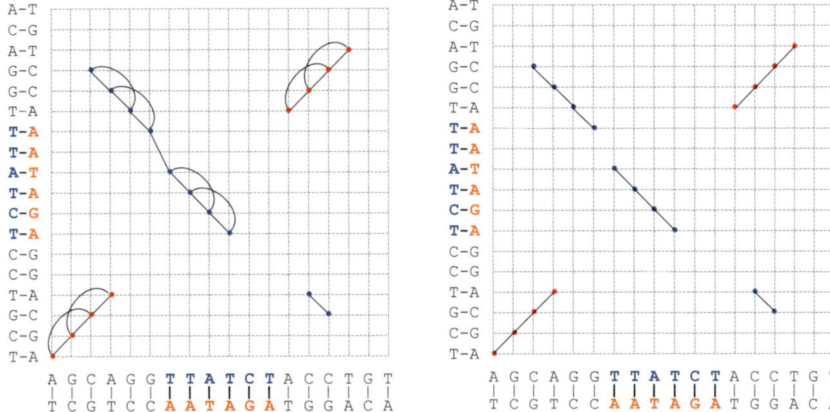

FIGURE 6.23 (Left) The graph SYNTENYGRAPH(*DotPlot*, 3) constructed from the genomic dot-plot of AGCAGG**TTATCT**CCCTGT and AGCAGG**AGATAA**CCCTGT for $k = 3$. Note that the spurious synteny block in the top left has unfortunately been combined with a correct synteny block. (Right) The graph SYNTENYGRAPH(*DotPlot*, 2), which allows us to distinguish between the incorrect and correct synteny blocks that were previously concatenated. However, note that even if we set *minSize* equal to 3, we will be able to discard the small spurious synteny block in the bottom right but not the larger one in the top left, since it has the same size as the correct synteny blocks.

Open Problem: Can Rearrangements Shed Light on Bacterial Evolution?

Although there exist efficient algorithms for analyzing *pairwise* genome rearrangements, constructing rearrangement scenarios for *multiple* genomes remains an open problem. For example, we now know how to find a most parsimonious rearrangement scenario transforming the mouse X chromosome into the human X chromosome. However, the problem of finding a most parsimonious rearrangement scenario for the human, mouse, and rat X chromosomes — let alone for their entire genomes — is a more difficult problem. The difficulties further amplify when we attempt to reconstruct a rearrangement history for dozens of mammalian genomes. To address this challenge, we start from the simpler (but still unsolved) case of bacterial genomes.

Let *Tree* be a tree (i.e., a connected acyclic undirected graph) with nodes labeled by some genomes. In the case of bacterial genomes, we assume that every node (genome) is labeled by a circular permutation on n elements. Given an edge e connecting nodes v and w in *Tree*, we define $\text{DISTANCE}(v, w)$ as the 2-break distance between genomes v and w. The **tree distance** $\text{DISTANCE}(\textit{Tree})$ is the sum

$$\sum_{\text{all edges } (v,w) \text{ in } \textit{Tree}} \text{DISTANCE}(v, w).$$

Given a set of genomes $P_1, \ldots, P_n$ and an evolutionary tree *Tree* with n leaves labeled by $P_1, \ldots, P_n$, the Ancestral Genome Reconstruction Problem attempts to reconstruct genomes at the internal nodes of the tree such that $\text{DISTANCE}(\textit{Tree})$ is minimized across all possible reconstructions of genomes at internal nodes.

Ancestral Genome Reconstruction Problem:
Given a tree with leaves labeled by genomes, reconstruct ancestral genomes that minimize the tree distance.

Input: A tree *Tree* with each leaf labeled by a genome.
Output: Genomes *AncestralGenomes* assigned to the internal nodes of *Tree* such that $\text{DISTANCE}(\textit{Tree})$ is minimized across all possible choices of *AncestralGenomes*.

In the case when *Tree* is not given, we need to infer it from the genomes.

Multiple Genome Rearrangement Problem:
Given a set of genomes, reconstruct a tree with leaves labeled by these genomes and minimum tree distance.

Input: A set of genomes.
Output: A tree *Tree* with leaves labeled by these genomes and internal nodes labeled by (unknown) genomes *AncestralGenomes* such that DISTANCE(*Tree*) is minimal among all possible choices of *Tree* and *AncestralGenomes*.

Although many heuristics have been proposed for the Multiple Genome Rearrangement Problem, they have mainly been applied to analyze mammalian evolution. However, there have been hardly any applications of the Multiple Genome Rearrangement Problem for analyzing bacterial evolution. The fact that bacterial genomes are approximately 1,000 times smaller than mammalian genomes does not make this problem 1,000 times easier. In fact, there are unique challenges and opportunities in bacterial evolutionary research.

Consider 100 genomes from three closely related bacterial genera, *Salmonella*, *Shigella*, and *Escherichia*, whose various species are responsible for dysentery, typhoid fever, and a variety of foodborne illnesses. After you construct synteny blocks shared by all these genomes, you will see that there are relatively few (usually fewer than 10) rearrangements between every pair of genomes. However, solving the Multiple Genome Rearrangement Problem even in the case of closely related genomes presents a formidable challenge, and nobody has been able to construct a rearrangement scenario for more than a couple dozen — let alone 100! — species yet.

After you solve this puzzle, you will be able to address the question of whether there are rearrangement hotspots in bacterial genomes. Answering this question for a pair of bacterial genomes, like we did for the human and mouse genomes, may not be possible because there are typically fewer than 10 rearrangements between them. But answering this question for 100 bacterial genomes may be possible if we witness the same breakage occurring independently on many branches of the evolutionary tree. However, you will need to develop algorithms to analyze rearrangement hotspots in multiple (rather than pairwise) genomes.

After you construct the evolutionary tree, you will also be in a position to analyze the question of what triggers rearrangements. While many authors have discussed the causes of fragility, this question remains open, with no shortage of hypotheses. Many rearrangements are flanked by **matching duplications**, a pair of long similar regions located within a pair of breakpoint regions corresponding to a rearrangement event. However, it remains unclear what triggers rearrangements in bacteria; can you answer this question?

Charging Stations

From genomes to the breakpoint graph

Our goal is to count the number of cycles in the breakpoint graph and therefore solve the 2-Break Distance Problem. First, however, we will need to obtain a convenient graph representation of genomes. In the main text, we represented a circular chromosome by converting each synteny block into a directed edge, and then connected adjacent synteny blocks in the chromosome with red edges. Although this provided us with a way of visualizing genomes, it is not immediately clear how to represent this graph with an adjacency list.

Given a genome P, we will represent its synteny blocks not as letters but as integers from 1 to $n = |P|$. For example, $(+a\ -b\ -c\ +d)$ will be represented as $(+1\ -2\ -3\ +4)$ (Figure 6.24 (left)). Then, we will convert the directed gray edges of P into undirected edges as follows. Given a directed edge labeled by integer x, we assign the node at the "head" of this edge as x_h and the node at the "tail" of this edge as x_t. For example, we replace the directed edge labeled "2" in Figure 6.24 (left) with an undirected edge connecting nodes 2_t and 2_h (Figure 6.24 (middle)). This results in the cyclic sequence of nodes $(1_t, 1_h, 2_h, 2_t, 3_h, 3_t, 4_t, 4_h)$. Finally, to simplify analysis of this graph even further,

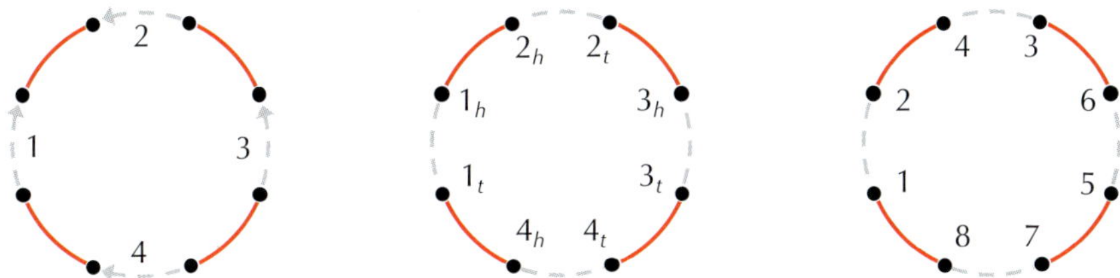

FIGURE 6.24 (Left) The circular chromosome $(+a\ -b\ -c\ +d)$ can be represented as $(+1\ -2\ -3\ +4)$ using integers. (Middle) Representing this chromosome by replacing the gray directed edges with undirected edges connecting "heads" and "tails" of each synteny block. (Right) Encoding head and tail nodes as integers. 1_t and 1_h are converted into 1 and 2; 2_t and 2_h are converted into 3 and 4; and so on. The original chromosome has been converted into the alternating cycle $(1, 2, 4, 3, 6, 5, 7, 8)$.

instead of using x_h and x_t to denote the head and tail of synteny block x, we will use the integers $2x$ and $2x - 1$, respectively (Figure 6.24 (right)). With this encoding, the original genome $(+1\ -2\ -3\ +4)$ is transformed into the cyclic sequence of nodes $(1, 2, 4, 3, 6, 5, 7, 8)$.

STOP and Think: Is the transformation illustrated in Figure 6.24 invertible? In other words, if we were to give you a cyclic sequence of nodes labeled from 1 to $2n$, could you reconstruct the chromosome with n synteny blocks from which it derives?

The following pseudocode bypasses the intermediate step in Figure 6.24 (middle) of assigning "head" and "tail" nodes in order to transform a single circular chromosome $Chromosome = (Chromosome_1, \ldots, Chromosome_n)$ into a cycle represented as a sequence of integers $Nodes = (Nodes_1, \ldots, Nodes_{2n})$.

6F

```
CHROMOSOMETOCYCLE(Chromosome)
    for j ← 1 to |Chromosome|
        i ← Chromosome_j
        if i > 0
            Nodes_{2j−1} ← 2i − 1
            Nodes_{2j} ← 2i
        else
            Nodes_{2j−1} ← −2i
            Nodes_{2j} ← −2i − 1
    return Nodes
```

This process is in fact invertible, as described by the following pseudocode.

```
CYCLETOCHROMOSOME(Nodes)
    for j ← 1 to |Nodes|/2
        if Nodes_{2j−1} < Nodes_{2j}
            Chromosome_j ← Nodes_{2j}/2
        else
            Chromosome_j ← −Nodes_{2j−1}/2
    return Chromosome
```

ChromosomeToCycle generates the sequence of nodes of a chromosome, but it does not explicitly add the edges. Any genome P with n synteny blocks will have the gray undirected edges BlackEdges(P) = $(1,2), (3,4), \ldots, (2n-1, 2n)$.

We now define ColoredEdges(P) as the set of colored edges in the graph of P. For the example in Figure 6.24, the set ColoredEdges(P) contains the edges $(2,4)$, $(3,6)$, $(5,7)$, and $(8,1)$.

The following algorithm constructs ColoredEdges(P) for a genome P. In this pseudocode, we will assume that an n-element array $(a_1, \ldots, a_n)$ has an invisible $(n+1)$-th element that is equal to its first element, i.e., $a_{n+1} = a_1$.

```
ColoredEdges(P)
    Edges ← an empty set
    for each chromosome Chromosome in P
        Nodes ← ChromosomeToCycle(Chromosome)
        for j ← 1 to |Chromosome|
            add the edge (Nodes_2j, Nodes_2j+1) to Edges
    return Edges
```

The colored edges in the breakpoint graph of P and Q are given by ColoredEdges(P) together with ColoredEdges(Q). Note that some edges in these two sets may connect the same two nodes, which results in trivial cycles.

Although we are now ready to solve the 2-Break Distance Problem, we will later find it helpful to implement a function converting a genome graph back into a genome.

```
GraphToGenome(GenomeGraph)
    P ← an empty set of chromosomes
    for each cycle in GenomeGraph
        Nodes ← sequence of nodes in this cycle (starting from node 1)
        Chromosome ← CycleToChromosome(Nodes)
        add Chromosome to P
    return P
```

Solving the 2-Break Sorting Problem

Note: This Charging Station uses some notation from **Charging Station: From Genomes to the Breakpoint Graph**.

Figure 6.25 (top) illustrates how a 2-break replaces colored edges $(1,6)$ and $(3,8)$ in a genome graph with two new colored edges $(1,3)$ and $(6,8)$. We will denote this operation as 2-BREAK$(1,6,3,8)$. Note that the order of the nodes in this function matter, since the operation 2-BREAK$(1,6,8,3)$ would represent a different 2-break that replaces $(1,6)$ and $(3,8)$ with $(1,8)$ and $(6,3)$ (Figure 6.25 (bottom)).

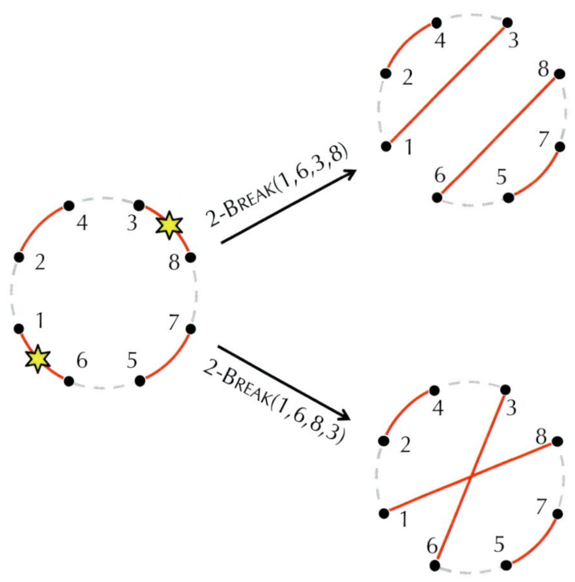

FIGURE 6.25 Operations 2-BREAK$(1, 6, 3, 8)$ (top) and 2-BREAK$(1, 6, 8, 3)$ (bottom) on the genome $(+1\ -2\ -3\ +4)$ from Figure 6.24.

The following pseudocode describes how 2-BREAK(i_1, i_2, i_3, i_4) transforms a genome graph.

```
2-BREAKONGENOMEGRAPH(GenomeGraph, i1, i2, i3, i4)
    remove colored edges (i1, i2) and (i3, i4) from GenomeGraph
    add colored edges (i1, i3) and (i2, i4) to GenomeGraph
    return GenomeGraph
```

We can extend this pseudocode to a 2-break defined on genome P.

```
2-BREAKONGENOME(P, i1, i2, i3, i4)
    GenomeGraph ← BLACKEDGES(P) and COLOREDEDGES(P)
    GenomeGraph ← 2-BREAKONGENOMEGRAPH(GenomeGraph, i1, i2, i3, i4)
    P ← GRAPHTOGENOME(GenomeGraph)
    return P
```

We are now ready to find a series of intermediate genomes in a shortest transformation of P into Q by 2-breaks. The idea of our algorithm is to find a 2-break that will increase the number of red-blue cycles in the breakpoint graph by 1. To do so, as illustrated in Figure 6.26, we select an arbitrary blue edge in a non-trivial alternating red-blue cycle and perform the 2-break on the two red edges flanking this blue edge in order to split the red-blue cycle into two cycles (at least one of which is trivial). The following pseudocode uses the concept of an edge being **incident** to a node v if v is one of the edge's endpoints.

```
SHORTESTREARRANGEMENTSCENARIO(P, Q)
    output P
    RedEdges ← COLOREDEDGES(P)
    BlueEdges ← COLOREDEDGES(Q)
    BreakpointGraph ← the graph formed by RedEdges and BlueEdges
    while BreakpointGraph has a non-trivial cycle Cycle
        (i1, i2, i3, i4) ← path starting at arbitrary blue edge in nontrivial red-blue cycle
        RedEdges ← RedEdges with edges (i1, i2) and (i3, i4) removed
        RedEdges ← RedEdges with edges (i1, i4) and (i2, i3) added
        BreakpointGraph ← the graph formed by RedEdges and BlueEdges
        P ← 2-BREAKONGENOME(P, i1, i2, i4, i3)
        output P
```

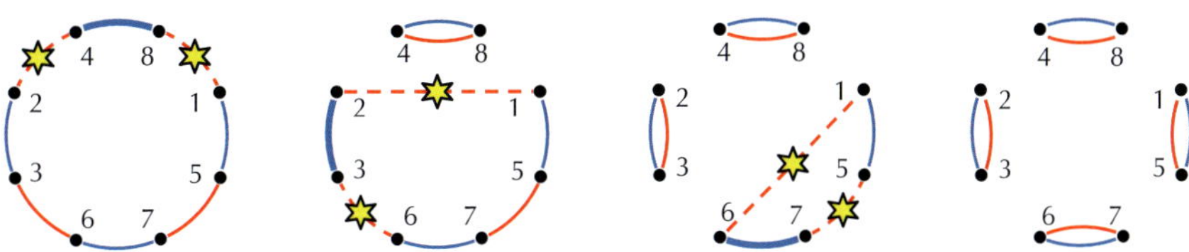

FIGURE 6.26 A shortest 2-break transformation (from left to right) of the breakpoint graph of $P = (+a\ -b\ -c\ +d)$ and $Q = (+a\ +b\ -d\ -c)$. The arbitrary blue edge selected by **SHORTESTREARRANGEMENTSCENARIO** at each step is shown in bold, and the red edges on either side of it are dashed (with stars indicating 2-breaks).

Exercise Break: Classify each of three 2-breaks shown in Figure 6.26 as a reversal, fusion, or fission.

Detours

Why is the gene content of mammalian X chromosomes so conserved?

Whereas mammalian X chromosomes are enriched in genes related to sexual reproduction, most of the approximately 1,000 genes on the X chromosome have nothing to do with gender. Ideally, they should be expressed (i.e., transcribed and eventually translated) in roughly the same quantities in females and males. But since females have two X chromosomes and males have only one, it would seem that all the X chromosome genes should have twice the expression level in females. This imbalance would cause a problem in the complex system of checks and balances underlying gene expression.

The need to balance gene expression in males and females led to the evolution of **dosage compensation**, or the inactivation of one X chromosome in females to equalize gene expression between the sexes. Because of dosage compensation, the gene content of the X chromosome is highly conserved between mammalian species because if a gene jumps off the X chromosome to any other chromosome, it will be present on two chromosomes in both males and females. As a result, its expression may double compared to its normal level, thus creating a gene expression imbalance.

Discovery of genome rearrangements

After Sturtevant discovered genome rearrangements in *Drosophila* in 1921, another breakthrough occurred with the discovery that the salivary glands of *Drosophila* contain **polytene cells**. In normal cellular division, each daughter cell receives one copy of the genome. However, in the nuclei of polytene cells, DNA replication occurs repeatedly in the absence of cell division. The resulting chromosomes then knit themselves together into much larger "superchromosomes" called **polytene chromosomes**.

Polytene chromosomes serve a practical purpose for the fruit fly, which uses the extra DNA to boost the production of gene transcripts, producing lots of sticky saliva. But the human value of polytene chromosomes is perhaps greater. When Sturtevant and his collaborator, Theodosius Dobzhansky, looked at polytene chromosomes under a microscope, they were able to witness the work of rearrangements firsthand in tangled mutant chromosomes. In 1938, they published a milestone paper with an evolutionary tree presenting a rearrangement scenario with seventeen reversals for various species of *Drosophila*, the first evolutionary tree ever to be constructed based on molecular data.

The exponential distribution

A **Bernoulli trial** is a random experiment with two possible outcomes, "success" (having probability p) and "failure" (having probability $1 - p$). The **geometric distribution** is the probability distribution underlying the random variable X representing the number of Bernoulli trials needed to obtain the first success:

$$\Pr(X = k) = (1 - p)^{k-1} p.$$

A **Poisson process** is a continuous-time probabilistic process counting the number of events in a given time interval, if we assume that the events occur independently and at a constant rate. For example, the Poisson process offers a good model of time points for passengers arriving to a large train station. If we assume that the number of passengers arriving during a very small time interval ϵ is $\lambda \cdot \epsilon$ (where λ is a constant), then we are interested in the probability $F(X)$ that nobody will arrive to the station during a time interval X. The **exponential distribution** describes the time between events in a Poisson process.

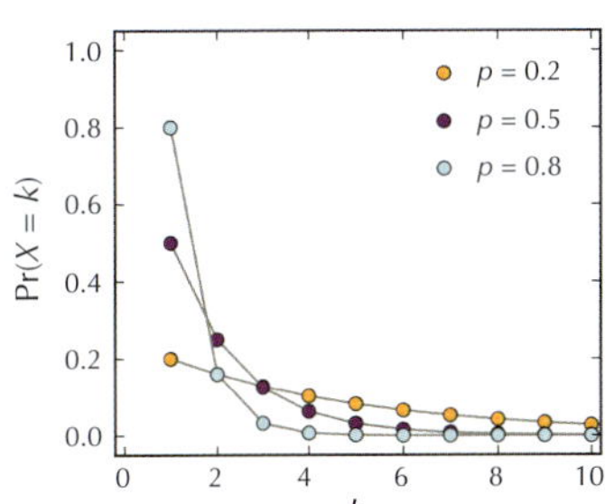

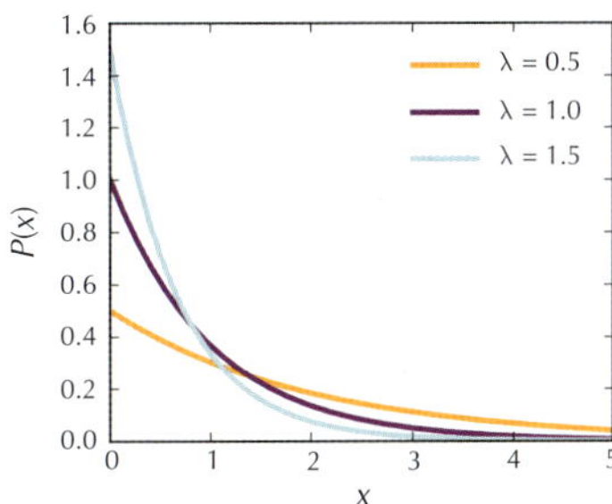

FIGURE 6.27 The probability density functions of the geometric (left) and exponential (right) distributions, each provided for three different parameter values.

STOP and Think: Do you see any similarities between the Poisson process and the Bernoulli trials or between the exponential and geometric distributions?

The exponential distribution is the continuous analogue of the geometric distribution. Precisely, a Poisson process is characterized by a **rate parameter** λ, such that the number of events k in a time interval of duration ϵ follows the **Poisson probability distribution**:

$$e^{-\lambda\cdot\epsilon}(\lambda\cdot\epsilon)^k \Big/ k!$$

The probability density function of the exponential distribution is $\lambda e^{-\lambda \cdot X}$ (compare with the geometric distribution shown in Figure 6.27).

Bill Gates and David X. Cohen flip pancakes

Before biologists faced genome rearrangement problems, mathematicians posed the **Pancake Flipping Problem**, arising from the following waiter's conundrum.

> *The chef in our place is sloppy, and when he prepares a stack of pancakes they come out all different sizes. Therefore, when I deliver them to a customer, on the way to a table I rearrange them (so that the smallest winds up on top, and so on, down to the largest at the bottom) by grabbing several from the top and flipping them over, repeating this (varying the number I flip) as many times as necessary. If there are n pancakes, what is the maximum number of flips that I will ever have to use to rearrange them?*

Formally, a **prefix reversal** is a reversal that flips a prefix, or initial interval, of a permutation. The Pancake Flipping Problem corresponds to sorting unsigned permutations by prefix reversals. For example, the series of prefix reversals shown below ignores signs and represents the sorting of the **unsigned permutation,** (1 7 6 10 9 8 2 11 3 5 4), into the **identity unsigned permutation,** (1 2 3 4 5 6 7 8 9 10 11). The inverted interval is shown in red, and sorted intervals at the end of the permutation are shown in blue.

(1 7 6 10 9 8 2 11 3 5 4)
(11 2 8 9 10 6 7 1 3 5 4)
(4 5 3 1 7 6 10 9 8 2 11)
(10 6 7 1 3 5 4 9 8 2 11)
(2 8 9 4 5 3 1 7 6 10 11)
(9 8 2 4 5 3 1 7 6 10 11)
(6 7 1 3 5 4 2 8 9 10 11)
(7 6 1 3 5 4 2 8 9 10 11)
(2 4 5 3 1 6 7 8 9 10 11)
(5 4 2 3 1 6 7 8 9 10 11)
(1 3 2 4 5 6 7 8 9 10 11)
(3 1 2 4 5 6 7 8 9 10 11)
(2 1 3 4 5 6 7 8 9 10 11)
(1 2 3 4 5 6 7 8 9 10 11)

When we search for a shortest series of prefix reversals sorting a *signed* permutation, the problem is called the **Burnt Pancake Flipping Problem** (each pancake is "burnt" on one side, giving it two possible orientations).

STOP and Think: Prove that every unsigned permutation of length n can be sorted using at most $2 \cdot (n-1)$ prefix reversals. Prove that every signed permutation of length n can be sorted using at most $3 \cdot (n-1)+1$ prefix reversals.

In the mid-1970s, Bill Gates, an undergraduate student at Harvard, and Christos Papadimitriou, Gates's professor, made the first attempt to solve the Pancake Flipping Problem and proved that any permutation of length n can be sorted with at most $5/3 \cdot (n+1)$ prefix reversals, a result that would not be improved for three decades. David X. Cohen worked on the Burnt Pancake Flipping Problem at Berkeley before he left computer science to become a writer for *The Simpsons* and eventually producer of *Futurama*. Along with Manuel Blum, he demonstrated that the Burnt Pancake Flipping Problem can be solved with at most $2 \cdot (n-1)$ prefix reversals.

Sorting linear permutations by reversals

In the main text, we defined the breakpoint graph for circular chromosomes, but this structure can easily be extended to linear chromosomes. Figure 6.28 depicts the human and the mouse X chromosomes as alternating red-gray and blue-gray paths (first and second panels). These two paths are superimposed in the third panel to form the breakpoint graph, which has five alternating red-blue cycles.

STOP and Think: Prove the following analogue of the Cycle Theorem for permutations: Given permutations P and Q, any reversal applied to P can increase $\text{CYCLES}(P, Q)$ by at most 1.

Whereas the number of trivial cycles is equal to $\text{BLOCKS}(Q, Q)$ in the trivial breakpoint graph of a circular permutation, the trivial breakpoint graph of a linear permutation has $\text{BLOCKS}(Q, Q) + 1$ trivial cycles. Since the Cycle Theorem holds for linear permutations, perhaps the reversal distance $d_{\text{rev}}(P, Q)$ is equal to $\text{BLOCKS}(P, Q) + 1 - \text{CYCLES}(P, Q)$ for linear chromosomes? After all, for the human and mouse X chromosomes, $\text{BLOCKS}(P, Q) + 1 - \text{CYCLES}(P, Q)$ is equal to $11 + 1 - 5 = 7$, which we already know to be the reversal distance between the human and mouse X chromosomes.

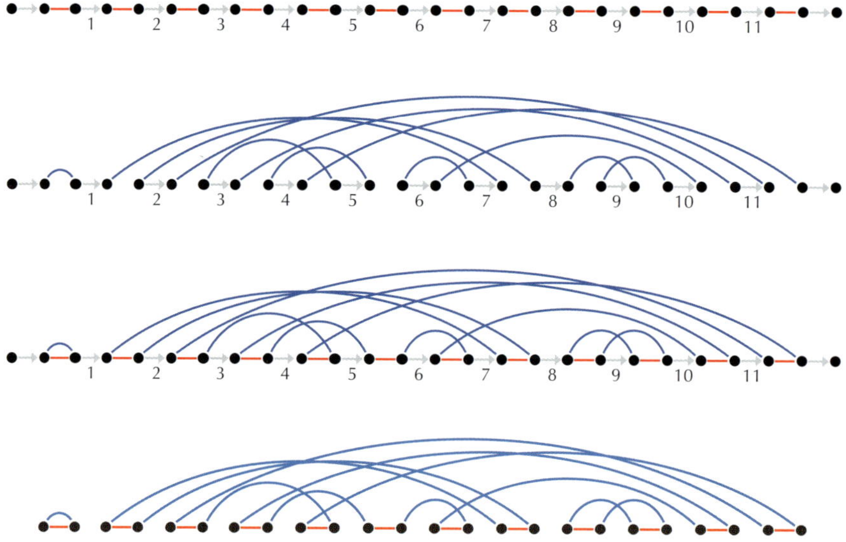

FIGURE 6.28 (1st panel) An alternating red-gray path representing the human X chromosome (+1 +2 +3 +4 +5 +6 +7 +8 +9 +10 +11). (2nd panel) An alternating blue-gray path representing the mouse X chromosome (+1 −7 +6 −10 +9 −8 +2 −11 −3 +5 +4). (3rd panel) The breakpoint graph of the mouse and human X chromosomes is obtained by superimposing red-gray and blue-gray paths from the first two panels. (4th panel) To highlight the five alternating red-blue cycles in the breakpoint graph, gray edges are removed.

STOP and Think: Can you modify the proof of the 2-Break Distance Theorem to prove that $d_{\text{rev}}(P, Q) = \text{BLOCKS}(P, Q) + 1 - \text{CYCLES}(P, Q)$ for linear permutations P and Q?

You can verify that if $P = (+2\ +1)$ and $Q = (+1\ +2)$, then $d_{\text{rev}}(P, Q) \neq \text{BLOCKS}(P, Q) + 1 - \text{CYCLES}(P, Q)$. This makes it unlikely that we will be able to develop a simple algorithm for the computation of reversal distance.

However, the lower bound $d_{\text{rev}}(P, Q) \geq \text{BLOCKS}(P, Q) + 1 - \text{CYCLES}(P, Q)$ approximates the reversal distance between linear permutations extremely well. This intriguing performance raised the question of whether this bound is close to an exact formula. In 1999, Hannenhalli and Pevzner found this formula by defining two special types of breakpoint graph structures called "hurdles" and "fortresses". Denoting the number of hurdles and fortresses in BREAKPOINTGRAPH(P, Q) by HURDLES(P, Q)

and $\textsc{Fortresses}(P, Q)$, respectively, they proved that the reversal distance $d_{\text{rev}}(P, Q)$ is given by

$$\textsc{Blocks}(P, Q) + 1 - \textsc{Cycles}(P, Q) + \textsc{Hurdles}(P, Q) + \textsc{Fortresses}(P, Q).$$

Using this formula, they developed a polynomial algorithm for computing $d_{\text{rev}}(P, Q)$. Nevertheless, $\textsc{Hurdles}(P, Q)$ and $\textsc{Fortresses}(P, Q)$ are small for the vast majority of permutations, and so the lower bound $\textsc{Blocks}(P, Q) + 1 - \textsc{Cycles}(P, Q)$ is a good approximation of the reversal distance in practice.

Bibliography Notes

Alfred Sturtevant was the first to discover rearrangements while comparing gene orders in fruit flies (Sturtevant, 1921). Together with Theodosius Dobzhansky, Sturtevant pioneered the analysis of genome rearrangements, publishing a milestone paper that presented a rearrangement scenario for many fruit fly species (Sturtevant and Dobzhansky, 1936). The Random Breakage Model was proposed by Ohno, 1973, further developed by Nadeau and Taylor, 1984, and refuted by Pevzner and Tesler, 2003b.

The notion of the breakpoint graph was proposed by Bafna and Pevzner, 1996. The polynomial algorithm for sorting by reversals was developed by Hannenhalli and Pevzner, 1999. The synteny block construction algorithm presented in this chapter was described by Pevzner and Tesler, 2003a. The 2-break operation was introduced in Yancopoulos, Attie, and Friedberg, 2005 under the name of "double cut and join".

The first algorithmic analysis of the Pancake Flipping problem was described by Gates and Papadimitriou, 1979. The first algorithmic analysis of the Burnt Pancake Flipping problem was described by Cohen and Blum, 1995.

The Multiple Genome Rearrangement problem was addressed by Ma et al., 2008 and Alekseyev and Pevzner, 2009. Zhao and Bourque, 2009 observed that matching duplications may trigger genome rearrangements.

Which Animal Gave Us SARS?

Evolutionary tree reconstruction

The Fastest Outbreak

Trouble at the Metropole Hotel

On February 21, 2003, a Chinese doctor named Liu Jianlun flew to Hong Kong to attend a wedding and checked into Room 911 of the Metropole Hotel. The next day, he became too ill to attend the wedding and was admitted to a hospital. Two weeks later, Dr. Liu was dead.

On his deathbed, Liu told doctors that he had recently treated sick patients in Guangdong Province, China where a deadly, highly contagious respiratory illness had infected hundreds of people. The Chinese government had made brief mention of this incident to the World Health Organization but had concluded that the likely culprit was a common bacterial infection.

By the time anyone realized the severity of the disease, it was already too late to stop the outbreak. On February 23, a man who had stayed across the hall from Dr. Liu at the Metropole traveled to Hanoi and died after infecting 80 people. On February 26, a woman checked out of the Metropole, traveled back to Toronto, and died after initiating an outbreak there. On March 1, a third guest was admitted to a hospital in Singapore, where sixteen additional cases of the illness arose within two weeks.

Consider that it took four years for the Black Death, which killed over a third of all Europeans in the 14th Century, to travel from Constantinople to Kiev. Or that HIV took two decades to circle the globe. In contrast, this mysterious new disease had crossed the Pacific Ocean within a week of entering Hong Kong.

As health officials braced for the impact of the fastest-traveling pandemic in human history, panic set in. Businesses were closed, sick passengers were removed from airplanes, and Chinese officials threatened to execute infected patients who violated quarantine.

International travel may have helped the disease spread rapidly, but international collaboration would eventually contain it. In a matter of a few weeks, biologists identified a virus that had caused the epidemic and sequenced its genome. In the process, the mysterious new disease earned a name: **Severe Acute Respiratory Syndrome**, or **SARS**.

The evolution of SARS

The virus causing SARS belongs to a family of viruses called **coronaviruses**, which are named after the Latin *corona* (meaning "crown") because the virus particle resembles the sun's corona (Figure 7.1). Coronaviruses infect the respiratory tracts of mammals and birds but typically cause only minor problems, like the common cold. Before SARS, no one believed that a coronavirus could wreak such havoc.

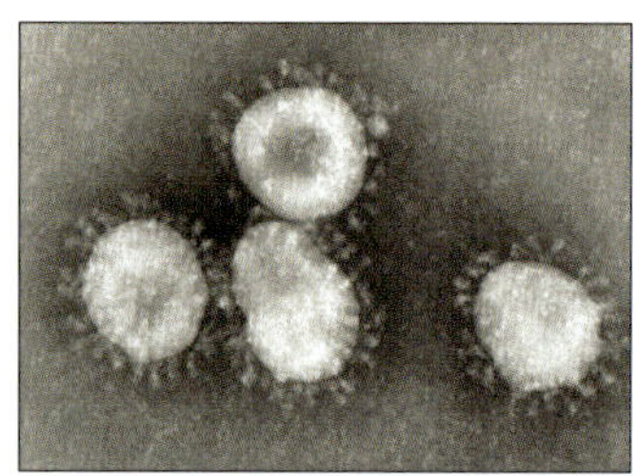

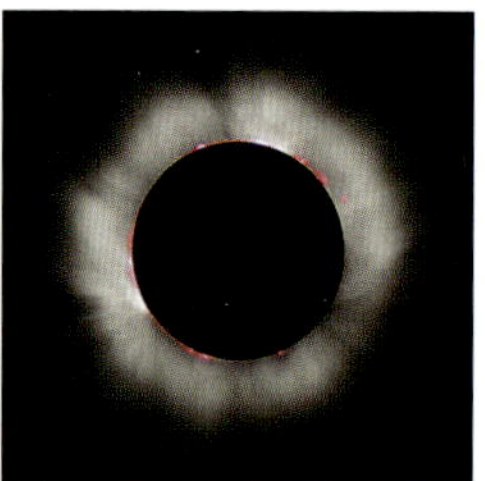

FIGURE 7.1 (Left) Coronavirus particles. (Right) A solar eclipse with the sun's corona visible.

Coronaviruses, influenza viruses, and HIV are all **RNA viruses**, meaning that they possess RNA instead of DNA. RNA replication has a higher error rate than DNA replication, and so RNA viruses are capable of mutating more quickly into divergent strains. The rapid mutation of RNA viruses explains why the flu shot changes from year to year and why there are many different subtypes of HIV.

SARS researchers initially hypothesized that, like HIV and influenza, the **SARS coronavirus** (abbreviated as **SARS-CoV**) had jumped from animals to humans. They first named birds as the likely suspect because of the similarities between SARS and "bird flu", a form of influenza originating in chickens that is difficult to transmit to humans but is even deadlier than SARS, killing over half of the people it infects. Yet when researchers sequenced the 29,751 nucleotide-long SARS-CoV genome in April 2003, it became evident that SARS did not come from birds because its genome did not resemble avian coronaviruses.

By fall 2003, researchers had sequenced many SARS-CoV strains from patients in various countries, but many questions still remained unanswered. How did SARS-CoV cross the species barrier to humans? When and where did it happen? How did SARS spread around the world, and who infected whom?

Each of these questions about SARS is ultimately related to the problem of constructing **evolutionary trees** (also known as **phylogenies**). For another example, by constructing an evolutionary tree of primate viruses related to HIV (Figure 7.2), scientists inferred that HIV was transmitted to humans on five separate occasions (see PAGE 401 DETOUR: When Did HIV Jump From Primates to Humans?). But what algorithm did they use to construct this phylogeny?

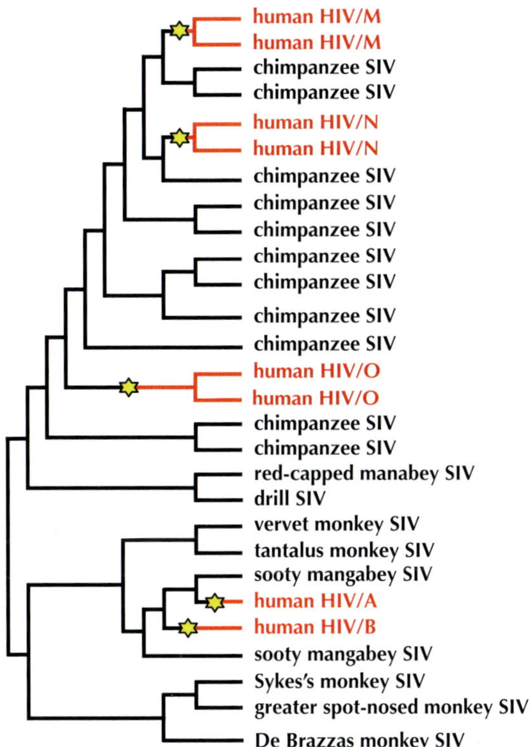

FIGURE 7.2 HIV comprises five different viral families, denoted as A, B, M, N, and O, with the M family responsible for 95% of all HIV infections. The five families are different offshoots of the evolutionary tree for Simian Immunodeficiency Virus (SIV), which infects primates. Stars indicate viruses transitioning from primates to humans. The A and B families originated in sooty mangabey monkeys, whereas the M, N, and O families originated in chimpanzees.

Transforming Distance Matrices into Evolutionary Trees

Constructing a distance matrix from coronavirus genomes

To determine how SARS jumped from animals to humans, scientists started sequencing coronaviruses from various species in order to determine which one is the most similar to SARS-CoV. However, constructing a multiple alignment of entire viral genomes is tricky because viral genes are often rearranged, inserted, and deleted. For this reason, scientists focused on only one of the six genes in SARS-CoV. This gene encodes the **Spike protein**, which identifies and binds to receptor sites on the host's cell membrane.

In SARS-CoV, the Spike protein is 1,255 amino acids long and has rather weak similarity with Spike proteins in other coronaviruses. However, even these subtle similarities turned out to be sufficient for constructing a multiple alignment of Spike proteins across various coronaviruses.

After constructing a multiple alignment of genes from n different species, biologists often transform this alignment into an $n \times n$ **distance matrix** D. In many cases, $D_{i,j}$ represents the number of differing symbols between the genes representing rows i and j of the alignment (Figure 7.3). However, distance matrices can be constructed using a variety of different distance functions in order to suit different applications. For example, $D_{i,j}$ could represent the edit distance between genes from the i-th and j-th species. Or, a distance matrix for n genomes could be constructed from the 2-break distances between each pair of genomes.

SPECIES	ALIGNMENT	DISTANCE MATRIX			
		Chimp	**Human**	**Seal**	**Whale**
Chimp	ACGTAGGCCT	0	3	6	4
Human	ATGTAAGACT	3	0	7	5
Seal	TCGAGAGCAC	6	7	0	2
Whale	TCGAAAGCAT	4	5	2	0

FIGURE 7.3 A multiple alignment of hypothetical DNA sequences from four species, along with the distance matrix produced by counting the number of differing symbols between each pair of rows in this multiple alignment.

Regardless of which distance function we use, in order to be a distance matrix, D must satisfy three properties. It must be **symmetric** (for all i and j, $D_{i,j} = D_{j,i}$), **non-negative** (for all i and j, $D_{i,j} \geq 0$) and satisfy the **triangle inequality** (for all i, j, and k, $D_{i,j} + D_{j,k} \geq D_{i,k}$).

Exercise Break: Prove that if $D_{i,j}$ is equal to the number of differing symbols between rows i and j of a multiple alignment, then D is symmetric, non-negative, and satisfies the triangle inequality.

By the end of 2003, bioinformaticians had sequenced many coronaviruses taken from a variety of animals and SARS patients and then computed the associated distance matrix. They needed to use this information in order to construct a coronavirus phylogeny and understand the origin and spread of the SARS epidemic.

Evolutionary trees as graphs

You may have noticed that the HIV tree in Figure 7.2 has the structure of a graph. Furthermore, Figure 7.4 (top) shows the representation of the phylogeny of all life as a graph.

Graphs that are used to model phylogenies share two properties. They are connected (i.e. it is possible to reach any node from any other node), and they contain no cycles. For this reason, we will define a **tree** as a connected graph without cycles (see Figure 7.4 (bottom) for a few additional examples).

Take another look at Figure 7.4 (top). You will see that present-day species have been assigned to the **leaves** of the tree, or nodes having degree 1 (in Chapter 3, we defined the degree of a node as the number of edges connected to that node). Nodes with degree larger than 1 are called **internal nodes** and represent unknown ancestor species. Given a leaf j, there is only one node connected to j by an edge, which we call the **parent** of j, denoted $\text{PARENT}(j)$. An edge connecting a leaf to its parent is called a **limb**.

Exercise Break: Prove the following statements:

- Every tree with at least two nodes has at least two leaves.
- Every tree with n nodes has $n - 1$ edges.

In a **rooted tree**, one node is designated as a special node called the **root**, and the edges in the tree automatically inherit an implicit orientation away from the root, which is placed at the top or left of the tree (Figure 7.5). This edge orientation models time: the ancestor of all species in the tree is found at the root, and evolution proceeds from the root outward through the tree. Trees without a designated root are called **unrooted**.

STOP and Think: Where would you place the root in the phylogeny in Figure 7.2?

We will analyze rooted trees when we attempt to infer the node corresponding to the ancestor of all species in the tree; otherwise, we will analyze unrooted trees. Figure 7.6 shows an unrooted tree of HIV viruses produced from a different dataset than the one used to create Figure 7.2. By proposing two additional subtypes of HIV, it illustrates that the classification of HIV into five families shown in Figure 7.2 is not written in stone.

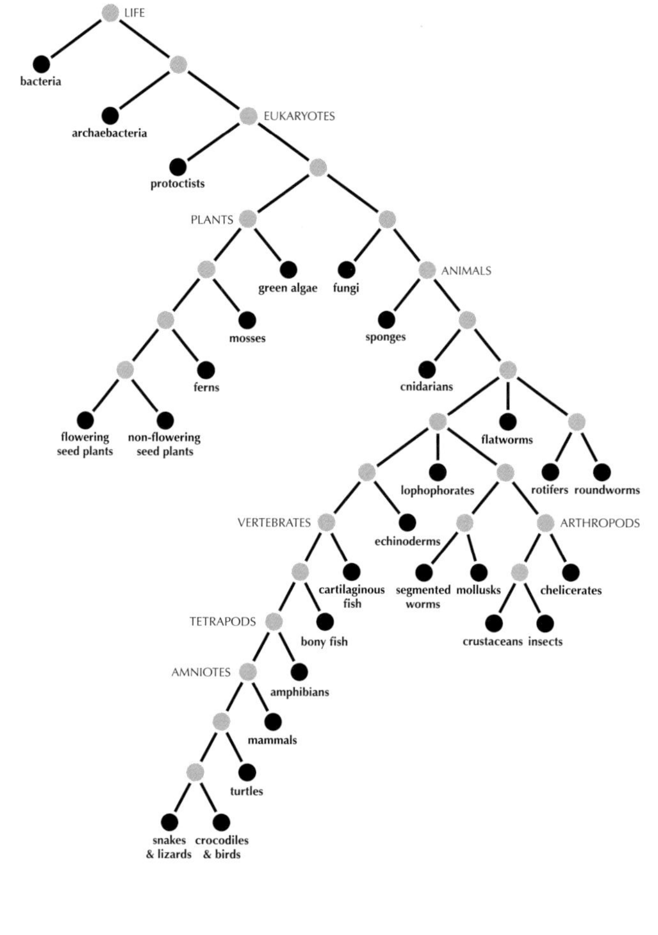

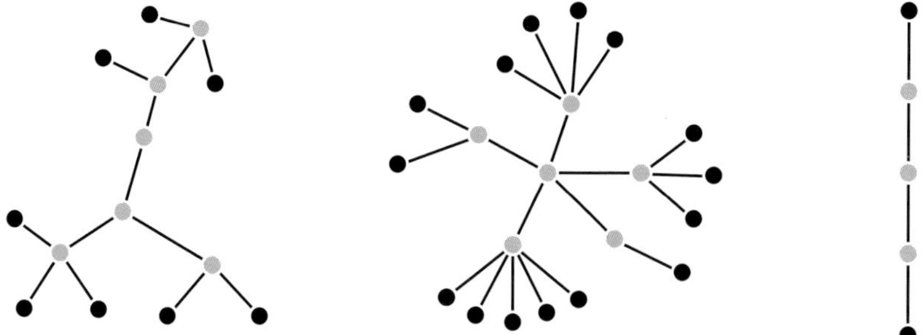

FIGURE 7.4 (Top) A connected acyclic graph that models an evolutionary tree of life on Earth. Present-day species are shown as darker nodes (leaves). (Bottom) Trees come in a variety of different shapes. In each of the three trees shown, leaves (i.e., nodes of degree 1) have been drawn darker than internal nodes (i.e., nodes of larger degree).

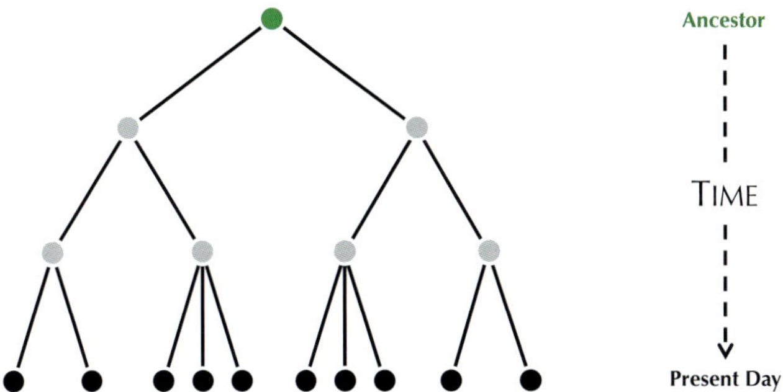

FIGURE 7.5 A rooted tree, with the root (representing an ancestor of all species in the tree) indicated in green at the top of the tree. The presence of the root implies an orientation of edges in the tree away from the root.

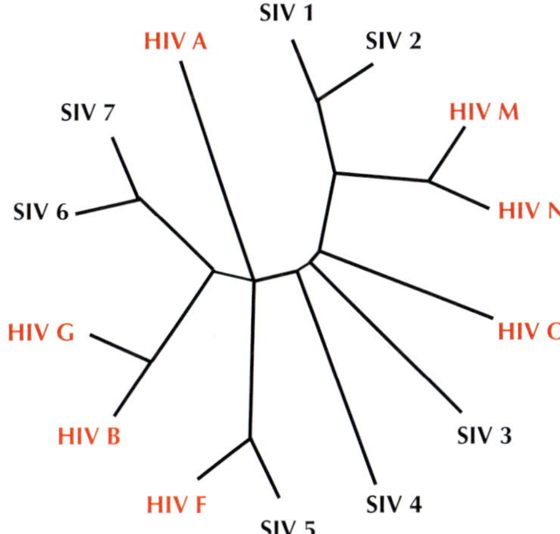

FIGURE 7.6 An unrooted tree of HIV and SIV viruses that suggests additional viral families F and G in addition to the viral families A, B, M, N, and O shown in Figure 7.2.

Distance-based phylogeny construction

We will first focus on deriving an unrooted tree from a distance matrix. The leaves of this tree should correspond to the species represented by the matrix (with internal nodes corresponding to unknown ancestral species). To reflect the evolutionary distance between species in a tree, we assign each edge a non-negative length representing the distance between the organisms that the edge connects, as shown in Figure 7.7.

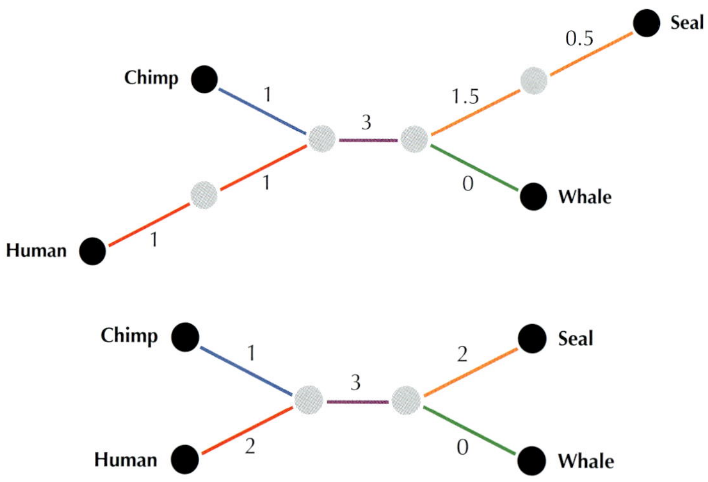

FIGURE 7.7 Two unrooted trees fitting the distance matrix in Figure 7.3. Each of the five maximal non-branching paths in the tree on the top is shown using a different color. Replacing each maximal non-branching path in this tree with a single edge (of length equal to the total length of edges) results in the simple tree shown on the bottom.

Exercise Break: Prove that there exists exactly one path connecting every pair of nodes in a tree. Hint: what would happen if there were two different paths connecting a pair of nodes? What would happen if there were no paths connecting a pair of nodes?

In this chapter, we define the length of a path in a tree as the sum of the lengths of its edges (rather than the number of edges on the path). As a result, the evolutionary distance between two present-day species corresponding to leaves i and j in a tree T is equal to the length of the unique path connecting i and j, denoted $d_{i,j}(T)$.

Distances Between Leaves Problem:
Compute the distances between leaves in a weighted tree.

Input: A weighted tree with n leaves.
Output: An $n \times n$ matrix $(d_{i,j})$, where $d_{i,j}$ is the length of the path between leaves i and j.

One way to find the distance between any pair of leaves i and j in a tree is to first construct a topological ordering of the tree that starts with i and ends with j (see **DETOUR: Constructing a Topological Ordering**). Then, we can apply our algorithm for finding the Longest Path in a DAG from Chapter 5 to find the length of the path connecting i to j.

PAGE 287

Yet we would like to solve the reverse problem, in which we must construct an unrooted tree that models a given distance matrix. We say that a weighted unrooted tree T **fits** a distance matrix D if $d_{i,j}(T) = D_{i,j}$ for every pair of leaves i and j.

Distance-Based Phylogeny Problem:
Reconstruct an evolutionary tree fitting a distance matrix.

Input: A distance matrix.
Output: A tree fitting this distance matrix.

STOP and Think: Does the Distance-Based Phylogeny Problem always have a solution?

Not every distance matrix has a tree fitting it (see **DETOUR: Searching for a Tree Fitting a Distance Matrix**). We therefore call a distance matrix **additive** if there exists a tree that fits this matrix and **non-additive** otherwise. The term "additive" is used because the lengths of all edges along the path between leaves i and j in a tree fitting the matrix D add to $D_{i,j}$.

Note that both trees in Figure 7.7 fit the distance matrix from Figure 7.3, so it would be nice to have a notion of a "canonical" tree fitting a distance matrix. Extending definitions introduced in Chapter 3 to undirected graphs, we say that a path in a tree is **non-branching** if every node other than the beginning and ending node of the path has degree equal to 2. A non-branching path is **maximal** if it is not a subpath of an even longer non-branching path. If we substitute every maximal non-branching path by a single edge whose length is equal to the length of the path, then the tree in Figure 7.7 (top) becomes the tree in Figure 7.7 (bottom). In general, after such a transformation, there are no nodes of degree 2; a tree satisfying this property is called a **simple tree**. It turns out that if a matrix is additive, then there exists a *unique* simple tree fitting this matrix. In the Distance-Based Phylogeny Problem, we will therefore use the terminology TREE(D) to denote the simple tree fitting the additive distance matrix D. Our question, then, is how to construct TREE(D) from D.

Exercise Break: Prove that every simple tree with n leaves has at most $n-2$ internal nodes.

Toward An Algorithm for Distance-Based Phylogeny Construction

A quest for neighboring leaves

A natural first step for solving the Distance-Based Phylogeny Problem would be to ensure that the two closest species with respect to the distance matrix D correspond to **neighbors** in TREE(D). In other words, the minimum value $D_{i,j}$ should correspond to leaves i and j having the same parent. In the rest of this chapter, when we refer to the minimum element of a matrix, we are referring to a minimum **off-diagonal** element, i.e., a value $D_{i,j}$ such that $i \neq j$.

Theorem. *Every simple tree with at least three nodes has a pair of neighboring leaves.*

Proof. Given a simple tree T with at least three nodes, consider a path $P = (v_1, \ldots, v_k)$ that has the maximum number of nodes of any path in T. Because T has at least three nodes, k must be at least 3. Furthermore, nodes v_1 and v_k must be leaves, since otherwise we could extend P into a longer path. Because T is simple, each internal node of T has degree at least 3. Thus, node v_2, which is the parent of v_1, must have at least three adjacent nodes: v_1, v_3, and yet another node w.

We claim that w is a leaf, which would imply that leaves v_1 and w are neighbors. We will proceed by contradiction: if w were not a leaf, then since T is simple, w would be adjacent to another node u. As a result, we could form the path $P' = (u, w, v_2, v_3, \ldots, v_k)$, which contains $k+1$ nodes and contradicts our original assumption that P has the maximum number of nodes. Thus, w must be a leaf, implying that v_1 and w are neighbors. □

Figure 7.8 (top) illustrates that for neighboring leaves i and j sharing a parent node m, the following equality holds for every other leaf k in the tree:

$$d_{k,m} = \frac{(d_{i,m} + d_{k,m}) + (d_{j,m} + d_{k,m}) - (d_{i,m} + d_{j,m})}{2} = \frac{d_{i,k} + d_{j,k} - d_{i,j}}{2}.$$

Since i, j, and k are leaves, we can compute the distance $d_{k,m}$ between nodes k and m in terms of elements of the additive distance matrix D,

$$d_{k,m} = \frac{(D_{i,k} + D_{j,k} - D_{i,j})}{2}.$$

In the case when the parent m has degree 3 (as in Figure 7.8 (top)), removing leaves i and j from the tree turns m into a leaf and thus reduces the total number of leaves (Figure 7.8 (bottom)). This operation is equivalent to removing rows i and j as well as columns i and j from D, then adding a new row and column corresponding to their parent m, where the distances from m to other leaves are computed according to the above formula.

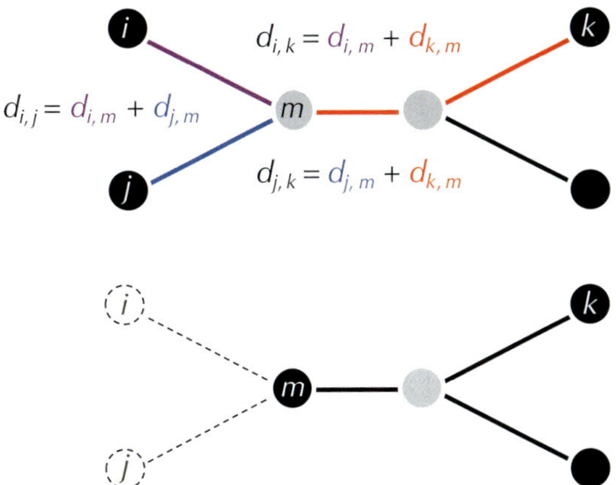

FIGURE 7.8 (Top) For neighboring leaves i and j and their parent node m, $d_{k,m} = (d_{i,k} + d_{j,k} - d_{i,j})/2$ for every other leaf k in the tree. (Bottom) Removing leaves i and j from the tree turns m into a leaf (we assume that m has degree 3). The distances from this new leaf to any other leaf k can be recomputed as $d_{k,m} = (D_{i,k} + D_{j,k} - D_{i,j})/2$.

Exercise Break: We have just described how to reduce the size of the tree as well as the dimension of the distance matrix D if the parent node (m) has degree 3. Design a similar approach in the case that the degree of m is larger than 3.

This discussion implies a recursive algorithm for the Distance-Based Phylogeny Problem:

- find a pair of neighboring leaves i and j by selecting the minimum element $D_{i,j}$ in the distance matrix;
- replace i and j with their parent, and recompute the distances from this parent to all other leaves as described above;

- solve the Distance-Based Phylogeny problem for the smaller tree;
- add the previously removed leaves i and j back to the tree.

Exercise Break: Apply this recursive approach to the distance matrix shown in Figure 7.9 (left). (Solve this exercise by hand.)

	v_1	v_2	v_3	v_4
v_1	0	13	21	22
v_2	13	0	12	13
v_3	21	12	0	13
v_4	22	13	13	0

	v_1	v_2	v_3	v_4
v_1	0	3	4	3
v_2	3	0	4	5
v_3	4	4	0	2
v_4	3	5	2	0

FIGURE 7.9 (Left) An additive 4 × 4 distance matrix. (Right) A non-additive 4 × 4 distance matrix.

Computing limb lengths

If you attempted the preceding exercise, then you were likely driven crazy. The reason why is that in the first step of our proposed algorithm, we assumed that a minimum element of an additive distance matrix corresponds to neighboring leaves. Yet as illustrated in Figure 7.10, this assumption is not necessarily true! Thus, we need a new approach to the Distance-Based Phylogeny Problem, as finding the animal coronavirus that is the smallest distance from SARS-CoV may not be the best way to identify the animal reservoir of SARS.

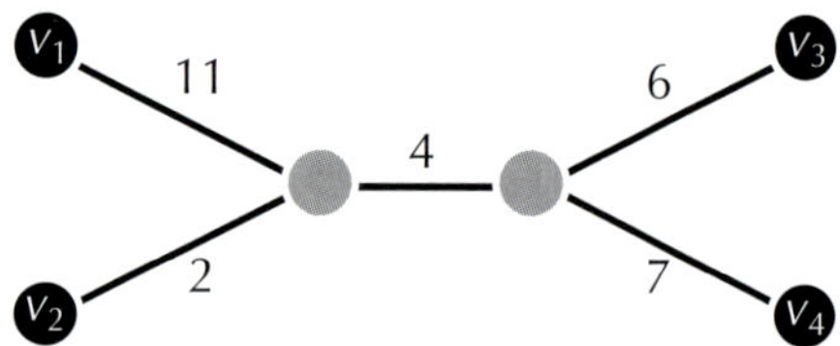

FIGURE 7.10 The simple tree fitting the distance matrix from Figure 7.9 (left). The two closest leaves in this tree (j and k) are not neighbors.

Our proposed recursive approach may have failed, but using recursion was a good idea, and so we will explore a different recursive algorithm. Rather than looking for a *pair* of neighbors in TREE(D), we will instead reduce the size of the tree by trimming its leaves *one at a time*. Of course, we don't know TREE(D), and so we must somehow trim leaves in TREE(D) by analyzing the distance matrix.

As a first step toward constructing TREE(D), we will address the more modest goal of computing the lengths of limbs in TREE(D). So, given a leaf j in a tree, we denote the length of the limb connecting j with its parent as LIMBLENGTH(j). Edges that are not limbs must connect two internal nodes and are therefore called **internal edges**.

Limb Length Problem:
Compute the length of a limb in the simple tree fitting an additive distance matrix.

Input: An additive distance matrix D and an integer j.
Output: LIMBLENGTH(j), the length of the limb connecting leaf j to its parent in TREE(D).

To compute LIMBLENGTH(j) for a given leaf j, note that because TREE(D) is simple, we know that PARENT(j) has degree at least 3 (unless TREE(D) has only two nodes). We can therefore think of PARENT(j) as dividing the other nodes of TREE(D) into at least three **subtrees**, or smaller trees that would remain if we were to remove PARENT(j) along with any edges connecting it to other nodes (Figure 7.11). Because j is a leaf, it must belong to a subtree by itself; we call this subtree T_j. This brings us to the following result.

Limb Length Theorem: *Given an additive matrix D and a leaf j,* LIMBLENGTH(j) *is equal to the minimum value of* $(D_{i,j} + D_{j,k} - D_{i,k})/2$ *over all leaves i and k.*

Proof. A given pair of leaves can belong to the same subtree or to different subtrees. So first assume that leaves i and k belong to different subtrees T_i and T_k (Figure 7.11). Because PARENT(j) is on the path connecting i to k, it follows that

$$d_{i,j} = d_{i,\,\text{PARENT}(j)} + \text{LIMBLENGTH}(j)$$
$$d_{j,k} = d_{k,\,\text{PARENT}(j)} + \text{LIMBLENGTH}(j)$$

Adding these two equations yields

$$d_{i,j} + d_{j,k} = d_{i,\,\text{PARENT}(j)} + d_{k,\,\text{PARENT}(j)} + 2 \cdot \text{LIMBLENGTH}(j).$$

Because $d_{i,\,\text{PARENT}(j)} + d_{k,\,\text{PARENT}(j)}$ is equal to $d_{i,k}$, it follows that

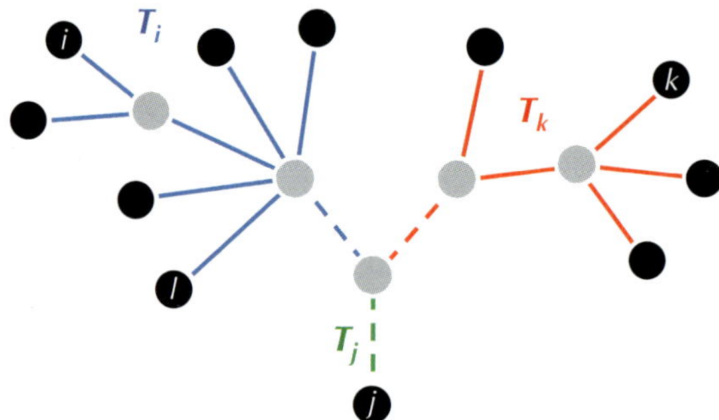

FIGURE 7.11 A simple tree with selected leaves i, j, k, and l. Removing the parent of j (along with the three dashed edges connecting it to other nodes) would separate this tree into three subtrees, whose edges are shown in different colors. Leaves i and l belong to T_i, whereas leaf k belongs to T_k. Leaf j belongs to T_j, which contains a single node.

$$\text{LIMBLENGTH}(j) = \frac{d_{i,j} + d_{j,k} - d_{i,k}}{2} = \frac{D_{i,j} + D_{j,k} - D_{i,k}}{2}.$$

On the other hand, assume that leaves i and l belong to the same subtree (Figure 7.11). Then the path from i to l does not pass through $\text{PARENT}(j)$, and so we have the *inequality*

$$d_{i,\text{PARENT}(j)} + d_{l,\text{PARENT}(j)} \geq d_{i,l}.$$

Combining this with the equation

$$d_{i,j} + d_{j,l} = d_{i,\text{PARENT}(j)} + d_{l,\text{PARENT}(j)} + 2 \cdot \text{LIMBLENGTH}(j)$$

yields that

$$\begin{aligned}\text{LIMBLENGTH}(j) &= \frac{d_{i,j} + d_{j,l} - (d_{i,\text{PARENT}(j)} + d_{l,\text{PARENT}(j)})}{2} \\ &\leq \frac{d_{i,j} + d_{j,l} - d_{i,l}}{2} = \frac{D_{i,j} + D_{j,l} - D_{i,l}}{2}.\end{aligned}$$

As a result of this discussion, $\text{LIMBLENGTH}(j)$ must be less than or equal to $(D_{i,j} + D_{j,k} - D_{i,k})/2$ for any choice of leaves i and k. Because we can always find leaves i and k belonging to different subtrees (why?), it follows that $\text{LIMBLENGTH}(j)$ is equal to the minimum value of $(D_{i,j} + D_{j,k} - D_{i,k})/2$ over all choices of i and k. □

We now have an algorithm for solving the Limb Length Problem. For each *j*, we can compute LIMBLENGTH(j) by finding the minimum value of $(D_{i,j} + D_{j,k} - D_{i,k})/2$ over all pairs of leaves *i* and *k*.

Exercise Break: The proposed algorithm computes LIMBLENGTH(j) in $\mathcal{O}(n^2)$ time (for an $n \times n$ distance matrix). Design an algorithm that computes LIMBLENGTH(j) in $\mathcal{O}(n)$ time.

Additive Phylogeny

Trimming the tree

Since we now know how to find the length of any limb in TREE(D), we can construct TREE(D) recursively using the algorithm illustrated in Figure 7.12.

First, imagine that we already know TREE(D), and pick an arbitrary leaf *j*. We will trim the limb of *j* by reducing its length by LIMBLENGTH(j). Because we do not know TREE(D), we need to represent trimming the leaf *j* in terms of the distance matrix *D*. To do so, we first subtract LIMBLENGTH(j) from each off-diagonal element in row *j* and column *j* of *D* to obtain a matrix D^{bald} for which the limb of *j* has become a **bald limb**, or a limb of length 0 (Figure 7.12). We will further assume that a bald limb has disappeared from the tree entirely. In terms of the distance matrix, ignoring a bald limb means removing row *j* and column *j* from *D* to produce a smaller $(n-1) \times (n-1)$ distance matrix D^{trimmed}. We can now recursively find TREE(D) in four steps:

- pick an arbitrary leaf *j*, compute LIMBLENGTH(j), and construct the distance matrix D^{trimmed};
- solve the Distance-Based Phylogeny Problem for D^{trimmed};
- identify the point in TREE(D^{trimmed}) where leaf *j* should be attached in TREE(D);
- add a limb of length LIMBLENGTH(j) growing from this attachment point in TREE(D^{trimmed}) to form TREE(D).

STOP and Think: When adding leaf *j* back to TREE(D^{trimmed}), how would you find its attachment point?

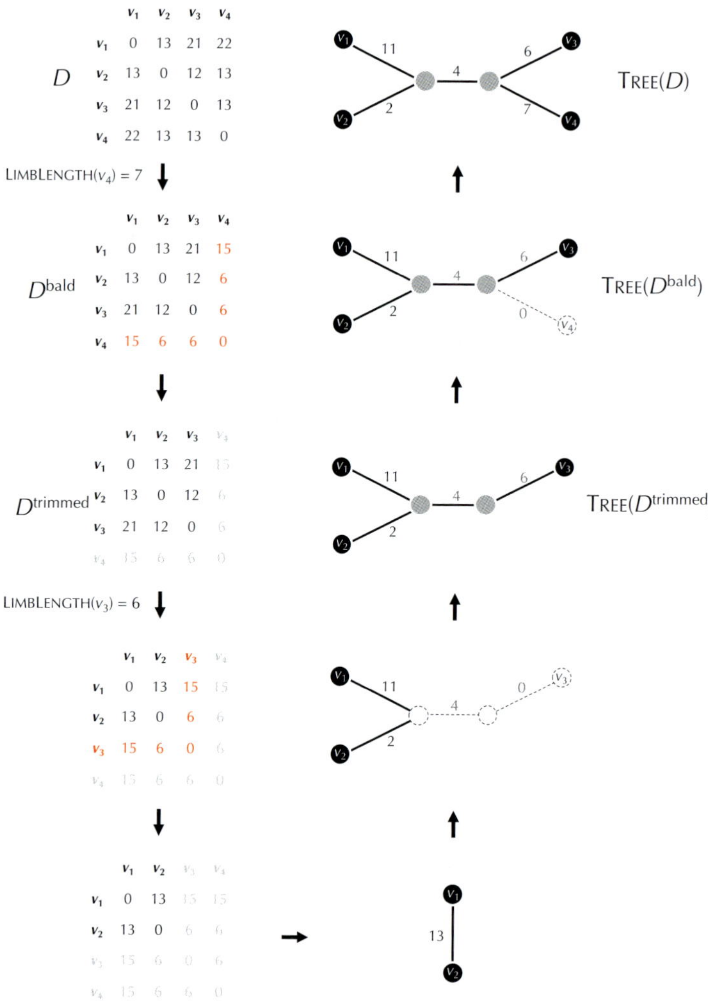

FIGURE 7.12 Converting the additive distance matrix from Figure 7.9 (left) into the simple tree fitting this matrix from Figure 7.10. On the left side, we first compute LIMBLENGTH$(v_4) = 7$, and then subtract 7 from the non-diagonal elements in the final row and column of D to obtain D^{bald} (updated values are shown in red). Removing this row and column yields a 3×3 distance matrix D^{trimmed}. We find that LIMBLENGTH$(v_3) = 6$ in D^{trimmed} and subtract 6 from the non-diagonal elements in the third row and column. Graying out this row and column yields a 2×2 distance matrix. On the right side, we can fit this 2×2 distance matrix to a tree consisting of a single edge. By finding the attachment points of removed limbs (shown on the left), we reconstruct TREE(D^{trimmed}), TREE(D^{bald}), and then TREE(D).

Attaching a limb

To find the attachment point of a leaf j in $\text{TREE}(D^{\text{trimmed}})$, consider $\text{TREE}(D^{\text{bald}})$, which is the same as $\text{TREE}(D)$ except that $\text{LIMBLENGTH}(j) = 0$. From the Limb Length Theorem, we know that there must be leaves i and k in $\text{TREE}(D^{\text{bald}})$ such that

$$\frac{D^{\text{bald}}_{i,j} + D^{\text{bald}}_{j,k} - D^{\text{bald}}_{i,k}}{2} = 0\,,$$

which implies that

$$D^{\text{bald}}_{i,k} = D^{\text{bald}}_{i,j} + D^{\text{bald}}_{j,k}\,.$$

Thus, the attachment point for leaf j must be located at distance $D^{\text{bald}}_{i,j}$ from leaf i on the path connecting i and k in the trimmed tree. This attachment point may occur at an existing node, in which case we connect j to this node. On the other hand, the attachment point for j may occur along an edge, in which case we place a new node at the attachment point and connect j to it.

An algorithm for distance-based phylogeny construction

The preceding discussion results in the recursive algorithm below, which we call **ADDITIVEPHYLOGENY**, for finding the simple tree fitting an $n \times n$ additive distance matrix D. We assume that you have already implemented a program **LIMB**(D, j) that computes $\text{LIMBLENGTH}(j)$ for a leaf j based on the distance matrix D. Rather than selecting an arbitrary leaf j from $\text{TREE}(D)$ for trimming, **ADDITIVEPHYLOGENY** selects leaf n (corresponding to the last row and column of D).

STOP and Think: Consider these questions about **ADDITIVEPHYLOGENY**.

- What is its running time?
- Although it may seem that **ADDITIVEPHYLOGENY** would construct a tree for any matrix, this is not the case. What goes wrong if you apply **ADDITIVEPHYLOGENY** to the non-additive distance matrix in Figure 7.9 (right)?
- Modify **ADDITIVEPHYLOGENY** to develop an algorithm that checks whether a given distance matrix is additive. Then, apply this test to the distance matrix for coronavirus Spike proteins shown in Figure 7.13 (top). Is this matrix additive?

```
ADDITIVEPHYLOGENY(D, n)
    if n = 2
        return the tree consisting of a single edge of length D_{1,2}
    limbLength ← LIMB(D, n)
    for j ← 1 to n − 1
        D_{j,n} ← D_{j,n} − limbLength
        D_{n,j} ← D_{j,n}
    (i, k) ← two leaves such that D_{i,k} = D_{i,n} + D_{n,k}
    x ← D_{i,n}
    remove row n and column n from D
    T ← ADDITIVEPHYLOGENY(D, n − 1)
    v ← the (potentially new) node in T at distance x from i on the path between i and k
    add leaf n back to T by creating a limb (v, n) of length limbLength
    return T
```

7C

Although the previous question suggests that **ADDITIVEPHYLOGENY** can be modified to determine whether a given distance matrix is additive, there exists an even simpler way to check for additivity (see **DETOUR: The Four Point Condition**).

PAGE 403

Constructing an evolutionary tree of coronaviruses

By the end of 2003, bioinformaticians had sequenced many coronaviruses from a variety of birds and mammals, from which we obtain the distance matrix in Figure 7.13 (top) based on a multiple alignment of Spike proteins.

Although you now understand the perils of concluding that the minimum element of the distance matrix corresponds to a pair of neighbors, common sense tells us with a glance at Figure 7.13 (top) that the civet must be the animal reservoir of SARS. This information led researchers to hypothesize that inadequate preparation of meat from palm civets (Figure 7.14) in the Guangdong region of China may have caused the SARS outbreak.

PAGE 404

Yet before rushing to this conclusion, you may like to read **DETOUR: Did Bats Give Us SARS?** to see why the history of interspecies viral transfer is often difficult to trace. In fact, some studies have suggested that humans first received SARS from bats, which later gave the virus to palm civets, which then transmitted the disease back to humans. The civet was identified as the animal reservoir of SARS in 2003 in part because SARS viruses from other potential suspects, including bats, had not yet been sequenced.

Since the distance matrix for SARS-like coronaviruses is non-additive, we will cheat a bit and slightly modify it to make it additive so you can apply **ADDITIVEPHYLOGENY** to it (Figure 7.13 (bottom)).

Exercise Break: Construct the simple tree fitting the distance matrix in Figure 7.13 (bottom).

	Cow	Pig	Horse	Mouse	Dog	Cat	Turkey	Civet	Human
Cow	0	295	300	524	1077	1080	978	941	940
Pig	295	0	314	487	1071	1088	1010	963	966
Horse	300	314	0	472	1085	1088	1025	965	956
Mouse	524	487	472	0	1101	1099	1021	962	965
Dog	1076	1070	1085	1101	0	818	1053	1057	1054
Cat	1082	1088	1088	1098	818	0	1070	1085	1080
Turkey	976	1011	1025	1021	1053	1070	0	963	961
Civet	941	963	965	962	1057	1085	963	0	16
Human	940	966	956	965	1054	1080	961	16	0

	Cow	Pig	Horse	Mouse	Dog	Cat	Turkey	Civet	Human
Cow	0	295	306	497	1081	1091	1003	956	954
Pig	295	0	309	500	1084	1094	1006	959	957
Horse	306	309	0	489	1073	1083	995	948	946
Mouse	497	500	489	0	1092	1102	1014	967	965
Dog	1081	1084	1073	1092	0	818	1056	1053	1051
Cat	1091	1094	1083	1102	818	0	1066	1063	1061
Turkey	1003	1006	995	1014	1056	1066	0	975	973
Civet	956	959	948	967	1053	1063	975	0	16
Human	954	957	946	965	1051	1061	973	16	0

FIGURE 7.13 (Top) The distance matrix based on pairwise alignment of Spike proteins from coronaviruses extracted from various animals. The distance between each pair of sequences was computed as the total number of mismatches and indels in their optimal alignment. (Bottom) A modification of the distance matrix to make it additive.

FIGURE 7.14 The palm civet.

Using Least Squares to Construct Approximate Distance-Based Phylogenies

If an $n \times n$ distance matrix D is non-additive, then we will instead look for a weighted tree T whose distances between leaves approximate the entries in D. To this end, we would like for T to minimize the **sum of squared errors** DISCREPANCY(T, D), which is given by the formula

$$\text{DISCREPANCY}(T, D) = \sum_{1 \leq i < j \leq n} (d_{i,j}(T) - D_{i,j})^2 \,.$$

Least Squares Distance-Based Phylogeny Problem:
Given a distance matrix, find the tree that minimizes the sum of squared errors.

Input: An $n \times n$ distance matrix D.
Output: A weighted tree T minimizing DISCREPANCY(T, D) over all weighted trees with n leaves.

Exercise Break: Let T be the tree in Figure 7.10 with all edge lengths removed. Given the non-additive 4×4 distance matrix D in Figure 7.9 (right), find the lengths of edges in this tree that minimize DISCREPANCY(T, D).

It turns out that for a specific tree T, it is easy to find edge weights in T minimizing DISCREPANCY(T, D). Yet our ability to minimize the sum of squared errors for a *specific* tree does not imply that we can efficiently solve the Least Squares Distance-Based Phylogeny Problem, since the number of different trees grows very quickly as the number of leaves in the tree increases. In fact, the Least Squares Distance-Based Phylogeny winds up being *NP*-Complete, and so we must abandon the hope of designing a fast algorithm to find a tree that best fits a non-additive matrix. In the next two sections, we will explore heuristics for constructing trees from non-additive matrices that solve this problem approximately.

Ultrametric Evolutionary Trees

Biologists often assume that every internal node in an evolutionary tree corresponds to a species that underwent a **speciation event**, splitting one ancestral species into two descendants. Note that every internal node in the tree in Figure 7.15 (top) (corresponding to a speciation event) has degree 3. We therefore define an **unrooted binary tree** as a tree where every node has degree equal to either 1 or 3.

Exercise Break: Prove that every unrooted binary tree with n leaves has $n-2$ internal nodes (and thus $2n-3$ edges).

A **rooted binary tree** is an unrooted binary tree that has a root (of degree 2) placed on one of its edges. In other words, we replace an edge (v, w) with a root and draw edges connecting the root to each of v and w (Figure 7.15 (bottom)).

If we had a **molecular clock** measuring evolutionary time, then we could assign an **age** to every node v in a rooted binary tree (denoted $\text{AGE}(v)$), where all of the leaves of the tree have age 0 because they correspond to present-day species. We could then define the weight of an edge (v, w) in the tree as the difference $\text{AGE}(v) - \text{AGE}(w)$. Consequently, the length of a path between the root and any node would be equal to the difference between their ages. Such a tree, in which the distance from the root to any leaf is the same, is called **ultrametric** (Figure 7.16 (bottom right)).

Our aim is to derive an ultrametric tree that explains a given distance matrix (even if it does so only approximately). **UPGMA** (which stands for **U**nweighted **P**air **G**roup **M**ethod with **A**rithmetic Mean) is a simple clustering heuristic that introduces a hypothetical molecular clock for constructing an ultrametric evolutionary tree. You can learn more about clustering in Chapter 8.

Given an $n \times n$ matrix D, **UPGMA** (which is illustrated in Figure 7.16) first forms n trivial clusters, each containing a single leaf. The algorithm then finds a pair of "closest" clusters. To clarify the notion of closest clusters, **UPGMA** defines the distance between clusters C_1 and C_2 as the average pairwise distance between elements of C_1 and C_2,

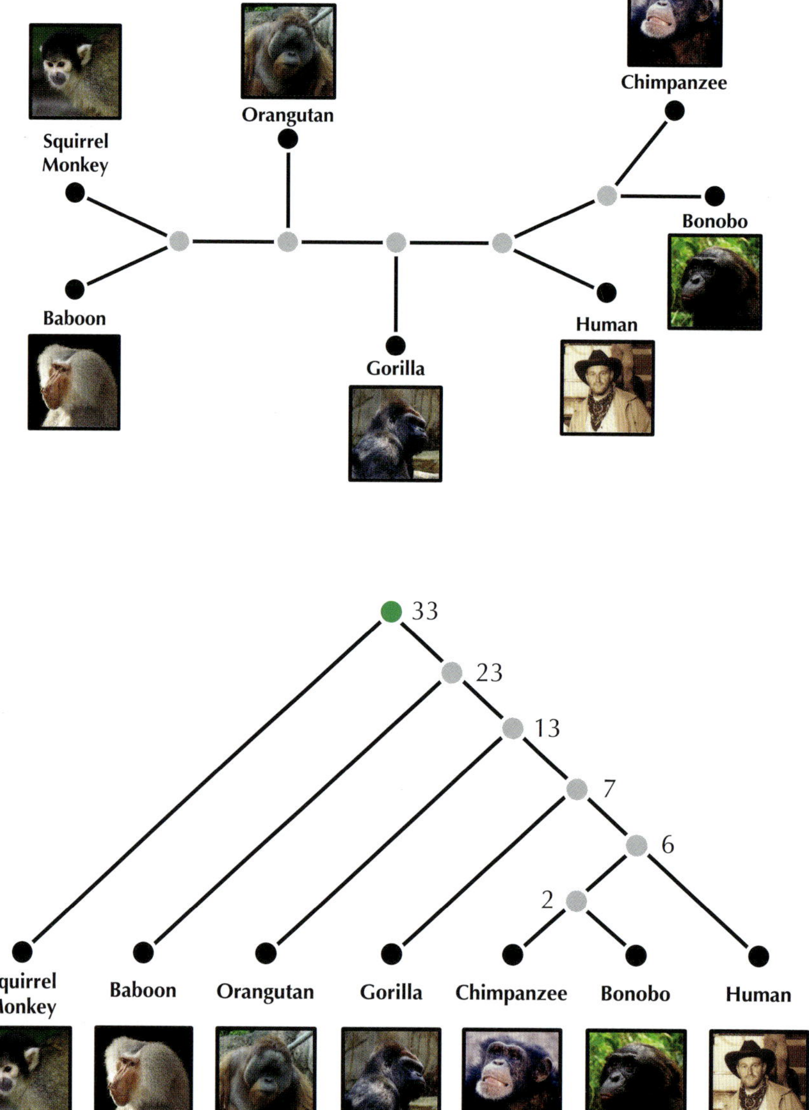

FIGURE 7.15 (Top) An unrooted binary tree representing a phylogeny of primates. (Bottom) Placing a root on the squirrel monkey's limb results in a rooted binary tree. The number at each node corresponds to the number of million years ago that the divergence at this node occurred.

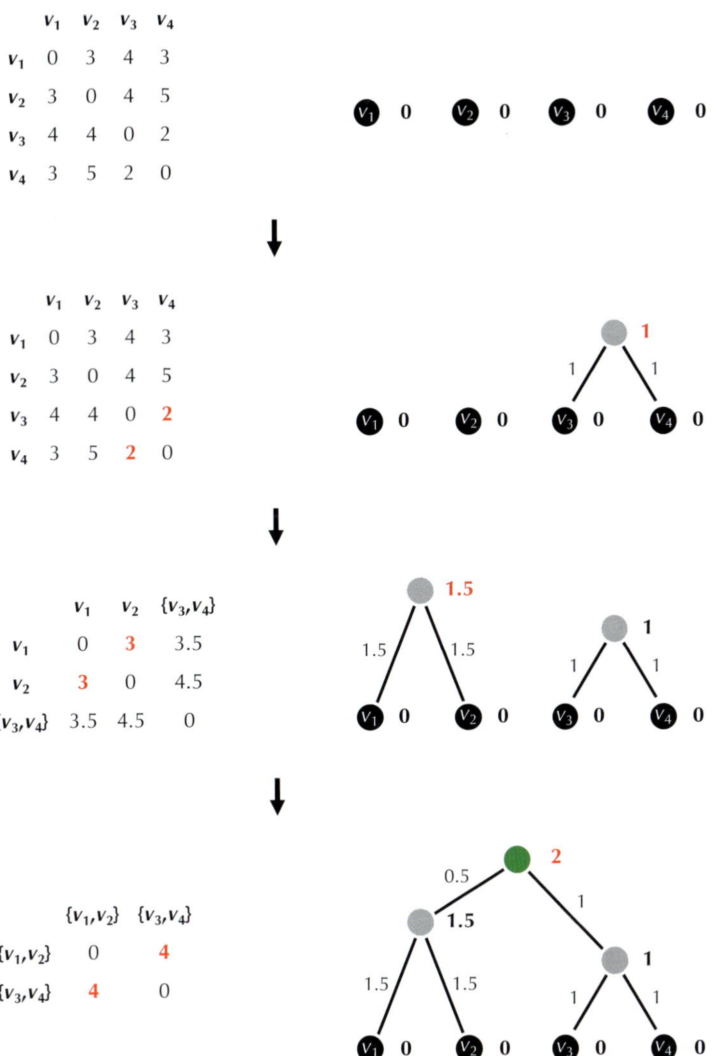

FIGURE 7.16 Tree reconstruction with **UPGMA** for the non-additive distance matrix from Figure 7.9 (right). **UPGMA** begins with forming one cluster for each leaf. In each step, it identifies the two closest clusters C_1 and C_2, merge them into a new node C, and connect C to C_1 and C_2 by directed edges. The age of C is set equal to $D_{C_1,C_2}/2$. We then iterate this process until only a single cluster remains, which must be the root. The resulting tree is ultrametric (i.e., the distance from the root to any leaf is the same).

$$D_{C_1,C_2} = \frac{\sum_{i \in C_1} \sum_{j \in C_2} D_{i,j}}{|C_1| \cdot |C_2|}.$$

In this equation, the notation $|C|$ denotes the number of leaves in cluster C.

Once **UPGMA** has identified a pair of closest clusters C_1 and C_2, it **merges** them into a cluster C with $|C_1| + |C_2|$ elements and then creates a node for C, which it connects to each of C_1 and C_2 by a directed edge. The age of C is set to be $D_{C_1,C_2}/2$. **UPGMA** then iterates this process of merging the two closest clusters until only a single cluster remains, which corresponds to the root.

```
UPGMA(D, n)
    Clusters ← n single-element clusters labeled 1, . . . , n
    construct a graph T with n isolated nodes labeled by single elements 1, . . . , n
    for every node v in T
        AGE(v) ← 0
    while there is more than one cluster
        find the two closest clusters C_i and C_j (break ties arbitrarily)
        merge C_i and C_j into a new cluster C_new with |C_i| + |C_j| elements
        add a new node labeled by cluster C_new to T
        connect node C_new to C_i and C_j by directed edges
        AGE(C_new) ← D_{C_i,C_j} / 2
        remove the rows and columns of D corresponding to C_i and C_j
        remove C_i and C_j from Clusters
        add a row/column to D for C_new by computing D(C_new, C) for each C in Clusters
        add C_new to Clusters
    root ← the node in T corresponding to the remaining cluster
    for each edge (v, w) in T
        length of (v, w) ← AGE(v) − AGE(w)
    return T
```

Exercise Break: Prove that after merging clusters C_i and C_j into a cluster C_{new}, the distance between C_{new} and another cluster C_m is equal to $(D_{C_i,C_m} \cdot |C_i| + D_{C_j,C_m} \cdot |C_j|) / (|C_i| + |C_j|)$.

UPGMA offers a step forward from **ADDITIVEPHYLOGENY**, since it can analyze non-additive distance matrices. Figure 7.17 shows the result of applying UPGMA to the coronavirus distance matrix from Figure 7.13 (top). However, the first step that **UPGMA**

takes is to merge the two leaves i and j with minimum distance $D_{i,j}$ into a single cluster. And we have already seen that the smallest element in the distance matrix does not necessarily correspond to a pair of neighboring leaves! This is a concern, since if **UPGMA** generates incorrect trees from additive matrices, then it is not an ideal heuristic for evolutionary tree construction from non-additive matrices. Can we find an algorithm that always identifies neighboring leaves in an additive distance matrix but also performs well on a non-additive distance matrix?

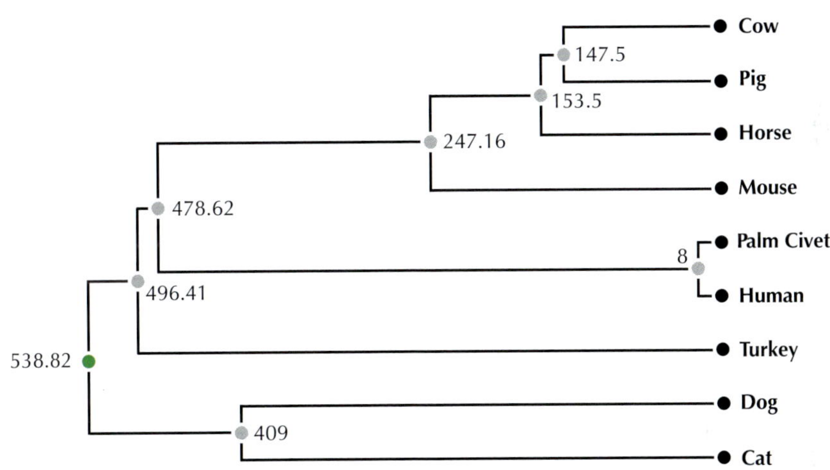

FIGURE 7.17 The ultrametric tree of coronaviruses created by **UPGMA** using the distance matrix in Figure 7.13 (top). The root is shown in green.

The Neighbor-Joining Algorithm

Transforming a distance matrix into a neighbor-joining matrix

In 1987, Naruya Saitou and Masatoshi Nei developed the **neighbor-joining algorithm** for evolutionary tree reconstruction. Given an additive distance matrix, this algorithm, which we call **NEIGHBORJOINING**, finds a pair of neighboring leaves and substitutes them by a single leaf, thus reducing the size of the tree. **NEIGHBORJOINING** can thus recursively construct a tree fitting the additive matrix. This algorithm also provides a heuristic for non-additive distance matrices that performs well in practice.

The central idea of **NEIGHBORJOINING** is that although finding a minimum element in a distance matrix D is not guaranteed to yield a pair of neighbors in TREE(D), we can transform D into a different matrix whose minimum element does yield a pair of neighbors. First, given an $n \times n$ distance matrix D, we define $\text{TOTALDISTANCE}_D(i)$ as the sum $\sum_{1 \leq k \leq n} D_{i,k}$ of distances from leaf i to all other leaves. The **neighbor-joining matrix** D^* is defined such that for all i and j, $D^*_{i,i} = 0$ and

$$D^*_{i,j} = (n-2) \cdot D_{i,j} - \text{TOTALDISTANCE}_D(i) - \text{TOTALDISTANCE}_D(j)\,.$$

NEIGHBORJOINING, which is illustrated in Figure 7.18, is a widely used method for evolutionary tree reconstruction; the paper that introduced it is one of the most cited in all of science, with over 30,000 citations. Yet this algorithm is non-intuitive: the above formula for computing the matrix D^* probably looks like witchcraft to you. In fact, despite having flawless intuition, Saitou and Nei *never proved* that their algorithm correctly solves the Distance-Based Phylogeny Problem for additive matrices! However, it took researchers another year to prove the following theorem, whose proof we have passed to **DETOUR: Why Does the Neighbor-Joining Algorithm Find Neighboring Leaves?**

PAGE 406

Neighbor-Joining Theorem: *Given an additive matrix D, the smallest element $D^*_{i,j}$ of its neighbor-joining matrix D^* corresponds to a pair of neighboring leaves i and j in* $\text{TREE}(D)$.

If $n = 2$, then **NEIGHBORJOINING**(D, n) returns the tree consisting of a single edge of length $D_{1,2}$. If $n > 2$, then it selects the minimum element in the neighbor-joining matrix, replaces the neighboring leaves i and j with a new leaf m, and then computes the distance from m to any other leaf k according to the formula

$$D_{k,m} = \tfrac{1}{2}(D_{k,i} + D_{k,j} - D_{i,j}),$$

which is motivated by Figure 7.8. This equation allows us to replace an $n \times n$ matrix D with an $(n-1) \times (n-1)$ matrix D' in which i and j have been replaced by m. By recursively applying **NEIGHBORJOINING** to D', we obtain an evolutionary tree with $n-1$ leaves. We then add two limbs starting at node m, one ending in leaf i and the other ending in leaf j. We set

$$\Delta_{i,j} = \frac{\text{TOTALDISTANCE}_D(i) - \text{TOTALDISTANCE}_D(j)}{n-2}$$

and assign

$$\text{LIMBLENGTH}(i) = \frac{1}{2}\left(D_{i,j} + \Delta_{i,j}\right)$$

$$\text{LIMBLENGTH}(j) = \frac{1}{2}\left(D_{i,j} - \Delta_{i,j}\right)$$

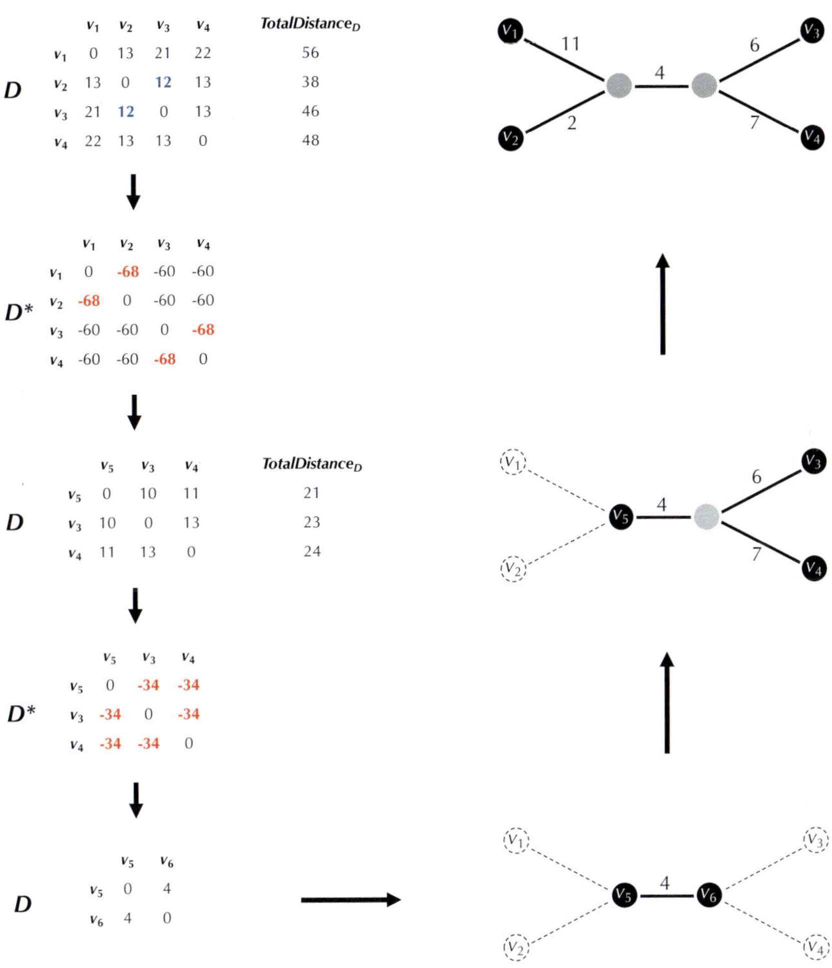

FIGURE 7.18 (Top left) The additive distance matrix D from Figure 7.9 (left) along with the array $TotalDistance_D$. D_{v_2,v_3} (shown in blue) is a minimum element of D, but as it turns out, leaves v_2 and v_3 are not neighbors in TREE(D). Moving down the left side, we construct the neighbor-joining matrix D^* and find that $D^*_{v_1,v_2}$ (red) is a minimum element of D^*. We transform the initial 4×4 distance matrix into a 3×3 distance matrix by replacing v_1 and v_2 with a single leaf v_5 and updating distances from v_5 to other leaves as $D_{v_3,v_5} = \frac{1}{2}(D_{v_3,v_1} + D_{v_3,v_2} - D_{v_1,v_2}) = \frac{1}{2}(21 + 12 - 13) = 10$ and $D_{v_4,v_5} = \frac{1}{2}(D_{v_4,v_1} + D_{v_4,v_2} - D_{v_1,v_2}) = \frac{1}{2}(22 + 13 - 13) = 11$. We select $D^*_{v_3,v_5}$ as a minimum element of D^* and replace leaves v_3 and v_4 with a single leaf v_6. The resulting 2×2 matrix corresponds to a tree with a single edge connecting v_5 and v_6. We then work our way up the right side, adding pairs of neighbors back into the tree at each step using formulas for limb lengths. (Top right) The tree TREE(D) fitting the original matrix D.

To see where these formulas come from, see **DETOUR: Computing Limb Lengths in the Neighbor-Joining Algorithm**.

Exercise Break: Prove that if D is additive, then for any i and j between 1 and n, both $\frac{1}{2}(D_{i,j} + \Delta_{i,j})$ and $\frac{1}{2}(D_{i,j} - \Delta_{i,j})$ are non-negative.

The following pseudocode summarizes the neighbor-joining algorithm.

```
NEIGHBORJOINING(D, n)
    if n = 2
        T ← the tree consisting of a single edge of length D_{1,2}
        return T
    D* ← the neighbor-joining matrix constructed from the distance matrix D
    find elements i and j such that D*_{i,j} is a minimum non-diagonal element of D*
    Δ ← (TOTALDISTANCE_D(i) − TOTALDISTANCE_D(j))/(n − 2)
    limbLength_i ← ½(D_{i,j} + Δ)
    limbLength_j ← ½(D_{i,j} − Δ)
    add a new row/column m to D so that D_{k,m} = D_{m,k} = ½(D_{k,i} + D_{k,j} − D_{i,j})
        for any k
    remove rows i and j from D
    remove columns i and j from D
    T ← NEIGHBORJOINING(D, n − 1)
    add two new limbs (connecting node m with leaves i and j) to the tree T
    assign length limbLength_i to LIMB(i)
    assign length limbLength_j to LIMB(j)
    return T
```

7E

Exercise Break: Before implementing **NEIGHBORJOINING**, apply it to the additive and non-additive distance matrices from Figure 7.9.

Exercise Break: Apply **NEIGHBORJOINING** to the coronavirus distance matrix from Figure 7.13 (top).

Analyzing coronaviruses with the neighbor-joining algorithm

Figure 7.19 shows the neighbor-joining tree of coronaviruses isolated from different animals based on the distance matrix in Figure 7.13 (top). We can also apply **NEIGHBORJOINING** to the distance matrix of SARS-CoV variants isolated from various human carriers, in addition to a coronavirus from palm civet (Figure 7.20).

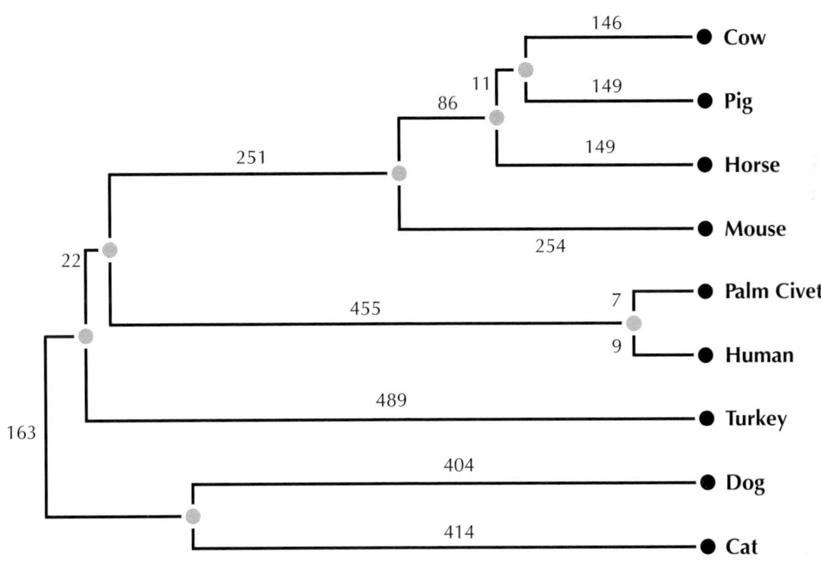

FIGURE 7.19 The neighbor-joining tree of coronaviruses isolated from different animals, based on the non-additive distance matrix in Figure 7.13 (top).

The SARS-CoV strain labeled "Hanoi" in Figure 7.20 was taken from Carlo Urbani, an Italian physician who worked for the World Health Organization. In February 2003, Urbani was called into a Hanoi hospital to examine a patient who had fallen ill with what local doctors believed to be a bad case of influenza; these doctors were afraid that it might be bird flu. In fact, the patient was the American man who had stayed across the hall from Liu Jianlun at the Metropole Hotel just one week earlier.

Fortunately, Urbani was quick to realize that the disease was not influenza, and he became the first physician to raise the alarm to public health officials. Yet rather than leaving Hanoi, he demanded that he remain there in order to oversee quarantine procedures. In an argument with his wife, who scolded him for risking his life to treat sick patients. Urbani replied, "What am I here for? Answering e-mails, going to cocktail

	Guangzhou Dec. 16, 2002	Zhongshan Dec. 16, 2002	Guangzhou Jan. 24, 2003	Guangzhou Jan. 31, 2003	Guangzhou Feb. 18, 2003	Hong Kong Feb. 18, 2003	Hanoi Feb. 26, 2003	Toronto Feb. 17, 2003	Hong Kong Mar. 15, 2003	Palm Civet
Guangzhou	0	4	12	8	9	9	12	12	11	3
Zhongshan	4	0	10	6	7	7	10	10	9	3
Guangzhou	12	10	0	4	5	3	2	2	1	11
Guangzhou	8	6	4	0	3	1	4	4	3	7
Guangzhou	9	7	5	3	0	2	5	5	4	8
Hong Kong	9	7	3	1	2	0	3	3	2	8
Hanoi	12	10	2	4	5	3	0	2	1	11
Toronto	12	10	2	4	5	3	2	0	1	11
Hong Kong	11	9	1	3	4	2	1	1	0	10
Palm Civet	3	3	11	7	8	8	11	11	10	0

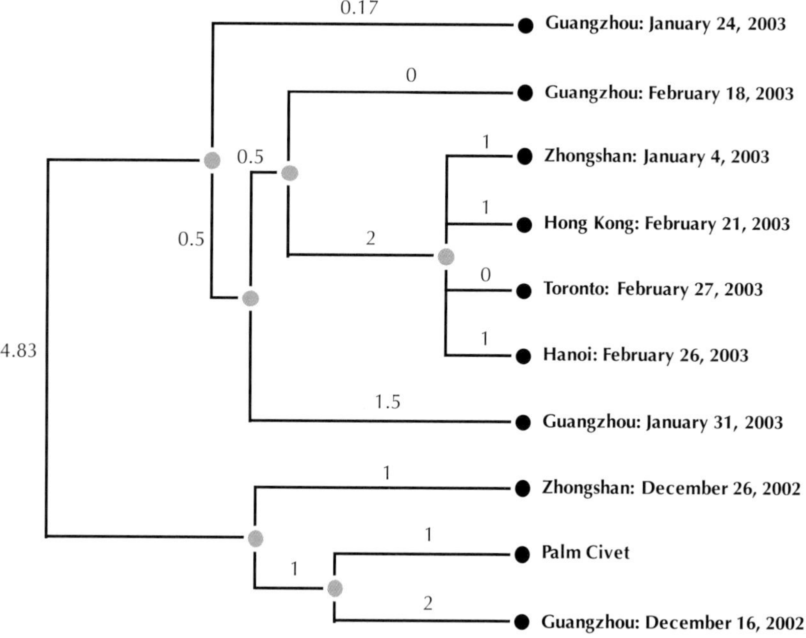

FIGURE 7.20 (Top) The distance matrix based on pairwise alignment of Spike proteins from SARS-CoV strains extracted from various patients as well as a coronavirus taken from a palm civet. The distance between each pair of sequences was computed as the total number of mismatches and indels in their optimal pairwise alignment. (Bottom) The evolutionary tree of these viruses constructed by the neighbor-joining algorithm.

parties and pushing paper?" Urbani would ultimately lose his life to SARS a month later. But his sacrifice also helped begin the massive worldwide action against this disease that may well have saved millions of lives.

Limitations of distance-based approaches to tree construction

Although distance-based tree reconstruction has successfully resolved questions about the origin and spread of SARS, many evolutionary controversies cannot be resolved by using a distance matrix. For example, when we convert each pair of rows of a multiple alignment into a distance value, we lose information contained in the alignment. As a result, distance-based methods do not allow us to reconstruct ancestral sequences of Spike proteins (corresponding to internal nodes in Figure 7.20), which could lead us to believe that such molecular paleontology is impossible. Therefore, a superior approach to evolutionary tree reconstruction would be to somehow use the alignment directly, without first converting it to a distance matrix.

Character-Based Tree Reconstruction

Character tables

Fifty years ago, biologists constructed phylogenies not from DNA and protein sequences but from anatomical and physiological features called **characters**. For example, when analyzing invertebrate evolution, one commonly used character is the presence or absence of wings, while another is the number of legs (varying from 0 to over 300 in some centipedes). These two characters result in the 3×2 **character table** shown in Figure 7.21 for three species.

In general, every row in an $n \times m$ character table represents a **character vector,** holding the values of m characters corresponding to one of n existing species. Our goal, roughly stated, is to construct an evolutionary tree in which leaves corresponding to present-day species with similar character vectors occur near each other in the tree. We would also like to assign m character values to each internal node in the tree in order to best explain the characters of ancestral species.

STOP and Think: Can you transform the preceding vague description into a well-formulated computational problem that models tree reconstruction based on character tables?

	wings	legs
winged stick insect	Yes	6
wingless stick insect	No	6
giant centipede	No	42

FIGURE 7.21 (Top panel) Winged (left) and wingless (middle) stick insects, each having six legs, and the giant centipede (right), which has 42 legs. (Bottom panel) A 3×2 character table describes two characters (wings and legs) in these three invertebrates.

From anatomical to genetic characters

In 1965, Emile Zuckerkandl and Linus Pauling published "Molecules as documents of evolutionary history", arguing that DNA sequences offer a much more informative source of data than anatomical and physiological characters. The idea may seem obvious today, especially after we have spent half of a chapter constructing evolutionary trees from distance matrices generated from DNA sequences. However, Zuckerkandl and Pauling's proposal was initially met with skepticism by many biologists, who felt that DNA analysis could not offer the same power as anatomical comparison. A now-famous argument was initiated when Zuckerkandl and Pauling found that the amino acid sequence of human beta-hemoglobin is very similar to that of gorillas, prompting Zuckerkandl to write in 1963:

> *From the point of view of hemoglobin structure, it appears that gorilla is just an abnormal human.*

Yet the surprising similarity between hemoglobin proteins in various primates flew in the face of clear anatomical differences between primates. As a result, the leading evolutionary biologist Gaylord Simpson immediately responded to Zuckerkandl:

> *...that is of course nonsense. What the comparison really indicates is that hemoglobin is a bad choice and has nothing to tell us about attributes, or indeed tells us a lie.*

Despite such vehement initial criticisms, genetic analysis had become the dominant technique in evolutionary studies by the 1970s. Indeed, the analysis of DNA sequences answered evolutionary questions that previous analysis of anatomical characters had failed to resolve. Early examples include the classification of the giant panda (see **DETOUR: Giant Panda: Bear or Raccoon?**) and the identification of human origins (see **DETOUR: Where Did Humans Come From?**). As a result, most evolutionary experts had no choice but to adapt their views and become experts on molecular evolution.

PAGE 412

PAGE 413

Ironically, modern evolutionary studies often view an $n \times m$ multiple alignment as an $n \times m$ character table, with each column representing a character of its own. Our goal is to construct a tree whose leaves correspond to the rows of this alignment and whose internal nodes correspond to ancestral sequences in accordance with the most parsimonious evolutionary scenario. Before rigorously defining a most parsimonious scenario, we will describe one example of how this algorithmic framework solved a longstanding puzzle in insect evolution.

How many times has evolution invented insect wings?

Wings provided a revolutionary adaptation for insects, allowing them to escape predators and disperse into new territories, thus leading to many new insect species. Yet despite the evolutionary advantages provided by wings, some insects are apparently better equipped for survival without them. In fact, nearly all winged species are related to many wingless counterparts belonging to the same genus, and some entire orders of insects are wingless, including fleas and lice.

The acquisition of wings would seem to pose an evolutionary challenge because complex physiological interactions are required to accommodate flight. As a result, we would be led to believe that wings evolved only once in insects. This argument parallels **Dollo's principle of irreversibility**, a hypothesis proposed by 19th Century paleontologist Louis Dollo. According to this principle, when a species loses a complex organ, such as wings, the organ will not reappear in exactly the same form in the species's descendants.

STOP and Think: What do you think about this argument as it pertains to insect wings?

STOP

Until recently, biologists followed Dollo's principle with respect to insect wings, believing that re-evolution of wings was essentially impossible because unused flight genes in wingless insects would be free to accumulate mutations, eventually eroding into non-functional pseudogenes. However, in 2003, Michael Whiting studied various winged and wingless stick insects from around the world and refuted this argument. He sequenced an approximately 2,000 nucleotide-long segment (the **18S ribosomal RNA** gene) from these stick insects and constructed an evolutionary tree based on these sequences. From this phylogeny, he inferred that wings were re-invented at least three times and lost at least four times during stick insect evolution (Figure 7.22).

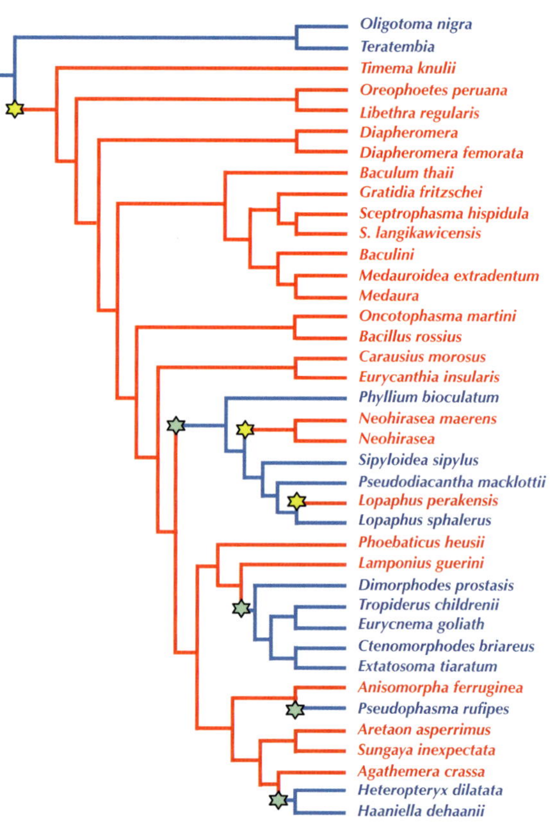

FIGURE 7.22 Evolutionary tree of winged (blue) and wingless (red) stick insects constructed from 18S ribosomal RNA genes. Transitions from winged to wingless species are shown by yellow stars; transitions from wingless to winged species are shown by green stars. 18S ribosomal RNAs are slow-evolving, making them ideal for reconstructing ancient divergences.

STOP and Think: Is it possible to infer an evolutionary scenario from the tree in Figure 7.22 that re-invents wings fewer than four times?

STOP

Whiting's work indicates the inherent complications of trying to infer evolutionary trees from anatomical characters. If you were asked to design a character-based phylogeny algorithm for the collection of all stick insects, then one of the first steps you would probably take would be to cluster all the winged insects on the opposite side of the tree from all of the wingless insects. However, the re-evolution of insect wings means that such an approach is flawed. Anatomical characters have even less power when we move into the microscopic world: just imagine trying to construct a phylogeny from direct observation of coronaviruses!

The Small Parsimony Problem

We will label each leaf of a tree by a row of a multiple alignment, and we will attempt to infer strings labeling internal nodes and corresponding to candidate ancestral sequences. However, we will need to develop a scoring function in order to quantify how well such a labeled tree fits the given multiple alignment. In what follows, we assume for simplicity that the multiple alignment contains only substitutions and no indels. In practice, researchers may start from a multiple alignment containing indels and then remove all columns containing indels.

An intuitive score of an evolutionary tree is the total number of mutations required to explain the strings at all nodes of the tree. Given a tree T with every node labeled by a string of length m, we will therefore set the length of edge (v, w) equal to the number of substitutions (Hamming distance) between the strings labeling v and w. The **parsimony score** of T is the sum of the lengths of its edges (Figure 7.23).

We will first assume that we are given the structure of a rooted binary tree in advance, in which case we only need to assign strings to the internal nodes in order to minimize the parsimony score.

Small Parsimony Problem:
Find the most parsimonious labeling of the internal nodes of a rooted tree.

> **Input**: A rooted binary tree with each leaf labeled by a string of length m.
> **Output**: A labeling of all other nodes of the tree by strings of length m that minimizes the tree's parsimony score.

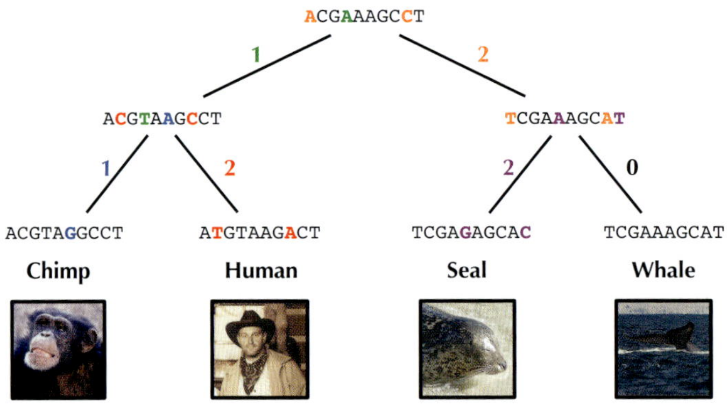

FIGURE 7.23 An evolutionary tree with parsimony score 8 whose leaves are the DNA strings in the multiple alignment from Figure 7.3. Colored letters indicate mismatches in strings connected by an edge.

Given a tree T for which every node v is labeled by a string of length m, we define labeled trees $T_1, \ldots, T_m$, where T_i has the same structure as T, and where a given node is labeled by the i-th symbol of the corresponding node in T. Since the parsimony score of T is the sum of parsimony scores of the trees $T_1, \ldots, T_m$, the Small Parsimony Problem can be solved independently for each column of the alignment. This observation lets us assume that every leaf is labeled by a single symbol rather than by a string. Thus, the weight of an edge connecting two nodes should be either 0 or 1, depending on whether these nodes are labeled by the same symbol or different symbols; given symbols i and j, we define $\delta_{i,j} = 0$ if $i = j$ and $\delta_{i,j} = 1$ if $i \neq j$.

We will describe a dynamic programming algorithm, called **SMALLPARSIMONY**, for solving the "single character" version of the Small Parsimony Problem. Recall that a rooted tree T can be viewed as a directed tree with all of its edges directed away from the root toward the leaves. Thus, every node v in T defines a subtree T_v formed by the nodes "beneath" v and consisting of all the nodes that can be reached by moving down from v (Figure 7.24).

Let k be a symbol in a given alphabet and v be a node in a tree T. Define $s_k(v)$ as the minimum parsimony score of the subtree T_v over all possible labelings of the nodes of T_v such that v is labeled by k. The initial conditions for **SMALLPARSIMONY** must assign scores to leaves. If leaf v is labeled by k, then the only character we are allowed to assign to this leaf is k. Therefore, $s_k(v) = 0$ if leaf v is labeled by symbol k, and $s_k(v) = \infty$ otherwise (Figure 7.25).

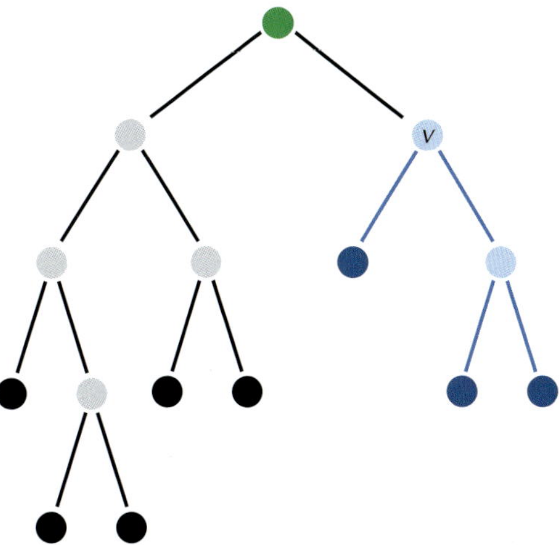

FIGURE 7.24 The (blue) subtree T_v of a node v within a larger rooted binary tree T.

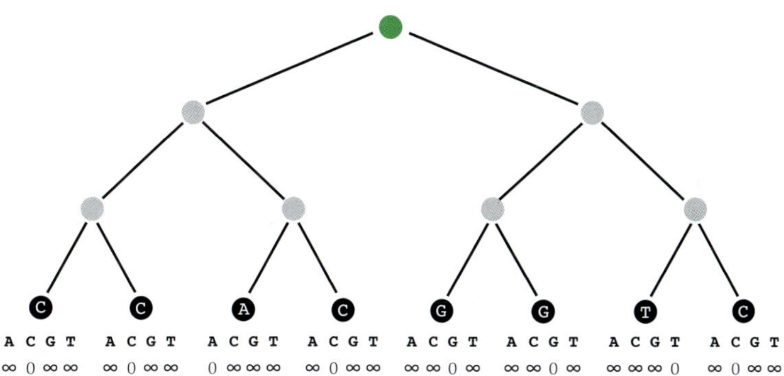

FIGURE 7.25 Initializing values $s_k(v)$ for all leaves v, represented above as an array for each leaf. We set $s_k(v)$ equal to zero if the leaf is labeled by symbol k; otherwise, we set $s_k(v)$ equal to infinity.

If v is an internal node of T, then v is connected to two "children" nodes (the nodes beneath v in T) that we arbitrarily denote as DAUGHTER(v) and SON(v). The score $s_k(v)$ can be computed as the minimum of $s_i(\text{DAUGHTER}(v)) + \delta_{i,k}$ over all possible symbols i, plus the minimum of $s_j(\text{SON}(v)) + \delta_{j,k}$ over all possible symbols j:

$$s_k(v) = \min_{\text{all symbols } i} \{s_i(\text{DAUGHTER}(v)) + \delta_{i,k}\} + \min_{\text{all symbols } j} \{s_j(\text{SON}(v)) + \delta_{j,k}\}$$

This equation allows us to compute all values $s_k(v)$ by working our way upward in T from the leaves to the root, denoted *root* (Figure 7.26). The subtree of the root is the entire tree T, and so the minimum parsimony score is given by the smallest score $s_k(root)$ over all symbols k,

$$\min_{\text{all symbols } k} s_k(root).$$

The pseudocode for **SMALLPARSIMONY** is shown below. It returns the parsimony score for a rooted binary tree T whose leaves are labeled by symbols stored in an array CHARACTER (i.e., CHARACTER(v) is the label of leaf v). At each iteration, it selects a node v and computes $s_k(v)$ for each symbol k in the alphabet. For each node v, **SMALLPARSIMONY** maintains a value TAG(v), which indicates whether the node has been processed (i.e., TAG$(v) = 1$ if the array $s_k(v)$ has been computed and TAG$(v) = 0$ otherwise). We call an internal node of T **ripe** if its tag is 0 but its children's tags are both 1. **SMALLPARSIMONY** works upward from the leaves, finding a ripe node v at which to compute $s_k(v)$ at each step.

7F

```
SMALLPARSIMONY(T, CHARACTER)
    for each node v in tree T
        TAG(v) ← 0
        if v is a leaf
            TAG(v) ← 1
            for each symbol k in the alphabet
                if CHARACTER(v) = k
                    s_k(v) ← 0
                else
                    s_k(v) ← ∞
    while there exist ripe nodes in T
        v ← a ripe node in T
        TAG(v) ← 1
        for each symbol k in the alphabet
            s_k(v) ← min_{all symbols i} {s_i(DAUGHTER(v)) + δ_{i,k}} + min_{all symbols j} {s_j(SON(v)) + δ_{j,k}}
    return min_{all symbols k} s_k(v)
```

STOP and Think: What is the final ripe node processed by **SMALLPARSIMONY**, regardless of the order in which we process ripe nodes?

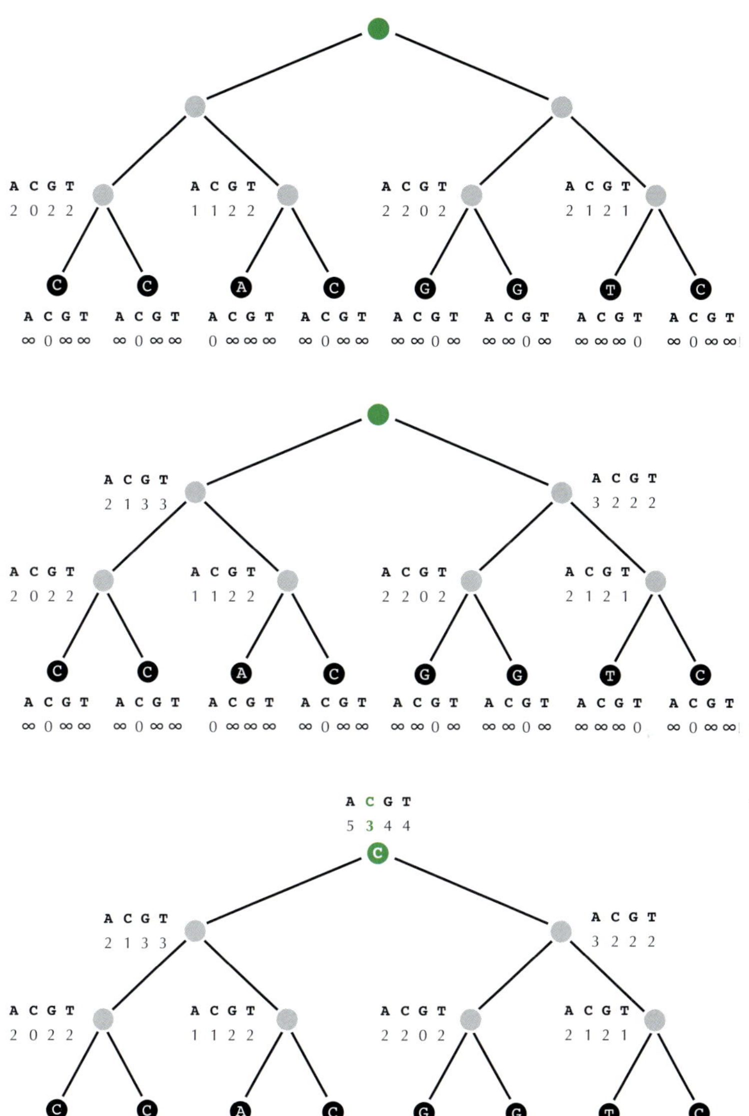

FIGURE 7.26 An illustration of **SMALLPARSIMONY** after initialization in Figure 7.25. The parsimony score is equal to the minimum score at the root, which for this tree is equal to 3. This value corresponds to symbol C, and so when we begin backtracking to assign symbols to internal nodes, we assign nucleotide C to the root.

Once we compute the parsimony score of a tree T, we also need some way of assigning symbols to internal nodes of T. The minimum value of $s_k(root)$ in Figure 7.26 (bottom) is equal to 3, which is achieved when $k = \text{C}$. Assigning symbols to the remaining internal nodes is similar to the backtracking approach that we used for sequence alignment. Because you are already a pro at dynamic programming, we leave this task to you as an exercise.

Exercise Break: Assign nucleotides to all nodes of the tree in Figure 7.26 in order to solve the Small Parsimony Problem.

When the position of the root in the tree is unknown, we can simply assign the root to any edge that we like, apply **SMALLPARSIMONY** to the resulting tree, and then remove the root. It can be shown that this method provides a solution to the following problem.

Small Parsimony in an Unrooted Tree Problem:
Find the most parsimonious labeling of the internal nodes of an unrooted tree.

7G

Input: An unrooted binary tree with each leaf labeled by a string of length m.
Output: A labeling of all other nodes of the tree by strings of length m that minimizes the tree's parsimony score.

Exercise Break: Estimate the runtime of the proposed algorithm for solving the Small Parsimony in an Unrooted Tree Problem.

Exercise Break: Given a multiple alignment of SARS viruses, reconstruct the most parsimonious amino acid sequence of the Spike protein in the ancestral SARS virus (the root of the tree) under the assumption that the evolutionary tree in Figure 7.17 is correct.

The Large Parsimony Problem

SMALLPARSIMONY does not help us if we do not know the evolutionary tree in advance. In this case, we must find a binary tree as well as assign ancestral strings to all internal nodes of this tree in order to minimize the parsimony score. Figure 7.27 presents the solutions of the Small Parsimony in an Unrooted Tree Problem for the three different unrooted binary trees with four leaves, where the leaves are assigned the strings from the toy multiple alignment in Figure 7.3.

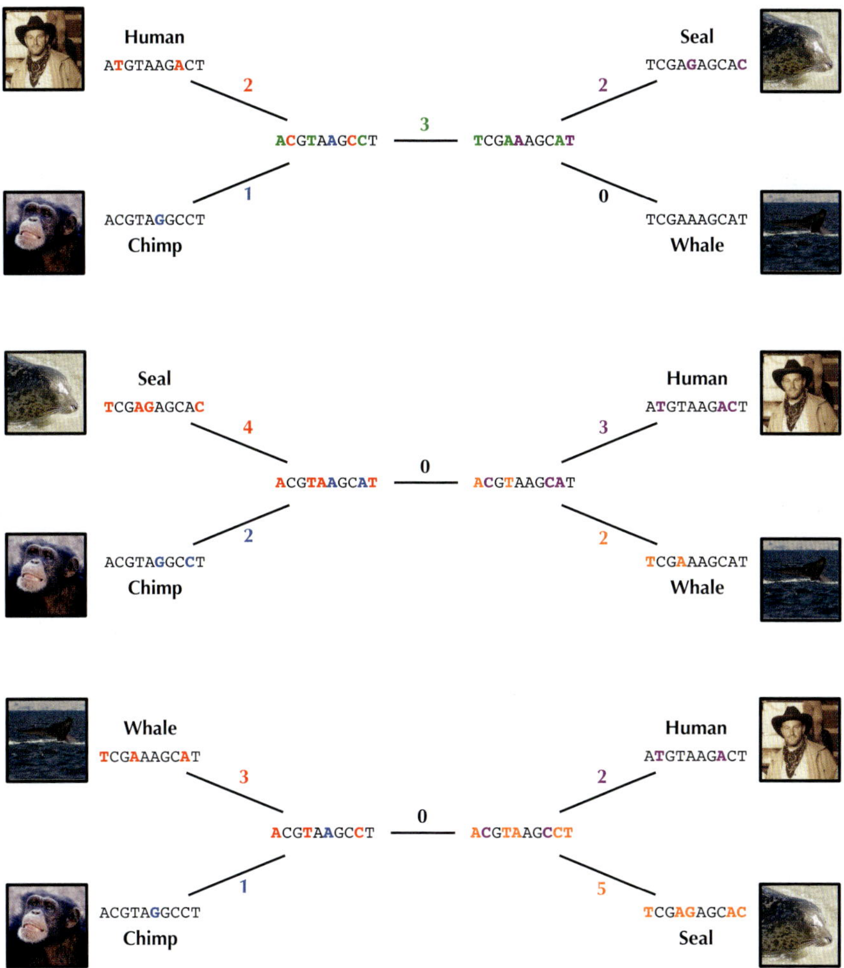

FIGURE 7.27 The three unrooted binary tree structures for the four species from Figure 7.3, with internal nodes labeled by a solution of the Small Parsimony in an Unrooted Tree Problem. The first tree solves the Large Parsimony Problem because it has smaller parsimony score (8) than the other two trees (11 each).

Large Parsimony Problem:
Given a set of strings, find a tree — with leaves labeled by all these strings — having minimum parsimony score.

> **Input**: A collection of strings of equal length.
> **Output**: An unrooted binary tree T that minimizes the parsimony score among all possible unrooted binary trees with leaves labeled by these strings.

Unfortunately, the Large Parsimony problem is *NP*-Complete, in part because the number of different trees grows very quickly with respect to the number of leaves. As a workaround, we will use a greedy heuristic that explores some but not all trees. First, note that in any unrooted binary tree, the removal of an internal edge, along with the two internal nodes that this edge connects, results in four subtrees, which we will call W, X, Y, and Z (Figure 7.28). These four subtrees can be combined into a tree in three different ways, which we denote $WX|YZ$, $WY|XZ$, and $WZ|XY$. These three trees are called **nearest neighbors**; a **nearest neighbor interchange** operation replaces a tree with one of its nearest neighbors.

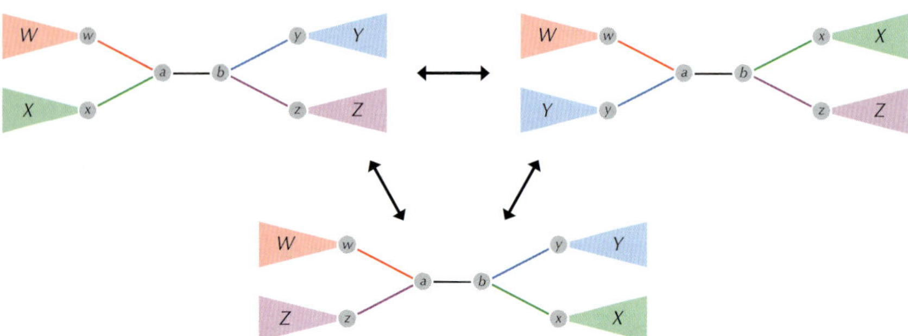

FIGURE 7.28 A nearest neighbor interchange on the internal edge (a, b), shown in black, results from rearranging the four colored subtrees W, X, Y, and Z, which are rooted at w, x, y, and z, respectively. The nearest neighbor interchange operation removes one edge connected to a and another edge connected to b, then replaces these edges with two new edges. The three possible tree structures resulting from nearest neighbor interchanges on (a, b) can be represented as $WX|YZ$ (top left), $WY|XZ$ (top right), and $WZ|XY$ (bottom).

Exercise Break: Find the two nearest neighbors of the tree in Figure 7.15 for the internal edge connecting the parents of gorilla and human.

Like the 2-break operation in Chapter 6, a nearest neighbor interchange corresponds to replacing two edges in the tree by two new edges. For example, denote the internal edge of a nearest neighbor interchange as (a, b); denote the remaining nodes adjacent to a as w and x; and denote the remaining nodes adjacent to b as y and z. The tree on the right in Figure 7.28 is obtained from the tree on the left by removing edges (a, x) and (b, y) and replacing them with (a, y) and (b, x). The tree on the bottom in Figure 7.28 is obtained from the tree on the left by removing edges (a, x) and (b, z) and substituting them with (a, z) and (b, x).

Exercise Break: Figure 7.29 shows all possible unrooted binary trees with five leaves. Find a pair of these trees that are the "farthest apart" in that they require the maximum number of nearest neighbor interchanges to transform one tree into the other.

Nearest Neighbors of a Tree Problem:
Given an edge in a binary tree, generate the tree's nearest neighbors.

Input: An internal edge in a binary tree.
Output: The two nearest neighbors of this tree (with respect to the given internal edge).

The **nearest neighbor interchange heuristic** for the Large Parsimony Problem starts from an arbitrary unrooted binary tree. It assigns input strings to arbitrary leaves of this tree, assigns strings to the internal nodes of the tree by solving the Small Parsimony Problem in an Unrooted Tree, and then moves to a nearest neighbor that provides the best improvement in the parsimony score. At each iteration, the algorithm explores all internal edges of a tree and generates all nearest neighbor interchanges for each internal edge. For each of these nearest neighbors, the algorithm solves the Small Parsimony Problem to reconstruct the labels of the internal nodes and computes the parsimony score. If a nearest neighbor with smaller parsimony score is found, then the algorithm. selects the one with smallest parsimony score (ties are broken arbitrarily) and iterates again; otherwise, the algorithm terminates. This is achieved by the following pseudocode.

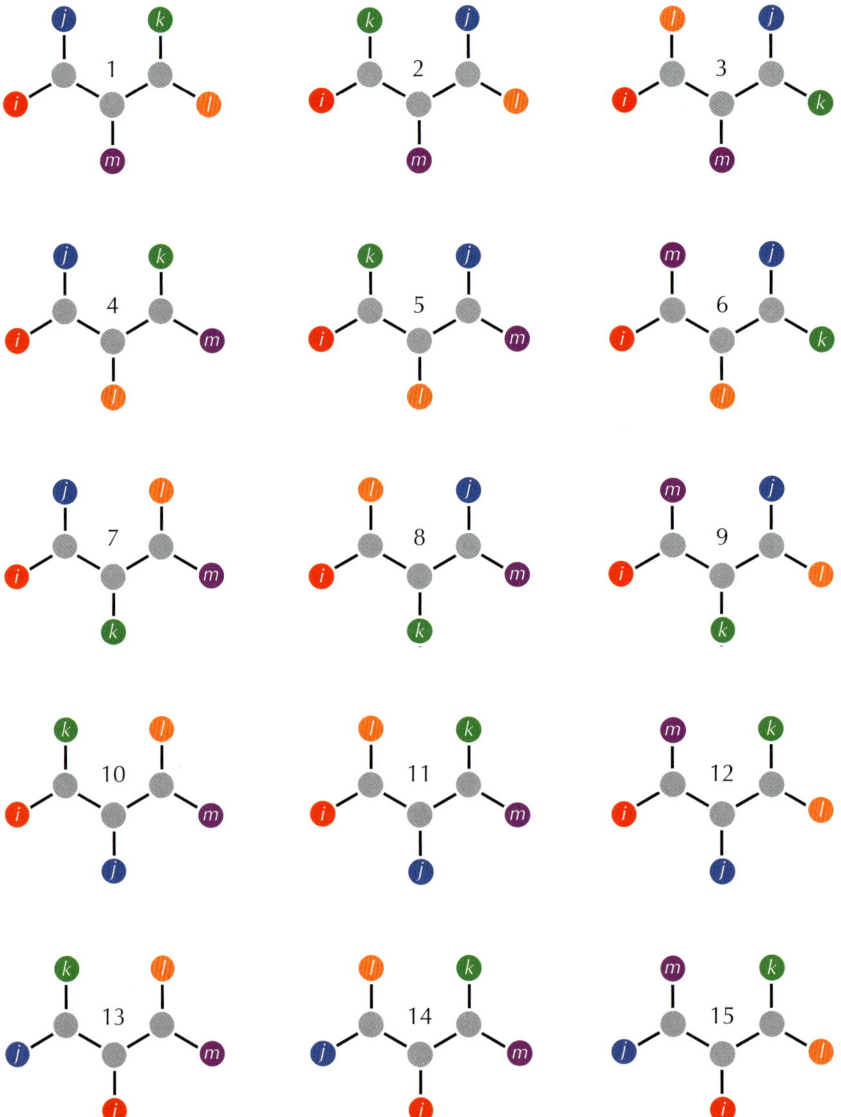

FIGURE 7.29 All fifteen unrooted binary trees with five labeled leaves. Tree 1 can be transformed into trees 4, 7, 12, and 15 by a single nearest neighbor interchange. Note that every tree has the same structure; this is not the case for trees containing more than five leaves.

```
NEARESTNEIGHBORINTERCHANGE(Strings)
    score ← ∞
    generate an arbitrary unrooted binary tree Tree with |Strings| leaves
    label the leaves of Tree by arbitrary strings from Strings
    solve the Small Parsimony in an Unrooted Tree Problem for Tree
    label the internal nodes of Tree according to a most parsimonious labeling
    newScore ← the parsimony score of Tree
    newTree ← Tree
    while newScore < score
        score ← newScore
        Tree ← newTree
        for each internal edge e in Tree
            for each nearest neighbor NeighborTree of Tree with respect to the edge e
                solve the Small Parsimony in an Unrooted Tree Problem for NeighborTree
                neighborScore ← the minimum parsimony score of NeighborTree
                if neighborScore < newScore
                    newScore ← neighborScore
                    newTree ← NeighborTree
    return newTree
```

71

Exercise Break: Chapter 8 describes how the alcohol dehydrogenase (Adh) gene helps yeast produce alcohol. To study the evolution of this gene, biologists constructed a multiple alignment of the Adh genes from various yeast species. Use this alignment to reconstruct the evolutionary tree of various yeast species and an ancient yeast Adh ancestor gene.

We have now encountered a number of algorithms for constructing evolutionary trees, but this does not mean that we can easily resolve various evolutionary controversies. For example, the identity of the chimpanzee's closest relative remained undecided until the mid-1990s (Figure 7.30), and the question of whether mice are closer to humans than to dogs is still the subject of debate (Figure 7.31).

FIGURE 7.30 (Left) Analysis of beta-globin genes in human, chimpanzee, and gorilla suggests a human-chimpanzee split. (Right) Analysis of dopamine D4 receptor gene suggests a gorilla-chimpanzee split.

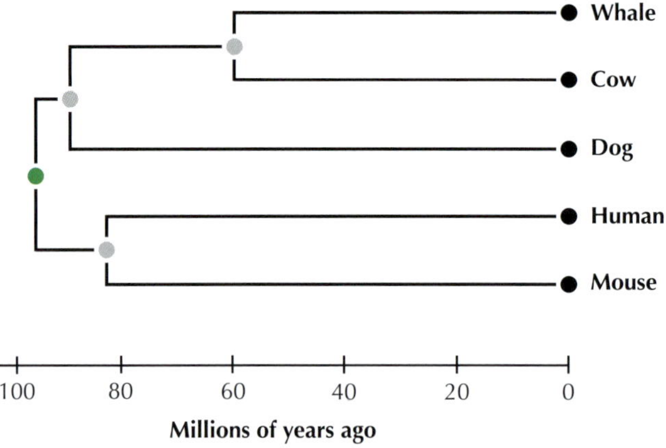

FIGURE 7.31 At the start of the 21st Century, biologists believed that dogs are evolutionarily closer to humans than mice are. However, recent studies suggest otherwise, as shown in the above phylogeny.

Epilogue: Evolutionary Trees Fight Crime

Janice Trahan met Dr. Richard Schmidt in 1982 when she began working as a nurse in Lafayette, Louisiana. Both Janice and Richard were married with children, but they fell in love. Janice soon divorced her husband, but although Richard promised that he would divorce his wife, he never did. After twelve years, she got tired of waiting, and broke off the relationship. Two weeks later, she awoke in the middle of the night to see Richard standing over her, a syringe in his hand.

Although Janice had broken up with Richard, she was not surprised to see him. She had even left her door unlocked for him because he had been giving her vitamin B-12 injections for her chronic fatigue. Since Janice had B-12 injections before, she knew what to expect. This time, however, she experienced scorching pain as Richard squeezed the syringe.

A few months later, Janice tested positive for HIV, and she accused Richard of infecting her via an injection. The police detective in Lafayette had never heard anything as bizarre as this revenge story — a syringe tainted with HIV had never been used as a murder weapon. At first, he suspected that Janice had fabricated the story to tarnish the reputation of her former lover. Nevertheless, he started an investigation by collecting samples from several HIV patients in Lafayette.

After investigating hospital records, the detective found that Richard had taken blood from Donald McClelland, an HIV patient, on the same day that he had injected Janice. Now the forensic challenge was to determine whether the HIV strain taken from McClelland was similar to Janice's strain. After HIV DNA was collected from Janice, McClelland, and many other unrelated HIV-infected patients from Lafayette, scientists constructed the evolutionary tree of these HIV viruses and found that the viruses sampled from Janice and McClelland formed a subtree of this tree (Figure 7.32).

The case "State of Louisiana vs. Richard Schmidt" went to trial in 1998. The prominent evolutionary biologist David Hillis presented the evolutionary tree as evidence of the crime, demonstrating that Janice's HIV sequence had been derived (with some small variations) from McClelland's HIV sequence. Richard Schmidt was then sentenced to 50 years in prison for attempted murder.

STOP and Think: If you had been Richard Schmidt's attorney, how would you have argued for his innocence?

STOP

Challenge Problem: Given HIV sequences from AIDS patients in Lafayette, construct the evolutionary tree for other HIV proteins. Does each tree support conviction of Dr. Schmidt? Reconstruct the ancestral HIV sequences at the internal nodes of the resulting trees.

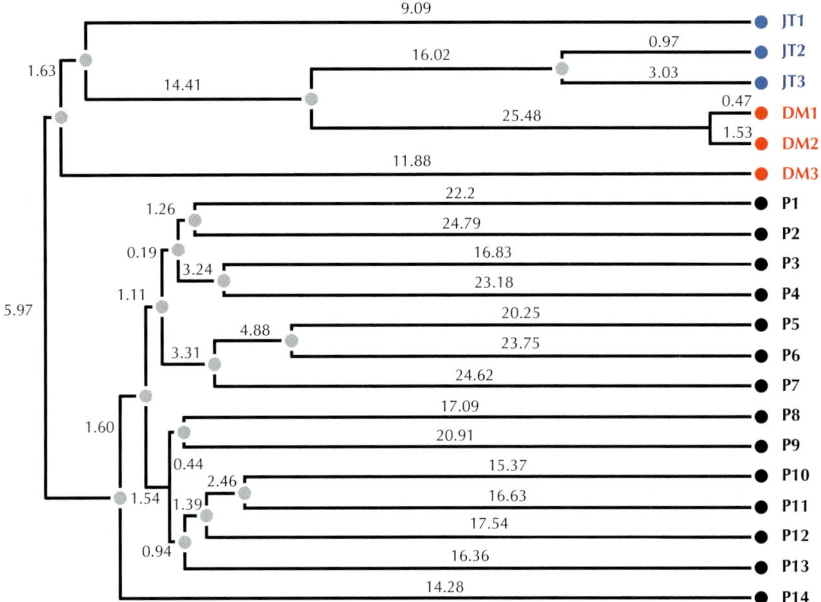

FIGURE 7.32 An evolutionary tree of HIV viruses taken from various patients in Lafayette. Samples from the victim, Janice Trahan (blue leaves JT1, JT2, and JT3), and Richard Schmidt's patient, Donald McClelland (red leaves DM1, DM2, DM3), are clustered together and are rather different than sequences from other patients from Lafayette (labeled P1 to P14).

Detours

When did HIV jump from primates to humans?

Scientists became aware that HIV causes AIDS in the 1980s, at a time when the virus was still rare. They immediately started looking for earlier cases of HIV infection in various medical records and found HIV in a blood sample taken from a Congolese patient in 1959. Genetic studies then revealed that HIV is closely related to Simian Immunodeficiency Virus (SIV), which infects primates, but it remained unclear how and when SIV entered the human population and evolved into HIV. The most popular current hypothesis is that SIV evolved into HIV when hunters killed monkeys to sell their meat and were exposed to the animals' blood. The virus circulating in the blood had further entered cuts in the hunters' skin, mutated, and later adapted to humans.

By finding regions in the viral genome that mutate at a roughly constant rate over time, researchers can infer a timeline of HIV evolution. Using this "molecular clock", biologists estimated the time points when the various subtypes of SIV jumped from primates to human: for HIV groups A, B, M, and O, these transitions have been estimated at 1940, 1945, 1908, and 1920, respectively (the timing of the jump for group N is currently unknown). By sequencing fecal samples of wild primates, biologists have even been able to locate populations of chimpanzees and sooty mangabeys whose SIVs are direct ancestors of HIV groups.

Searching for a tree fitting a distance matrix

Every 3×3 matrix D is additive. To see why, consider the tree in Figure 7.33 with three leaves labeled 1, 2, and 3, as well as an internal node labeled as c. As illustrated in Figure 7.33, the lengths of edges in this tree must satisfy the following three equations:

$$d_{1,c} + d_{2,c} = D_{1,2} \qquad d_{1,c} + d_{3,c} = D_{1,3} \qquad d_{2,c} + d_{3,c} = D_{2,3}$$

Solving this system of equations yields the following formulas for the lengths of edges in terms of values of the matrix D:

$$d_{1,c} = \frac{D_{1,2} + D_{1,3} - D_{2,3}}{2} \qquad d_{2,c} = \frac{D_{2,1} + D_{2,3} - D_{1,3}}{2} \qquad d_{3,c} = \frac{D_{3,1} + D_{3,2} - D_{1,2}}{2}$$

Figure 7.34 illustrates an attempt to fit the distance matrix in Figure 7.9 (right) to all possible unrooted trees with four leaves. Each such tree leads to a system of six linear equations (in either four or five variables) that does not have a solution. Thus, this distance matrix must be non-additive.

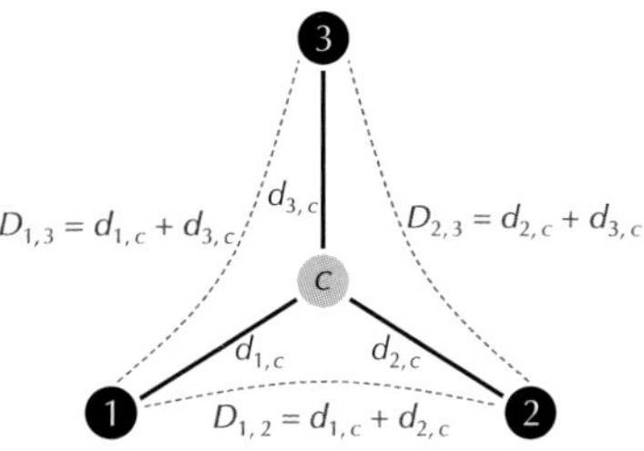

FIGURE 7.33 The tree with three leaves 1, 2, and 3 in addition to an internal node c. The distances between leaves ($D_{1,2}$, $D_{1,3}$, and $D_{2,3}$) uniquely define the lengths of edges ($d_{1,c}$, $d_{2,c}$, and $d_{3,c}$).

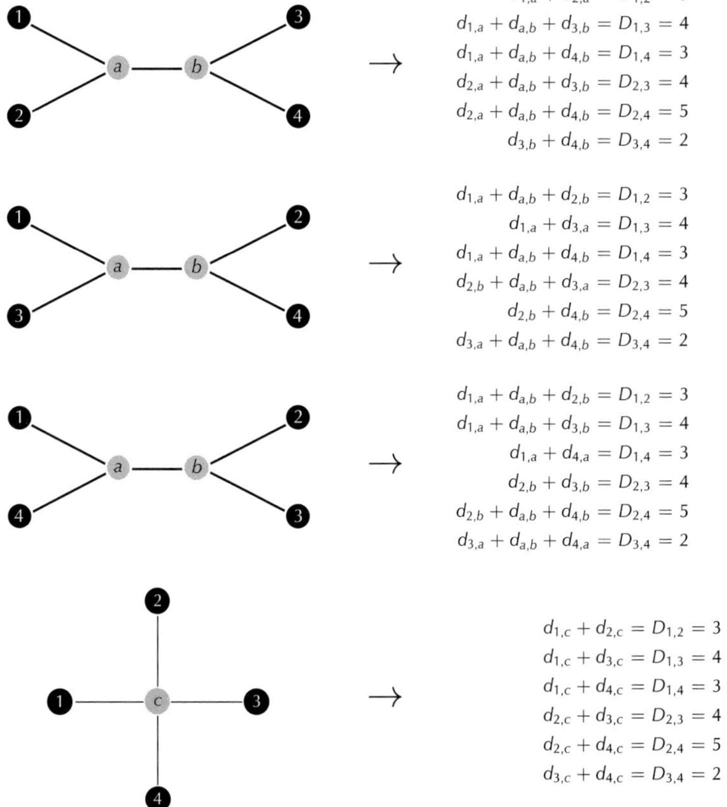

FIGURE 7.34 (Left) All four trees with four leaves. (Right) Each attempt to fit the distance matrix in Figure 7.9 to a tree results in a system of six linear equations. Because none of these systems has a solution, this matrix must be non-additive.

The reason we failed to fit a tree to the 4×4 distance matrix in Figure 7.34 is partly because the number of equations was larger than the number of variables ($d_{i,j}$) for each tree (a fact that will also hold for larger values of n). If the number of equations in a linear system of equations is smaller than or equal to the number of variables in the system, then a solution usually exists, whereas if the number of equations exceeds the number of variables, there is usually no solution. However, there are exceptions in both directions. Consult an introductory linear algebra text for more details.

The four point condition

The **four point condition** gives an alternative way to determine if a matrix is additive. Consider the tree containing only four leaves in Figure 7.35. For this tree, observe that

$$d_{i,j} + d_{k,l} \leq d_{i,k} + d_{j,l} = d_{i,l} + d_{j,k}$$

because the first sum is the sum of lengths of all edges in the tree *minus* the length of the internal edge, while the last two sums are equal to the sum of lengths of all edges in the tree *plus* the length of the internal edge.

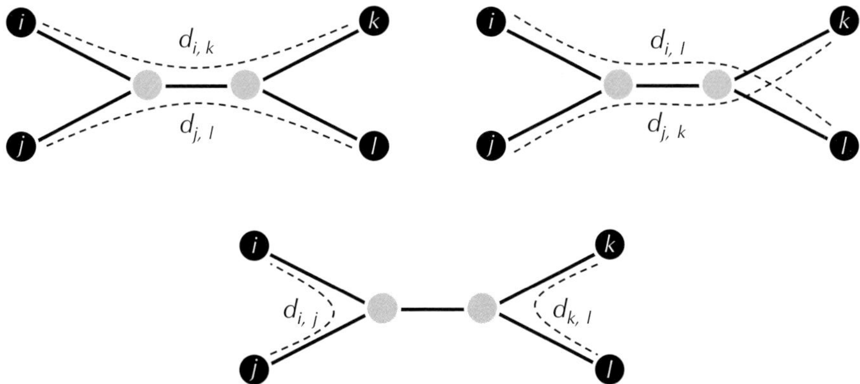

FIGURE 7.35 Three pairs of paths through a tree with four leaves. The paths in the top left and top right traverse the same edges, and so $d_{i,k} + d_{j,l} = d_{i,l} + d_{j,k}$. Furthermore, $d_{i,j} + d_{k,l}$ must be less than or equal to these two sums, since it does not traverse the internal edge of the tree, as shown in the tree on the bottom.

Indeed, for any quartet of leaves (i, j, k, l) in an arbitrary tree, if we compute the three sums

$$d_{i,j} + d_{k,l} \quad d_{i,k} + d_{j,l} \quad d_{i,l} + d_{j,k}$$

then we will find that two of the sums are equal and that the third sum is less than or equal to the other two sums. In terms of an $n \times n$ distance matrix, we say that a quartet of indices (i, j, k, l) satisfies the four point condition if two of the following sums are equal, and the third sum is less than or equal to the other two sums:

$$D_{i,j} + D_{k,l} \quad D_{i,k} + D_{j,l} \quad D_{i,l} + D_{j,k}$$

Four Point Theorem: *A distance matrix is additive if and only if the four point condition holds for every quartet (i, j, k, l) of indices of this matrix.*

Exercise Break: Prove the Four Point Theorem.

The Four Point Theorem provides us with an alternative way of determining whether a given distance matrix is additive, since we can simply check whether the four point condition holds for each quartet of indices of the matrix.

Exercise Break: Compare the running time of this proposed method with that of a variant of **ADDITIVEPHYLOGENY** deciding whether a given distance matrix is additive.

Exercise Break: Find a quartet of indices from the distance matrix in Figure 7.9 (right) that violates the four points condition.

Did bats give us SARS?

During the search for the animal reservoir of the SARS virus, biologists discovered infected palm civets at a live animal market in China. Meat from these animals is often added to "dragon-tiger-phoenix soup", an expensive Cantonese dish. The discovery did not greatly change the infected civets' fate: instead of ending up in soup, they were made into SARS scapegoats and slaughtered anyway.

Yet when further searches failed to identify more SARS-infected civets, biologists started to wonder whether palm civets really were the original source of SARS. In 2005, they discovered a SARS-like virus in Chinese horseshoe bats (Figure 7.36). The bats turned out to be SARS-CoV carriers, but they can probably only pass the virus to humans through intermediate hosts. Since bat meat is considered a delicacy and is also used in traditional Chinese medicine, bats had plenty of chances to come in close contact with civets at overcrowded live animal markets.

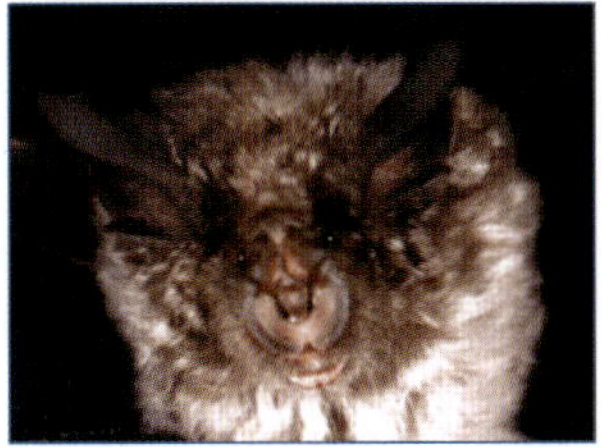

FIGURE 7.36 The horseshoe bat.

When biologists constructed the evolutionary tree of coronaviruses from bats, civets, and humans (Figure 7.37), they found that both the civet and human variants of SARS-CoV are nested within a bat virus phylogeny.

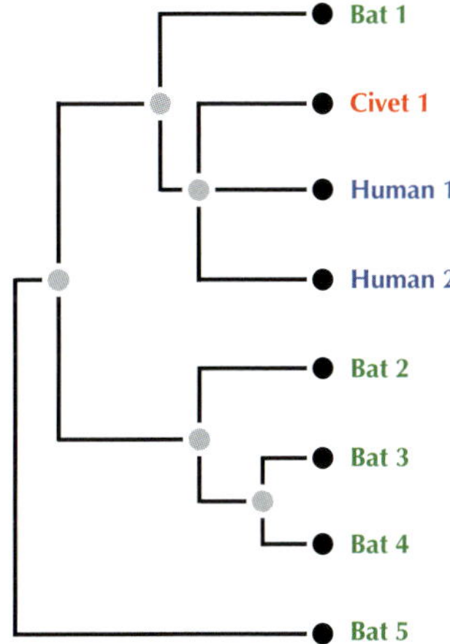

FIGURE 7.37 An evolutionary tree of coronaviruses from bats, civets, and humans.

Even before sequencing SARS-CoV, biologists knew of other human coronaviruses but considered them harmless. However, a new deadly SARS-like coronavirus emerged in 2012 in Saudi Arabia, This virus (causing **Middle East Respiratory Syndrome (MERS)**) generated international headlines as it quickly spread to other countries.

Although researchers initially believed that camels were to blame for MERS — many Saudis consume unpasteurized camel milk — most afflicted patients had not come in contact with camels. Yet when researchers tested a coronavirus taken from a bat found just a few miles from one of the first MERS patient's homes, they found this virus to be a nearly perfect match with the patient's virus sample.

Why does the neighbor-joining algorithm find neighboring leaves?

We stated in the main text that if the distance matrix D is additive, then there exists a unique simple tree TREE(D) fitting this matrix. This is not quite true, since if a simple tree fits D and has an internal edge of weight zero, then we can easily remove this edge by "gluing" together the nodes that it connects (Figure 7.38). Thus, we will make a further assumption that not only is TREE(D) simple, but that it contains no internal edges of length zero.

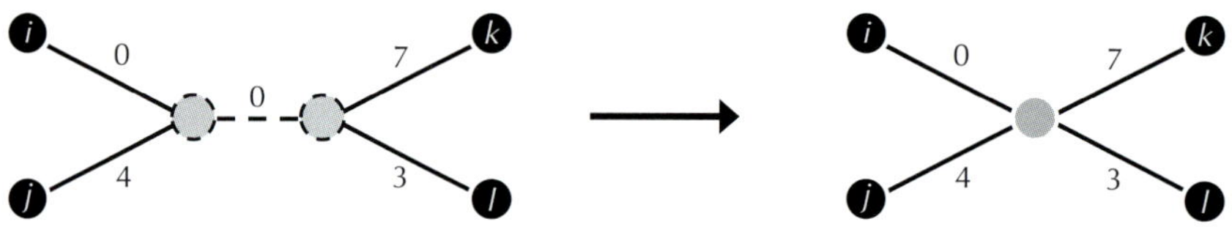

FIGURE 7.38 Gluing the nodes at the endpoints of an internal edge having zero length (shown by the dashed edge).

Now, we can rewrite the neighbor-joining matrix formula as follows:

$$\begin{aligned} D^*_{i,j} &= (n-2) \cdot D_{i,j} - \text{TOTALDISTANCE}_D(i) - \text{TOTALDISTANCE}_D(j) \\ &= (n-2) \cdot D_{i,j} - \sum_{1 \le k \le n} D_{i,k} - \sum_{1 \le k \le n} D_{j,k} \,. \end{aligned}$$

Each element in this formula further breaks down as a sum of edge weights in TREE(D). For example, for the first tree in Figure 7.34,

$$\begin{aligned} D^*_{1,2} &= 2 \cdot d_{1,2} - (d_{1,3} + d_{1,4} + d_{1,2}) - (d_{2,3} + d_{2,4} + d_{1,2}) \\ &= 2 \cdot (d_{1,a} + d_{a,2}) - ((d_{1,a} + d_{a,b} + d_{b,3}) + (d_{1,a} + d_{a,b} + d_{b,4}) + \\ &\quad (d_{1,a} + d_{a,2})) - ((d_{2,a} + d_{a,b} + d_{b,3}) + (d_{2,a} + d_{a,b} + d_{b,4}) + (d_{1,a} + d_{a,2})) \\ &= -2 \cdot d_{1,a} - 2 \cdot d_{2,a} - 2 \cdot d_{b,3} - 2 \cdot d_{b,4} - 4 \cdot d_{a,b} \,, \end{aligned}$$

and

$$\begin{aligned} D^*_{1,3} &= 2 \cdot d_{1,3} - (d_{1,2} + d_{1,4} + d_{1,3}) - (d_{3,4} + d_{3,2} + d_{3,1}) \\ &= 2 \cdot (d_{1,a} + d_{a,b} + d_{b,3}) - ((d_{1,a} + d_{a,2}) + (d_{1,a} + d_{a,b} + d_{b,4}) + \\ &\quad (d_{1,a} + d_{a,b} + d_{b,3})) - ((d_{3,b} + d_{b,4}) + (d_{3,b} + d_{b,a} + d_{b,2}) + (d_{3,b} + d_{b,a} + d_{a,1})) \\ &= -2 \cdot d_{1,a} - 2 \cdot d_{2,a} - 2 \cdot d_{b,3} - 2 \cdot d_{b,4} - 2 \cdot d_{a,b} . \end{aligned}$$

Note that the expressions for $D^*_{1,2}$ and $D^*_{1,3}$ are nearly identical, with only the coefficients of $d_{a,b}$ differing (highlighted in red above). Since $D^*_{1,2} - D^*_{1,3} = -2 \cdot d_{a,b} < 0$, the neighbor-joining algorithm will prefer the smaller $D^*_{1,2}$ over $D^*_{1,3}$.

Given an edge e in TREE(D), the **multiplicity** of this edge in $D^*_{i,j}$ is the coefficient of d_e in $D^*_{i,j}$, denoted MULTIPLICITY$_{i,j}(e)$,. For example, for the edge $e = (a, b)$ in the first tree in Figure 7.34, MULTIPLICITY$_{1,2}(e) = -4$, and MULTIPLICITY$_{1,3}(e) = -2$. The following result shows that all limbs in TREE(D) have the same multiplicity.

Lemma. *For an additive distance matrix D and any pair of leaves i and j in* TREE(D), MULTIPLICITY$_{i,j}(e)$ *is equal to -2 for any limb e in* TREE(D).

Proof. If a limb e is not the limb of leaf i or leaf j, then it is counted zero times in $(n-2) \cdot D_{i,j}$, once in TOTALDISTANCE$_D(i)$ and once in TOTALDISTANCE$_D(j)$, making its multiplicity -2. On the other hand, if e is the limb of i or j (say, i), then it is counted $n-2$ times in $(n-2) \cdot D_{i,j}$, $n-1$ times in TOTALDISTANCE$_D(i)$, and once in TOTALDISTANCE$_D(j)$. Therefore, its multiplicity is $n - 2 - (n-1) - 1 = -2$. □

This lemma implies that regardless of which pair of leaves i and j we choose, the limbs of TREE(D) will all have the same contribution to the computation of $D^*_{i,j}$. As a result, only the multiplicities of internal edges of TREE(D) differentiate values of the matrix D^*. Can we determine these multiplicities?

Exercise Break: Prove that for any i and j, and for any internal edge e in TREE(D), MULTIPLICITY$_{i,j}(e) \leq -2$.

Not only does the multiplicity of each internal edge not exceed -2, but we also have a condition determining when an internal edge will have multiplicity equal to -2. To derive this condition, first note that the removal of an internal edge e disconnects any tree into two subtrees. If e lies on the (unique) path connecting leaves i and j in TREE(D), denoted PATH(i, j), then i and j belong to different subtrees, denoted T_i and T_j, respectively; otherwise, i and j belong to the same subtree (Figure 7.39). In the latter case, we denote the number of leaves in the subtree that *does not* include leaves i and j as $\text{LEAVES}_{i,j}(e)$.

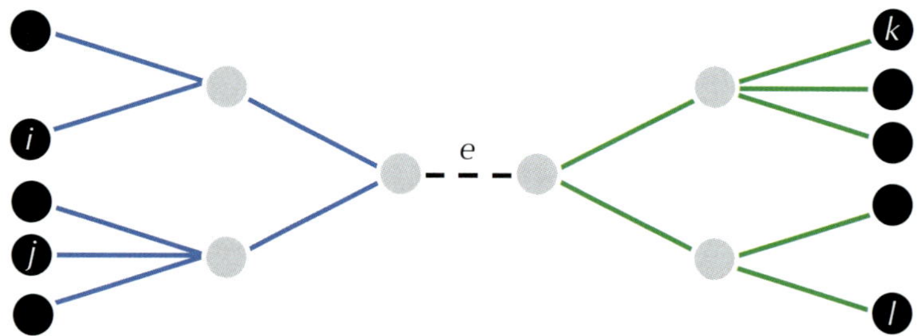

FIGURE 7.39 If an internal edge e lies on the unique path connecting two leaves in the original tree (such as j and k), then these leaves are separated into different subtrees (shown as blue and green) after removing e. If e does not lie on the unique path connecting two leaves (such as k and l), then these leaves belong to the same subtree after removing e.

Edge Multiplicity Theorem: *Given an additive matrix D, the multiplicity of an internal edge e is equal to* -2 *if e lies on* $\text{PATH}(i,j)$ *in* $\text{TREE}(D)$ *and is equal to* $-2 \cdot \text{LEAVES}_{i,j}(e)$ *otherwise.*

Proof. If an internal edge e lies on $\text{PATH}(i,j)$ in $\text{TREE}(D)$, then e has a coefficient of $n-2$ in $(n-2) \cdot D_{i,j}$ term in $D^*_{i,j}$. To compute $\text{MULTIPLICITY}_{i,j}(e)$, consider the subtrees T_i and T_j formed by the removal of e. For each leaf k in T_i, $\text{PATH}(j,k)$ passes through e, thus contributing 1 to the coefficient of e in $\text{TOTALDISTANCE}_D(j)$, but $\text{PATH}(i,k)$ does not pass through e, thus contributing 0 to the coefficient of e in $\text{TOTALDISTANCE}_D(i)$).

Likewise, for each leaf k in T_j, $\text{PATH}(i,k)$ passes through e, contributing 1 to the coefficient of e in $\text{TOTALDISTANCE}_D(i)$, but $\text{PATH}(j,k)$ does not pass through e, contributing 0 to the coefficient of e in $\text{TOTALDISTANCE}_D(i)$). As a result, every leaf k will contribute 1 to the coefficient of e either in $\text{TOTALDISTANCE}_D(i)$ or in $\text{TOTALDISTANCE}_D(j)$. The coefficient of e in $\text{TOTALDISTANCE}_D(i) + \text{TOTALDISTANCE}_D(j)$ is n, which means that

$$\text{MULTIPLICITY}_{i,j}(e) = (n-2) - n = -2\,.$$

On the other hand, if e does not lie on $\text{PATH}(i,j)$, then e has a coefficient of 0 in $(n-2) \cdot D_{i,j}$. And if k is a leaf in the subtree not containing i and j, then to reach k, we must pass through e. Thus, the coefficient of e in each of $\text{TOTALDISTANCE}_D(i)$ and $\text{TOTALDISTANCE}_D(j)$ is equal to $\text{LEAVES}_{i,j}(e)$. It follows that

$$\text{MULTIPLICITY}(e) = 0 - 2 \cdot \text{LEAVES}_{i,j}(e) = -2 \cdot \text{LEAVES}_{i,j}(e)\,.$$

□

We can interpret the Edge Multiplicity Theorem as stating that internal edges on the path PATH(i,j) have large multiplicities (-2) and other internal edges have small multiplicities (less than -2). Thus, if we are trying to minimize $D^*_{i,j}$ (among all possible choices of i and j), then we should look for a pair of leaves (i,j) having few internal edges on PATH(i,j). Neighbors have no internal edges connecting them, which makes them attractive candidates. The following exercise will get us part of the way toward proving that neighbors indeed minimize $D^*_{i,j}$.

Exercise Break: Show that if leaves i and j are neighbors in TREE(D), and leaf k is not a neighbor of i, then $D^*_{i,j} < D^*_{i,k}$.

Neighbor-Joining Theorem: *Given an additive distance matrix D, a minimum element $D^*_{i,j}$ of the neighbor-joining matrix D^* corresponds to neighboring leaves i and j in* TREE(D).

Proof. Assume that $D^*_{i,j}$ is a minimum element of D^* but that i and j are not neighbors in TREE(D). We aim to reach a contradiction by finding a pair of neighbors k and l such that $D^*_{k,l} < D^*_{i,j}$. By the preceding exercise, neither i nor j can have a neighbor if $D^*_{i,j}$ is a minimum element of D^*. Thus, i and j are the only leaves connected to PARENT(i) and PARENT(j), respectively. Because TREE(D) is simple, PARENT(i) and PARENT(j) have degree at least equal to 3, meaning that each of these nodes is connected to at least two other nodes of TREE(D), one of which lies on PATH(i,j). The other node is part of its own subtree; we call these subtrees T_1 and T_2 (Figure 7.40).

Without loss of generality, we will assume that the number of leaves in T_1 does not exceed the number of leaves in T_2. Because i and j are not in T_1 or T_2, T_1 must therefore contain fewer than $n/2$ leaves, and the rest of TREE(D) must contain more than $n/2$ leaves (recall that n refers to the total number of leaves in TREE(D)). Because i has no neighbors, T_1 must have at least two leaves, which implies that T_1 has a *pair* of neighbors, which we denote as (k,l). We will show that $D^*_{k,l} < D^*_{i,j}$.

Consider an internal edge e of TREE(D). We will first show that $D^*_{k,l} \leq D^*_{i,j}$ by showing that the multiplicity of e in $D^*_{k,l}$ does not exceed the multiplicity of e in $D^*_{i,j}$. There are three possibilities.

- If e lies on PATH(i,j), then by the Edge Multiplicity Theorem, MULTIPLICITY$_{i,j}(e) = -2$, and the result follows.

- If e lies on PATH(i, k), then the removal of e breaks TREE(D) into two subtrees, one containing k and l (with number of leaves equal to LEAVES$_{i,j}(e) < n/2$), and the other containing i and j (with number of leaves equal to LEAVES$_{k,l}(e) > n/2$). Thus, by the Edge Multiplicity Theorem, MULTIPLICITY$_{k,l}(e) = -2 \cdot$ LEAVES$_{k,l}(e) <$ MULTIPLICITY$_{i,j}(e) = -2 \cdot$ LEAVES$_{i,j}(e)$.

- If e lies on neither PATH(i, j) nor PATH(i, k), then the subtree not containing i and j when e is removed is the same as the subtree not containing k and l. As a result, we have that LEAVES$_{i,j}(e) =$ LEAVES$_{k,l}(e)$, which in turn implies that MULTIPLICITY$_{i,j}(e) =$ MULTIPLICITY$_{k,l}(e)$ (the Edge Multiplicity Theorem).

To prove that $D^*_{k,l}$ is in fact less than $D^*_{i,j}$, note that PATH(i, k) must contain an internal edge e because i and k are not neighbors. By the middle case above, we know that MULTIPLICITY$_{k,l}(e) <$ MULTIPLICITY$_{i,j}(e)$. Our assumption that no internal edges of TREE(D) have length zero means that d_e must be positive, which yields the result. □

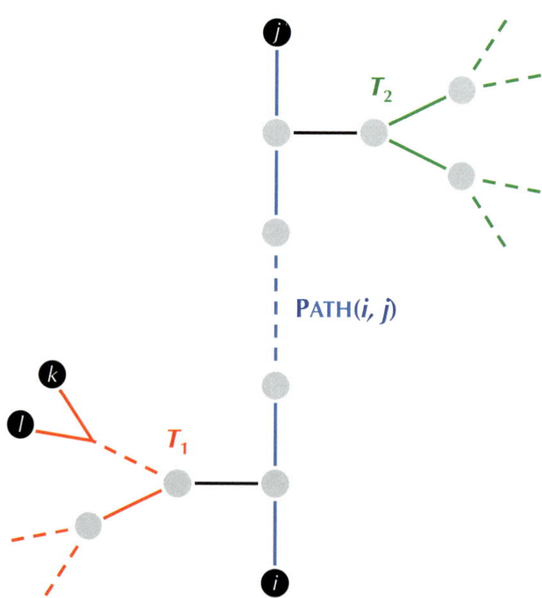

FIGURE 7.40 Leaves i and j are not neighbors. The unique path connecting them in TREE(D), PATH(i, j), is shown in blue. Because TREE(D) is simple, PARENT(i) and PARENT(j) must be connected to at least two other internal nodes, thus forming subtrees T_1 and T_2 (shown in red and green).

Computing limb lengths in the neighbor-joining algorithm

In the main text, we were attempting to assign limb lengths to leaves of a tree constructed from an arbitrary distance matrix. We set the limb length of i equal to $\frac{1}{2}(D_{i,j} + \Delta_{i,j})$ and the limb length of j equal to $\frac{1}{2}(D_{i,j} - \Delta_{i,j})$, where

$$\Delta_{i,j} = \frac{\text{TOTALDISTANCE}_D(i) - \text{TOTALDISTANCE}_D(j)}{n-2}\,.$$

Where do these formulas come from?

Assume for a moment that D is an additive matrix, and select some leaf k that is not equal to i or j. If m is the parent of i and j, then we already have the limb length formula

$$\text{LIMBLENGTH}(i) = \frac{D_{i,j} + D_{i,k} - D_{j,k}}{2}\,.$$

As a result, it might seem like we should use this formula in the neighbor-joining algorithm for an arbitrary distance matrix D. However, if D is non-additive, then the expression $(D_{i,j} + D_{i,k} - D_{j,k})/2$ will vary depending on how we select k. So we need a formula that still computes $\text{LIMBLENGTH}(i)$ when D is additive but that also gives us a single value when D is non-additive. To this end, we can compute the average of the above formula over all choices of $n-2$ leaves k:

$$\frac{1}{n-2} \cdot \sum_{\text{all leaves } k \neq i,j} \frac{D_{i,j} + D_{i,k} - D_{j,k}}{2}$$

If D is additive, then the sum above contains $n-2$ terms, all of which are equal to $\text{LIMBLENGTH}(i)$. Furthermore, if D is non-additive, then this formula provides us with an estimate of the limb length. Note also that the sum above has $n-2$ occurrences of $D_{i,j}$. If we separate these out, then we obtain

$$\begin{aligned}
\text{LIMBLENGTH}(i) &= \frac{D_{i,j}}{2} + \frac{1}{n-2} \cdot \sum_{\text{all leaves } k \neq i,j} \frac{D_{i,k} - D_{j,k}}{2} \\
&= \frac{D_{i,j}}{2} + \frac{1}{n-2} \cdot \left(\sum_{\text{all leaves } k \neq i,j} \frac{D_{i,k}}{2} - \sum_{\text{all leaves } k \neq i,j} \frac{D_{j,k}}{2} \right) \\
&= \frac{1}{2} \cdot \left(D_{i,j} + \frac{1}{n-2} \cdot \left(\sum_{\text{all leaves } k \neq i,j} D_{i,k} - \sum_{\text{all leaves } k \neq j} D_{j,k} \right) \right) \\
&= \frac{1}{2} \cdot \left(D_{i,j} + \frac{\text{TOTALDISTANCE}_D(i) - \text{TOTALDISTANCE}_D(j)}{n-2} \right) \\
&= \frac{1}{2} \cdot (D_{i,j} + \Delta_{i,j})\,,
\end{aligned}$$

which is the formula in the main text that we used for computing LIMBLENGTH(i).

Giant panda: bear or raccoon?

For many years, biologists could not agree on whether the giant panda should be classified as a bear or as a raccoon. Although giant pandas look like bears, they have features that are unusual for bears and typical of raccoons: they do not hibernate in the winter, and their male genitalia are tiny and backward-pointing. As a result, Edwin Colbert wrote in 1938:

> *So the quest has stood for many years with the bear proponents and the raccoon adherents and the middle-of-the-road group advancing their several arguments with the clearest of logic, while in the meantime the giant panda lives serenely in the mountains of Szechuan with never a thought about the zoological controversies he is causing by just being himself.*

Whereas analysis of anatomical and behavioral characters only led to unsettled debates, the analysis of genetic characters by Stephen O'Brien in 1985 demonstrated that giant pandas are indeed more closely related to bears than raccoons (Figure 7.41).

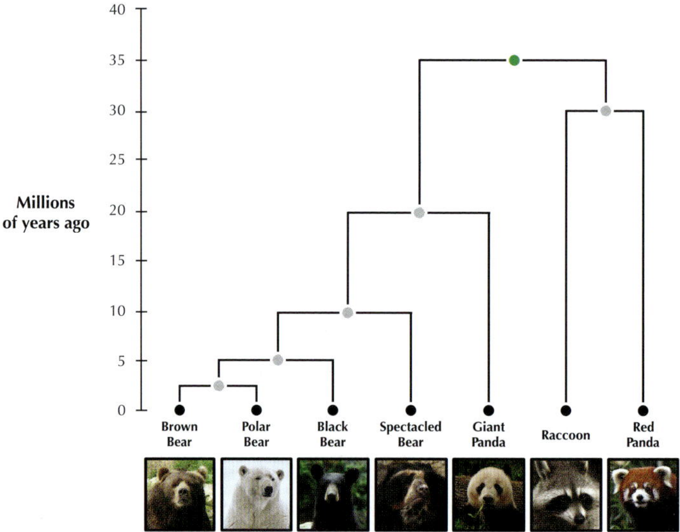

FIGURE 7.41 An evolutionary tree of bears and raccoons.

Where did humans come from?

In 1987, Rebecca Cann, Mark Stoneking and Allan Wilson constructed an evolutionary tree of **mitochondrial DNA (mtDNA)** of 133 people representing African, Asian, Australian, Caucasian, and New Guinean ethnic groups. This tree led to the **Out of Africa** hypothesis, which claims that humans have a common ancestor who lived in Africa. This study turned the question of human origins into an algorithmic problem.

The mtDNA evolutionary tree showed a trunk splitting into two major branches (Figure 7.42). One branch, containing the bottom five individuals in Figure 7.42, consisted only of Africans, whereas the other branch included some modern Africans as well as all people belonging to other ethnic groups.

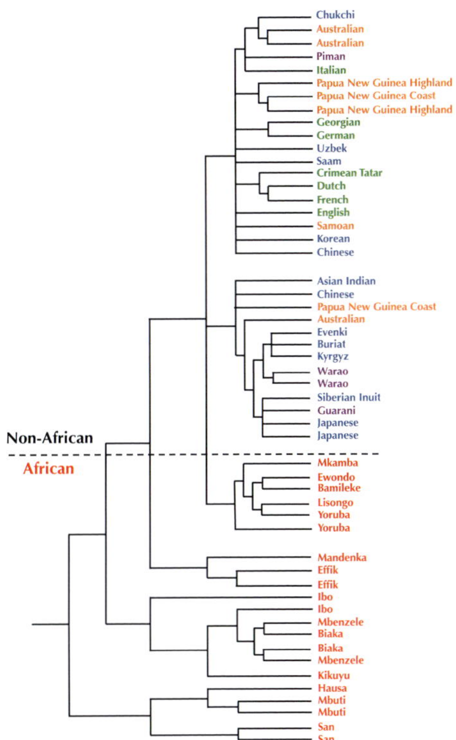

FIGURE 7.42 Evolutionary tree constructed for various human mitochondrial genomes from Africa (red), Asia (blue), North and South America (purple), Europe (green), and Oceania (orange). The dashed line divides African from non-African genomes.

STOP and Think: After looking at the tree in Figure 7.42, where do you think all humans came from?

If humans populated Africa before Asia, then African genomes started to diverge from each other earlier than Asian genomes. Thus, we would expect to find that African genomes, which had more time to diverge from each other, have more mutations (compared to each other and other genomes) than Asian genomes. This reasoning gives us a hint on how to check whether the human race spread from Africa.

African genomes are indeed more diverse than genomes from other continents, which led Wilson and colleagues to conclude that the African lineage is the oldest and that modern humans trace their roots back to Africa. Thus, a population of Africans, the first modern humans, forms one subtree, whereas another subtree represents a subgroup that left Africa and later spread out to the rest of the world. Using the mitochondrial tree, Wilson and colleagues further estimated that humans emerged from Africa 130,000 years ago, with racial differences arising only 50,000 years ago. Figure 7.43 shows putative human migration patterns derived from genomic data.

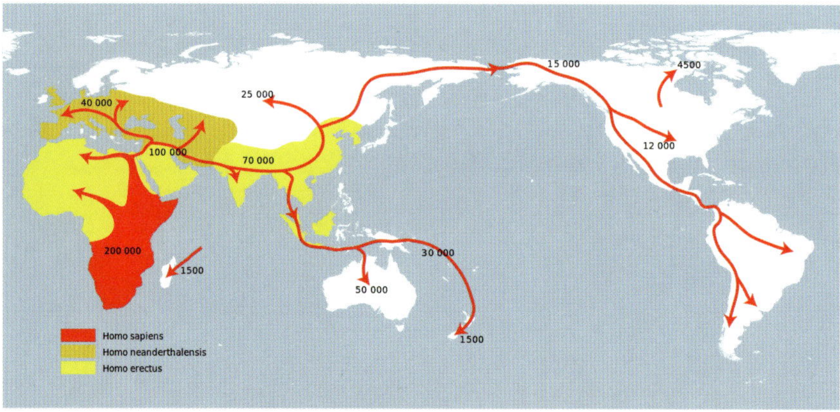

FIGURE 7.43 Putative routes of human migration derived from genomic data labeled by the number of years ago these migrations occurred.

Bibliography Notes

Zuckerkandl and Pauling, 1965 published "Molecules as documents of evolutionary history". The evolutionary tree that resolved the giant panda riddle was constructed by O'Brien et al., 1985. The *Out of Africa* hypothesis was proposed by Cann, Stoneking, and Wilson, 1987. The studies of the origins of HIV and the primate reservoir of the HIV virus were initiated by Gao et al., 1999. The first molecular evidence of HIV transmission in a criminal case was presented by Metzker et al., 2002. Whiting, Bradler, and Maxwell, 2003 published the study of loss and recovery of wings in stick insects.

The UPGMA approach to evolutionary tree reconstruction was developed by Sokal and Michener, 1958. The neighbor joining algorithm was developed by Saitou and Nei, 1987. Studier and Keppler, 1988 proved that this algorithm solves the Distance-Based Phylogeny Problem for additive trees. The dynamic programming algorithm solving the Small Parsimony problem was developed by Sankoff, 1975. The four point condition was formulated by Zaretskii, 1965. The nearest neighbor interchange approach to exploring trees was proposed by Robinson, 1971. Felsenstein, 2004 provides excellent coverage of various algorithms for tree reconstruction.

HOW DID YEAST BECOME A WINEMAKER?
Clustering Algorithms
XXX

An Evolutionary History of Wine Making

How long have we been addicted to alcohol?

One of the first organisms that humans domesticated was yeast. In 2011, while excavating an old graveyard in an Armenian cave, scientists discovered a 6,000 year-old winery, complete with a wine press, fermentation vessels, and even drinking cups. This winery was a major technological innovation that required understanding how to control *Saccharomyces*, the genus of yeast used in alcohol and bread production.

Yet our interest in alcohol may date back much farther than 6,000 years. In 2008, scientists discovered that pen-tailed tree shrews (Figure 8.1), which are similar to the ancient ancestors of all primates, are alcoholics. Their beverage of choice? A "palm wine" produced by the flowers of Bertram palms and naturally fermented by *Saccharomyces* yeast living on the flowers. This finding suggests that our own taste for alcohol may have a genetic basis that predates the Armenian wine cave by millions of years!

FIGURE 8.1 The pen-tailed tree shrew.

Pound for pound, the amount of palm wine that tree shrews consume would be lethal to most mammals. Fortunately, the shrews have developed efficient ways of metabolizing alcohol, and so they avoid inebriation, which would increase their risk of being killed by a predator. Because of tree shrews' alcohol tolerance, scientists believe that alcohol probably offers the shrews some evolutionary advantages, such as protection against heart attack. It is also possible that our more recent primate ancestors were also heavy drinkers — after all, chimpanzees drink naturally brewed fruit nectar — and that we may have inherited an ancestral association of alcohol intake with caloric gain.

The diauxic shift

The species of yeast that we will consider in this chapter is *Saccharomyces cerevisiae*, which can brew wine because it converts the **glucose** found in fruit into **ethanol**. We will therefore begin with a simple question: if *S. cerevisiae* often lives on grapevines, why must crushed grapes be stored in tightly sealed barrels in order to make wine?

Once its supply of glucose runs out, *S. cerevisiae* must do something to survive. It therefore inverts its metabolism, with the ethanol that it just produced becoming its new food supply. This metabolic inversion, called the **diauxic shift**, can only occur in the presence of oxygen. Without oxygen, *S. cerevisiae* hibernates until either glucose or oxygen becomes available. In other words, if winemakers don't seal their barrels, then the yeast in the barrel will metabolize the ethanol that it just produced, ruining the wine.

The diauxic shift is a complex process that affects the expression of many genes. Accordingly, it must derive from a major evolutionary event that equipped the *Saccharomyces* ancestor with a formidable advantage over its competitors: not only can *Saccharomyces* kill its competitors by producing ethanol, which is toxic to most bacteria and other yeasts, but it can then use the accumulated ethanol as an energy source. But how and when did *Saccharomyces* invent the diauxic shift? And which genes does it involve?

Identifying Genes Responsible for the Diauxic Shift

Two evolutionary hypotheses with different fates

Remember Susumu Ohno and his Random Breakage Model? Ohno also hypothesized that there exist rare evolutionary events called **whole genome duplications**, or **WGDs**, which duplicate an entire genome. He proposed this **WGD Model** in 1970, when there was absolutely no evidence to support it. Yet he believed that a WGD would be needed at the time of a critical evolutionary innovation for a species to implement some revolutionary new function, such as the diauxic shift.

For example, imagine the time, millions of years ago, when the first fruit-bearing plants had evolved, but no organisms could metabolize the glucose produced by these fruits and take advantage of the produced ethanol. The first species to do so would have possessed an enormous evolutionary advantage, but metabolizing glucose — let alone ethanol — is not a simple task. Rather than creating a new gene here or there, the diauxic shift would have required creating new metabolic pathways with many genes working together. Ohno argued that a WGD would provide a platform for such a revolutionary innovation, since every duplicated gene would have two copies. One copy would be free to evolve without compromising the gene's existing function, which would be carried out by the remaining copy.

The Random Breakage Model and the WGD Model had very different fates. From the time of its proposal, the Random Breakage Model was embraced by biologists, and it became dogma until its refutation in 2003. In contrast, the WGD Model was initially met with skepticism (because only 13% of *S. cerevisiae* genes are duplicated) and did not gain traction for 25 years.

In 1997, Wolfe and Shields provided the first computational arguments in favor of a WGD in *S. cerevisiae*. They argued that the fact that only 13% of *S. cerevisiae* genes are duplicated is not surprising because even if hundreds of genes contribute to an evolutionary innovation after a WGD, most genes are not needed for this innovation. Thus, unneeded duplicate genes will be bombarded by mutations until they turn into pseudogenes and eventually disappear from the genome after millions of years. See **DETOUR: Whole Genome Duplication or a Series of Single-Gene Duplications?** to learn more about the arguments surrounding a WGD in *S. cerevisiae*.

Which yeast genes drive the diauxic shift?

One of the many steps that yeast performs during fermentation is the conversion of **acetaldehyde** into ethanol. If oxygen becomes subsequently available, then the accumulated ethanol is converted back into acetaldehyde. Both the acetaldehyde-to-ethanol and ethanol-to-acetaldehyde conversions are catalyzed by an enzyme called **alcohol dehydrogenase (Adh)**. In *S. cerevisiae*, Adh activity is encoded by two genes, Adh_1 and Adh_2, which arose from the duplication of a single ancestral gene. The enzyme encoded by Adh_1 has an elevated ability of producing ethanol, whereas the enzyme encoded by Adh_2 has an elevated ability of consuming ethanol.

In 2005, Michael Thomson used a multiple alignment of *Adh* genes from various yeast species to reconstruct an ancient *Saccharomyces* gene. This gene showed a preference to convert acetaldehyde to ethanol, which resembled the behavior of Adh_1. Thomson therefore concluded that before the WGD in *Saccharomyces*, alcohol dehydrogenase was mainly involved in the generation, not consumption, of ethanol. After the WGD, Adh_1 carried out its original function, whereas Adh_2 was free to help power the diauxic shift.

STOP and Think: How would you find the rest of the genes in *S. cerevisae* that work together to accomplish the diauxic shift?

Imagine that you were able to monitor all n yeast genes at m time checkpoints on either side of the diauxic shift, resulting in an $n \times m$ **gene expression matrix** E, where $E_{i,j}$ is a number representing the expression level of gene i at checkpoint j. The i-th row of E is called the **expression vector** of gene i. Just by looking at the yeast genes' expression vectors, you would observe different patterns of gene behavior with respect to the diauxic shift. You would see genes whose expression hardly changes, genes whose expression rapidly increases before the diauxic shift and decreases afterwards, genes whose expression suddenly increases after the diauxic shift, and so on.

Although this chapter focuses on gene expression with respect to the diauxic shift, expression matrices are commonplace in biological analysis. For example, if the expression vector of a newly sequenced gene is similar to the expression vector of a gene with known function, a biologist may suspect that these genes perform related functions. Also, genes with similar expression vectors may imply that the genes are co-regulated, meaning that their expression is controlled by the same transcription factor. This suggests a "guilt by association" strategy for inferring gene functions by starting from a few genes with known functions and potentially propagating the functions of these genes to other genes with similar expression vectors.

Finally, gene expression analysis is important in biomedical studies such as analyzing tissues before and after a drug is administered or contrasting cancerous and non-cancerous cells. For example, expression analysis led to **MammaPrint**, a diagnostic test that determines the likelihood of breast cancer recurrence based on the expression analysis of 70 human genes associated with tumor activation and suppression.

Yet across all these applications, the question remains: what methods do biologists use to analyze gene expression data?

Introduction to Clustering

Gene expression analysis

In 1997, Joseph DeRisi conducted the first massive gene expression experiment by sampling an *S. cerevisiae* culture every two hours for the six hours before and after the diauxic shift. Since there are approximately 6,400 genes in *S. cerevisiae*, and there were seven time points, this experiment resulted in a $6{,}400 \times 7$ gene expression matrix.

STOP and Think: What technology would you use to generate this matrix?

We have already encountered three technologies that could be used to generate this matrix (see **DETOUR: Measuring Gene Expression**), but none of these technologies had matured by 1997! For this reason, DeRisi had to use **microarrays**, which differ from the DNA arrays that we discussed in Chapter 2 (see **DETOUR: Microarrays**). Microarrays are rarely used today, but the algorithmic approaches deRisi used for microarray analysis work equally well for modern gene expression technologies.

PAGE 460

PAGE 461

STOP and Think: Figure 8.2 visualizes the expression vectors of three yeast genes. Which of these genes do you think are involved in the diauxic shift?

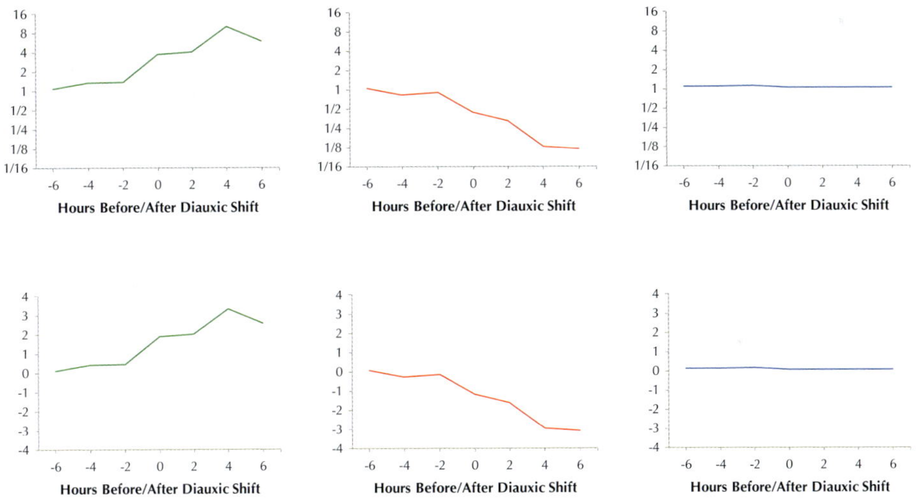

FIGURE 8.2 (Top) Expression vectors (1.07, 1.35, 1.37, 3.70, 4.00, 10.00, 5.88), (1.06, 0.83, 0.90, 0.44, 0.33, 0.13, 0.12), and (1.11, 1.11, 1.12, 1.06, 1.05, 1.06, 1.05) of three yeast genes (YLR258W, YPL012W, and YPR055W, respectively) visualized as plots. Each expression vector $(e_1, \ldots, e_m)$ is represented as a collection of line segments connecting points (j, e_j) to $(j+1, e_{j+1})$ for each j between 1 and $m-1=6$. In the DeRisi experiment, the expression level at the initial checkpoint corresponds to the base level of expression; note that it is close to 1 for the three genes. Values above 1 in expression vectors correspond to increased expression, while values below 1 correspond to decreased expression. (Bottom) The expression vectors of the same three genes with expression levels substituted by their base-2 logarithms: (0.11, 0.43, 0.45, 1.89, 2.00, 3.32, 2.56), (0.09, -0.28, -0.15, -1.18, -1.59, -2.96, -3.08), and (0.15, 0.15, 0.17, 0.09, 0.07, 0.09, 0.07).

Note that the pattern of the expression vector of gene YPR055W (Figure 8.2 (top right)) remains flat during the diauxic shift. We therefore conclude that this gene is probably not involved in the diauxic shift. On the other hand, the expression of gene YLR258W (Figure 8.2 (top left)) significantly changes during the diauxic shift, leading us to hypothesize that this gene is involved in the diauxic shift. Indeed, checking the *Saccharomyces* Genome Database (http://yeastgenome.org) reveals that YLR258W is **glycogen synthase**. This enzyme controls the production of **glycogen**, a glucose polysaccharide that is the main storage vessel for glucose in yeast cells.

STOP and Think: Do you think that gene YPL012W is involved in the diauxic shift?

In practice, biologists often take the logarithm of expression values (Figure 8.2 (bottom)). After this transformation, positive values of a gene's expression vector correspond to increased expression, and negative values correspond to decreased expression. Figure 8.3 shows an expression matrix of ten yeast genes after taking logarithms.

Gene	Expression Vector						
YLR361C	0.14	0.03	-0.06	0.07	-0.01	-0.06	-0.01
YMR290C	0.12	-0.23	-0.24	-1.16	-1.40	-2.67	-3.00
YNR065C	-0.10	-0.14	-0.03	-0.06	-0.07	-0.14	-0.04
YGR043C	-0.43	-0.73	-0.06	-0.11	-0.16	3.47	2.64
YLR258W	0.11	0.43	0.45	1.89	2.00	3.32	2.56
YPL012W	0.09	-0.28	-0.15	-1.18	-1.59	-2.96	-3.08
YNL141W	-0.16	-0.04	-0.07	-1.26	-1.20	-2.82	-3.13
YJL028W	-0.28	-0.23	-0.19	-0.19	-0.32	-0.18	-0.18
YKL026C	-0.19	-0.15	0.03	0.27	0.54	3.64	2.74
YPR055W	0.15	0.15	0.17	0.09	0.07	0.09	0.07

FIGURE 8.3 A 10 × 7 submatrix of DeRisi's 6,400 × 7 gene expression matrix for ten yeast genes (after taking the base-2 logarithm of each expression value). The genes from Figure 8.2 are colored appropriately.

Clustering yeast genes

Our goal is to **partition** the set of all yeast genes into k disjoint **clusters** so that genes in the same cluster have similar expression vectors. In practice, the number of clusters is not known *a priori*, and so biologists typically apply clustering algorithms to gene expression data for various values of k, selecting the value of k that makes sense biologically. For simplicity, we will assume that k is fixed. Figure 8.4 shows a partition of the genes from Figure 8.3 into three clusters indicating increased, decreased, and flat expression during the diauxic shift.

Gene	Expression Vector						
YLR361C	**0.14**	0.03	-0.06	0.07	-0.01	-0.06	-0.01
YMR290C	0.12	-0.23	-0.24	-1.16	-1.40	-2.67	**-3.00**
YNR065C	-0.10	**-0.14**	-0.03	-0.06	-0.07	-0.14	-0.04
YGR043C	-0.43	-0.73	-0.06	-0.11	-0.16	**3.47**	2.64
YLR258W	0.11	0.43	0.45	1.89	2.00	**3.32**	2.56
YPL012W	0.09	-0.28	-0.15	-1.18	-1.59	-2.96	**-3.08**
YNL141W	-0.16	-0.04	-0.07	-1.26	-1.20	-2.82	**-3.13**
YJL028W	-0.28	-0.23	-0.19	-0.19	**-0.32**	-0.18	-0.18
YKL026C	-0.19	-0.15	0.03	0.27	0.54	**3.64**	2.74
YPR055W	0.15	0.15	**0.17**	0.09	0.07	0.09	0.07

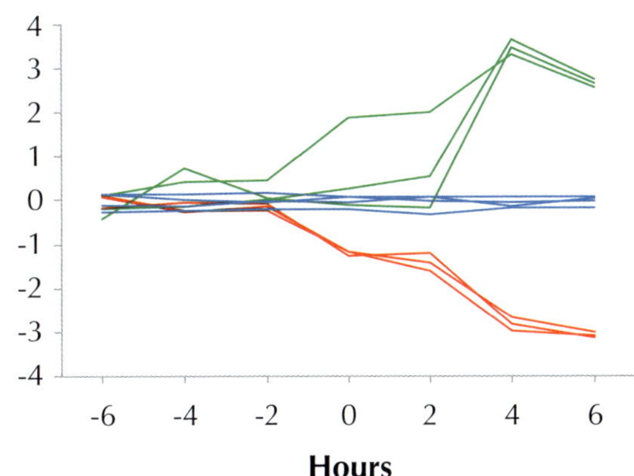

FIGURE 8.4 The rows of the gene expression matrix from Figure 8.3 partitioned into three clusters. Green genes exhibit increased expression, red genes exhibit decreased expression, and blue genes exhibit flat behavior and are unlikely to be associated with the diauxic shift. The element with the largest absolute value in each expression vector is shown in bold. (Bottom) The rows of the matrix visualized as plots.

Although the diauxic shift is an important event in the life of *S. cerevisiae*, it has no bearing on most of the yeast's functions. We therefore suspect that most *S. cerevisiae* genes have flat expression during the diauxic shift, and we would like to exclude these genes from further consideration, thus reducing the size of the expression matrix.

The expression levels of most yeast genes hardly change before and after the diauxic shift (the blue genes in Figure 8.4). These genes also have the property that all of their expression vector values are very close to zero. In our analysis, we will exclude genes with expression vectors whose values are all between $-\delta$ and δ for some parameter δ (we have selected $\delta = 2.3$ in our example). Doing so reduces the original 6,400 yeast gene dataset to a dataset containing 230 genes whose expression changes significantly around the diauxic shift.

You can see in Figure 8.4 that gene YLR258W has a different pattern of change than gene YGR043C, indicating that dividing the 230 yeast genes into just two clusters (i.e., those with increasing and decreasing expression levels) may be too simplistic. Our goal is to cluster these genes based on similar patterns of behavior.

The Good Clustering Principle

To identify groups of genes with similar expression patterns, we will think of an expression vector of length m as a point in m-dimensional space; genes with similar expression vectors will therefore form clusters of nearby points. Ideally, clusters should satisfy the following common-sense principle, which is illustrated in Figure 8.5 for $m = 2$.

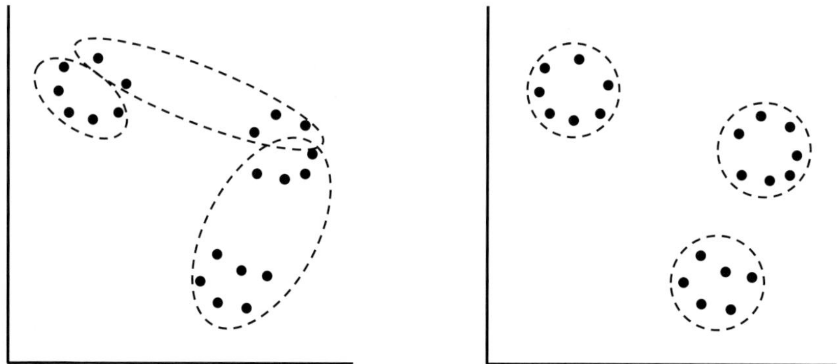

FIGURE 8.5 (Left) A partition of twenty points into three clusters that do not satisfy the Good Clustering Principle. (Right) A different partition of these points that does satisfy the Good Clustering Principle.

Good Clustering Principle: *Every pair of points from the same cluster should be closer to each other than any pair of points from different clusters.*

We have therefore embedded gene expression analysis within the algorithmic problem of partitioning a collection of n points in m-dimensional space into k clusters, which will be our focus in this chapter.

Good Clustering Problem:
Partition a set of points into clusters.

> **Input**: A set of n points in m-dimensional space and an integer k.
> **Output**: A partition of the n points into k clusters satisfying the Good Clustering Principle.

Exercise Break: Form ten points in two-dimensional space by taking the fourth and seventh columns of the matrix in Figure 8.4. How should these points be partitioned into three clusters?

Exercise Break: Compute the number of partitions of n points into two non-empty clusters.

The eye naturally divides the points in Figure 8.6 (left) into two clusters. Unfortunately, these clusters do not satisfy the Good Clustering Principle; in fact, no such partition of these points into two clusters exists! As a result, we will need to take a different approach in order to devise a well-defined computational problem for clustering.

Exercise Break: Design a polynomial algorithm to check whether there is a solution of the Good Clustering Problem.

STOP and Think: Figure 8.6 (right) shows eight data points in two-dimensional space. How would you partition these points into three clusters? How can we transform the Good Clustering Problem into a well-defined computational problem?

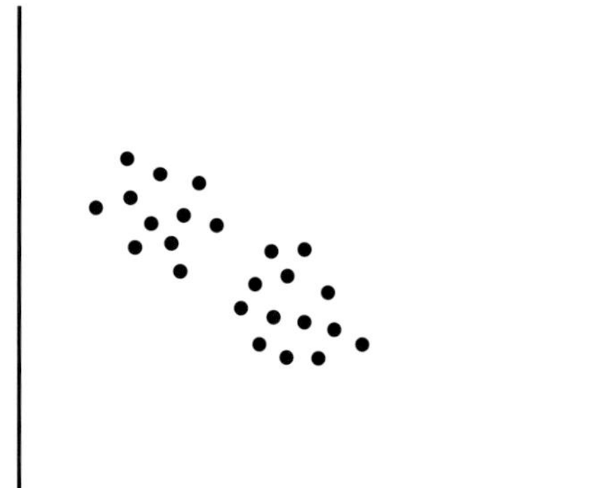

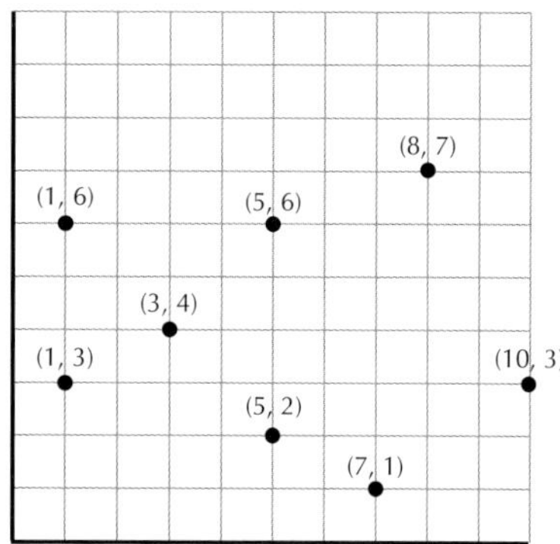

FIGURE 8.6 (Left) A collection of points that obviously form two clusters but for which no partition into two clusters satisfies the Good Clustering Principle. (Right) Eight points in two-dimensional space.

Clustering as an Optimization Problem

Rather than thinking about clustering as dividing data points *Data* into k clusters, we will instead try to select a set *Centers* of k points that will serve as the **centers** of these clusters. We would like to choose *Centers* so that they minimize some distance function between *Centers* and *Data* over all possible choices of centers. But how should this distance function be defined?

First, we define the **Euclidean distance** between points $v = (v_1, \ldots, v_m)$ and $w = (w_1, \ldots, w_m)$ in m-dimensional space, denoted $d(v, w)$, as the length of the line segment connecting these points,

$$d(v,w) = \sqrt{\sum_{i=1}^{m}(v_i - w_i)^2}\,.$$

Next, given a point *DataPoint* in multi-dimensional space and a set of k points *Centers*, we define the distance from *DataPoint* to *Centers*, denoted $d(DataPoint, Centers)$, as the Euclidean distance from *DataPoint* to its closest center,

$$d(DataPoint, Centers) = \min_{\text{all points } x \text{ from } Centers} d(DataPoint, x)\,.$$

The length of the segments in Figure 8.7 correspond to $d(DataPoint, Centers)$ for each point *DataPoint*.

We now define the distance between all data points *Data* and centers *Centers*. This distance, denoted MAXDISTANCE$(Data, Centers)$, is the maximum of $d(DataPoint, Centers)$ among all data points *DataPoint*,

$$\text{MAXDISTANCE}(Data, Centers) = \max_{\text{all points } DataPoint \text{ from } Data} d(DataPoint, Centers).$$

In Figure 8.7, this distance corresponds to the length of the red segment.

Exercise Break: Compute MAXDISTANCE(*Data*, *Centers*) for *Data* shown in Figure 8.6 (right) and *Centers* (2,4), (6,7), and (7,3).

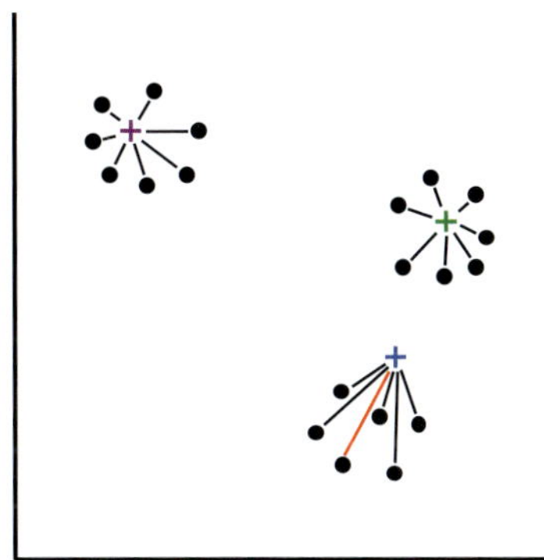

FIGURE 8.7 The collection of points *Data* from Figure 8.5 (shown as black points) along with three centers forming a set *Centers* (shown as colored points). For each point *DataPoint* in *Data*, $d(DataPoint, Centers)$ is equal to the length of the segment connecting it to its nearest center. MAXDISTANCE(*Data*, *Centers*) is equal to the length of the longest such segment, which is shown in red.

We can now formulate a well-defined clustering problem.

***k*-Center Clustering Problem**:
Given a set of data points, find k centers minimizing the maximum distance between these data points and centers.

Input: A set of points *Data* and an integer *k*.
Output: A set *Centers* of *k* centers that minimize the distance MAXDISTANCE(*DataPoints*, *Centers*) over all possible choices of *k* centers.

Exercise Break: How would you select a center in the case of only a single cluster (i.e., when $k = 1$)?

Farthest First Traversal

Although the k-Center Clustering Problem is easy to state, it is NP-Hard. The **Farthest First Traversal** heuristic, whose pseudocode is shown below, selects centers from the points in *Data* (instead of from all m-dimensional points). It first selects an arbitrary point in *Data* as the first center and iteratively adds a new center as the point in *Data* that is farthest from the centers chosen so far, with ties broken arbitrarily (Figure 8.8).

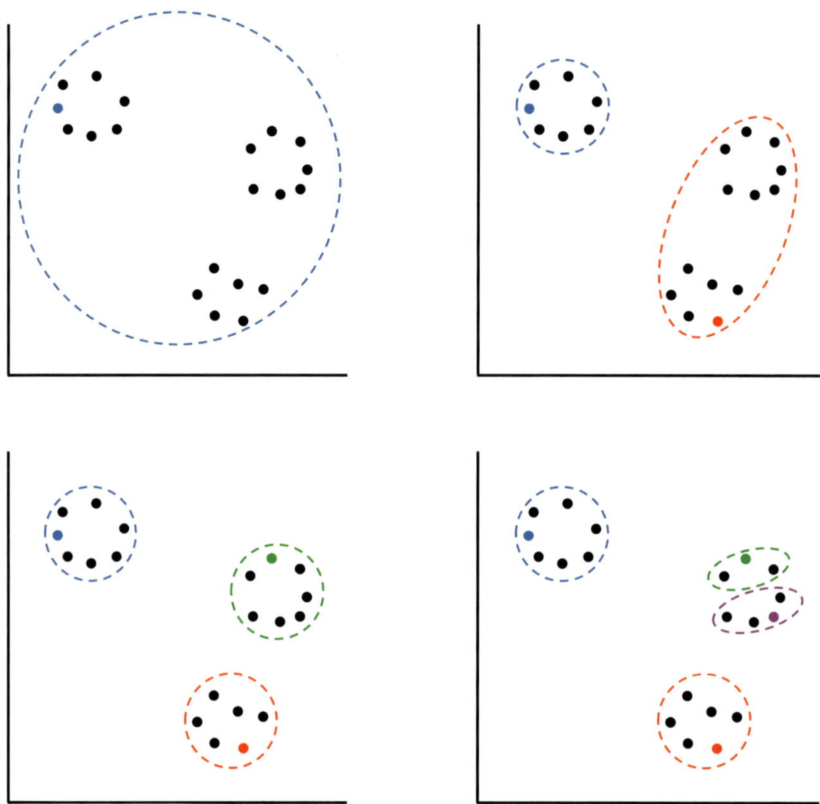

FIGURE 8.8 Applying **FARTHESTFIRSTTRAVERSAL** to the data in Figure 8.5. (Top left) An arbitrary point from the dataset (shown in blue) is selected as the first center. All points belong to a single cluster. (Top right) The red point is selected as the second center, since it is the farthest from the blue point. (Bottom left) After computing each data point's minimum distance to each of the first two centers, we find that the point with the largest such distance is the green point, which becomes the third center. (Bottom right) The fourth center is shown in purple.

```
FARTHESTFIRSTTRAVERSAL(Data, k)
    Centers ← the set consisting of a single randomly chosen point from Data
    while |Centers| < k
        DataPoint ← the point in Data maximizing d(DataPoint, Centers)
        add DataPoint to Centers
    return Centers
```

Exercise Break: Apply **FARTHESTFIRSTTRAVERSAL** to the eight points in Figure 8.6 (right) with $k = 3$. How does the result change if you change the point that is selected first?

Exercise Break: Let *Centers* be the set of centers returned by **FARTHESTFIRSTTRAVERSAL**, and let $Centers_{opt}$ be a set of centers corresponding to an optimal solution of the *k*-Center Clustering Problem. Prove that

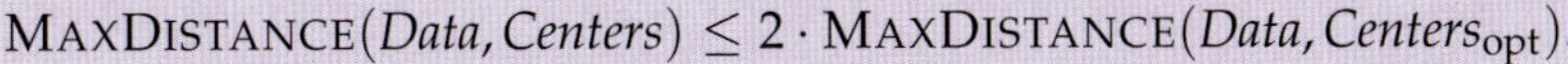

$$\text{MAXDISTANCE}(Data, Centers) \leq 2 \cdot \text{MAXDISTANCE}(Data, Centers_{opt}).$$

Can you find a collection of data points such that the centers returned by **FARTHESTFIRSTTRAVERSAL** are suboptimal?

FARTHESTFIRSTTRAVERSAL is fast, and according to the preceding exercise, its solution approximates the optimal solution of the *k*-Center Clustering Problem; however, this algorithm is rarely used for gene expression analysis. In *k*-Center Clustering, we selected *Centers* so that these points would minimize MAXDISTANCE(*Data*, *Centers*), the maximum distance between any point in *Data* and its nearest center. But biologists are usually interested in analyzing *typical* rather than *maximum* deviations, since the latter may correspond to outliers representing experimental errors (Figure 8.9).

STOP and Think: Can you devise an alternative scoring function that is more biologically appropriate than MAXDISTANCE(*Data*, *Centers*)?

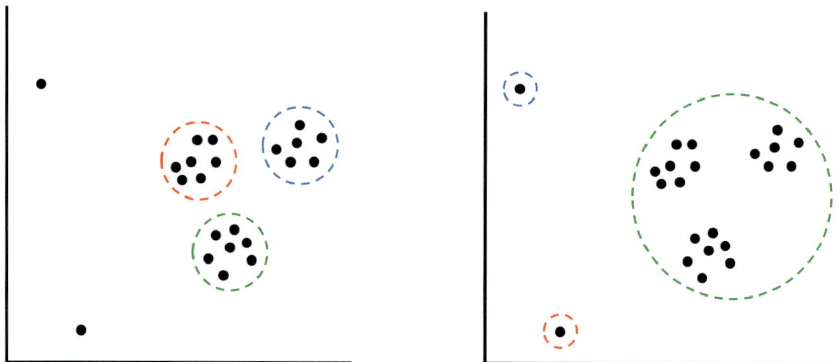

FIGURE 8.9 (Left) A set of data points with three clearly seen clusters and two outliers. (Right) Because **FARTHESTFIRSTTRAVERSAL** relies on MAXDISTANCE to compute new centers, if we attempt to cluster the data points into three clusters, then regardless of which point is selected as the first center, the two outliers on the left will be selected as centers of single-element clusters, and all remaining data points will be assigned to a single cluster.

k-Means Clustering

Squared error distortion

To address limitations of MAXDISTANCE, we will introduce a new scoring function. Given a set *Data* of n data points and a set *Centers* of k centers, the **squared error distortion** of *Data* and *Centers*, denoted DISTORTION(*Data*, *Centers*), is defined as the mean squared distance from each data point to its nearest center,

$$\text{DISTORTION}(Data, Centers) = \frac{1}{n} \sum_{\text{all points } DataPoint \text{ in } Data} d(DataPoint, Centers)^2 \,.$$

Note that whereas MAXDISTANCE(*Data*, *Centers*) only accounts for the length of the single red segment in Figure 8.7, the squared error distortion accounts for the length of all segments in this figure.

Exercise Break: Compute the values of MAXDISTANCE(*Data*, *Centers*) and DISTORTION(*Data*, *Centers*) for the eight data points and the three centers shown in Figure 8.10 (left). How do these values differ for the centers in Figure 8.10 (right)?

Squared Error Distortion Problem:
Compute the squared error distortion of a set of data points with respect to a set of centers.

Input: A set of points *Data* and a set of centers *Centers*.
Output: The squared error distortion DISTORTION(*Data*, *Centers*).

The squared error distortion leads us to the following modification of the k-Centers Clustering Problem.

---

k-Means Clustering Problem:
Given a set of data points, find k center points minimizing the squared error distortion.

Input: A set of points *Data* and an integer k.
Output: A set *Centers* of k centers that minimize DISTORTION(*Data*, *Centers*) over all possible choices of k centers.

Although the k-Centers and k-Means Clustering Problems look similar, they may produce different results (Figure 8.10). The key difference between the k-Centers and k-Means Clustering Problems is that in the latter, the placement of a center is far less affected by outliers (Figure 8.11).

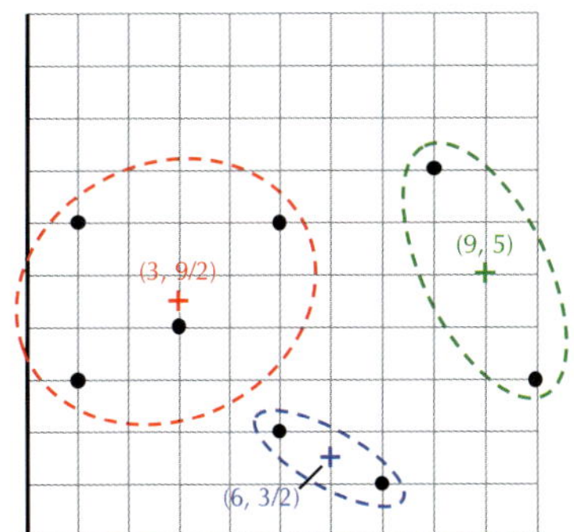

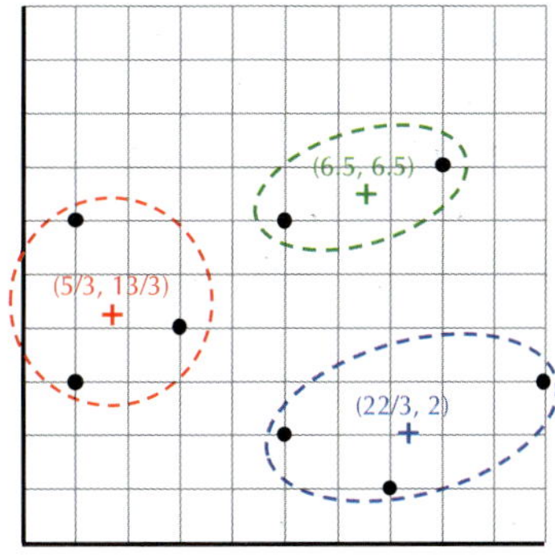

FIGURE 8.10 (Left) The three colored centers solving the k-Centers Clustering Problem for the eight black data points from Figure 8.6 (right), along with the clusters that they form. (Right) The three colored centers solving the k-Means Clustering Problem for these points, along with the clusters that they form.

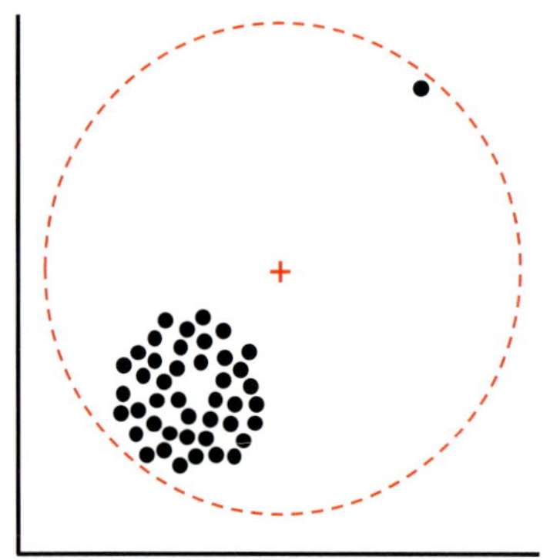

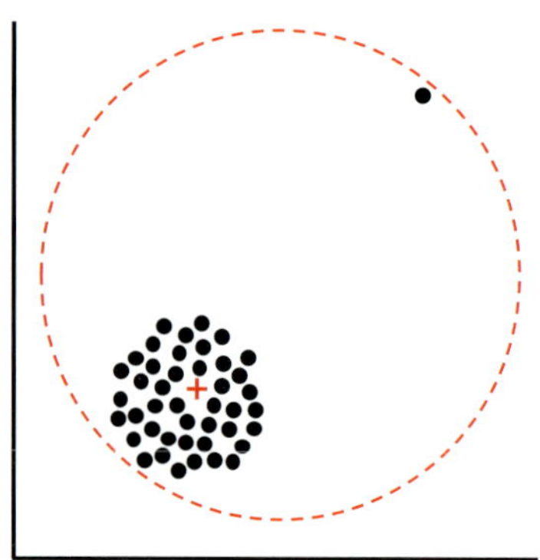

FIGURE 8.11 Center placement varies in different clustering problem formulations. (Left) In the k-Center Clustering Problem, a cluster's center is chosen so that the maximum distance between the center and any point in the cluster is minimized. As a result, the position of the center can be greatly influenced by outliers. (Right) In the k-Means Clustering Problem, the outlier's influence over the placement of the center is much smaller. This behavior is preferable when analyzing biological datasets, in which outliers often correspond to erroneous data.

k-means clustering and the center of gravity

It turns out that the k-Means Clustering Problem is *NP*-Hard when $k > 1$. However, when $k = 1$, the k-Means Clustering Problem amounts to finding a single center point x that minimizes the squared error distortion. Although we acknowledge that partitioning a set of data points into a single cluster is trivial, it remains unclear how to find a single center minimizing the squared error distortion. We would like to solve this simpler problem because it will help us design a heuristic for the case when $k > 1$.

We define the **center of gravity** of *Data* as the point whose i-th coordinate is the average of the i-th coordinates of all points from *Data*. For example, the center of gravity of the points $(3, 8)$, $(8, 0)$, and $(7, 4)$ is

$$\left(\frac{3+8+7}{3}, \frac{8+0+4}{3}\right) = (6, 4).$$

Center of Gravity Theorem: *The center of gravity of a set of points Data is the unique point solving the k-Means Clustering Problem for $k = 1$.*

PAGE 462 For a proof of this theorem, see **DETOUR: Proof of the Center of Gravity Theorem**.

Exercise Break: Although the k-Means Clustering Problem is NP-hard for $k > 1$, it can be solved in polynomial time for any value of k in the case of clustering in one-dimensional space, i.e., when all data points fall on a line. Design an algorithm for solving the k-Means Clustering Problem in this case.

Exercise Break: Prove that the centers in Figure 8.10 (right) solve the k-Means Clustering Problem for $k = 3$.

The Lloyd Algorithm

From centers to clusters and back again

The **Lloyd algorithm** is one of the most popular clustering heuristics for the k-Means Clustering Problem. It first chooses k arbitrary distinct points *Centers* from *Data* as centers and then iteratively performs the following two steps (Figure 8.12):

- **Centers to Clusters:** After centers have been selected, assign each data point to the cluster corresponding to its nearest center; ties are broken arbitrarily.
- **Clusters to Centers:** After data points have been assigned to clusters, assign each cluster's center of gravity to be the cluster's new center.

STOP and Think: Is it possible for the Lloyd algorithm to produce two centers that coincide (thus resulting in fewer than k clusters)?

In Figure 8.12, the centers appear to be moving less and less between iterations. We say that the Lloyd algorithm has **converged** if the centers (and therefore their clusters) stop changing between iterations.

STOP and Think: Can you find a set of data points for which the Lloyd algorithm does not converge?

If the Lloyd algorithm has not converged, the squared error distortion must decrease in any step, according to the following reasoning:

- In a "Centers to Clusters" step, if a data point is assigned to a new center, then this point must be closer to the new center than its previous center. Thus, the squared error distortion must decrease.

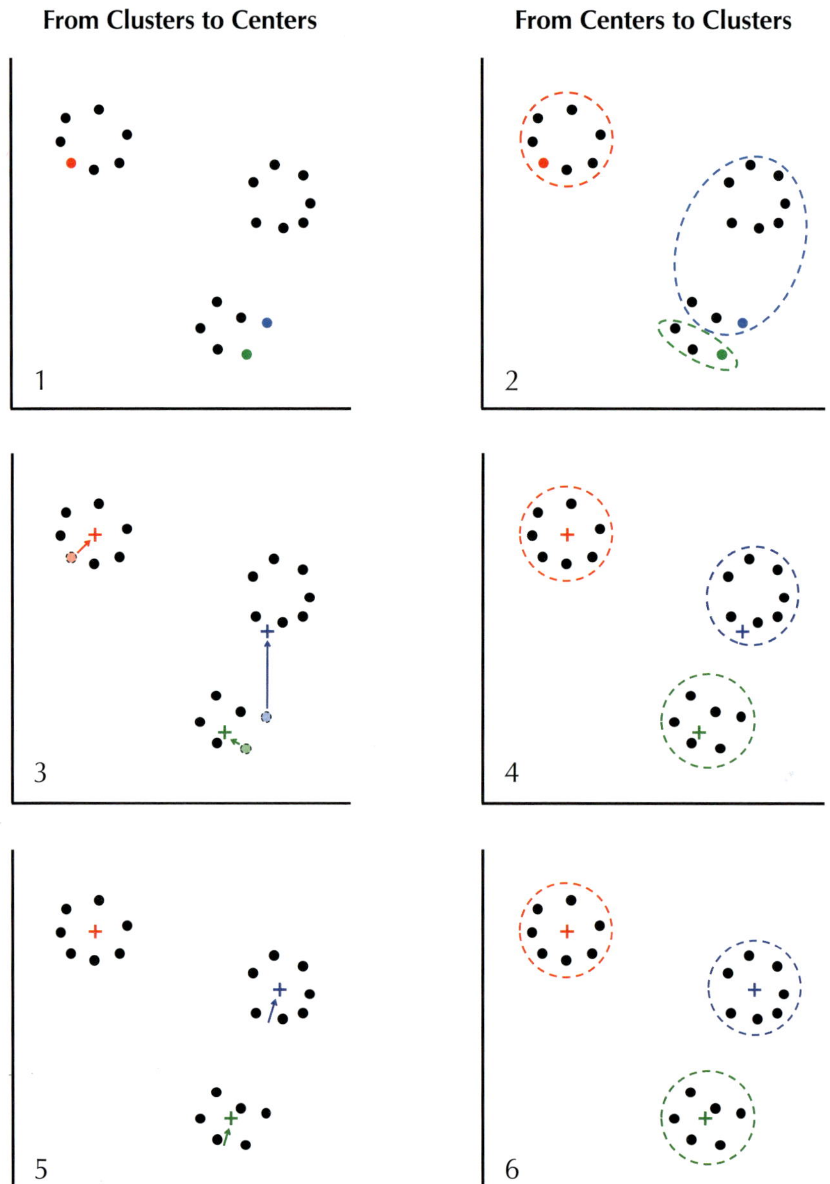

FIGURE 8.12 The Lloyd algorithm in action for $k = 3$. In the top left panel, we select three arbitrary data points as centers, shown as differently colored points. In subsequent panels, we iterate the "Centers to Clusters" step followed by the "Clusters to Centers" step. In the bottom right panel, the Lloyd algorithm has converged.

- In a "Clusters to Center" step, if a center is updated as a cluster's center of gravity, then by the Center of Gravity Theorem, the new center is the *only* point minimizing the squared error distortion for the points in its cluster. Thus, the squared error distortion must decrease.

STOP and Think: Does this reasoning imply that the Lloyd algorithm must converge?

It is not true that the Lloyd algorithm must converge just because the squared error distortion decreases at each step. For example, it could be the case that subsequent decreases in squared error distortion become smaller and smaller, leading to an infinite process (e.g., if the error distortion decreases by $1/2$, then $1/4$, then $1/8$, and so on).The following exercise ensures that such a scenario cannot occur.

Exercise Break: Prove that the number of iterations of the Lloyd algorithm does not exceed the number of partitions of the data points into k clusters.

Just because the Lloyd algorithm converges does not imply that it converges to an optimal solution to the k-Means Clustering Problem. To see why, try the following exercise.

Exercise Break: Run the Lloyd algorithm for the set of four one-dimensional data points $\{0, 1, 1.9, 3\}$ with the two initial centers $\{1, 3\}$. Does the Lloyd algorithm result in a solution to the k-Means Clustering Problem?

Exercise Break: Pick your favorite parameter k and run the Lloyd algorithm 1,000 times on the 230-gene diauxic shift dataset, each time initialized with a new set of k randomly chosen centers. Construct a histogram of the squared error distortions of the resulting 1,000 outcomes. How many times did you have to run the Lloyd algorithm before finding the run that scored highest among your 1,000 runs?

Initializing the Lloyd algorithm

Figure 8.13 illustrates that things can go horribly wrong if we do not pay attention to the Lloyd algorithm's initialization step. In Figure 8.13 (top), we select no centers from clump 1, two centers from clump 3, and one center from each of clumps 2, 4, and 5. As shown in Figure 8.13 (bottom), after the first iteration of the Lloyd algorithm, all points in clumps 1 and 2 will be assigned to the red center, which will move approximately halfway between clumps 1 and 2. The two centers in clump 3 will divide the points in that clump into two clusters. And the centers in clumps 4 and 5 will move toward the middle of these clumps. The Lloyd algorithm will then quickly converge, resulting in an incorrect clustering.

Exercise Break: Compute the probability that at least one of the five clumps in Figure 8.13 will have no centers if five centers are chosen randomly from the data (like in the Lloyd algorithm).

STOP and Think: How would you change the Lloyd algorithm's initialization step to improve the clusters it finds?

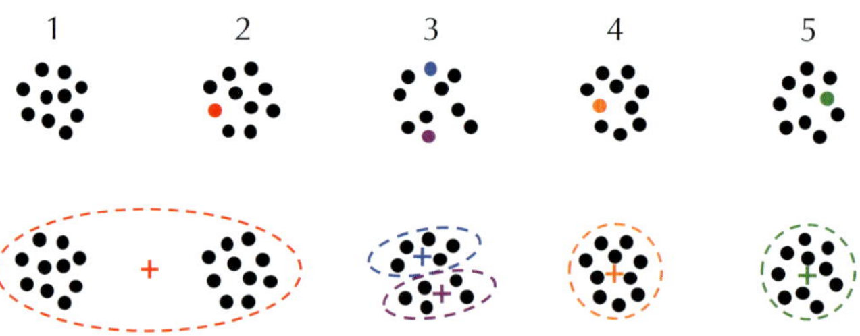

FIGURE 8.13 (Top) Five clumps of ten points in two-dimensional space. The Lloyd algorithm is initialized so that clump 1 contains no centers, clump 3 contains two centers (blue and purple), and each of the other three clumps contains one center (red, orange, and green). (Bottom) The Lloyd algorithm has combined the points in clumps 1 and 2 into a single cluster and split clump 3 into two clusters.

k-means++ Initializer

We have thus far not paid much attention to how initial centers are chosen in the Lloyd algorithm, which selects them randomly. Similarly to **FARTHESTFIRSTTRAVERSAL**, ***k*-MEANS++INITIALIZER** picks k centers one at a time, but instead of choosing the point farthest from those picked so far, it chooses each point at random in such a way that distant points are more likely to be chosen than nearby points. Specifically, the probability of selecting a center *DataPoint* from *Data* is proportional to the squared distance of *DataPoint* from the centers already chosen, i.e., to $d(DataPoint, Centers)^2$.

For a simple example, say that we have just three data points, and that the squared distances from these points to the existing centers *Centers* are equal to 1, 4, and 5. Then the probability of **k-MEANS++INITIALIZER** selecting each of these points as the next center is $1/10$, $4/10$, and $5/10$, respectively.

```
k-MEANS++INITIALIZER(Data, k)
    Centers ← the set consisting of a single randomly chosen point from Data
    while |Centers| < k
        randomly select DataPoint from Data with probability proportional to
            d(DataPoint, Centers)^2
        add DataPoint to Centers
    return Centers
```

STOP and Think: Although **k-MEANS++INITIALIZER** may also fall into the trap in Figure 8.13, it does so rarely. Why?

Exercise Break: Power up your implementation of the Lloyd algorithm using **k-MEANS++INITIALIZER**, and apply it to the 230-gene diauxic shift dataset for varying values of k.

Clustering Genes Implicated in the Diauxic Shift

Since selecting the most biologically relevant value of k can be challenging, we will (somewhat arbitrarily) choose to cluster the 230 yeast genes into six clusters (Figure 8.14).

The plots in Figure 8.14 reveal six patterns of behavior of genes involved in the diauxic shift and raise questions for further biological studies beyond the focus of this chapter. For example, what regulatory mechanisms force the genes in the first cluster to increase their expression? What mechanisms cause the genes in the fourth cluster to decrease their expression? And how do these changes contribute to the diauxic shift?

Exercise Break: The clustering of the entire 6,400-gene yeast dataset into six clusters implies a clustering of the abridged set containing 230 genes. How does this implied clustering compare to the clustering in Figure 8.14?

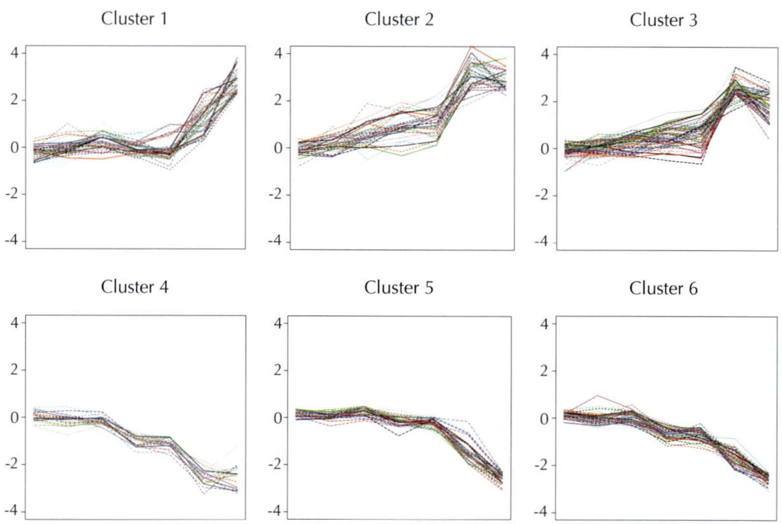

FIGURE 8.14 Applying the Lloyd algorithm (with $k = 6$) to the abridged yeast dataset containing 230 genes results in six clusters revealing six different types of regulatory behavior and containing 37, 36, 58, 19, 36, and 44 genes. Expression vectors for all genes in each of these six clusters are visualized as separate plots.

Limitations of *k*-Means Clustering

After seeing the Lloyd algorithm in action, it may seem that clustering is easy. If you think so, consider the following question.

STOP and Think: How would you cluster the points in Figure 8.15?

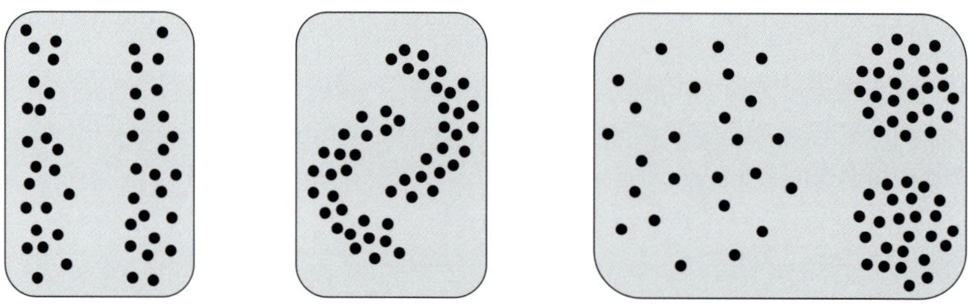

FIGURE 8.15 Difficult clustering problems for $k = 2$ (left and middle) and $k = 3$ (right).

In the case of challenging clustering problems, the Lloyd algorithm sometimes fails to identify what may seem like obvious clusters (Figure 8.16).

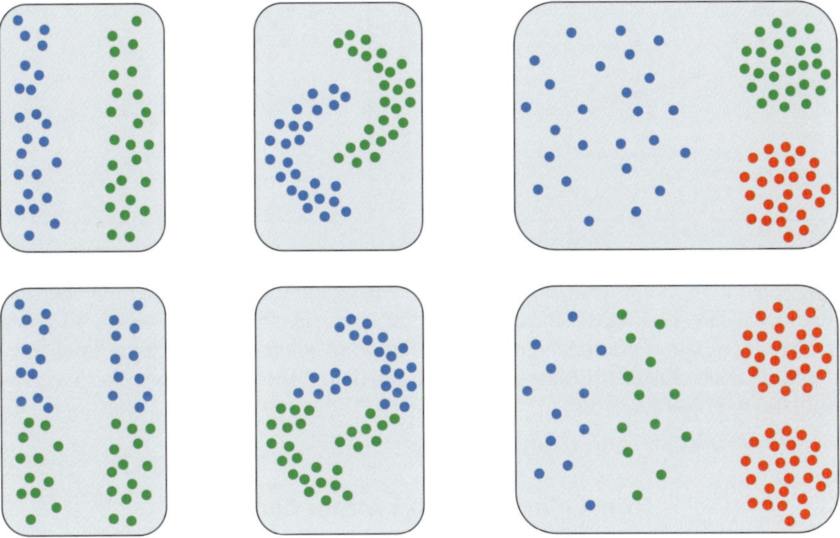

FIGURE 8.16 The human eye (top) and the Lloyd algorithm (bottom) often disagree in the case of elongated clusters (left), clusters with non-globular shapes (middle), and clusters with widely different data point densities (right).

STOP and Think: The Lloyd algorithm assigns each data point to its closest center, with ties broken arbitrarily. What are the negative effects of this assignment?

STOP

One weakness with our formulation of the *k*-Means Clustering Problem is that it forces us to make a "hard" assignment of each point to only one cluster. This strategy makes little sense for **midpoints**, or points that are approximately equidistant from two centers. To deal with midpoints, our goal is to transition away from a rigid assignment of a data point to a single cluster (Figure 8.17 (left)) and toward a "soft" assignment (Figure 8.17 (right)).

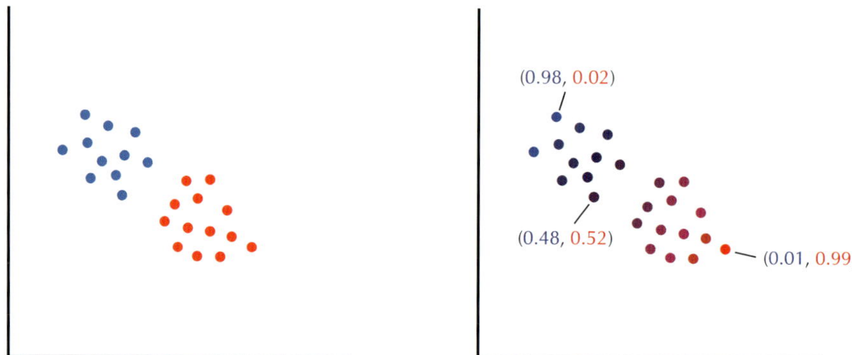

FIGURE 8.17 (Left) The points from Figure 8.6 (left) partitioned into two clusters by the Lloyd algorithm; points are colored red or blue depending on their cluster membership. (Right) We can visualize a soft clustering of the same data into two clusters as assigning each point a pair of numbers representing the point's percentage of "blue" and "red" based on each cluster's "responsibility" for this point. The colors mix to form a color from the red-blue spectrum.

From Coin Flipping to *k*-Means Clustering

Flipping coins with unknown biases

To develop an algorithm for soft clustering, we will introduce a seemingly unrelated analogy. Tom Stoppard's play *Rosencrantz and Guildenstern Are Dead* opens with the title characters flipping a coin over and over, finding that it results in "heads" 157 consecutive times. Rosencrantz and Guildenstern question whether they have become detached from the laws of probability, but if you witnessed such an event, you would probably guess that the coin was biased. Say that your friend flips a biased coin n times, and you would like to estimate the probability θ that a single coin flip results in heads.

STOP and Think: For each value of θ between 0 and 1, you can compute the probability of a given sequence of flips. How would you estimate the value of θ maximizing this probability after watching a series of flips where the coin lands on heads i out of n times?

It seems like the best estimate of θ should be the number of occurrences of heads divided by the total number of coin flips. But how can we prove this? Given a sequence of n coin flips containing i heads, the probability that a coin with bias θ generated this sequence is $f(\theta) = \theta^i \cdot (1-\theta)^{n-i}$. Since the most likely coin bias is the value of θ that maximizes this probability, we will set the derivative of $f(\theta)$ equal to zero,

$$\begin{aligned} f'(\theta) &= i \cdot \theta^{i-1} \cdot (1-\theta)^{n-i} - \theta^i \cdot (n-i) \cdot (1-\theta)^{n-i-1} \\ &= [i \cdot (1-\theta) - \theta \cdot (n-i)] \cdot \theta^{i-1} \cdot (1-\theta)^{n-i-1} \\ &= (i - \theta \cdot n) \cdot \theta^{i-1} \cdot (1-\theta)^{n-i-1} = 0\,. \end{aligned}$$

Other than $\theta = 0$ and $\theta = 1$, the only solution of this equation is $\theta = i/n$, implying that the observed proportion of heads provides the best estimate for θ.

To make the coin flipping problem a bit more interesting, suppose that your friend secretly switches between two coins A and B that look identical but have unknown biases θ_A and θ_B. After observing a sequence of coin flips, your goal is to estimate θ_A and θ_B, which we collectively denote by *Parameters*.

We will simplify the problem by assuming that every n flips, your friend secretly decides to either keep the same coin or switch coins. Five sequences of $n = 10$ flips are shown in Figure 8.18; we will represent the proportion of heads in each of these sequences as a vector,

$$Data = (Data_1, Data_2, Data_3, Data_4, Data_5) = (0.4, 0.9, 0.8, 0.3, 0.7)\,.$$

	Data
H T T T H T T H T H	0.4
H H H H T H H H H H	0.9
H T H H H H H T H H	0.8
H T T T T T H H T T	0.3
T H H H T H H H T H	0.7

FIGURE 8.18 Five sequences of ten coin flips result in *Data* = (0.4, 0.9, 0.8, 0.3, 0.7). "H" denotes heads and "T" denotes tails.

If you knew that your friend used coin A in the first and fourth sequences of flips, then you would estimate θ_A by computing the proportion of heads in these sequences,

$$\theta_A = \frac{Data_1 + Data_4}{2} = \frac{0.4 + 0.3}{2} = 0.35\,.$$

You would then estimate θ_B as the proportion of heads in the remaining three sequences,

$$\theta_B = \frac{Data_2 + Data_3 + Data_5}{3} = \frac{0.9 + 0.8 + 0.7}{3} = 0.8\,.$$

We will represent this choice of coins as a binary vector *HiddenVector* = (1, 0, 0, 1, 0), where a 1 in the k-th position denotes that coin A was used to generate the k-th sequence of flips, and a 0 denotes that coin B was used. This notation allows us to rewrite the equations for *Parameters* in terms of *Data* and *HiddenVector*:

$$\theta_A = \frac{\sum_i HiddenVector_i \cdot Data_i}{\sum_i HiddenVector_i} = \frac{1 \cdot 0.4 + 0 \cdot 0.9 + 0 \cdot 0.8 + 1 \cdot 0.3 + 0 \cdot 0.7}{1+0+0+1+0} = 0.35$$

$$\theta_B = \frac{\sum_i (1 - HiddenVector_i) \cdot Data_i}{\sum_i (1 - HiddenVector_i)} = \frac{0 \cdot 0.4 + 1 \cdot 0.9 + 1 \cdot 0.8 + 0 \cdot 0.3 + 1 \cdot 0.7}{0+1+1+0+1} = 0.80$$

where i runs over all data points.

The expression $\sum_i HiddenVector_i \cdot Data_i$ is the **dot product** of vectors *HiddenVector* and *Data*, written *HiddenVector* · *Data*. Define the **all ones vector**, written $\overrightarrow{1}$, as the vector consisting of all ones and whose length is equal to that of *HiddenVector*. This allows us to write $\sum_i HiddenVector_i$ as the dot product $HiddenVector \cdot \overrightarrow{1}$ and $\sum_i (1 - HiddenVector_i)$ as the dot product $(\overrightarrow{1} - HiddenVector) \cdot \overrightarrow{1}$. As a result, the above equations become

$$\theta_A = \frac{HiddenVector \cdot Data}{HiddenVector \cdot \overrightarrow{1}}$$

$$\theta_B = \frac{(\overrightarrow{1} - HiddenVector) \cdot Data}{(\overrightarrow{1} - HiddenVector) \cdot \overrightarrow{1}}$$

STOP and Think: We just saw that given *Data* and *HiddenVector*, we can find *Parameters* = (θ_A, θ_B). If you are given *Data* and *Parameters*, can you find the most likely choice of *HiddenVector*?

If we know *Parameters*, then deciding on the most likely choice of *HiddenVector* corresponds to determining whether coin A or coin B was more likely to have generated the n observed flips in each of the five coin flipping sequences. For example, suppose we know that $Parameters = (\theta_A, \theta_B) = (0.6, 0.82)$. If coin A was used to generate the fifth sequence of flips, then the probability that it generated the outcome in Figure 8.18 is

$$\theta_A^7 (1 - \theta_A)^3 = 0.6^7 \cdot 0.4^3 \approx 0.00179\,.$$

If coin B was used to generate the fifth sequence, then the probability that it generated this outcome is

$$\theta_B^7 (1 - \theta_B)^3 = 0.82^7 \cdot 0.18^3 \approx 0.00145\,.$$

Since $0.00179 > 0.00145$, we would set $HiddenVector_5$ equal to 1.

Exercise Break: Determine the rest of the entries in *HiddenVector* for *Parameters* = (0.6, 0.82) and the sequences of coin flips in Figure 8.18.

More generally, let $\Pr(Data_i|\theta)$ denote the **conditional probability** of generating the outcome $Data_i$ given a coin with bias θ,

$$\Pr(Data_i|\theta) = \theta^{n \cdot Data_i}(1-\theta)^{n \cdot (1-Data_i)} .$$

If $\Pr(Data_i|\theta_A) > \Pr(Data_i|\theta_B)$, then coin A is more likely to have generated the i-th sequence of flips, and we set $HiddenVector_i$ equal to 1. If $\Pr(Data_i|\theta_A) < \Pr(Data_i|\theta_B)$, then coin B is more likely, and we set $HiddenVector_i$ equal to 0. Ties are broken arbitrarily.

In summary, if *HiddenVector* is known and *Parameters* is unknown, then we can reconstruct the most likely *Parameters* = (θ_A, θ_B):

$$(Data, HiddenVector, ?) \rightarrow Parameters$$

Likewise, if *Parameters* is known and *HiddenVector* is unknown, then we can reconstruct the most likely *HiddenVector*:

$$(Data, ?, Parameters) \rightarrow HiddenVector$$

Our original problem, however, was that both *HiddenVector* and *Parameters* are unknown:

$$(Data, ?, ?) \rightarrow ???$$

Where is the computational problem?

You may have noticed that we have not formulated the computational problem that we are trying to solve. So define the conditional probability of generating a sequence of coin flips $Data_i$ given *HiddenVector* and *Parameters* as

$$\Pr(Data_i|HiddenVector, Parameters) = \begin{cases} \Pr(Data_i|\theta_A) & \text{if } HiddenVector_i = 1 \\ \Pr(Data_i|\theta_B) & \text{if } HiddenVector_i = 0 \end{cases}$$

Furthermore, define the conditional probability of generating *Data* given *HiddenVector* and *Parameters* as

$$\Pr(Data|HiddenVector, Parameters) = \prod_{i=1}^{n} \Pr(Data_i|HiddenVector, Parameters) .$$

Given *Data*, the computational problem we are trying to solve is to find *HiddenVector* and *Parameters* maximizing $\Pr(Data|HiddenVector, Parameters)$.

From coin flipping to the Lloyd algorithm

Identifying *HiddenVector* and *Parameters* from *Data* may appear hopeless, but we have already learned that starting from a random guess is not necessarily a bad idea. We will therefore start from an arbitrary choice of $Parameters = (\theta_A, \theta_B)$ and immediately reconstruct the most likely *HiddenVector*:

$$(Data, ?, Parameters) \rightarrow HiddenVector$$

As soon as we know *HiddenVector*, we will question the wisdom of our initial choice of *Parameters* and re-estimate *Parameters'*:

$$(Data, HiddenVector, ?) \rightarrow Parameters'$$

As illustrated in Figure 8.19 for the initial choice of *Parameters* = (0.6, 0.82), we repeat these two steps and hope that *Parameters* and *HiddenVector* are moving closer to the values that maximize $\Pr(Data|HiddenVector, Parameters)$,

$$\begin{aligned}(Data, ?, Parameters) &\rightarrow (Data, HiddenVector, Parameters)\\ &\rightarrow (Data, HiddenVector, \quad ? \quad)\\ &\rightarrow (Data, HiddenVector, Parameters')\\ &\rightarrow (Data, \quad ? \quad , Parameters')\\ &\rightarrow (Data, HiddenVector', Parameters')\\ &\rightarrow \cdots\end{aligned}$$

Exercise Break: Prove that this process terminates, i.e., that *HiddenVector* and *Parameters* eventually stop changing between iterations.

STOP and Think: If *HiddenVector* consists of all zeroes, then there are no flips with coin A, and the formula for computing θ_A is invalid. What would you do to address this complication?

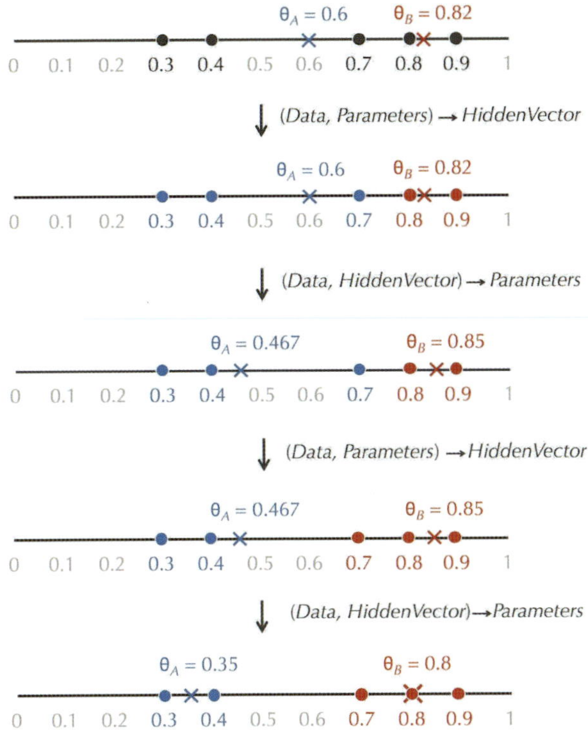

FIGURE 8.19 Starting with *Parameters* = (0.6, 0.82) results in *HiddenVector* = (1, 0, 0, 1, 1) for *Data* = (0.4, 0.9, 0.8, 0.3, 0.7). We then update *Parameters* as (0.467, 0.85), which in turn results in *HiddenVector* = (1, 0, 0, 1, 0). This new vector leads to the assignment of *Parameters* as (0.35, 0.8), at which point the process has terminated because *HiddenVector* will not change in the next step.

Return to clustering

Figure 8.19 has revealed that although our coin flipping analogy seems like a different problem, it is really just a one-dimensional clustering problem in disguise!

STOP and Think: What are *Data*, *HiddenVector*, and *Parameters* in the Lloyd algorithm?

Given n data points in m-dimensional space $Data = (Data_1, \ldots, Data_n)$, we represent their assignment to k clusters as an n-dimensional vector

$$HiddenVector = (HiddenVector_1, \ldots, HiddenVector_n),$$

where each $HiddenVector_i$ can take integer values from 1 to k. We will then represent the k centers as k points in m-dimensional space, $Parameters = (\theta_1, \ldots, \theta_k)$.

In k-means clustering, similarly to the coin flipping analogy, we are given *Data*, but *HiddenVector* and *Parameters* are unknown. The Lloyd algorithm starts from randomly chosen *Parameters*, and we can now rewrite its two main steps as follows:

- Centers to Clusters: $(Data, ?, Parameters) \rightarrow HiddenVector$
- Clusters to Centers: $(Data, HiddenVector, ?) \rightarrow Parameters$

The only difference between the coin flipping algorithm and the Lloyd algorithm for k-means clustering is how they execute the "Centers to Clusters" step. In the former, we compute $HiddenVector_i$ by comparing $\Pr(Data_i|\theta_A)$ with $\Pr(Data_i|\theta_B)$, whereas in the latter, we assign a point to the cluster containing the center nearest to that point.

STOP and Think: Consider the following questions regarding coin flipping and clustering.

- Is it fair to always select coin A if $\Pr(Data_i|\theta_A)$ is only slightly larger than $\Pr(Data_i|\theta_B)$?
- Is it fair to always assign a point to a center if this center is only slightly closer to the point than another center?

Making Soft Decisions in Coin Flipping

Expectation maximization: the E-step

We will now use our coin flipping analogy to motivate a soft version of k-means clustering. Given $Parameters = (\theta_A, \theta_B)$, we can make hard decisions for *HiddenVector* by comparing $\Pr(Data_i|\theta_A)$ with $\Pr(Data_i|\theta_B)$. But this does not mean that we are certain which coin was used. If $\Pr(Data_i|\theta_B)$ were approximately equal to $\Pr(Data_i|\theta_A)$, then our confidence that coin B was used would be approximately 50%. On the other hand, if $\Pr(Data_i|\theta_B)$ were much larger than $\Pr(Data_i|\theta_A)$, then we would be almost positive that coin B was used. More generally, we can speak of our confidence that a coin was used as the "responsibility" of this coin for a given sequence of flips. (The responsibilities should sum to 1.)

In terms of k-means clustering, if a data point is a midpoint between two centers, then each of these centers should have about the same responsibility for attracting it to their clusters. As with coin flipping, the responsibilities of all centers for a given data point should sum to 1.

STOP and Think: Given *Parameters* = (θ_A, θ_B) = (0.6, 0.82) and the sequence of coin flips "THHHTHHHTH", how would you compute responsibilities for coins A and B?

STOP

To answer the preceding question, we have seen that $\Pr(0.7|\theta_A) = 0.6^7 \cdot 0.4^3 \approx 0.00179$ and that $\Pr(0.7|\theta_B) = 0.82^7 \cdot 0.18^3 \approx 0.00145$. Before, we rigidly concluded that coin A was more likely. Now, since coin A is more likely to generate seven heads in a sequence of ten coin flips, we should assign a larger responsibility to coin A than to coin B. One possible way to assign these responsibilities is given by the formulas

$$\frac{\Pr(0.7|\theta_A)}{\Pr(0.7|\theta_A) + \Pr(0.7|\theta_B)} = \frac{0.00179}{0.00179 + 0.00145} \approx 0.55$$

$$\frac{\Pr(0.7|\theta_B)}{\Pr(0.7|\theta_A) + \Pr(0.7|\theta_B)} = \frac{0.00145}{0.00179 + 0.00145} \approx 0.45$$

As a result, instead of a vector *HiddenVector*, we now have a 2×5 **responsibility profile** *HiddenMatrix* that can be constructed from *Data* and *Parameters*,

$$(Data, ?, Parameters) \rightarrow HiddenMatrix\,.$$

We call this transition the **E-step** (Figure 8.20).

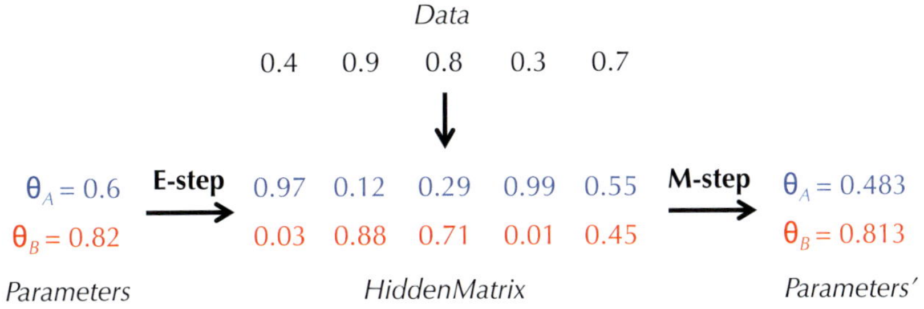

				Data				
		0.4	0.9	0.8	0.3	0.7		
				↓				
$\theta_A = 0.6$	**E-step** →	0.97	0.12	0.29	0.99	0.55	**M-step** →	$\theta_A = 0.483$
$\theta_B = 0.82$		0.03	0.88	0.71	0.01	0.45		$\theta_B = 0.813$
Parameters				*HiddenMatrix*				*Parameters′*

FIGURE 8.20 In the E-step, we compute *HiddenMatrix* from *Data* = (0.4, 0.9, 0.8, 0.3, 0.7) and *Parameters* = (0.6, 0.82). In the M-step, we compute an updated *Parameters′* from *HiddenMatrix* and *Data*.

For an arbitrary set of data points $Data = (Data_1, \ldots, Data_n)$ and a set of k coins with biases $Parameters = (\theta_1, \ldots, \theta_k)$, the $k \times n$ matrix *HiddenMatrix* is defined as

$$HiddenMatrix_{i,j} = \frac{\Pr(Data_j|\theta_i)}{\sum_{1 \le t \le k} Pr(Data_j|\theta_t)} \ .$$

Expectation maximization: the M-step

When making hard assignments, we computed *Parameters* from *Data* and *HiddenVector* as follows:

$$\theta_A = \frac{HiddenVector \cdot Data}{HiddenVector \cdot \overrightarrow{1}}$$

$$\theta_B = \frac{(\overrightarrow{1} - HiddenVector) \cdot Data}{(\overrightarrow{1} - HiddenVector) \cdot \overrightarrow{1}} \ .$$

To make soft assignments, note that the hard assignment of outcomes to the two coins can be represented by the binary responsibility matrix below. An occurrence of 1 in the i-th position of the first row means that we conclude that coin A generated the i-th sequence of flips, and an occurrence of 1 in the second row means that we conclude that coin B generated the i-th sequence of flips:

$$HiddenMatrix \quad \begin{matrix} 1 & 0 & 0 & 1 & 0 \\ 0 & 1 & 1 & 0 & 1 \end{matrix}$$

Thus, the first row of *HiddenMatrix*, denoted $HiddenMatrix_A$, is just *HiddenVector*, and the second row of *HiddenMatrix*, denoted $HiddenMatrix_B$, is just $\overrightarrow{1} - HiddenVector$. We can therefore rewrite the previous formulas for θ_A and θ_B in terms of *HiddenMatrix*:

$$\theta_A = \frac{HiddenMatrix_A \cdot Data}{HiddenMatrix_A \cdot \overrightarrow{1}}$$

$$\theta_B = \frac{HiddenMatrix_B \cdot Data}{HiddenMatrix_B \cdot \overrightarrow{1}}$$

For the responsibility matrix in Figure 8.20, we can now recompute *Parameters* as follows:

$$\theta_A = \frac{0.97 \cdot 0.4 + 0.12 \cdot 0.9 + 0.29 \cdot 0.8 + 0.99 \cdot 0.3 + 0.55 \cdot 0.7}{0.97 + 0.12 + 0.29 + 0.99 + 0.55} = \frac{1.41}{2.92} \approx 0.483$$

$$\theta_B = \frac{0.03 \cdot 0.4 + 0.88 \cdot 0.9 + 0.71 \cdot 0.8 + 0.01 \cdot 0.3 + 0.45 \cdot 0.7}{0.03 + 0.88 + 0.71 + 0.01 + 0.45} = \frac{1.69}{2.08} \approx 0.813$$

STOP and Think: Notice that the soft parameter choices $\theta_A = 0.483$ and $\theta_B = 0.813$ are a little closer to each other than the hard parameter choices $\theta_A = 0.467$ and $\theta_B = 0.85$. Why do you think that this is the case?

In general, the transition

$$(Data, HiddenMatrix, ?) \rightarrow Parameters$$

is called the **M-step**.

The expectation maximization algorithm

The **expectation maximization algorithm** starts with a random choice of *Parameters*. It then alternates between the E-step, in which we compute a responsibility matrix *HiddenMatrix* for *Data* given *Parameters*:

$$(Data, ?, Parameters) \rightarrow HiddenMatrix$$

and the M-step, in which we re-estimate *Parameters* using *HiddenMatrix*:

$$(Data, HiddenMatrix, ?) \rightarrow Parameters$$

Exercise Break: Carry out a few more steps of the expectation maximization algorithm for the data in Figure 8.20. When should we stop the algorithm?

Soft *k*-Means Clustering

Applying expectation maximization to clustering

We are now ready to use the expectation maximization algorithm to modify the Lloyd algorithm into a **soft *k*-means clustering algorithm**. This algorithm starts from randomly chosen centers and iterates the following two steps:

- **Centers to Soft Clusters (E-step):** After centers have been selected, assign each data point a "responsibility" for each cluster, where higher responsibilities correspond to stronger cluster membership.

- **Soft Clusters to Centers (M-step):** After data points have been assigned to soft clusters, compute new centers.

Centers to soft clusters

We begin with the "Centers to Soft Clusters" step for k centers $Centers = (x_1, \ldots, x_k)$ and n points $Data = (Data_1, \ldots, Data_n)$. When we introduced the E-step, we applied the formula

$$HiddenMatrix_{i,j} = \frac{\Pr(Data_j|x_i)}{\sum_{1 \le t \le k} Pr(Data_j|x_t)} .$$

However, this formula only works for the coin flipping problem when we know the probabilities $\Pr(Data_j|x_i)$. Since it is not clear how to define this probability in the case of an arbitrary set of points *Data*, we will now describe a different formula for computing *HiddenMatrix*.

We have already used the term "center of gravity" when computing centers; if we think about the centers as stars and the data points as planets, then the closer a point is to a center, the stronger that center's "pull" should be on the point. Given k centers $Centers = (x_1, \ldots x_k)$ and n points $Data = (Data_1, \ldots, Data_n)$, we therefore need to construct a $k \times n$ responsibility matrix *HiddenMatrix* for which $HiddenMatrix_{i,j}$ is the pull of center i on data point j. This pull can be computed according to the Newtonian inverse-square law of gravitation,

$$HiddenMatrix_{i,j} = \frac{1/d(Data_j, x_i)^2}{\sum_{\text{all centers } x_t} 1/d(Data_j, x_t)^2}.$$

Unfortunately for Newton fans, the following **partition function** from statistical physics often works better in practice:

$$HiddenMatrix_{i,j} = \frac{e^{-\beta \cdot d(Data_j, x_i)}}{\sum_{\text{all centers } x_t} e^{-\beta \cdot d(Data_j, x_t)}}$$

In this formula, e is the base of the natural logarithm ($e \approx 2.718$), and β is a parameter reflecting the amount of flexibility in our soft assignment and called the **stiffness parameter**. Figure 8.21 illustrates different approaches for computing *HiddenMatrix* when *Data* represents points in one-dimensional space and should give you an idea of why β is called the "stiffness" parameter.

STOP and Think: How does the assignment of the points in Figure 8.21 to soft clusters change as $\beta \to -\infty$?

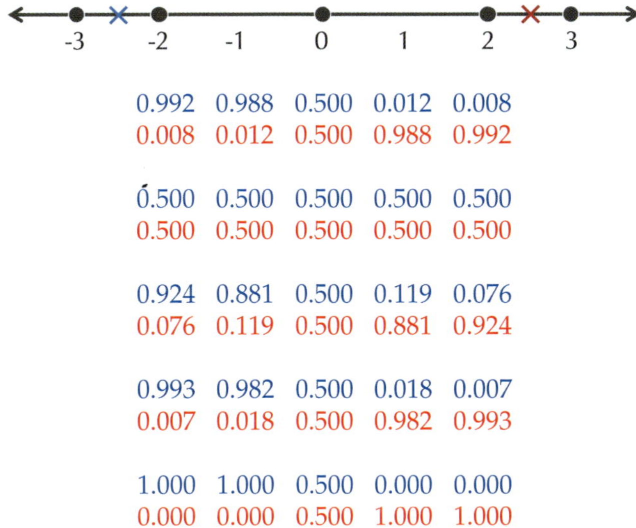

FIGURE 8.21 (Top) Five one-dimensional points *Data* = (-3, -2, 0, +2, +3) with two centers (shown in blue and red) *Centers* = {-2.5, +2.5}. (Bottom) Five versions of *HiddenMatrix* constructed for *Data* and *Centers*, using the Newtonian inverse-square law (first matrix) and the partition function with stiffness $\beta = 0$ (second matrix), $\beta = 0.5$ (third matrix), $\beta = 1$ (fourth matrix), and $\beta = 1000$ (fourth matrix).

Exercise Break: Compute *HiddenMatrix* using the Newtonian inverse-square law for the three centers and eight data points shown in Figure 8.10 (left).

Soft clusters to centers

When we implemented the M-step for coin flipping, we obtained the following formulas for θ_A and θ_B:

$$\theta_A = \frac{HiddenMatrix_A \cdot Data}{HiddenMatrix_A \cdot \vec{1}}$$

$$\theta_B = \frac{HiddenMatrix_B \cdot Data}{HiddenMatrix_B \cdot \vec{1}}$$

In soft k-means clustering, if we let $HiddenMatrix_i$ denote the i-th row of *HiddenMatrix*, then we can update center x_i using an analogue of the above formulas. Specifically, we will define the j-th coordinate of center x_i, denoted $x_{i,j}$, as

$$x_{i,j} = \frac{HiddenMatrix_i \cdot Data^j}{HiddenMatrix_i \cdot \overrightarrow{1}}$$

Here, $Data^j$ is the n-dimensional vector holding the j-th coordinates of the n points in *Data*. The updated center x_i is called a **weighted center of gravity** of the points *Data*.

Computing weighted centers of gravity for the *HiddenMatrix* at the bottom of Figure 8.21 produces the following updated centers:

$$x_1 = \frac{0.993 \cdot (-3) + 0.982 \cdot (-2) + 0.500 \cdot (0) + 0.018 \cdot (2) + 0.007 \cdot (3)}{0.993 + 0.982 + 0.500 + 0.018 + 0.007} = -1.955$$

$$x_2 = \frac{0.007 \cdot (-3) + 0.018 \cdot (-2) + 0.500 \cdot (0) + 0.982 \cdot (2) + 0.993 \cdot (3)}{0.007 + 0.018 + 0.500 + 0.982 + 0.993} = 1.955$$

You are now ready to implement the expectation maximization algorithm for soft k-means clustering.

STOP and Think: Does the soft k-means clustering algorithm terminate? If not, how would you modify it to ensure that it does not run forever?

Exercise Break: Recompute centers using the data points from the exercise on page 449 along with the responsibility matrix obtained as the result of this algorithm.

Exercise Break: Apply soft k-means clustering to the abridged yeast diauxic shift gene expression data, and compare the results with those of the Lloyd algorithm.

Hierarchical Clustering

Introduction to distance-based clustering

In Chapter 7, we discussed two approaches to evolutionary tree reconstruction with different strengths and weaknesses: distance-based algorithms (including the neighbor-joining algorithm), and alignment-based algorithms (including the algorithm for the Small Parsimony Problem). Similarly, biologists do not always analyze the $n \times m$ gene expression matrix directly. Instead, they sometimes first transform this matrix into an $n \times n$ **distance matrix** D, where $D_{i,j}$ indicates the *distance* between the expression vectors for genes i and j (Figure 8.22 (top right)). In this section, we will see how to

use a distance matrix to partition genes into clusters (see **DETOUR: Transforming an Expression Matrix into a Distance/Similarity Matrix** for more details). 
PAGE 463

	1 hr	2 hr	3 hr
g_1	10.0	8.0	10.0
g_2	10.0	0.0	9.0
g_3	4.0	8.5	3.0
g_4	9.5	0.5	8.5
g_5	4.5	8.5	2.5
g_6	10.5	9.0	12.0
g_7	5.0	8.5	11.0
g_8	3.7	8.7	2.0
g_9	9.7	2.0	9.0
g_{10}	10.2	1.0	9.2

	g_1	g_2	g_3	g_4	g_5	g_6	g_7	g_8	g_9	g_{10}
g_1	0.0	8.1	9.2	7.7	9.3	2.3	5.1	10.2	6.1	7.0
g_2	8.1	0.0	12.0	0.9	12.0	9.5	10.1	12.8	2.0	1.0
g_3	9.2	12.0	0.0	11.2	0.7	11.1	8.1	1.1	10.5	11.5
g_4	7.7	0.9	11.2	0.0	11.2	9.2	9.5	12.0	1.6	1.1
g_5	9.3	12.0	0.7	11.2	0.0	11.2	8.5	1.0	10.6	11.6
g_6	2.3	9.5	11.1	9.2	11.2	0.0	5.6	12.1	7.7	8.5
g_7	5.1	10.1	8.1	9.5	8.5	5.6	0.0	9.1	8.3	9.3
g_8	10.2	12.8	1.1	12.0	1.0	12.1	9.1	0.0	11.4	12.4
g_9	6.1	2.0	10.5	1.6	10.6	7.7	8.3	11.4	0.0	1.1
g_{10}	7.0	1.0	11.5	1.1	11.6	8.5	9.3	12.4	1.1	0.0

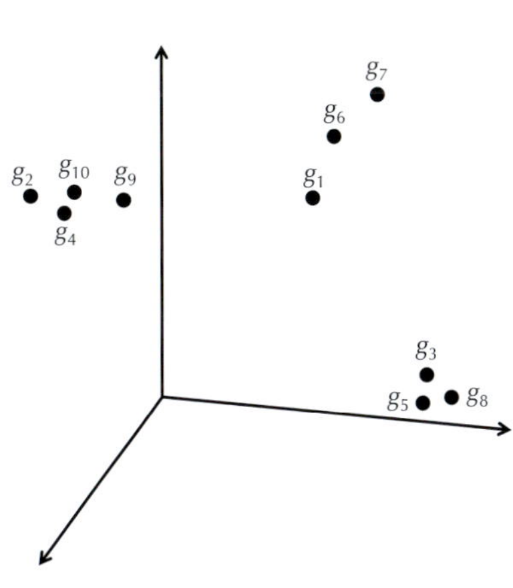

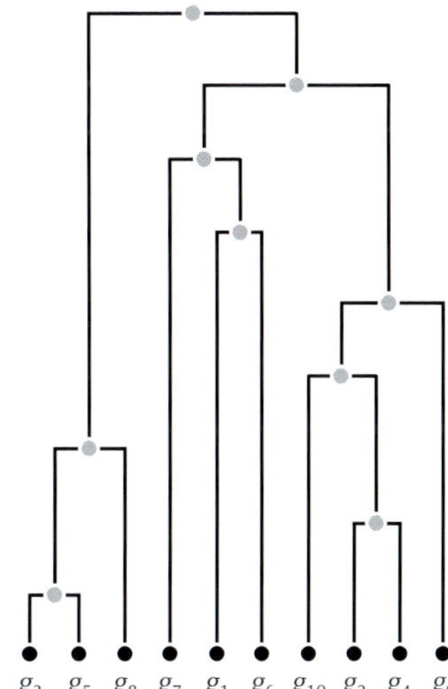

FIGURE 8.22 A toy gene expression matrix of ten genes measured at three time points (top left), the distance matrix based on Euclidean distance (top right), the gene expression vectors as points in three-dimensional space (bottom left), and the tree produced from the distance matrix by the hierarchical clustering algorithm (bottom right). Leaves correspond to genes; internal nodes correspond to clusters of genes.

In previous sections, we assumed that we were working with a fixed number of clusters k. But in practice, clusters often have subclusters, which have subsubclusters, and so on. To capture this cluster stratification, the **hierarchical clustering** algorithm uses an $n \times n$ distance matrix D to organize n data points into a tree (Figure 8.22 (bottom

right)). As shown in Figure 8.23, a horizontal line crossing the tree in i places divides the n genes into i clusters.

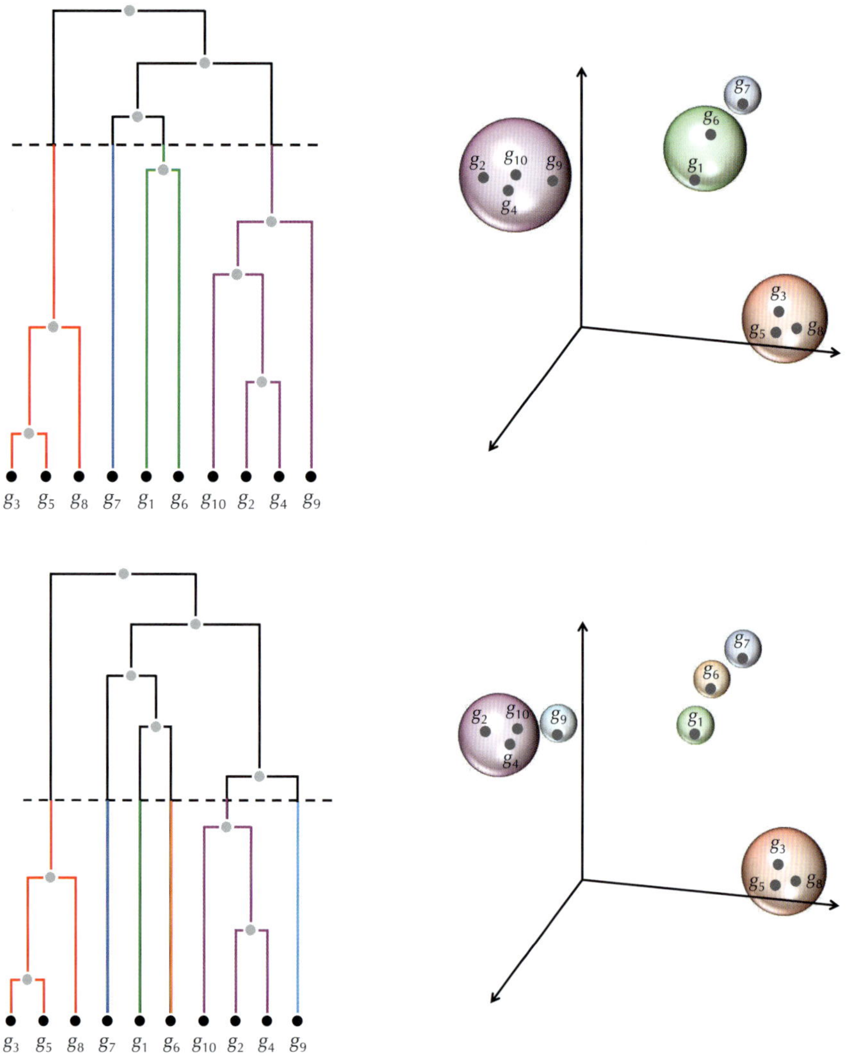

FIGURE 8.23 A tree with n leaves imposes n different ways of partitioning the data into clusters. (Top) The horizontal line through the tree (left) crosses the data in four places and partitions the data into four clusters (right). (Bottom) The same tree with a different horizontal line (left) partitions the data into six clusters (right).

Exercise Break: Figure 8.23 illustrates two ways of clustering the data from Figure 8.22 using a tree. Find the remaining eight ways of clustering these data using the same tree.

Inferring clusters from a tree

HIERARCHICALCLUSTERING, whose pseudocode is shown below, progressively generates n different partitions of the underlying data into clusters, all represented by a tree in which each node is labeled by a cluster of genes. The first partition has n single-element clusters represented by the leaves of the tree, with each element forming its own cluster. The second partition merges the two "closest" clusters into a single cluster consisting of two elements. In general, the i-th partition merges the two closest clusters from the (i - 1)-th partition and has $n - i + 1$ clusters. We hope this algorithm looks familiar — it is **UPGMA** (from Chapter 7) in disguise.

```
HIERARCHICALCLUSTERING(D, n)
    Clusters ← n single-element clusters labeled 1, . . . , n
    construct a graph T with n isolated nodes labeled by single elements 1, . . . , n
    while there is more than one cluster
        find the two closest clusters C_i and C_j (break ties arbitrarily)
        merge C_i and C_j into a new cluster C_new with |C_i| + |C_j| elements
        add a new node labeled by cluster C_new to T
        connect node C_new to C_i and C_j by directed edges
        remove the rows and columns of D corresponding to C_i and C_j
        remove C_i and C_j from Clusters
        add a row/column to D for C_new by computing D(C_new, C) for each C in Clusters
        add C_new to Clusters
    root ← the node in T corresponding to the remaining cluster
    return T
```

Note that we have not yet defined how **HIERARCHICALCLUSTERING** computes the distance $D(C_{\text{new}}, C)$ between a newly formed cluster C_{new} and each old cluster C. In practice, clustering algorithms vary in how they compute these distances, with results that can vary greatly. One commonly used approach (Figure 8.24) defines the distance between clusters C_1 and C_2 as the smallest distance between any pair of elements from these clusters,

$$D_{\min}(C_1, C_2) = \min_{\text{all points } i \text{ in cluster } C_1, \text{ all points } j \text{ in cluster } C_2} D_{i,j} .$$

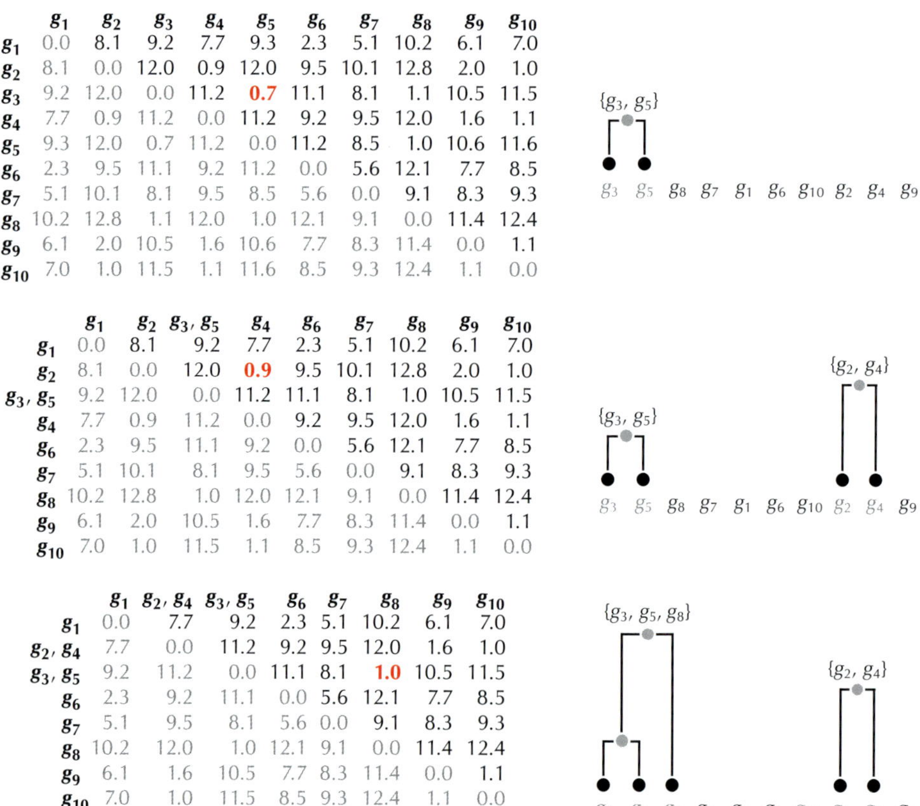

	g_1	g_2	g_3	g_4	g_5	g_6	g_7	g_8	g_9	g_{10}
g_1	0.0	8.1	9.2	7.7	9.3	2.3	5.1	10.2	6.1	7.0
g_2	8.1	0.0	12.0	0.9	12.0	9.5	10.1	12.8	2.0	1.0
g_3	9.2	12.0	0.0	11.2	**0.7**	11.1	8.1	1.1	10.5	11.5
g_4	7.7	0.9	11.2	0.0	11.2	9.2	9.5	12.0	1.6	1.1
g_5	9.3	12.0	0.7	11.2	0.0	11.2	8.5	1.0	10.6	11.6
g_6	2.3	9.5	11.1	9.2	11.2	0.0	5.6	12.1	7.7	8.5
g_7	5.1	10.1	8.1	9.5	8.5	5.6	0.0	9.1	8.3	9.3
g_8	10.2	12.8	1.1	12.0	1.0	12.1	9.1	0.0	11.4	12.4
g_9	6.1	2.0	10.5	1.6	10.6	7.7	8.3	11.4	0.0	1.1
g_{10}	7.0	1.0	11.5	1.1	11.6	8.5	9.3	12.4	1.1	0.0

	g_1	g_2	g_3, g_5	g_4	g_6	g_7	g_8	g_9	g_{10}
g_1	0.0	8.1	9.2	7.7	2.3	5.1	10.2	6.1	7.0
g_2	8.1	0.0	12.0	**0.9**	9.5	10.1	12.8	2.0	1.0
g_3, g_5	9.2	12.0	0.0	11.2	11.1	8.1	1.0	10.5	11.5
g_4	7.7	0.9	11.2	0.0	9.2	9.5	12.0	1.6	1.1
g_6	2.3	9.5	11.1	9.2	0.0	5.6	12.1	7.7	8.5
g_7	5.1	10.1	8.1	9.5	5.6	0.0	9.1	8.3	9.3
g_8	10.2	12.8	1.0	12.0	12.1	9.1	0.0	11.4	12.4
g_9	6.1	2.0	10.5	1.6	7.7	8.3	11.4	0.0	1.1
g_{10}	7.0	1.0	11.5	1.1	8.5	9.3	12.4	1.1	0.0

	g_1	g_2, g_4	g_3, g_5	g_6	g_7	g_8	g_9	g_{10}
g_1	0.0	7.7	9.2	2.3	5.1	10.2	6.1	7.0
g_2, g_4	7.7	0.0	11.2	9.2	9.5	12.0	1.6	1.0
g_3, g_5	9.2	11.2	0.0	11.1	8.1	**1.0**	10.5	11.5
g_6	2.3	9.2	11.1	0.0	5.6	12.1	7.7	8.5
g_7	5.1	9.5	8.1	5.6	0.0	9.1	8.3	9.3
g_8	10.2	12.0	1.0	12.1	9.1	0.0	11.4	12.4
g_9	6.1	1.6	10.5	7.7	8.3	11.4	0.0	1.1
g_{10}	7.0	1.0	11.5	8.5	9.3	12.4	1.1	0.0

FIGURE 8.24 HIERARCHICALCLUSTERING in action. (Top left) The distance matrix from Figure 8.22 (top left), with its minimum element shown in red, corresponding to genes g_3 and g_5. (Top right) Merging the single-element clusters containing g_3 and g_5. (Middle left) The updated distance matrix after computing $D_{\min}$ for the new cluster with respect to each other (single-element) cluster, with its minimum element shown in red. (Middle right) Merging the two clusters corresponding to the minimum element. (Bottom) Updating the distance matrix (left) and merging two additional clusters (right). Subsequent steps will reconstruct the tree from Figure 8.22.

The distance function that we encountered with **UPGMA** uses the average distance between elements in two clusters,

$$D_{\text{avg}}(C_1, C_2) = \frac{\sum_{\text{all points } i \text{ in cluster } C_1} \sum_{\text{all points } j \text{ in cluster } C_2} D_{i,j}}{|C_1| \cdot |C_2|}.$$

Exercise Break: Apply **HIERARCHICALCLUSTERING** to the distance matrix in Figure 8.22 using D_{avg} instead of D_{min}.

Exercise Break: Apply **HIERARCHICALCLUSTERING** (with D_{avg}) to the abridged 230-gene yeast dataset, and partition this dataset into six clusters. Do you expect these clusters to be roughly the same as the clusters shown in Figure 8.14? If not, should we be concerned?

Analyzing the diauxic shift with hierarchical clustering

Figure 8.25 visualizes expression vectors for each of the six clusters obtained after applying **HIERARCHICALCLUSTERING** (using D_{avg}) to the yeast dataset.

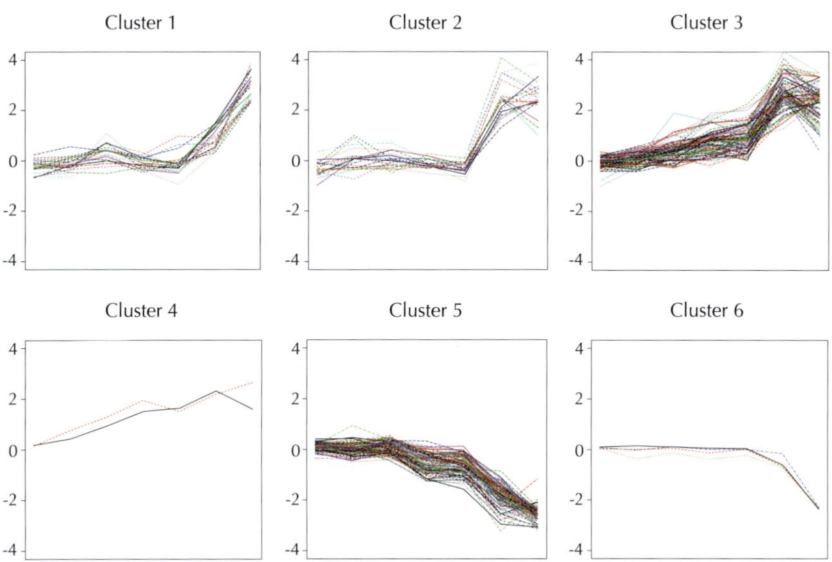

FIGURE 8.25 Applying **HIERARCHICALCLUSTERING** to the yeast dataset results in six clusters with 22, 20, 87, 2, 95, and 4 genes, respectively.

STOP and Think: **HIERARCHICALCLUSTERING** and the Lloyd algorithm (Figure 8.14) have produced different clusters. Should we be concerned?

Exercise Break: Implement **HIERARCHICALCLUSTERING** (with D_{min} rather than D_{avg}) and apply it to partition the abridged yeast gene expression dataset into six clusters. How does the result differ from Figure 8.25?

Biologists are not discouraged by the fact that different clustering approaches may produce different clusters because applying these clustering algorithms is often just the first step on the road to discovery (see DETOUR: Clustering and Corrupted Cliques for yet another clustering approach). For this reason, gene expression studies are typically followed by experimental work to confirm that derived clusters make sense biologically. Once clusters have been generated, further research often focuses on specific genes within these clusters.

For example, each of the clusters in Figure 8.25 can be further analyzed to reveal sub-clusters of genes with even more pronounced expression profiles than the genes in the entire cluster. In particular, cluster 1 contains seven genes exhibiting rather slight changes during the first six checkpoints and a surge in gene expression at the final checkpoint. Biologists discovered that six of these seven genes have the **carbon source response element (CSRE) regulatory motif** with consensus sequence `CATTCATCCG` in their upstream regions. Further analysis of the entire yeast genome revealed that only four other yeast genes have this motif in their upstream region, suggesting that it was a good idea to group these six genes into a sub-cluster within cluster 1.

The more important issue, however, is to understand *why* these six genes are related. Yeast prefers using glucose as an energy source compared to other compounds like ethanol, and so in the presence of glucose, the transcription of genes responsible for metabolizing these less tasty compounds is repressed. Researchers have thus concluded that the CSRE motif somehow helps yeast sense the presence of glucose and activates the six genes in question when the organism runs out of glucose, thus serving as an important component of the diauxic shift.

Finally, if you believe that we have exhausted every possible avenue of clustering, take another look at Figure 8.16 (top). Although **HIERARCHICALCLUSTERING** with D_{min} is able to find the clusters on the left and top middle, none of the clustering algorithms we have encountered can produce the clusters on the top right. Clustering seems like a straightforward problem in part because the human eye is so adept at grouping points into shapes. After all, computer vision researchers are still trying to teach computers to mimic our visual experience of the world, which is the outcome of millions of years of evolution.

Epilogue: Clustering Tumor Samples

As we mentioned earlier, gene expression analysis has a wide variety of applications, including cancer studies. In 1999, Uri Alon analyzed gene expression data for 2,000 genes from 40 colon tumor tissues and compared them with data from colon tissues belonging to 21 healthy individuals, all measured at a single time point. We can represent his data as a 2,000 $\times$ 61 gene expression matrix, where the first 40 columns describe tumor samples and the last 21 columns describe normal samples.

Now, suppose you performed a gene expression experiment with a colon sample from a new patient, corresponding to a 62nd column in an augmented gene expression matrix. Your goal is to predict whether this patient has a colon tumor. Since the partition of tissues into two clusters (tumor vs. healthy) is known in advance, it may seem that classifying the sample from a new patient is easy. Indeed, since each patient corresponds to a point in 2,000-dimensional space, we can compute the center of gravity of these points for the tumor sample and for the healthy sample. Afterwards, we can simply check which of the two centers of gravity is closer to the new tissue.

Alternatively, we could perform a blind analysis, pretending that we do not already know the classification of samples into cancerous vs. healthy, and analyze the resulting 2,000 $\times$ 62 expression matrix to divide the 62 samples into two clusters. If we obtain a cluster consisting predominantly of cancer tissues, this cluster may help us diagnose colon cancer.

Challenge Problem: These approaches may seem straightforward, but it is unlikely that either of them will reliably diagnose the new patient. Why do you think this is the case? Given Alon's 2,000 $\times$ 61 gene expression matrix and gene data from a new patient, derive a superior approach to evaluate whether this patient is likely to have a colon tumor.

Detours

Whole genome duplication or a series of duplications?

WGDs are quickly followed by massive gene loss and rearrangements, making it difficult to reconstruct the pre-duplicated genome. Indeed, as we mentioned in the main text, only 13% of genes have duplicates in modern-day *S. cerevisiae*. How, then, can we argue that *S. cerevisiae* has indeed undergone a WGD instead of a sequence of smaller duplications?

In 2004, Manolis Kellis analyzed *K. waltii*, a related yeast species. By aligning synteny blocks from *K. waltii* and *S. cerevisiae*, he discovered that nearly every synteny block of *K. waltii* aligns to *two* regions of *S. cerevisiae*. Because very few genes in the duplicated *S. cerevisiae* blocks occurred in both blocks (Figure 8.26), Kellis argued that there was indeed a WGD during yeast evolution.

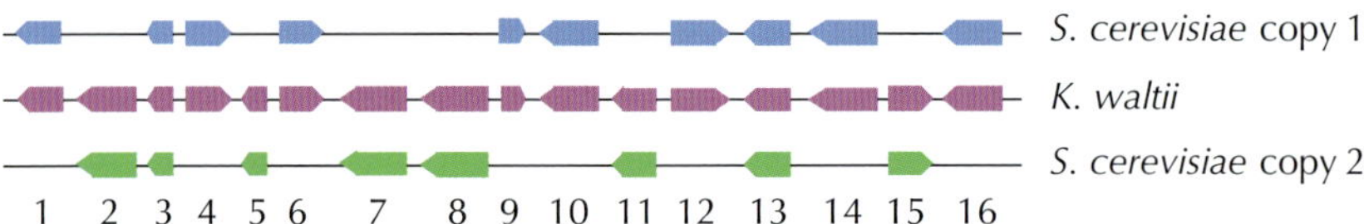

FIGURE 8.26 Two synteny blocks from *S. cerevisiae* (whose respective genes are shown in blue and green) aligned against a synteny block of *K. waltii* (whose genes are shown in purple). Although only three of the sixteen genes in this synteny block have two copies in *S. cerevisiae*, every gene in the *K. waltii* synteny block has a copy in at least one of the two synteny blocks in *S. cerevisiae*. In fact, most synteny blocks in *K. waltii* display this phenomenon, thus suggesting that there was indeed a WGD on the evolutionary path from the common ancestor of *S. cerevisiae* and *K. waltii* to *S. cerevisiae*.

STOP and Think: Although Manolis Kellis argued in 2004 that Figure 8.26 provides evidence for a WGD, three years later, Gustavo Caetano-Anollés raised doubts about Kellis's conclusion. Can you devise an alternative explanation for Figure 8.26 and propose a different evolutionary scenario that does not require a WGD?

Measuring gene expression

In the main text, we mentioned that we have encountered three technologies that could be used to measure gene expression. First, in a mass spectrometry experiment

(Chapter 4), biologists generate a set of spectra and match them against a proteome. The number of spectra matching peptides from a given protein offers a proxy for the expression level of this protein. To estimate the protein expression from this proxy, bioinformaticians must account for the varying length of proteins, poor fragmentation of some peptides (leading to difficulties in their identification), and other practical concerns.

Second, in an **RNA sequencing** experiment, we generate reads from a **transcriptome**, or all RNA transcripts present in a cell. By estimating the quantity of each protein-coding RNA transcript in a sample, we obtain a proxy for the expression of the resulting protein. A number of processes affect protein production in the cell in addition to transcription, such as translation, post-translational modifications, and protein degradation. These additional factors can muddle the correlation between the quantity of a transcript and the expression of its corresponding protein.

Third, we can use DNA arrays (Chapter 2) carrying probes (k-mers) aimed at each gene in a species of interest. Each probe is characterized by an intensity, which offers a proxy for the number of transcripts of a given gene present in a sample. A deficiency of DNA arrays is that they only target the identification of known transcripts and often fail to evaluate unknown transcripts. For example, many cancers are caused by rare mutations and would go undetected when using DNA arrays. As a result, RNA sequencing is often more attractive in cancer studies, and it is now the dominant technology for analyzing gene expression.

Microarrays

The microarrays that DeRisi used to study the diauxic shift were manufactured as follows. After capturing many RNA transcripts expressed in yeast cells, DeRisi converted each RNA transcript to **complementary DNA (cDNA)** using an enzyme called **reverse transcriptase**, and spotted these cDNAs on a glass slide. He then hybridized the cDNA against fluorescently labeled RNA from a sample of interest in order to measure the expression levels of various yeast genes.

The amount of cDNA printed on each spot of the microarray can vary greatly, a complication that DeRisi needed to address in order to ensure that fluorescence intensities could be compared across spots and across arrays. He therefore hybridized two samples corresponding to two different timestamps to each array (Figure 8.27). He then labeled the samples with different colors of fluorescent dyes so that the samples could be distinguished by image-processing software.

The expression values obtained from a microarray are represented as the ratio of fluorescent intensities of the two samples. Thus, expression is measured as relative changes in the expression of individual genes between samples and time points. For example, if a gene's expression value is 2, then this gene's expression is twice as large in the first sample; if the expression value is $1/3$, then the expression is three times as large in the second sample. Following DeRisi, researchers commonly take the logarithm of these expression ratios to generate expression matrices.

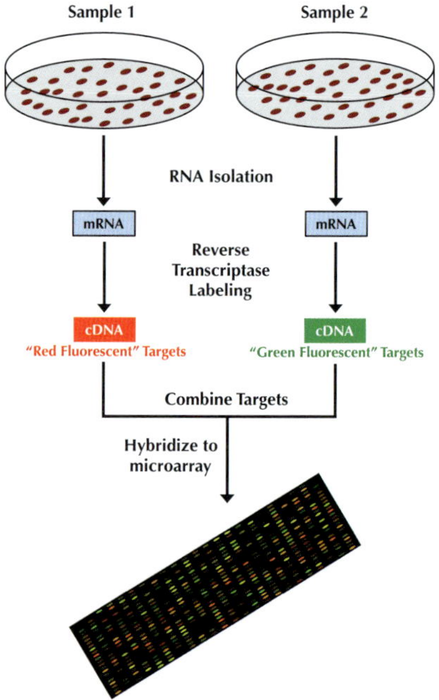

FIGURE 8.27 A microarray against which two samples of fluorescently labeled RNA (red and green) has been hybridized.

Proof of the Center of Gravity Theorem

Note that when $k = 1$, the k-Means Clustering Problem is equivalent to finding a single center point x that minimizes the sum of squared distances from x to points in *Data*.

Our goal is to show that the center of gravity of a set of points *Data* is the unique point that minimizes DISTORTION$(Data, x)$ over all possible centers x. Since the squared Euclidean distance between $DataPoint = (DataPoint_1, \ldots, DataPoint_m)$ and center $x =$

$(x_1, \ldots, x_m)$ is equal to $\sum_{1 \le j \le m}(DataPoint_j - x_j)^2$, we have that

$$\begin{aligned}\text{DISTORTION}(Data, x) &= \frac{1}{n} \sum_{\text{all points } DataPoint \text{ in } Data} d(DataPoint, x)^2 \\ &= \frac{1}{n} \sum_{\text{all points } DataPoint \text{ in } Data} \sum_{j=1}^{m} (DataPoint_j - x_j)^2 \\ &= \frac{1}{n} \sum_{j=1}^{m} \sum_{\text{all points } DataPoint \text{ in } Data} (DataPoint_j - x_j)^2 .\end{aligned}$$

The last line of this formula implies that we can independently minimize DISTORTION$(Data, x)$ in each of m dimensions by minimizing each of the m expressions

$$\sum_{\text{all points } DataPoint \text{ in } Data} (DataPoint_j - x_j)^2.$$

Each of these expressions is a concave-up quadratic function of a single variable x_j. Thus, we can find the minimum of this function by finding where its derivative is equal to zero:

$$\sum_{\text{all points } DataPoint \text{ in } Data} -2 \cdot (DataPoint_j - x_j) = 0 .$$

The only solution of this equation is given by

$$x_j = \frac{1}{n} \sum_{\text{all points } DataPoint \text{ in } Data} DataPoint_j ,$$

implying that the center's j-th coordinate is the mean value of the j-th coordinates of the data points. In other words, the unique solution of the k-Means Clustering Problem for $k = 1$ is simply the center of gravity of all data points. □

Transforming an expression matrix into a distance/similarity matrix

There are many ways to quantify the similarity between expression vectors $x = (x_1, \ldots, x_m)$ and $y = (y, \ldots, y_m)$. One possibility is the dot product, $\sum_{i=1,m} x_i \cdot y_i$. Another is the **Pearson correlation coefficient** PEARSONCORRELATION(x, y), where

$$\text{PEARSONCORRELATION}(x, y) = \frac{\sum_{i=1}^{m}(x_i - \mu(x)) \cdot (y_i - \mu(y))}{\sqrt{\sum_{i=1}^{m}(x_i - \mu(x))^2 \cdot \sum_{i=1}^{m}(y_i - \mu(y))^2}}.$$

In the above formula, $\mu(x)$ denotes the means of all coordinates of vector x.

STOP and Think: Given a vector x, which vectors y maximize and minimize PEARSONCORRELATION(x, y)?

The Pearson correlation coefficient varies between -1 and 1, where -1 indicates total negative correlation, 0 indicates no correlation, and 1 indicates total positive correlation. Based on the Pearson correlation coefficient, we can define the **Pearson distance** between vectors x and y as

$$\text{PEARSONDISTANCE}(x, y) = 1 - \text{PEARSONCORRELATION}(x, y).$$

Exercise Break: Compute the Pearson correlation coefficient for the following pairs of vectors:

1. $(\cos\alpha, \sin\alpha)$ and $(\sin\alpha, -\cos\alpha)$ for an arbitrary value of α;
2. $(\sqrt{0.75}, 0.5)$ and $(-\sqrt{0.75}, 0.5)$.

Clustering and corrupted cliques

In expression analysis studies, a similarity matrix R is often transformed into a **similarity graph** $G(R, \theta)$. The nodes of this graph represent genes, and an edge connects genes i and j if and only if the similarity between them ($R_{i,j}$) exceeds a threshold value θ.

STOP and Think: Consider a clustering of genes that satisfies the Good Clustering Principle: the similarity between any two genes within the same cluster exceeds θ, and the similarity between any two genes in different clusters is less than θ. What does the similarity graph $G(R, \theta)$ look like for these genes?

If clusters satisfy the Good Clustering Principle, then there should be some value of θ such that each connected component of $G(R, \theta)$ is a **clique**, or a graph in which every pair of nodes are connected by an edge (Figure 8.28). In general, a graph whose connected components are all cliques is called a **clique graph**.

Errors in expression data and the absence of a universal threshold θ often result in corrupted similarity graphs whose connected components are not cliques (Figure 8.29). Either genes from the same cluster may have a similarity value falling below θ, thus removing edges from a clique, or genes from different clusters may have a similarity value exceeding θ, thus adding edges between different cliques. This observation leads us to ask how to transform a corrupted similarity graph into a clique graph using the smallest number of edge additions and deletions.

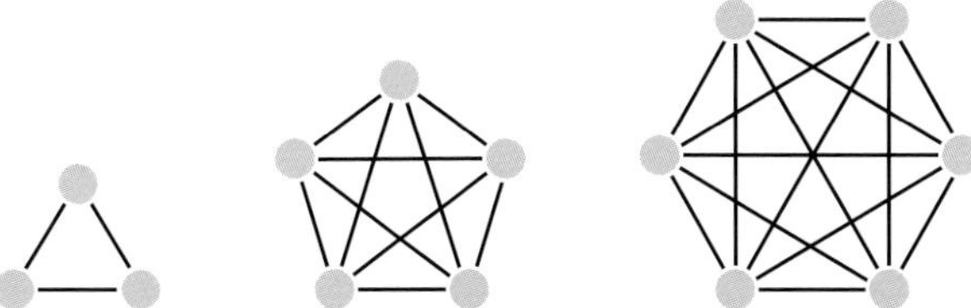

FIGURE 8.28 A clique graph consisting of three cliques.

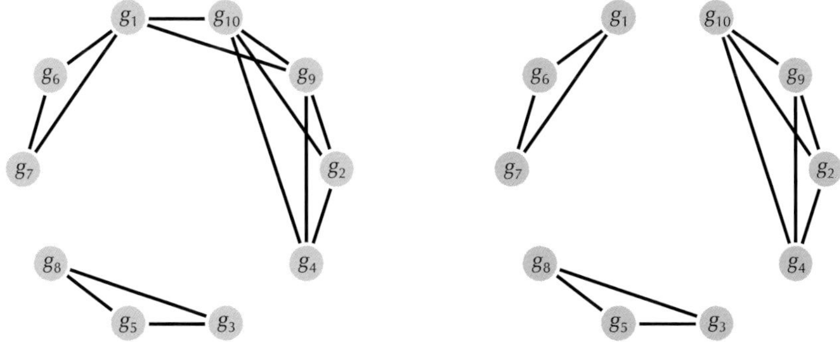

FIGURE 8.29 (Left) One possible similarity graph for the genes from Figure 8.22. (Right) The similarity graph can be transformed into a clique graph (right) by removing edges (g_1, g_{10}) and (g_1, g_9).

Corrupted Cliques Problem:
Find the minimum number of edges that need to be added or deleted to transform a graph into a clique graph.

> **Input**: A graph.
> **Output**: The minimum number of edge additions and deletions that transform this graph into a clique graph.

The Corrupted Cliques Problem is difficult to solve exactly, so some heuristics have been proposed. The **Cluster Affinity Search Technique algorithm (CAST)**, described below, performs remarkably well at clustering gene expression data.

Define the similarity between gene i and cluster C as the average similarity between i and all genes in C:

$$R_{i,C} = \sum_{\text{all elements } j \text{ in cluster } C} \frac{R_{i,j}}{|C|}$$

Given a threshold θ, a gene i is **θ-close** to cluster C if $R_{i,C} > \theta$ and **θ-distant** from C otherwise. A cluster is called **consistent** if all genes in C are θ-close to C and all genes not in C are θ-distant from C. **CAST** uses the similarity graph and the threshold θ to iteratively find consistent clusters by starting with a single-element cluster C and then adding the "closest" gene not in C and removing the "most distant" gene in C. After a consistent cluster is found, all nodes in cluster C are removed from the similarity graph, and **CAST** iterates over the resulting smaller graph.

```
CAST(R, ϑ)
    Graph ← G(R, ϑ)
    Clusters ← empty set
    while Graph is nonempty
        C ← a single-node cluster consisting of a node of maximal degree in Graph
        while there exists a ϑ-close gene i not in C or a ϑ-distant gene i in C
            find the nearest ϑ-close gene i not in C and add it to C
            find the farthest ϑ-distant gene i in C and remove it from C
        add C to the set Clusters
        remove the nodes of C from Graph
    return Clusters
```

Exercise Break: Implement **CAST** and use it to cluster the abridged gene expression dataset.

Bibliography Notes

The soft *k*-means algorithm for clustering, developed by Bezdek, 1981, is a variant of the expectation maximization algorithm, which was first proposed by Ceppellini, Siniscalco, and Smith, 1955 and rediscovered many times by various researchers. Do and Batzoglou, 2008 wrote an excellent primer on expectation maximization that inspired our discussion of coin flipping. The Lloyd algorithm for *k*-means clustering was introduced by Lloyd, 1982. Arthur and Vassilvitskii, 2007 developed *k*-means++ initialization for *k*-means clustering. The **CAST** algorithm was developed by Ben-Dor, Shamir, and Yakhini, 1999.

DeRisi, Iyer, and Brown, 1997 performed the first large-scale gene expression experiment to analyze the diauxic shift (see Cristianini and Hahn, 2007 for an excellent analysis of this experiment). Eisen et al., 1998 described the first applications of hierarchical clustering to gene expression analyses. Alon et al., 1999 analyzed patterns of gene expression in colon tumors.

Ohno, 1970 proposed the Whole Genome Duplication Model. Wolfe and Shields, 1997 provided the first convincing arguments in favor of a whole genome duplication in yeast. Kellis, Birren, and Lander, 2004 provided further evidence for a whole genome duplication by analyzing various yeast species. However, these arguments have not convinced Martin et al., 2007, who published a rebuttal. Thomson et al., 2005 resurrected the sequence of ancient alcohol dehydrogenases from yeast.

Burrows-Wheeler Transportation
ANN-B$AA
HOW DO WE
LOCATE
DISEASE-CAUSING MUTATIONS?
Combinatorial Pattern Matching

What Causes Ohdo Syndrome?

About 1% of babies are born with mental retardation, but this affliction remains poorly understood because it can be caused by a variety of different genetic disorders. One of these disorders is **Ohdo syndrome**, which causes an expressionless, "mask-like" face. In 2011, biologists solved the genetic puzzle underlying Ohdo syndrome by discovering a handful of mutations shared by multiple patients, which the researchers used to identify a single protein-truncating mutation responsible for Ohdo syndrome.

The discovery of Ohdo syndrome's root cause represents just one of many new discoveries arising from the use of **read mapping** to study genetic disorders. In read mapping, researchers compare sequenced DNA reads taken from an individual against a **reference human genome** (see DETOUR: The Reference Human Genome) in order to find which reads perfectly match the reference and which reads indicate mutations of one nucleotide into another (**single nucleotide polymorphisms**, or **SNPs**). The reference genome is a gross simplification of species identity, since in addition to about 3 million SNPs (0.1% of the human genome), humans differ by genome rearrangements, insertions, and deletions that can span thousands of nucleotides (see DETOUR: Rearrangements, Insertions, and Deletions in Human Genomes). However, in this chapter we will focus only on algorithms for finding SNPs.

PAGE 520

PAGE 520

But wait, you may say, *why not use one of the algorithms that we already covered?* After all, we could always sequence the entire genome of an individual and then compare it against the reference genome. However, sequencing methods are computationally intensive and not perfect, as they often generate error-prone contigs. As a result, it makes sense to map reads from an individual human to the reference human genome to find out the differences.

To see why read mapping should be easier than genome assembly, let us return to the analogy of a jigsaw puzzle, which is sold with a picture of the completed puzzle on its box. This photo makes reconstructing the puzzle far easier; for a simple example, if the completed puzzle shows a sun in a blue sky, then you can automatically move all of the bright yellow pieces and all of the light blue pieces to the top of the puzzle.

Yet aside from genome sequencing, two other methods come to mind for read mapping. First, you could align each read to the reference genome (using a fitting alignment, Chapter 5) to find the most similar region. Second, you could apply approximate pattern matching algorithms to match each read one at a time against the reference genome.

STOP and Think: What computational challenges might arise from using these methods to map millions of reads to a reference human genome?

Both of these methods are guaranteed to solve the problem of mapping reads to a reference genome, but their runtimes become a bottleneck when we scale to millions of reads. Therefore, our goal in this chapter is to figure out how to use the reference genome as a "photo on the box" shortcut to find SNPs.

Introduction to Multiple Pattern Matching

Recall from Chapter 3 that reads are typically a few hundred base pairs long. These reads will form a collection of strings *Patterns* that we wish to match against a genome *Text*. For each string in *Patterns*, we will first find all its *exact* matches as a substring of *Text* (or conclude that it does not appear in *Text*). When hunting for the cause of a genetic disorder, we can immediately eliminate from consideration areas of the reference genome where exact matches occur. In the epilogue, we will generalize this problem to find *approximate* matches, where single nucleotide substitutions in reads separate the individual from the reference genome (or represent errors in reads).

Multiple Pattern Matching Problem:
Find all occurrences of a collection of patterns in a text.

> **Input**: A string *Text* and a collection *Patterns* containing (shorter) strings.
> **Output**: All starting positions in *Text* where a string from *Patterns* appears as a substring.

A naive approach to the Multiple Pattern Matching Problem would attempt repeated applications of an algorithm for the (single) Pattern Matching Problem, which we encountered in Chapter 1. This algorithm, which we call **BRUTEFORCEPATTERNMATCHING**, would slide each *Pattern* along *Text*, checking whether the substring starting at each position of *Text* matches *Pattern*. Recall that the runtime of a naive algorithm for a single pattern is $\mathcal{O}(|Text| \cdot |Pattern|)$. Thus, the runtime of **BRUTEFORCEPATTERNMATCHING** for the Multiple Pattern Matching Problem is $\mathcal{O}(|Text| \cdot |Patterns|)$, where $|Text|$ is the length of *Text* and $|Patterns|$ is the sum of the lengths of all strings in *Patterns*.

The problem with applying **BRUTEFORCEPATTERNMATCHING** to read mapping is that $|Text|$ and $|Patterns|$ are both huge. In the case of the human genome (3 GB), the total length of all reads may exceed 1 TB; as a result, any algorithm with runtime $\mathcal{O}(|Text| \cdot |Patterns|)$ will be too slow.

STOP and Think: The estimate $\mathcal{O}(|Text| \cdot |Patterns|)$ presents the *worst-case* estimate of the runtime for **BRUTEFORCEPATTERNMATCHING**. What is the *average-case* estimate?

Herding Patterns into a Trie

Constructing a trie

The reason why the runtime of **BRUTEFORCEPATTERNMATCHING** is so high is that each string in *Patterns* must traverse all of *Text* independently. If you think about *Text* as a long road, then **BRUTEFORCEPATTERNMATCHING** is analogous to loading each pattern into its own car when driving down *Text*, an inefficient strategy. Instead, our goal is to herd the patterns onto a bus so that we only need to make one trip from the beginning to the end of *Text*. More formally, we would like to organize *Patterns* into a data structure to prevent multiple passes down *Text* and to reduce runtime. To this end, we will consolidate *Patterns* into a directed acyclic graph called a **trie** (pronounced "try"), which is written TRIE(*Patterns*) and has the following properties (Figure 9.1).

- The trie has a single root node with indegree 0, denoted *root*; all other nodes have indegree 1.
- Each edge of TRIE(*Patterns*) is labeled with a letter of the alphabet.
- Edges leading out of a given node have distinct labels.
- Every string in *Patterns* is spelled out by concatenating the letters along some path from the root downward.
- Every path from the root to a **leaf**, or node with outdegree 0, spells a string from *Patterns*.

Trie Construction Problem:
Construct a trie from a collection of patterns.

Input: A collection of strings *Patterns*.
Output: TRIE(*Patterns*).

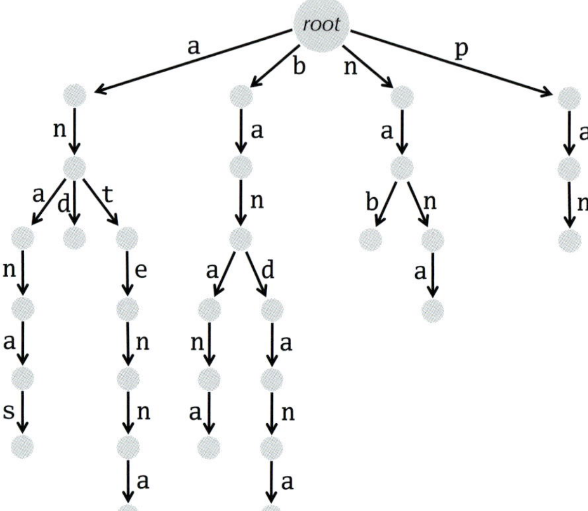

FIGURE 9.1 The trie for the following collection of strings *Patterns*: "ananas", "and", "antenna", "banana", "bandana", "nab", "nana", "pan".

The most obvious way to construct TRIE(*Patterns*) is by iteratively adding each string from *Patterns* to the growing trie, as implemented by the following algorithm.

9A

```
TRIECONSTRUCTION(Patterns)
    Trie ← a graph consisting of a single node root
    for each string Pattern in Patterns
        currentNode ← root
        for i ← 1 to |Pattern|
            currentSymbol ← i-th symbol of Pattern
            if there is an outgoing edge from currentNode with label currentSymbol
                currentNode ← ending node of this edge
            else
                add a new node newNode to Trie
                add a new edge from currentNode to newNode with label currentSymbol
                currentNode ← newNode
    return Trie
```

STOP and Think: How can we use the trie to solve the Multiple Pattern Matching Problem?

Applying the trie to multiple pattern matching

Given a string *Text* and TRIE(*Patterns*), we can quickly check whether any string from *Patterns* matches a *prefix* of *Text*. To do so, we start reading symbols from the beginning of *Text* and see what string these symbols "spell" as we proceed along the path downward from the root of the trie, as illustrated in Figure 9.2 (left). For each new symbol in *Text*, if we encounter this symbol along an edge leading down from the present node, then we continue along this edge; otherwise, we stop and conclude that no string in *Patterns* matches a prefix of *Text*. If we make it all the way to a leaf, then the pattern spelled out by this path matches a prefix of *Text*. This algorithm is called **PREFIXTRIEMATCHING**.

```
PREFIXTRIEMATCHING(Text, Trie)
    symbol ← first letter of Text
    v ← root of Trie
    while forever
        if v is a leaf in Trie
            output the pattern spelled by the path from the root to v
        else if there is an edge (v, w) in Trie labeled by symbol
            symbol ← next letter of Text
            v ← w
        else
            return "no matches found"
```

STOP and Think: For **PREFIXTRIEMATCHING** to work, we have made hidden assumption that no string in *Patterns* is a prefix of another string in *Patterns* (or, for that matter, longer than *Text*). How can this algorithm be modified when *Patterns* is an arbitrary collection of strings? Hint: consider adding "pantry" to the patterns in Figure 9.2 (left).

PREFIXTRIEMATCHING finds whether any strings in *Patterns* match a prefix of *Text*. To find whether any strings in *Patterns* match a substring of *Text* starting at position i, this algorithm must match a prefix of the suffix of *Text* starting at position i. We can therefore run **PREFIXTRIEMATCHING** on all suffixes of *Text*. This requires running **PREFIXTRIEMATCHING** a total of $|Text|$ times, chopping the first symbol off of *Text* before each new iteration (Figure 9.2 (right)).

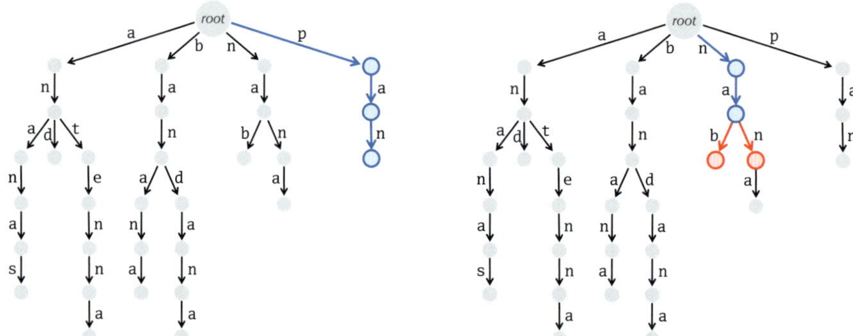

FIGURE 9.2 (Left) The pattern "pan" matches *Text* = "**pan**amabananas" when starting at the beginning of *Text*. (Right) No pattern match is found in TRIE(*Patterns*) for the strings *Patterns* from Figure 9.1 when starting at the third symbol of "pa**nam**abananas".

```
TRIEMATCHING(Text, Trie)
    while Text is nonempty
        PREFIXTRIEMATCHING(Text, Trie)
        remove first symbol from Text
```

We need |*Patterns*| steps to construct TRIE(*Patterns*), which contains at most |*Patterns*| + 1 nodes. Each iteration of **PREFIXTRIEMATCHING** takes at most |*LongestPattern*| steps, where *LongestPattern* is the longest string in *Patterns*. **TRIEMATCHING** makes |*Text*| total calls to **PREFIXTRIEMATCHING**, making the total number of steps equal to |*Patterns*| + |*Text*| · |*LongestPattern*|. This runtime offers a speed-up compared to the |*Text*| · |*Patterns*| steps required by **BRUTEFORCEPATTERNMATCHING**. The **Aho-Corasick algorithm**, developed in 1975, further reduces the number of steps required after constructing the trie from $\mathcal{O}(|Text| \cdot |LongestPattern|)$ steps to $\mathcal{O}(|Text|)$ steps (see **DETOUR: The Aho-Corasick Algorithm**).

STOP and Think: Do you see any computational challenges with using **TRIEMATCHING** to solve the Multiple Pattern Matching Problem?

Although **TRIEMATCHING** is fast, storing a trie consumes a lot of memory. Recall that **BRUTEFORCEPATTERNMATCHING** works with a single read at a time, which keeps the memory low because we only need to store the genome in memory. Yet **TRIEMATCHING** needs to store the entire trie in memory, which is proportional to

|*Patterns*|. Since a collection of reads for the human genome may consume upwards of 1 TB, the memory required to store the trie is prohibitive.

STOP and Think: How can we avoid multiple passes through the genome without needing to consolidate all the reads into a huge data structure?

Preprocessing the Genome Instead

Introduction to suffix tries

Since storing TRIE(*Patterns*) requires so much memory, let's process *Text* into a data structure instead. Our goal is to compare each string in *Patterns* against *Text* without needing to traverse *Text* from beginning to end. In more familiar terms, instead of packing *Patterns* onto a bus and riding the long distance down *Text*, our new data structure will be able to "teleport" each string in *Patterns* directly to its occurrences in *Text*.

A **suffix trie**, denoted SUFFIXTRIE(*Text*), is the trie formed from all suffixes of *Text* (Figure 9.3). From now on, we append the dollar-sign ("$") to *Text* in order to mark the end of *Text* (there is nothing special about this choice of symbol). We will also label each leaf of the resulting trie by the starting position of the suffix whose path through the trie ends at this leaf (using 0-based indexing). This way, when we arrive at a leaf, we will immediately know where this suffix came from in *Text*.

Exercise Break: Construct the suffix trie for papa without first appending the dollar-sign to mark the end of the text. Where are the paths corresponding to each suffix of papa? Do you now see why we first append "$" to the end of the text? (Hint: try to locate the suffix "pa" in your trie.)

STOP and Think: How can we use the suffix trie for pattern matching?

Using suffix tries for pattern matching

To match a single string *Pattern* to *Text*, note that if *Pattern* matches a substring of *Text* starting at position *i*, then *Pattern* must also appear at the beginning of the suffix of *Text* starting at position *i*. We can therefore determine whether *Pattern* occurs in SUFFIXTRIE(*Text*) by starting at the root and spelling symbols of *Pattern* downward. If

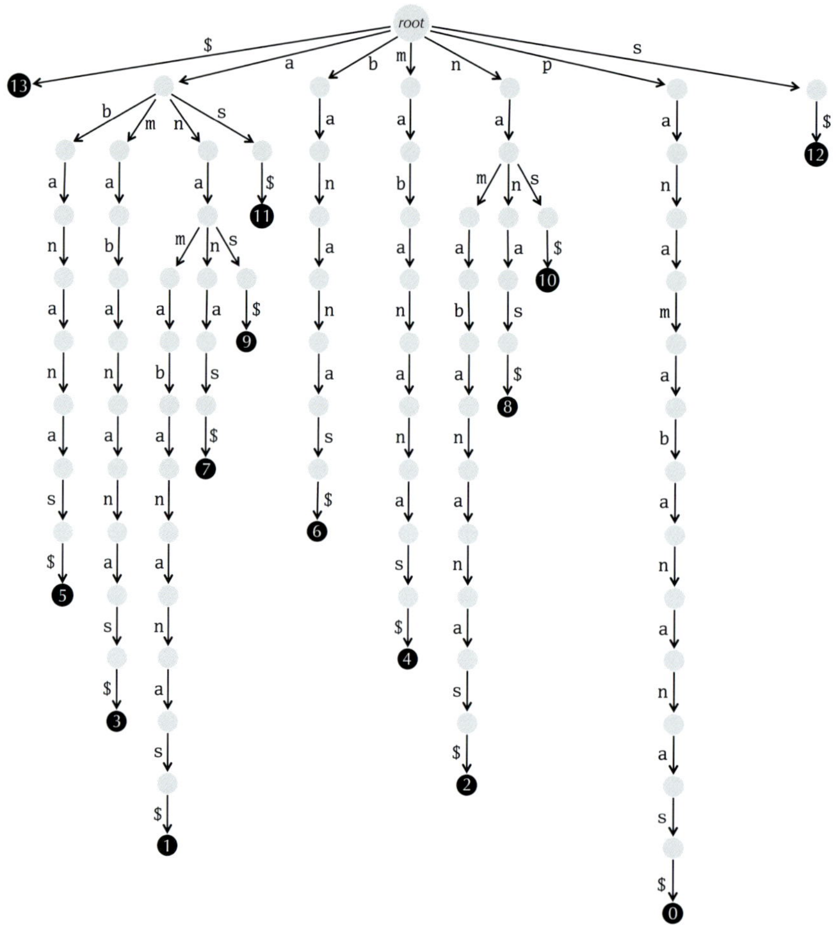

FIGURE 9.3 SuffixTrie("panamabananas$"), with leaf labels (corresponding to starting positions of suffixes) varying from 0 to 13.

we can find a path in the suffix trie spelling out *Pattern*, then we know that *Pattern* must occur in *Text* (Figure 9.4). We can then iterate over all strings in *Patterns*.

STOP and Think: Figure 9.4 illustrates how to find the pattern "nanas" in SuffixTrie("panamabananas$"), but it does not tell us where "nanas" occurs in *Text*. How can we obtain this information?

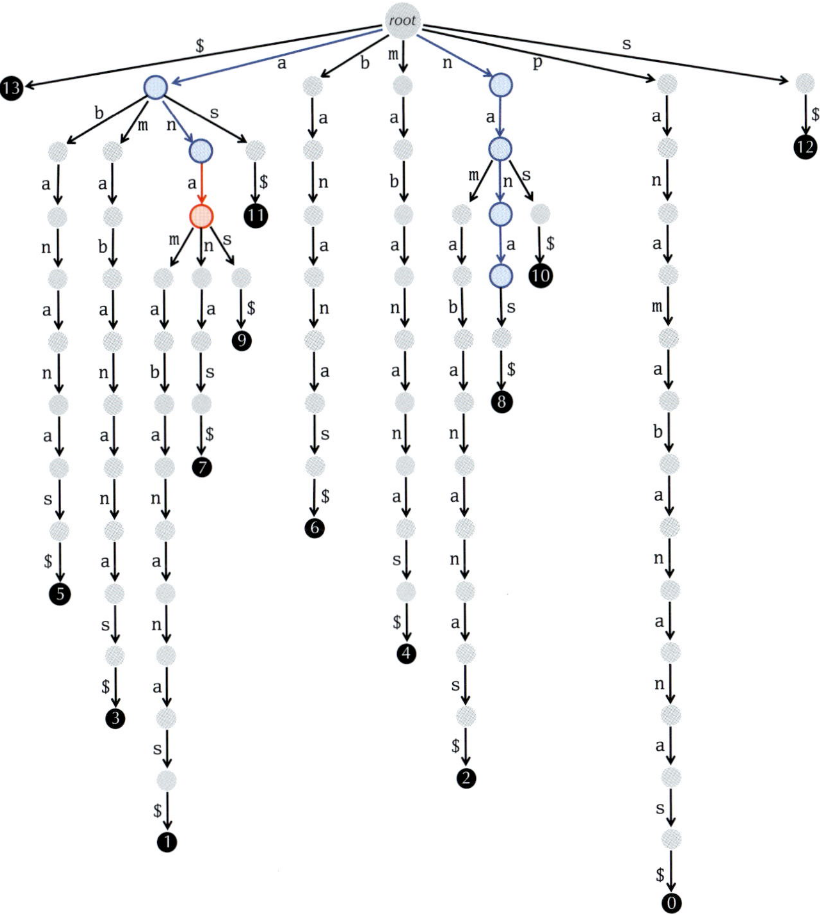

FIGURE 9.4 Threading "antenna" through SUFFIXTRIE("panamabananas$") fails to find a match because no suffix of "panamabananas$" begins with "**ant**"; however, threading "nana" through the suffix trie does find a match: "panamaba**nana**s$".

To determine where *Pattern* appears in *Text*, assume first that *Pattern* matches *Text* at a leaf of SUFFIXTRIE(*Text*). In this case, *Pattern* must appear in *Text* as a suffix, and we can consult the label at that leaf to determine the starting position of the suffix. For example, threading "nanas" into the suffix trie in Figure 9.4 shows that it matches the suffix starting at position 8 of "panamaba**nanas**$".

If the path spelling out *Pattern* stops before a leaf at some node v of SUFFIXTRIE(*Text*), then *Pattern* may occur more than once in *Text*. To locate these occurrences, follow all paths from v down to the leaves of SUFFIXTRIE(*Text*), which will indicate all starting positions of *Pattern* in *Text*. For example, as illustrated in Figure 9.5, the pattern "ana" corresponds to a path in SUFFIXTRIE("panamabananas$") that can be extended to three different leaves with labels 1, 7, and 9, corresponding to three occurrences of "ana": "p**ana**mabananas$", "panamab**ana**nas$", and "panamaban**ana**s$".

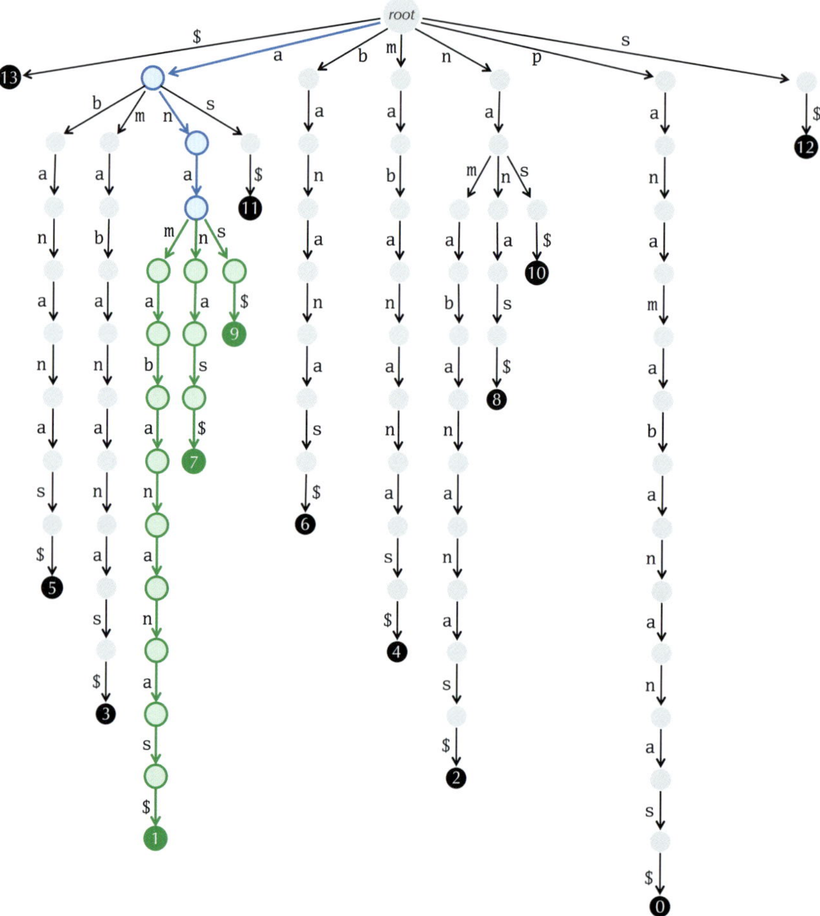

FIGURE 9.5 All paths starting with "ana" reveal the three occurrences of "ana" in "p**ana**mab**anana**s$" . Extending these paths to the leaves (shown in green) reveals that the starting positions of these occurrences are 1, 7, and 9.

STOP and Think: How much runtime and memory will it take to construct SUFFIXTRIE(*Text*)?

Recall that constructing TRIE(*Patterns*) required $\mathcal{O}(|Patterns|)$ runtime and memory. Accordingly, the runtime and memory required to construct SUFFIXTRIE(*Text*) are both equal to the combined length of all suffixes in *Text*. There are $|Text|$ suffixes of *Text*, ranging in length from 1 to $|Text|$ and having total length $|Text| \cdot (|Text| + 1)/2$, which is $\mathcal{O}(|Text|^2)$. Thus, we need to reduce both the construction time and memory requirements of suffix tries to make them practical.

Suffix Trees

Let's not give up hope on suffix tries. We can reduce the number of edges in the suffix trie by combining the edges on any non-branching path into a single edge. We then label this edge with the concatenation of symbols on the consolidated edges, as shown in Figure 9.6. The resulting data structure is called a **suffix tree**, written SUFFIXTREE(*Text*).

To match a single *Pattern* to *Text*, we thread *Pattern* into SUFFIXTREE(*Text*) by the same process used for a suffix trie. Similarly to the suffix trie, we can use the leaf labels to find starting positions of successfully matched patterns.

Exercise Break: Prove that SUFFIXTREE(*Text*) has exactly $|Text| + 1$ leaves and at most $|Text| + 1$ other nodes.

Compared to a suffix trie, which may have a quadratic number of nodes in the length of *Text*, the number of nodes in SUFFIXTREE(*Text*) does not exceed $2 \cdot |Text|$. Therefore, the memory required for SUFFIXTREE(*Text*) is $\mathcal{O}(|Text|)$.

STOP and Think: Wait a second. Suffix trees seem like a cosmetic modification of suffix tries. Since we still need to keep all concatenated edge labels in memory, why should suffix trees be more memory-efficient than suffix tries?

Suffix trees save memory because they do not need to store concatenated edge labels from each non-branching path. For example, a suffix tree does not need ten bytes to store the edge labeled "mabananas$" in Figure 9.6; instead, it suffices to store a **pointer** to position 4 of "panamabananas$", as well as the *length* of "mabananas$". Furthermore, suffix trees can be constructed in linear time, without having to first construct the suffix

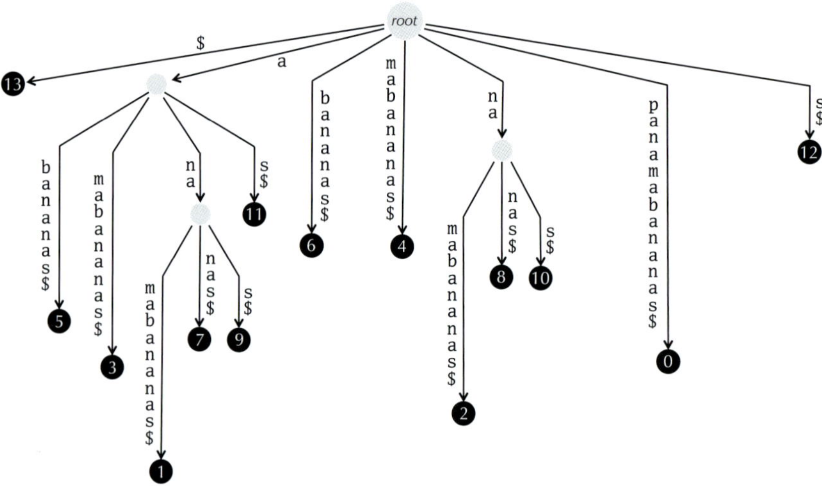

FIGURE 9.6 SUFFIXTREE("panamabananas$"), formed by compressing the edges on non-branching paths in Figure 9.3.

trie! We will not ask you to implement this fast suffix tree construction algorithm because it is quite complex.

Suffix Tree Construction Problem:
Construct the suffix tree of a string.

Input: A string *Text*.
Output: SUFFIXTREE(*Text*).

Charging Station (Constructing a Suffix Tree): We hope that you are wondering how the suffix tree can possibly save memory, since it needs to store potentially very long strings on each edge. It would be memory-inefficient to first construct the suffix trie and then consolidate each non-branching path into a single edge, storing its label in memory. In practice, researchers implement the suffix tree by storing *pointers* to locations in *Text* rather than storing potentially lengthy substrings. To see how this idea can be implemented, check out this Charging Station.

Although the suffix tree decreases memory requirements from $\mathcal{O}(|Text|^2)$ to $\mathcal{O}(|Text|)$, on average it still requires about 20 times as much memory as *Text* (when the length of *Text* is on the order of the human genome). For a 3 GB human genome, 60 GB of RAM is a huge improvement over the 1 TB that we needed to work with a trie constructed from all reads, but it still presents a memory challenge for most machines. This reveals a dark secret of big-O notation, which is that it ignores constant factors. For long strings such as the human genome, we will need to pay attention to this constant factor, since the expression $\mathcal{O}(|Text|)$ applies to both an algorithm with $2 \cdot |Text|$ memory and an algorithm with $1000 \cdot |Text|$ memory.

Yet before seeing how we can further reduce the memory needed for multiple pattern matching, we ask you to solve three problems for which suffix trees are useful.

Longest Repeat Problem:
Find the longest repeat in a string.

Input: A string *Text*.
Output: A longest substring of *Text* that appears in *Text* more than once.

Longest Shared Substring Problem:
Find the longest substring shared by two strings.

Input: Strings $Text_1$ and $Text_2$.
Output: The longest substring that occurs in both $Text_1$ and $Text_2$.

Shortest Non-Shared Substring Problem:
Find the shortest substring of one string that does not appear in another string.

Input: Strings $Text_1$ and $Text_2$.
Output: The shortest substring of $Text_1$ that does not appear in $Text_2$.

Charging Station (Solving the Longest Shared Substring Problem): One way of solving the Longest Shared Substring Problem is to construct two suffix trees, one for $Text_1$ and one for $Text_2$. Check out this Charging Station to learn about a more elegant solution.

PAGE 517

Suffix Arrays

Constructing a suffix array

In 1993, Udi Manber and Gene Myers introduced **suffix arrays** as a memory-efficient alternative to suffix trees. To construct SUFFIXARRAY(*Text*), we first sort all suffixes of *Text* lexicographically, assuming that "$" comes first in the alphabet (Figure 9.7). The suffix array is the list of starting positions of these sorted suffixes:

$$\text{SUFFIXARRAY}(\texttt{"panamabananas\$"}) = [13,\ 5,\ 3,\ 1,\ 7,\ 9,\ 11,\ 6,\ 4,\ 2,\ 8,\ 10,\ 0,\ 12].$$

Sorted Suffixes	Starting Positions
$	13
abananas$	5
amabananas$	3
anamabananas$	1
ananas$	7
anas$	9
as$	11
bananas$	6
mabananas$	4
namabananas$	2
nanas$	8
nas$	10
panamabananas$	0
s$	12

FIGURE 9.7 The sorted list of suffixes of *Text* = "panamabananas$", along with their starting positions in *Text*; these starting positions form the suffix array of *Text*.

Suffix Array Construction Problem:
Construct the suffix array of a string.

Input: A string *Text*.
Output: SUFFIXARRAY(*Text*).

The Suffix Array Construction Problem can easily be solved after sorting all suffixes of *Text*, but since even the fastest algorithms for sorting an array of n elements require $\mathcal{O}(n \cdot \log n)$ comparisons, sorting all suffixes takes $\mathcal{O}(|Text| \cdot \log(|Text|))$ comparisons, each one of which taking $\mathcal{O}(|Text|)$ time. However, there exists a faster algorithm that constructs suffix arrays in linear time (without constructing a suffix tree first) and requires only about a fifth as much memory as suffix trees, which knocks the 60 GB memory requirement for the human genome down to 12 GB.

STOP and Think: Given a suffix tree, can you quickly transform it into a suffix array? Given a suffix array, can you quickly transform it into a suffix tree?

As the preceding question suggests, suffix arrays and suffix trees are practically equivalent; every algorithm using suffix trees can be translated into an algorithm using suffix arrays (see **DETOUR: From Suffix Trees to Suffix Arrays**), and vice-versa (see **DETOUR: From Suffix Arrays to Suffix Trees**).

Pattern matching with the suffix array

Once we have constructed the suffix array of a string *Text*, we can use it to quickly locate every occurrence of a string *Pattern* in *Text*. First, recall that when pattern matching with the suffix trie, we observed that all matches of *Pattern* in *Text* must occur at the beginning of suffixes of *Text*. Second, note that after sorting the suffixes of *Text*, the suffixes beginning with *Pattern* clump together. For example, in Figure 9.7, *Pattern* = "ana" occurs at the beginning of the suffixes "anamabananas$", "ananas$", and "anas$" of *Text* = "panamabananas$"; these suffixes occur in three consecutive rows and correspond to the starting positions 1, 7, and 9 in *Text*.

The question is how to find these starting positions for an arbitrary string *Pattern* without needing to store the sorted suffixes of *Text*. The following algorithm, called PATTERNMATCHINGWITHSUFFIXARRAY, identifies the first and last index of the suffix array corresponding to suffixes beginning with *Pattern* (these indices are denoted *first* and *last*, respectively).

PATTERNMATCHINGWITHSUFFIXARRAY offers a variation of a general search technique called **binary search** that finds a data point in a sorted collection of data by iteratively dividing the data in half and determining the half in which the data point lies (see **DETOUR: Binary Search** for more details). The pseudocode below uses the notation $String_1 < String_2$ to indicate that $String_1$ is lexicographically smaller than $String_2$. We also say that *Pattern* matches a suffix of *Text* if the first $|Pattern|$ symbols of this suffix match *Pattern*. Note that all suffixes of *Text* are different from each other and from *Pattern* (which doesn't contain the "$" sign). As a result, the first while loop in the pseudocode below sets $minIndex = maxIndex + 1$ such that *Pattern* is "between" the

suffixes corresponding to SUFFIXARRAY(*maxIndex*) and SUFFIXARRAY(*minIndex*). That is, the following two properties hold:

- the suffix of *Text* starting at position SUFFIXARRAY(*maxIndex*) < *Pattern*;
- the suffix of *Text* starting at position SUFFIXARRAY(*minIndex*) > *Pattern*.

```
PATTERNMATCHINGWITHSUFFIXARRAY(Text, Pattern, SUFFIXARRAY)
    minIndex ← 0
    maxIndex ← |Text| − 1
    while minIndex ≤ maxIndex
        midIndex ← ⌊(minIndex + maxIndex)/2⌋
        if Pattern > suffix of Text starting at position SUFFIXARRAY(midIndex)
            minIndex ← midIndex + 1
        else
            maxIndex ← midIndex − 1
    if Pattern matches suffix of Text starting at position SUFFIXARRAY(midIndex)
        first ← minIndex
    else
        return Pattern does not appear in Text
    minIndex ← first
    maxIndex ← |Text| − 1
    while minIndex ≤ maxIndex
        midIndex ← ⌊(minIndex + maxIndex)/2⌋
        if Pattern matches suffix of Text starting at position SUFFIXARRAY(midIndex)
            minIndex ← midIndex + 1
        else
            maxIndex ← midIndex − 1
    last ← maxIndex
    return (first, last)
```

The Burrows-Wheeler Transform

Genome compression

Suffix arrays have greatly reduced the memory required for efficient text searches, and until the start of this century, they represented the state of the art in pattern matching.

Can we be so ambitious as to look for a data structure that would encode *Text* using memory approximately equal to the length of *Text* while still enabling fast pattern matching?

To answer this question, we will digress to consider the seemingly unrelated topic of **text compression**. In one simple compression technique called **run-length encoding**, we replace a **run** of k consecutive occurrences of symbol s with only two symbols: k, followed by s. For example, run-length encoding would compress the string `TTTTTGGGAAAACCCCCCA` into `5T3G4A6C1A`.

Run-length encoding works well for strings having lots of long runs, but real genomes do not have many runs. What they do have, as we saw in Chapter 3, are *repeats*. It would therefore be nice if we could first manipulate the genome to convert repeats into runs and then apply run-length encoding to the resulting string.

A naive way of creating runs in a string is to reorder the string's symbols lexicographically. For example, `TACGTAACGATACGAT` would become `AAAAACCCGGGTTTT`, which we could then compress into `5A3C3G4T`. This method would represent a 3 GB human genome file using just four numbers.

STOP and Think: What is wrong with applying this compression method to genomes?

STOP

Ordering a string's symbols lexicographically is not suitable for compression because many different strings will get compressed into the *same* string. For example, the DNA strings `GCATCATGCAT` and `ACTGACTACTG` — as well as any string with the same nucleotide counts — get reordered into `AAACCCGGTTT`. As a result, we cannot **decompress** the compressed string, i.e., invert the compression operation to produce the original string.

Constructing the Burrows-Wheeler transform

Let's consider a different method of converting the repeats of a string into runs that was proposed by Michael Burrows and David Wheeler in 1994. First, form all possible **cyclic rotations** of *Text*; a cyclic rotation is defined by chopping off a suffix from the end of *Text* and appending this suffix to the beginning of *Text*. Next — similarly to suffix arrays — order all the cyclic rotations of *Text* lexicographically to form a $|Text| \times |Text|$ matrix of symbols that we call the **Burrows-Wheeler matrix** and denote by M(*Text*) (Figure 9.8).

Notice that the first column of M(*Text*) contains the symbols of *Text* ordered lexicographically, which is just the naive rearrangement of *Text* that we already described.

Cyclic Rotations	*M*("panamabananas$")
panamabananas$	$ p a n a m a b a n a n a **s**
$panamabananas	a b a n a n a s $ p a n a **m**
s$panamabanana	a m a b a n a n a s $ p a **n**
as$panamabanan	a n a m a b a n a n a s $ **p**
nas$panamabana	a n a n a s $ p a n a m a **b**
anas$panamaban	a n a s $ p a n a m a b a **n**
nanas$panamaba	a s $ p a n a m a b a n a **n**
ananas$panamab	b a n a n a s $ p a n a m **a**
bananas$panama	m a b a n a n a s $ p a n **a**
abananas$panam	n a m a b a n a n a s $ p **a**
mabananas$pana	n a n a s $ p a n a m a b **a**
amabananas$pan	n a s $ p a n a m a b a n **a**
namabananas$pa	p a n a m a b a n a n a s **$**
anamabananas$p	s $ p a n a m a b a n a n **a**

FIGURE 9.8 All cyclic rotations of "panamabananas$" (left) and the Burrows-Wheeler matrix M("panamabananas$") of all lexicographically ordered cyclic rotations (right). BWT("panamabananas$") is the last column of M("panamabananas$"): "**smnpbnnaaaaa$a**".

In turn, the second column of M(*Text*) contains the second symbols of all cyclic rotations of *Text*, and so it too represents a (different) rearrangement of symbols from *Text*. The same reasoning applies to show that any column of M(*Text*) is some rearrangement of the symbols of *Text*. We are interested in the last column of M(*Text*), called the **Burrows-Wheeler transform** of *Text*, or BWT(*Text*), which is shown in red in Figure 9.8.

9I

Burrows-Wheeler Transform Construction Problem:
Construct the Burrows-Wheeler transform of a string.

Input: A string *Text*.
Output: BWT(*Text*).

STOP and Think: Figure 9.8 suggests a simple (but inefficient) algorithm for computing BWT(*Text*) based on constructing M(*Text*). Can you construct BWT(*Text*) using less memory given *Text* and SUFFIXARRAY(*Text*)?

It may be puzzling why we are so interested in the last column of M(*Text*). Why not the seventh column, or the penultimate column? Bear with us and you will see that the last column of this matrix has a unique property: you can reconstruct *Text* from BWT(*Text*). It turned out that no other column of M(*Text*) has this property.

From repeats to runs

If we re-examine the Burrows-Wheeler transform in Figure 9.8, we immediately notice that it has created the run "aaaaa" in BWT("panamabananas$") = "smnpbnnaaaaa$a".

STOP and Think: Why do you think that the Burrows-Wheeler Transform produced this run?

Imagine that we take the Burrows-Wheeler transform of Watson and Crick's 1953 paper on the double helix structure of DNA. The word "and" is repeated often in English, which means that when we form all possible cyclic rotations of the Watson & Crick paper, we will witness a large number of rotations beginning with "and..." In turn, we will observe many rotations that begin with "nd..." and end with "...a". When all the cyclic rotations of *Text* are sorted lexicographically to form M(*Text*), all rows that begin with "nd..." and end with "...a" will tend to clump together. As illustrated in Figure 9.9, this clumping produces runs of "a" in the final column of M(*Text*), which we know is BWT(*Text*).

The substring "ana" in "panamabananas$" plays the role of "and" in Watson and Crick's paper and explains three of the five occurrences of "a" in the repeat "aaaaa" in BWT("panamabananas$") = "smnpbnnaaaaa$a". When the Burrows-Wheeler transform is applied to a genome, it converts the genome's many repeats into runs. As we already suggested, after applying the Burrows-Wheeler transform, we can apply an additional compression method such as run-length encoding in order to further reduce the memory.

Exercise Break: There is only one run of length at least 10 in the *E. coli* genome. How many runs of length at least 10 do you find after applying the Burrows-Wheeler transform to the *E. coli* genome?

```
nd Corey (1).  They kindly made their manuscript availa ...... a
nd criticism, especially on interatomic distances.  We  ...... a
nd cytosine.  The sequence of bases on a single chain d ...... a
nd experimentally (3,4) that the ratio of the amounts o ...... u
nd for this reason we shall not comment on it.  We wish ...... a
nd guanine (purine) with cytosine (pyrimidine).  In oth ...... a
nd ideas of Dr.  M. H. F. Wilkins, Dr.  R. E. Franklin  ...... a
nd its water content is rather high.  At lower water co ...... a
nd pyrimidine bases.  The planes of the bases are perpe ...... a
nd stereochemical arguments.  It has not escaped our no ...... a
nd that only specific pairs of bases can bond together  ...... u
nd the atoms near it is close to Furberg's 'standard co ...... a
nd the bases on the inside, linked together by hydrogen ...... a
nd the bases on the outside.  In our opinion, this stru ...... a
nd the other a pyrimidine for bonding to occur.  The hy ...... a
nd the phosphates on the outside.  The configuration of ...... a
nd the ration of guanine to cytosine, are always very c ...... a
nd the same axis (see diagram).  We have made the usual ...... u
nd their co-workers at King's College, London.  One of  ...... a
```

FIGURE 9.9 A few consecutive rows selected from M(*Text*), where *Text* is Watson and Crick's 1953 paper on the double helix. Rows beginning with "nd..." often end with "...a" because of the common occurrence of the word "and" in English, which causes runs of "a" in BWT(*Text*). Each row in this figure contains a long string, indicated by "...", that does not fit on the page. Instances of "u" in the last column correspond to the words "round" and "found".

Inverting the Burrows-Wheeler Transform

A first attempt at inverting the Burrows-Wheeler transform

Before we get ahead of ourselves, remember that compressing a genome does not count for much if we cannot decompress it. In particular, if there exist a pair of genomes that the Burrows-Wheeler transform compresses into the same string, then we will not be able to decompress this string. But it turns out that the Burrows-Wheeler transform is reversible!

STOP and Think: Can you find the (unique) string whose Burrows-Wheeler transform is "enwvpeoseu$llt"? It could be "newtloveslupe$", "elevenplustwo$", "unwellpesovet$", or something else entirely.

Consider the toy example BWT(*Text*) = "ard$rcaaaabb". First, recall that the first column of M(*Text*) is the lexicographic rearrangement of symbols in BWT(*Text*), i.e., "$aaaaabbcdrr". For convenience, we will use the terms *FirstColumn* and *LastColumn* (i.e., BWT(*Text*)) when referring to the first and last columns of M(*Text*), respectively.

We know that the first row of M(*Text*) is the cyclic rotation of *Text* beginning with "$", which occurs at the end of *Text*. Thus, if we determine the first row of M(*Text*), then we can move the "$" to the end of this row and reproduce *Text*. But how do we determine the remaining symbols in this first row, if all we know is *FirstColumn* and *LastColumn*?

```
$ ? ? ? ? ? ? ? ? ? ? a
a ? ? ? ? ? ? ? ? ? ? r
a ? ? ? ? ? ? ? ? ? ? d
a ? ? ? ? ? ? ? ? ? ? $
a ? ? ? ? ? ? ? ? ? ? r
a ? ? ? ? ? ? ? ? ? ? c
b ? ? ? ? ? ? ? ? ? ? a
b ? ? ? ? ? ? ? ? ? ? a
c ? ? ? ? ? ? ? ? ? ? a
d ? ? ? ? ? ? ? ? ? ? a
r ? ? ? ? ? ? ? ? ? ? b
r ? ? ? ? ? ? ? ? ? ? b
```

STOP and Think: Using the first and last columns of the Burrows-Wheeler matrix shown above, can you find the first symbol of *Text*?

Note that the first symbol in *Text* must follow "$" in *any* cyclic rotation of *Text*. Because "$" occurs as the fourth symbol of *LastColumn* = "ard$rcaaaabb", we know that if we walk one symbol to the right from the end of the fourth row of M(*Text*), then we will "wrap around" and arrive at the fourth symbol of *FirstColumn*, which is "**a**" in "$aa**a**aabbcdrr". Therefore, this "**a**" belongs in the first position of *Text*:

```
$ a ? ? ? ? ? ? ? ? ? a
a ? ? ? ? ? ? ? ? ? ? r
a ? ? ? ? ? ? ? ? ? ? d
a ? ? ? ? ? ? ? ? ? ? $
a ? ? ? ? ? ? ? ? ? ? r
a ? ? ? ? ? ? ? ? ? ? c
b ? ? ? ? ? ? ? ? ? ? a
b ? ? ? ? ? ? ? ? ? ? a
c ? ? ? ? ? ? ? ? ? ? a
d ? ? ? ? ? ? ? ? ? ? a
r ? ? ? ? ? ? ? ? ? ? b
r ? ? ? ? ? ? ? ? ? ? b
```

STOP and Think: Which symbol is hiding in the second position of *Text*?

Following the same logic of "wrapping around", the next symbol of *Text* should be the first symbol in a row of M(*Text*) that ends in "a". The only trouble is that five rows end in "a", and we don't know which of them is the correct one! If we guess that this "a" is the seventh symbol of "ard$rcaaaabb", then we obtain "**b**" in the second position of *Text* (Figure 9.10 (left)). On the other hand, if we guess that this "a" is the ninth symbol of "ard$rcaaaabb", then we obtain "**c**" in the second position of *Text* (Figure 9.10 (middle)). Finally, if we guess that this "a" is the tenth symbol of "ard$rcaaaabb", then we obtain "**d**" in the second position of *Text* (Figure 9.10 (right)).

```
$ a b ? ? ? ? ? ? ? ? a      $ a c ? ? ? ? ? ? ? ? a      $ a d ? ? ? ? ? ? ? ? a
a ? ? ? ? ? ? ? ? ? ? r      a ? ? ? ? ? ? ? ? ? ? r      a ? ? ? ? ? ? ? ? ? ? r
a ? ? ? ? ? ? ? ? ? ? d      a ? ? ? ? ? ? ? ? ? ? d      a ? ? ? ? ? ? ? ? ? ? d
a ? ? ? ? ? ? ? ? ? ? $      a ? ? ? ? ? ? ? ? ? ? $      a ? ? ? ? ? ? ? ? ? ? $
a ? ? ? ? ? ? ? ? ? ? r      a ? ? ? ? ? ? ? ? ? ? r      a ? ? ? ? ? ? ? ? ? ? r
a ? ? ? ? ? ? ? ? ? ? c      a ? ? ? ? ? ? ? ? ? ? c      a ? ? ? ? ? ? ? ? ? ? c
b ? ? ? ? ? ? ? ? ? ? a      b ? ? ? ? ? ? ? ? ? ? a      b ? ? ? ? ? ? ? ? ? ? a
b ? ? ? ? ? ? ? ? ? ? a      b ? ? ? ? ? ? ? ? ? ? a      b ? ? ? ? ? ? ? ? ? ? a
c ? ? ? ? ? ? ? ? ? ? a      c ? ? ? ? ? ? ? ? ? ? a      c ? ? ? ? ? ? ? ? ? ? a
d ? ? ? ? ? ? ? ? ? ? a      d ? ? ? ? ? ? ? ? ? ? a      d ? ? ? ? ? ? ? ? ? ? a
r ? ? ? ? ? ? ? ? ? ? b      r ? ? ? ? ? ? ? ? ? ? b      r ? ? ? ? ? ? ? ? ? ? b
r ? ? ? ? ? ? ? ? ? ? b      r ? ? ? ? ? ? ? ? ? ? b      r ? ? ? ? ? ? ? ? ? ? b
```

FIGURE 9.10 The three possibilities ("**b**", "**c**", or "**d**") for the third element of the first row of M(*Text*) when BWT(*Text*) is "ard$rcaaaabb". One of these possibilities must correspond to the second symbol of *Text*.

STOP and Think: How would you choose among "**b**", "**c**", and "**d**" for the second symbol of *Text*?

The First-Last Property

To determine the remaining symbols of *Text*, we need to use a subtle property of M(*Text*) that may seem completely unrelated to inverting the Burrows-Wheeler transform. Below, we have indexed the occurrences of each symbol in *FirstColumn* with subscripts according to their order of appearance in this column. When *Text* = "panamabananas$", six instances of "a" appear in *FirstColumn*.

```
$  p a n a m a b a n a n a s
a1 b a n a n a s $ p a n a m
a2 m a b a n a n a s $ p a n
a3 n a m a b a n a n a s $ p
a4 n a n a s $ p a n a m a b
a5 n a s $ p a n a m a b a n
a6 s $ p a n a m a b a n a n
b  a n a n a s $ p a n a m a
m  a b a n a n a s $ p a n a
n  a m a b a n a n a s $ p a
n  a n a s $ p a n a m a b a
n  a s $ p a n a m a b a n a
p  a n a m a b a n a n a s $
s  $ p a n a m a b a n a n a
```

Consider "a_1" in *FirstColumn*, which occurs at the beginning of the cyclic rotation "$a_1$bananas$panam". Cyclically rotating this string yields "panama$_1$bananas$". Thus, "$a_1$" in *FirstColumn* is actually the third occurrence of "a" in "panamabananas$". We can now identify the positions of the other five instances of "a" in "panamabananas$":

$$pa_3na_2ma_1ba_4na_5na_6s\$$$

Exercise Break: Where are the three instances of "n" from *FirstColumn* (i.e., "n_1", "n_2", and "n_3") located in "panamabananas$"?

To locate "a_1" in *LastColumn*, we need to cyclically rotate the second row of the matrix M("panamabananas$"), transforming "$a_1$bananas$panam" into "bananas$panam$a_1$", corresponding to the eighth row of the matrix:

```
$  p a n a m a b a n a n a s
a1 b a n a n a s $ p a n a m
a2 m a b a n a n a s $ p a n
a3 n a m a b a n a n a s $ p
a4 n a n a s $ p a n a m a b
a5 n a s $ p a n a m a b a n
a6 s $ p a n a m a b a n a n
b  a n a n a s $ p a n a m a1
m  a b a n a n a s $ p a n a
n  a m a b a n a n a s $ p a
n  a n a s $ p a n a m a b a
n  a s $ p a n a m a b a n a
p  a n a m a b a n a n a s $
s  $ p a n a m a b a n a n a
```

Exercise Break: Where are the other five instances of "a" located in *LastColumn*?

You hopefully saw that *LastColumn* can be recorded as "smnpbnn$a_1a_2a_3a_4a_5$\$$a_6$", as shown in Figure 9.11. Note that the six instances of "a" appear in exactly the same order in *FirstColumn* and *LastColumn*. This observation is not a fluke. On the contrary, it is a principle that holds for any string *Text* and any symbol that we choose.

First-Last Property: *The k-th occurrence of a symbol in FirstColumn and the k-th occurrence of this symbol in LastColumn correspond to the same position of this symbol in Text.*

```
$  p a n a m a b a n a n a s
a1 b a n a n a s $ p a n a m
a2 m a b a n a n a s $ p a n
a3 n a m a b a n a n a s $ p
a4 n a n a s $ p a n a m a b
a5 n a s $ p a n a m a b a n
a6 s $ p a n a m a b a n a n
b  a n a n a s $ p a n a m a1
m  a b a n a n a s $ p a n a2
n  a m a b a n a n a s $ p a3
n  a n a s $ p a n a m a b a4
n  a s $ p a n a m a b a n a5
p  a n a m a b a n a n a s $
s  $ p a n a m a b a n a n a6
```

FIGURE 9.11 The six occurrences of "a" occur in the same order in *FirstColumn* as they do in *LastColumn*.

To see why the First-Last Property is true, consider the rows of M("panamabananas$") beginning with "a":

```
a1 b a n a n a s $ p a n a m
a2 m a b a n a n a s $ p a n
a3 n a m a b a n a n a s $ p
a4 n a n a s $ p a n a m a b
a5 n a s $ p a n a m a b a n
a6 s $ p a n a m a b a n a n
```

These rows are already ordered lexicographically, so if we chop off the "a" from the beginning of each row, then the remaining strings should still be ordered lexicographically:

```
b a n a n a s $ p a n a m
m a b a n a n a s $ p a n
n a m a b a n a n a s $ p
n a n a s $ p a n a m a b
n a s $ p a n a m a b a n
s $ p a n a m a b a n a n
```

Adding "a" back to the end of each row should not change the lexicographic ordering of these rows:

```
b a n a n a s $ p a n a m a1
m a b a n a n a s $ p a n a2
n a m a b a n a n a s $ p a3
n a n a s $ p a n a m a b a4
n a s $ p a n a m a b a n a5
s $ p a n a m a b a n a n a6
```

But these are just the rows of M("panamabananas$") containing "a" in *LastColumn*! As a result, the k-th occurrence of "a" in *FirstColumn* corresponds to the k-th occurrence of "a" in *LastColumn*. This argument generalizes for any *symbol* and any string *Text*, which establishes the First-Last property.

Using the First-Last property to invert the Burrows-Wheeler transform

The First-Last Property is interesting, but how can we use it to invert BWT(*Text*) = "ard$rcaaaabb"? Recalling Figure 9.10, let's return to where we were in our attempt to reconstruct the first row of M(*Text*) and index the occurrences of each symbol in *FirstColumn* and *LastColumn*:

```
$1 a ? ? ? ? ? ? ? ? ? a1
a1 ? ? ? ? ? ? ? ? ? ? r1
a2 ? ? ? ? ? ? ? ? ? ? d1
a3 ? ? ? ? ? ? ? ? ? ? $1
a4 ? ? ? ? ? ? ? ? ? ? r2
a5 ? ? ? ? ? ? ? ? ? ? c1
b1 ? ? ? ? ? ? ? ? ? ? a2
b2 ? ? ? ? ? ? ? ? ? ? a3
c1 ? ? ? ? ? ? ? ? ? ? a4
d1 ? ? ? ? ? ? ? ? ? ? a5
r1 ? ? ? ? ? ? ? ? ? ? b1
r2 ? ? ? ? ? ? ? ? ? ? b2
```

The First-Last Property reveals where "$\mathbf{a_3}$" is hiding in *LastColumn*:

```
$1 a ? ? ? ? ? ? ? ? ? a1
a1 ? ? ? ? ? ? ? ? ? ? r1
a2 ? ? ? ? ? ? ? ? ? ? d1
a3 ? ? ? ? ? ? ? ? ? ? $1
a4 ? ? ? ? ? ? ? ? ? ? r2
a5 ? ? ? ? ? ? ? ? ? ? c1
b1 ? ? ? ? ? ? ? ? ? ? a2
b2 ? ? ? ? ? ? ? ? ? ? a3
c1 ? ? ? ? ? ? ? ? ? ? a4
d1 ? ? ? ? ? ? ? ? ? ? a5
r1 ? ? ? ? ? ? ? ? ? ? b1
r2 ? ? ? ? ? ? ? ? ? ? b2
```

Since we know that "a_3" is located at the end of the eighth row, we can wrap around this row to determine that "b_2" follows "a_3" in *Text*. Thus, the second symbol of *Text* is "**b**", which we can now add to the first row of M(*Text*):

```
$1 a b ? ? ? ? ? ? ? ? a1
a1 ? ? ? ? ? ? ? ? ? ? r1
a2 ? ? ? ? ? ? ? ? ? ? d1
a3 ? ? ? ? ? ? ? ? ? ? $1
a4 ? ? ? ? ? ? ? ? ? ? r2
a5 ? ? ? ? ? ? ? ? ? ? c1
b1 ? ? ? ? ? ? ? ? ? ? a2
b2 ? ? ? ? ? ? ? ? ? ? a3
c1 ? ? ? ? ? ? ? ? ? ? a4
d1 ? ? ? ? ? ? ? ? ? ? a5
r1 ? ? ? ? ? ? ? ? ? ? b1
r2 ? ? ? ? ? ? ? ? ? ? b2
```

In Figure 9.12, we illustrate repeated applications of the First-Last Property to reconstruct more and more symbols from *Text*. Presto — the string that we have been trying to reconstruct is "abracadabra$".

Exercise Break: Reconstruct the string whose Burrows-Wheeler transform is "enwvpeoseu$llt".

STOP and Think: Can *any* string (having a single "$" symbol) be inverted using the inverse Burrows-Wheeler transform?

You are now ready to implement the inverse of the Burrows-Wheeler transform.

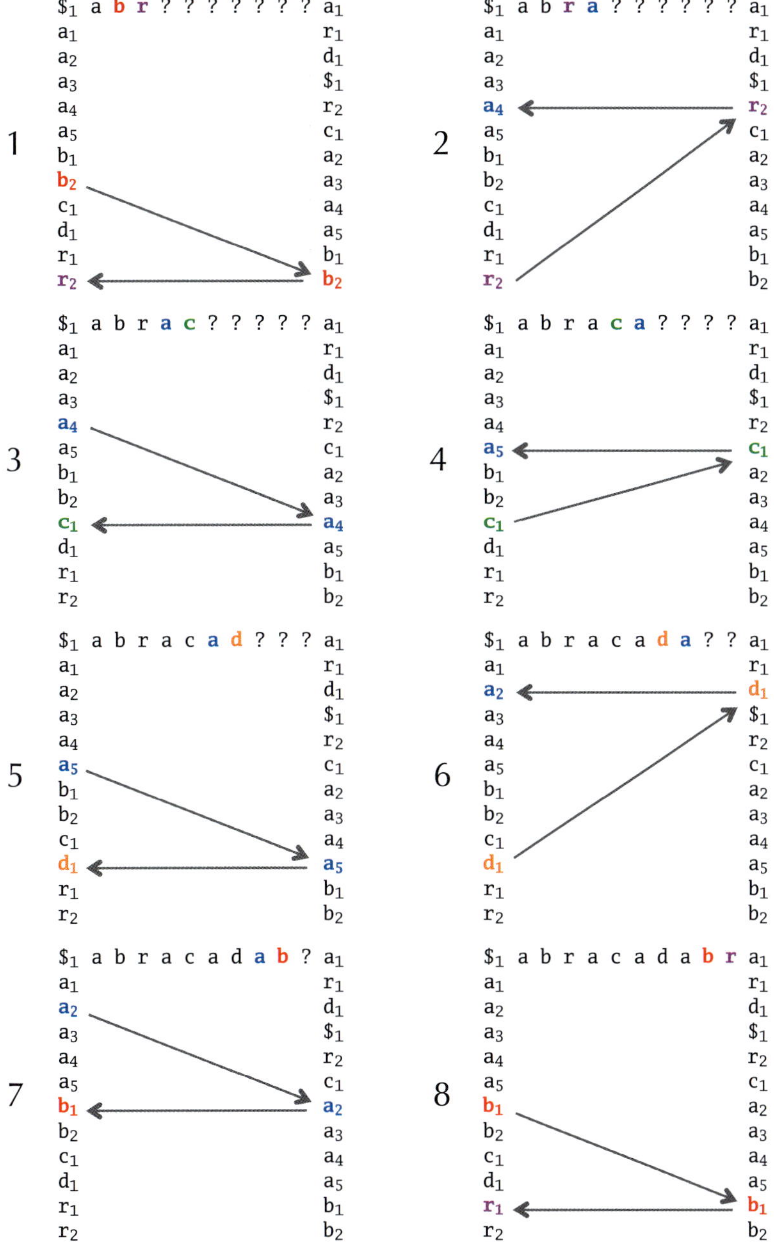

FIGURE 9.12 Repeated applications of the First-Last Property reconstruct the string "abracadabra$" from its Burrows-Wheeler transform "ard$rcaaaabb".

Inverse Burrows-Wheeler Transform Problem:
Reconstruct a string from its Burrows-Wheeler transform.

> **Input**: A string *Transform* (with a single "$" symbol).
> **Output**: The string *Text* such that BWT(*Text*) = *Transform*.

Pattern Matching with the Burrows-Wheeler Transform

A first attempt at Burrows-Wheeler pattern matching

The Burrows-Wheeler transform may be fascinating, but how could it possibly help us decrease the memory required for pattern matching? The idea motivating a Burrows-Wheeler-based approach to pattern matching relies on the observation that each row of M(*Text*) begins with a different suffix of *Text*. Since these suffixes are already ordered lexicographically, as we already noted when pattern matching with the suffix array, any matches of *Pattern* in *Text* will appear at the beginning of consecutive rows of M(*Text*), as shown in Figure 9.13.

M(*Text*)	SUFFIXARRAY(*Text*)
$ p a n a m a b a n a n a s	13
a b a n a n a s $ p a n a m	5
a m a b a n a n a s $ p a n	3
a n a m a b a n a n a s $ p	**1**
a n a n a s $ p a n a m a b	**7**
a n a s $ p a n a m a b a n	**9**
a s $ p a n a m a b a n a n	11
b a n a n a s $ p a n a m a	6
m a b a n a n a s $ p a n a	4
n a m a b a n a n a s $ p a	2
n a n a s $ p a n a m a b a	8
n a s $ p a n a m a b a n a	10
p a n a m a b a n a n a s $	0
s $ p a n a m a b a n a n a	12

FIGURE 9.13 (Left) Because the rows of M(*Text*) are ordered lexicographically, suffixes beginning with the same string ("**ana**") appear in consecutive rows of the matrix. (Right) The suffix array records the starting position of each suffix in *Text* and immediately tells us the locations of "**ana**".

We now have the outline of a method to match *Pattern* to *Text*. Construct M(*Text*), and then identify rows beginning with the first symbol of *Pattern*. Among these rows, determine which ones have a second element matching the second symbol of *Pattern*. Continue this process until we find which rows of M(*Text*) begin with *Pattern*.

STOP and Think: What is wrong with this approach?

Moving backward through a pattern

The issue with this proposed method for pattern matching is that we cannot afford storing the entire matrix M(*Text*), which has $|Text|^2$ entries. In an effort to reduce memory requirements, let's forbid ourselves from accessing any information in M(*Text*) other than *FirstColumn* and *LastColumn*. Using these two columns, we will try to match *Pattern* to *Text* by moving *backward* through *Pattern*. For example, if we want to match *Pattern* = "ana" to *Text* = "panamabananas$", then we will first identify rows of M(*Text*) beginning with "**a**", the last letter of "**ana**":

```
$1 p a n a m a b a n a n a s1
a1 b a n a n a s $ p a n a m1
a2 m a b a n a n a s $ p a n1
a3 n a m a b a n a n a s $ p1
a4 n a n a s $ p a n a m a b1
a5 n a s $ p a n a m a b a n2
a6 s $ p a n a m a b a n a n3
b1 a n a n a s $ p a n a m a1
m1 a b a n a n a s $ p a n a2
n1 a m a b a n a n a s $ p a3
n2 a n a s $ p a n a m a b a4
n3 a s $ p a n a m a b a n a5
p1 a n a m a b a n a n a s $1
s1 $ p a n a m a b a n a n a6
```

As we are moving backward through "**ana**", we will next look for rows of M(*Text*) beginning with "**na**". To do this without knowing the entire matrix M(*Text*), we again use the fact that a symbol in *LastColumn* must precede the symbol of *Text* found in the same row in *FirstColumn*. Thus, we only need to identify those rows of M(*Text*) beginning with "**a**" and ending with "**n**":

```
$1 p a n a m a b a n a n a s1
a1 b a n a n a s $ p a n a m1
a2 m a b a n a n a s $ p a n1
a3 n a m a b a n a n a s $ p1
a4 n a n a s $ p a n a m a b1
a5 n a s $ p a n a m a b a n2
a6 s $ p a n a m a b a n a n3
b1 a n a n a s $ p a n a m a1
m1 a b a n a n a s $ p a n a2
n1 a m a b a n a n a s $ p a3
n2 a n a s $ p a n a m a b a4
n3 a s $ p a n a m a b a n a5
p1 a n a m a b a n a n a s $1
s1 $ p a n a m a b a n a n a6
```

The First-Last Property tells us where to find the three highlighted "**n**" in *FirstColumn*, as shown below. All three rows end with "**a**", yielding three total occurrences of "**ana**" in *Text*.

```
$1 p a n a m a b a n a n a s1
a1 b a n a n a s $ p a n a m1
a2 m a b a n a n a s $ p a n1
a3 n a m a b a n a n a s $ p1
a4 n a n a s $ p a n a m a b1
a5 n a s $ p a n a m a b a n2
a6 s $ p a n a m a b a n a n3
b1 a n a n a s $ p a n a m a1
m1 a b a n a n a s $ p a n a2
n1 a m a b a n a n a s $ p a3
n2 a n a s $ p a n a m a b a4
n3 a s $ p a n a m a b a n a5
p1 a n a m a b a n a n a s $1
s1 $ p a n a m a b a n a n a6
```

The highlighted occurrences of "**a**" in *LastColumn* correspond to the third, fourth, and fifth occurrences of "**a**" in this column, and the First-Last Property tells us that they should correspond to the third, fourth, and fifth occurrences of "**a**" in *FirstColumn* as well, which identifies the three matches of "**ana**":

```
$1 p a n a m a b a n a n a s1
a1 b a n a n a s $ p a n a m1
a2 m a b a n a n a s $ p a n1
a3 n a m a b a n a n a s $ p1
a4 n a n a s $ p a n a m a b1
a5 n a s $ p a n a m a b a n2
a6 s $ p a n a m a b a n a n3
b1 a n a n a s $ p a n a m a1
m1 a b a n a n a s $ p a n a2
n1 a m a b a n a n a s $ p a3
n2 a n a s $ p a n a m a b a4
n3 a s $ p a n a m a b a n a5
p1 a n a m a b a n a n a s $1
s1 $ p a n a m a b a n a n a6
```

Exercise Break: Match *Pattern* = "banana" to *Text* = "panamabananas$" by walking backward through *Pattern* using the Burrows-Wheeler transform of *Text*.

The Last-to-First mapping

We now know how to use BWT(*Text*) to find all matches of *Pattern* in *Text* by walking backward through *Pattern*. However, every time we walk backward, we need to keep track of the rows of M(*Text*) where the matches of a suffix of *Pattern* are hiding. Fortunately, we know that at each step, the rows of M(*Text*) that match a suffix of *Pattern* clump together in consecutive rows of M(*Text*). This means that the collection of all matching rows is revealed by only two pointers, *top* and *bottom*: *top* holds the index of the first row of M(*Text*) that matches the current suffix of *Pattern*, and *bottom* holds the index of the last row of M(*Text*) that matches this suffix. Figure 9.14 shows the process of updating pointers; after walking backward through *Pattern* = "ana", we have that *top* = 3 and *bottom* = 5. After traversing *Pattern*, we can compute the total number of matches of *Pattern* in *Text* by calculating *bottom* − *top* + 1 (e.g., there are $5 - 3 + 1 = 3$ matches of "ana" in "panamabananas$").

Let's concentrate on how pointers are updated from one stage to the next. Consider the transition from the second to the third panel in Figure 9.14; how did we know to update the pointers (*top* = 1, *bottom* = 6) into (*top* = 9, *bottom* = 11)? We are looking for the first and last occurrence of "n" in the range of positions from *top* = 1 to *bottom* = 6 in *LastColumn*. The first occurrence of "n" in this range is "$\mathbf{n_1}$" (in position 2) and the last is "$\mathbf{n_3}$" (position 6).

In order to update the *top* and *bottom* pointers, we need to determine where "$\mathbf{n_1}$" and "$\mathbf{n_3}$" occur in *FirstColumn*. The **Last-to-First mapping**, denoted LASTTOFIRST(i),

answers the following question: given a symbol at position *i* in *LastColumn*, what is its position in *FirstColumn*?

For our ongoing example, LASTTOFIRST(2) = 9, since the symbol at position 2 of *LastColumn* ("n_1") occurs at position 9 in *FirstColumn*, as shown in Figure 9.15. Similarly, LASTTOFIRST(6) = 11, since the symbol at position 6 of *LastColumn* ("n_3") occurs at position 11 in *FirstColumn*. Therefore, with the help of the Last-to-First mapping, we can quickly update the pointers (*top* = 1, *bottom* = 6) into (*top* = 9, *bottom* = 11).

We are now ready to describe **BWMATCHING**, an algorithm that counts the total number of matches of *Pattern* in *Text*, where the only information that we are given is *FirstColumn* and *LastColumn* in addition to the Last-to-First mapping. The pointers *top* and *bottom* are updated by the green lines in the following pseudocode.

```
BWMATCHING(FirstColumn, LastColumn, Pattern, LASTTOFIRST)
    top ← 0
    bottom ← |LastColumn| − 1
    while top ≤ bottom
        if Pattern is nonempty
            symbol ← last letter in Pattern
            remove last letter from Pattern
            if positions from top to bottom in LastColumn contain symbol
                topIndex ← first position of symbol among positions from top to bottom
                           in LastColumn
                bottomIndex ← last position of symbol among positions from top to
                              bottom in LastColumn
                top ← LASTTOFIRST(topIndex)
                bottom ← LASTTOFIRST(bottomIndex)
            else
                return 0
        else
            return bottom − top + 1
```

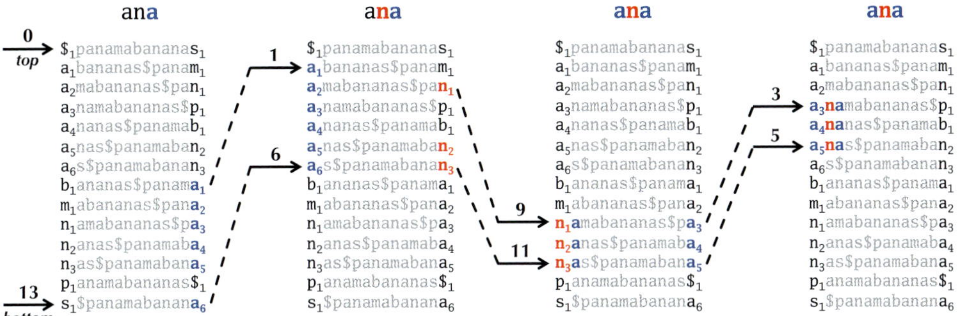

FIGURE 9.14 The pointers *top* and *bottom* hold the indices of the first and last rows of M(*Text*) matching the current suffix of *Pattern* = "ana". The above diagram shows how these pointers are updated when walking backwards through "ana" and looking for substring matches in "panamabananas$".

i	*FirstColumn*	*LastColumn*	LASTTOFIRST(i)	COUNT $	a	b	m	n	p	s
				$	a	b	m	n	p	s
0	$\$_1$	s_1	13	0	0	0	0	0	0	0
1	a_1	m_1	8	0	0	0	0	0	0	1
2	a_2	n_1	9	0	0	0	1	0	0	1
3	a_3	p_1	12	0	0	0	1	1	0	1
4	a_4	b_1	7	0	0	0	1	1	1	1
5	a_5	n_2	10	0	0	1	1	1	1	1
6	a_6	n_3	11	0	0	1	1	2	1	1
7	b_1	a_1	1	0	0	1	1	3	1	1
8	m_1	a_2	2	0	1	1	1	3	1	1
9	n_1	a_3	3	0	2	1	1	3	1	1
10	n_2	a_4	4	0	3	1	1	3	1	1
11	n_3	a_5	5	0	4	1	1	3	1	1
12	p_1	$\$_1$	0	0	5	1	1	3	1	1
13	s_1	a_6	6	1	5	1	1	3	1	1
				1	6	1	1	3	1	1

FIGURE 9.15 The Last-to-First mapping and count array for BWT("panamabananas$"). Precomputing the count array prevents time-consuming updates of the *top* and *bottom* pointers in **BWMATCHING**.

Speeding Up Burrows-Wheeler Pattern Matching

Substituting the Last-to-First mapping with count arrays

If you implemented **BWMATCHING** in the previous section, you probably found this algorithm to be slow. The reason for its sluggishness is that updating the pointers *top* and *bottom* is time-intensive, since it requires examining every symbol in *LastColumn* between *top* and *bottom* at each step. To improve **BWMATCHING**, we introduce a function $\text{COUNT}_{symbol}(i, LastColumn)$, which holds the number of occurrences of *symbol* in the first i positions of *LastColumn* (Figure 9.15). For example, $\text{COUNT}_{\text{"n"}}(10, \text{"smnpbnnaaaaa\$a"}) = 3$, and $\text{COUNT}_{\text{"a"}}(4, \text{"smnpbnnaaaaa\$a"}) = 0$. In Figure 9.15, we show arrays holding $\text{COUNT}_{symbol}(i, \text{"smnpbnnaaaaa\$a"})$ for every *symbol* occurring in "panamabananas$".

Exercise Break: Compute the arrays COUNT for BWT("abracadabra$").

We will now show how to update pointers without examining every symbol in *LastColumn* between *top* and *bottom* at each step. We say that the k-th occurrence of *symbol* in a column of a matrix has **rank** k in this column. For $Text$ = "panamabananas$", note that the first and last occurrences of *symbol* in the range of positions from *top* to *bottom* in *LastColumn* have respective ranks

$$\text{COUNT}_{symbol}(top, LastColumn) + 1$$

and

$$\text{COUNT}_{symbol}(bottom + 1, LastColumn).$$

As illustrated in Figure 9.15, when $top = 1$, $bottom = 6$, and $symbol$ = "n",

$$\text{COUNT}_{\text{"n"}}(top, LastColumn) + 1 = 1$$
$$\text{COUNT}_{\text{"n"}}(bottom + 1, LastColumn) = 3$$

The occurrences of "n" having ranks 1 and 3 are located at positions 2 and 6 of *LastColumn*, implying that we should update *top* to $\text{LASTTOFIRST}(2) = 9$ and *bottom* to $\text{LASTTOFIRST}(6) = 11$. Thus, the four green lines in the pseudocode for **BWMATCHING** can be rewritten as follows.

```
topIndex ← position of symbol with rank COUNT_symbol(top, LastColumn) + 1
               in LastColumn
bottomIndex ← position of symbol with rank COUNT_symbol(bottom + 1, LastColumn)
               in LastColumn
top ← LASTTOFIRST(topIndex)
bottom ← LASTTOFIRST(bottomIndex)
```

By eliminating the variables *topIndex* and *bottomIndex*, we can reduce these four lines of pseudocode to only two lines:

```
top ← LASTTOFIRST(position of symbol with rank COUNT_symbol(top, LastColumn) +
                   1 in LastColumn)
bottom ← LASTTOFIRST(position of symbol with rank COUNT_symbol(bottom + 1,
                   LastColumn) in LastColumn)
```

Note that these two lines of pseudocode merely compute the position of *symbol* with rank i in *FirstColumn* from its positions in *LastColumn*. This task can be compactly described without the Last-to-First mapping by the following two lines:

```
top ← position of symbol with rank COUNT_symbol(top, LastColumn) + 1 in
        FirstColumn
bottom ← position of symbol with rank COUNT_symbol(bottom + 1, LastColumn) in
        FirstColumn
```

For $top = 1$, $bottom = 6$, and $symbol =$ "n", the occurrences of "n" having ranks $\text{COUNT}_{\text{"n"}}(top, LastColumn) + 1 = 1$ and $\text{COUNT}_{\text{"n"}}(bottom + 1, LastColumn) = 3$ are located in positions 9 and 11 of *FirstColumn*, respectively. We therefore update $top = 9$ and $bottom = 11$ as before.

STOP and Think: Do we need to store all of *FirstColumn* in memory in order to execute the preceding two lines of pseudocode?

Getting rid of the first column of the Burrows-Wheeler matrix

BWMATCHING requires us to store *FirstColumn*. We can reduce the memory needed to store the information contained in *FirstColumn* by defining FIRSTOCCURRENCE(*symbol*) as the first position of *symbol* in *FirstColumn*. If *Text* = "panamabananas$", then *FirstColumn* is "$aaaaaabmnnnps", and the array holding all values of FIRSTOCCURRENCE is [0, 1, 7, 8, 9, 12, 13], as shown in Figure 9.16. The array FIRSTOCCURRENCE can be computed directly from *LastColumn*, and for DNA strings of any length, this array contains only five elements. The previous two lines of code can now be rewritten as follows:

```
top ← FIRSTOCCURRENCE(symbol) + COUNT_symbol(top, LastColumn)
bottom ← FIRSTOCCURRENCE(symbol) + COUNT_symbol(bottom + 1, LastColumn) − 1
```

i	*FirstColumn*	FIRSTOCCURRENCE
0	$\$_1$	0
1	a_1	1
2	a_2	
3	a_3	
4	a_4	
5	a_5	
6	a_6	
7	b_1	7
8	m_1	8
9	n_1	9
10	n_2	
11	n_3	
12	p_1	12
13	s_1	13

FIGURE 9.16 The array FIRSTOCCURRENCE has just seven elements, which is equal to the number of distinct symbols in "panamabananas$".

When *top* = 1, *bottom* = 6, and *symbol* = "n", we have that

$$\text{FIRSTOCCURRENCE}(\text{"n"}) = 9$$
$$\text{COUNT}_{\text{"n"}}(top, LastColumn) = 0$$
$$\text{COUNT}_{\text{"n"}}(bottom + 1, LastColumn) = 3$$

Recalling Figure 9.14, this again implies that $(top = 1, bottom = 6)$ will be updated as

$$top = 9 + 0 = 9$$
$$bottom = 9 + 3 - 1 = 11$$

In the process of simplifying the green lines of pseudocode from **BWMATCHING**, we have also substituted *FirstColumn* by FIRSTOCCURRENCE and LASTTOFIRST by COUNT, resulting in a more efficient algorithm called **BETTERBWMATCHING**, shown below.

You may be wondering why we call this algorithm "better", since on the one hand, if you have to compute the COUNT arrays as you go, then you will not obtain a runtime speedup. On the other hand, if you have to precompute these arrays, then you will need to store them in memory, which is space-intensive. Hold onto this thought.

```
BETTERBWMATCHING(FIRSTOCCURRENCE, LastColumn, Pattern, COUNT)
    top ← 0
    bottom ← |LastColumn| − 1
    while top ≤ bottom
        if Pattern is nonempty
            symbol ← last letter in Pattern
            remove last letter from Pattern
            top ← FIRSTOCCURRENCE(symbol) + COUNT_symbol(top, LastColumn)
            bottom ← FIRSTOCCURRENCE(symbol) + COUNT_symbol(bottom + 1,
                        LastColumn) − 1
        else
            return bottom − top + 1
    return
```

Where are the Matched Patterns?

We hope that you have noticed a limitation of **BETTERBWMATCHING** — even though this algorithm counts the *number* of occurrences of *Pattern* in *Text*, it does not tell us *where* these occurrences are located in *Text*! To locate pattern matches identified by the algorithm, we can once again use the suffix array, as shown in Figure 9.13 (right). In this figure, the suffix array immediately finds the three matches of "ana" in "panamabananas$".

The suffix array makes our job easy, but recall that our original motivation for using the Burrows-Wheeler transform was to *reduce* the amount of memory used by the suffix array for pattern matching. If we add the suffix array to Burrows-Wheeler-based pattern matching, then we are right back where we started!

The memory-saving device that we will employ is inelegant but useful. We will

build a **partial suffix array** of *Text*, denoted $\text{SUFFIXARRAY}_K(Text)$, which only contains values that are multiples of some positive integer K (Figure 9.17). In real applications, partial suffix arrays are often constructed for $K = 100$, thus reducing memory usage by a factor of 100 compared to a full suffix array. See **Charging Station: Partial Suffix Array Construction** for more details.

STOP and Think: The partial suffix array reduces the memory by a factor of K but also slows down the search for a pattern match due to "walking backwards", as illustrated in Figure 9.17. What is the worst-case scenario for the number of steps that must be taken when walking backwards? How does it affect the (big-O) runtime of our pattern matching algorithm?

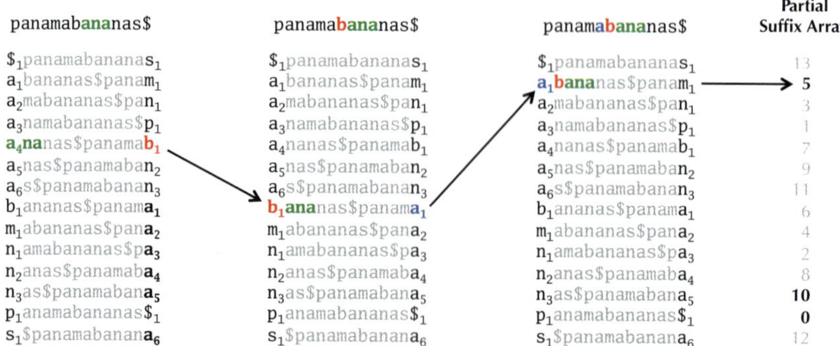

FIGURE 9.17 One of the matches of "ana" in "panamab**ana**nas$" is highlighted in the matrix on the left. By walking backward, we find that "**ana**" is preceded by "$\mathbf{b_1}$", which in turn is preceded by "$\mathbf{a_1}$". The partial suffix array above, generated for $K = 5$, indicates that "$\mathbf{a_1}$" occurs at position 5 of "panama**ba**nanas". Since it took us two steps to walk backward to "$\mathbf{a_1}$", we conclude that this occurrence of "ana" begins at position $5 + 2 = 7$.

Burrows and Wheeler Set Up Checkpoints

We will now discuss how to improve **BETTERBWMATCHING** by resolving the trade-off between precomputing the values of $\text{COUNT}_{symbol}(i, LastColumn)$ (requiring substantial memory) and computing these values as we go (requiring substantial runtime).

The balance that we strike is similar to the one used for the partial suffix array. Rather than storing $\text{COUNT}_{symbol}(i, LastColumn)$ for *all* positions i, we will only store the COUNT arrays when i is divisible by C, where C is a constant; these arrays are called **checkpoint arrays** (Figure 9.18). When C is large (C is typically equal to 100 in practice) and the alphabet is small (e.g., four nucleotides), checkpoint arrays require only a fraction of the memory used by BWT(*Text*).

i	*LastColumn*	COUNT						
		$	a	b	m	n	p	s
0	s_1	**0**	**0**	**0**	**0**	**0**	**0**	**0**
1	m_1	0	0	0	0	0	0	1
2	n_1	0	0	0	1	0	0	1
3	p_1	0	0	0	1	1	0	1
4	b_1	0	0	0	1	1	1	1
5	n_2	**0**	**0**	**1**	**1**	**1**	**1**	**1**
6	n_3	0	0	1	1	2	1	1
7	a_1	0	0	1	1	3	1	1
8	a_2	0	1	1	1	3	1	1
9	a_3	0	2	1	1	3	1	1
10	a_4	**0**	**3**	**1**	**1**	**3**	**1**	**1**
11	a_5	0	4	1	1	3	1	1
12	$\$_1$	0	5	1	1	3	1	1
13	a_6	1	5	1	1	3	1	1
		1	6	1	1	3	1	1

FIGURE 9.18 The COUNT checkpoint arrays for *Text* = "panamabananas$" and $C = 5$ are highlighted in bold. If we want to compute COUNT"a"(13, "smnpbnnaaaaa$a"), then the checkpoint array at position 10 tells us that there are 3 occurrences of "a" *before* position 10 of "smnpbnnaaaaa$a". We then check whether "a" is present at position **10** (yes), **11** (yes), and **12** (no) of *LastColumn* to conclude that COUNT"a"(13, "smnpbnnaaaaa$a") = **3** + **2** = 5.

What about runtime? Using checkpoint arrays, we can compute the *top* and *bottom* pointers in a constant number of steps (i.e., fewer than C). Since each string *Pattern* requires at most |*Pattern*| pointer updates, the modified **BETTERBWMATCHING** algorithm now requires $\mathcal{O}(|Patterns|)$ runtime, which is the same as using a trie or suffix array.

Furthermore, we now only have to store the following data in memory: BWT(*Text*), FIRSTOCCURRENCE, the partial suffix array, and the checkpoint arrays. Storing this data requires memory approximately equal to $1.5 \cdot |Text|$. Thus, we have finally knocked down the memory required for solving the Multiple Pattern Matching Problem for millions of sequencing reads into a workable range.

As the plummeting cost of DNA sequencing has made headlines, it is easy to fail to appreciate the progress that has been made in the computational side of read mapping. So, before continuing into approximate pattern matching, we would like to pause and reflect on just how far the algorithms for read mapping – and the hardware running them – have progressed. In 1975, the state-of-the-art Aho-Corasick algorithm was touted as requiring 15 minutes to map just 24 English words to a dictionary. Just a generation later, when Burrows-Wheeler based approaches were introduced into read mapping, the same 15 minutes was used to map almost *ten million* reads to a reference genome. Having seen how far we have come in the last four decades, we can only imagine where it will take us in the future.

Epilogue: Mismatch-Tolerant Read Mapping

Reducing approximate pattern matching to exact pattern matching

In this section, we will return to the goal of identifying SNPs in an individual genome when compared against the reference genome. To do so, we need to generalize the Approximate Pattern Matching Problem from Chapter 1 to the case of multiple patterns.

Multiple Approximate Pattern Matching Problem:
Find all approximate occurrences of a collection of patterns in a text.

> **Input**: A string *Text*, a collection of shorter strings *Patterns*, and an integer *d*.
> **Output**: All starting positions in *Text* where a string from *Patterns* appears as a substring with at most *d* mismatches.

We begin with the simple observation that if *Pattern* occurs in *Text* with a single mismatch, then we can divide *Pattern* into two halves, one of which occurs exactly in *Text*, as illustrated below.

```
Pattern        acttggct
   Text  ...ggcacactaggctcc...
```

Thus, we can find whether *Pattern* occurs with one mismatch in *Text* by dividing *Pattern* into two halves and then searching for exact matches of these shorter strings. If we find a match with one of these halves of *Pattern*, we then check if the entire string *Pattern* occurs with a single mutation.

This method can easily be generalized for approximate pattern matching with $d > 1$ mismatches; if *Pattern* approximately matches a substring of *Text* with at most d mismatches, then *Pattern* and *Text* must share at least one k-mer for a sufficiently large value of k. For example, if we are looking for a pattern of length 20 with at most $d = 3$ mismatches, then we can divide this pattern into four parts of length $20/(3+1) = 5$ and search for exact matches of these shorter substrings:

Pattern	`acttaggctcgggataatcc`
Text	`...actaagtctcgggataagcc...`

This observation is helpful because it reduces *approximate* pattern matching to the *exact* matching of shorter patterns, which allows us to use fast algorithms that are designed for exact pattern matching but are not applicable to approximate pattern matching, such as approaches based on suffix trees and suffix arrays.

STOP and Think: If *Pattern* has length 23 and appears in *Text* with 3 mismatches, can we conclude that *Pattern* shares a 6-mer with *Text*? Can we conclude that it shares a 5-mer with *Text*?

The question remains how to find the maximum value of k that guarantees that any *Pattern* of length n with d mismatches in *Text* will have a k-mer substring exactly matching *Text*.

Theorem. *If two strings of length n match with at most d mismatches, then they must share a k-mer of length $k = \lfloor n/(d+1) \rfloor$.*

Proof. Divide the first string into $d+1$ substrings, where the first d substrings have exactly k symbols and the final substring has at least k symbols. For the case $d = 3$, here is a subdivision of a 23-nucleotide string ($n = 23$) into $d + 1 = 4$ substrings, where the first 3 substrings have $k = \lfloor n/(d+1) \rfloor = \lfloor 23/4 \rfloor = 5$ symbols and the last substring has 8 symbols.

`acttaggctcgggataatccgga`

If you distribute d mismatches among the positions of the string, then the mismatches may affect at most d substrings, which leaves at least one substring (of length at least k) unchanged. This substring is shared by both strings. □

We now have the outline of an algorithm for matching a string *Pattern* of length n to *Text* with at most d mismatches. We first divide *Pattern* into $d + 1$ segments of length $k = \lfloor n/(d+1) \rfloor$, called **seeds**. After finding which seeds match *Text* exactly (**seed detection**), we attempt to extend seeds in both directions in order to verify whether *Pattern* occurs in *Text* with at most d mismatches (**seed extension**).

BLAST: Comparing a sequence against a database

Using shared k-mers to find similarities between biological sequences has some disadvantages. For example, two proteins may have similar functions but not share any k-mers, even for small values of k.

The **Basic Local Alignment Search Tool (BLAST)** is a heuristic that can find similarities between proteins, even if all amino acids in one protein have mutated compared to the other protein. BLAST is so fast that it is often used to query a protein against *all other* known proteins in protein databases. The paper introducing BLAST was released in 1990, and with over 40,000 citations, it has become one of the most cited scientific papers ever published.

To see how BLAST works, say we are given an integer k and strings $x = x_1 \dots x_n$ and $y = y_1 \dots y_m$ to compare (representing, say, two proteins). We define a **segment pair** of x and y as a pair formed by a k-mer from x and a k-mer from y. The score of the segment pair corresponding to the k-mers starting at position i in x and position j in y is

$$\sum_{t=0}^{k-1} \text{SCORE}(x_{i+t}, y_{j+t}),$$

where $\text{SCORE}(x_{i+t}, y_{j+t})$ is determined by a scoring matrix such as the PAM scoring matrix that we introduced in Chapter 5. A **locally maximal segment pair** is a segment pair whose score cannot be increased by extending or shortening both strings in the segment pair. BLAST attempts to find not the highest-scoring segment pair of x and y, but rather all locally maximal segment pairs in these strings with scores above some threshold.

For each k-mer in the query string, BLAST quickly finds the set of all k-mers that score above a given threshold against that k-mer, and combines these sets (for all k-mers in the query string) to obtain a list of potential seeds. If the score threshold is high, then the set of all resulting segment pairs formed between the query string and strings in the database is not too large. In this case, the database can be searched for exact occurrences of these high-scoring k-mers from this set, producing an initial set of seeds. This is an instance of the Multiple Pattern Matching Problem, which we have learned how to solve quickly.

Exercise Break: Given the PAM_{250} scoring matrix, only five 3-mers score higher than 23 against `CFC`: `CIC`, `CLC`, `CMC`, `CWC`, and `CYC`. How many 3-mers score higher than 20 against `CFC`?

After finding seeds, BLAST attempts to extend these seeds (allowing for insertions and deletions) in order to obtain locally maximal segment pairs.

Exercise Break: Given the PAM_{250} scoring matrix, an amino acid k-mer *Peptide*, and a threshold θ, develop an efficient algorithm for finding the exact number of k-mers scoring more than θ against *Peptide*.

Approximate pattern matching with the Burrows-Wheeler transform

To extend the Burrows-Wheeler approach to approximate pattern matching, we will not stop when we encounter a mismatch. On the contrary, we will proceed onward until we either find an approximate match or exceed the limit of d mismatches.

Figure 9.19 illustrates the search for "asa" in "panamabananas$" with at most 1 mismatch. Let's first proceed as in the case of exact pattern matching, threading "asa" backwards using the Burrows-Wheeler transform. After finding six occurrences of "a", we identify six inexact occurrences of "sa": "**p**a", "**m**a", "**b**a", and three occurrences of "**n**a". We note that these three strings have accumulated a mismatch and then continue with all six strings.

In the next step, five of the inexact occurrences of "sa" can be extended into inexact occurrences of "asa" with only a single mismatch: "a**m**a", "a**b**a", and three occurrences of "a**n**a". We fail to extend "**p**a", which we eliminate from consideration. Generally, we fail to extend a growing pattern either because we hit the end of *Text* (as illustrated by hitting "$" in the case of "**p**a") or because we exceed the maximum allowable number of mismatches.

In practice, this heuristic faces complications. We do not want to start allowing mismatched strings at the early stages of **BetterBWMatching**, or else we will have to consider too many frivolous candidate strings. We may therefore require that a suffix of *Pattern* of some threshold length matches *Text* exactly. Moreover, the method becomes time-intensive when using large values of d, as we must explore many inexact matches. Practical applications often limit the value of d to at most 3.

You should now be ready to design your own approach to solve the Multiple Approximate Pattern Matching Problem and use this solution to map real sequencing reads.

Challenge Problem: Given the Burrows-Wheeler transform and a partial suffix array of the bacterial genome *Mycoplasma pneumoniae* along with a collection of reads, find all reads occurring in the genome with at most one mismatch.

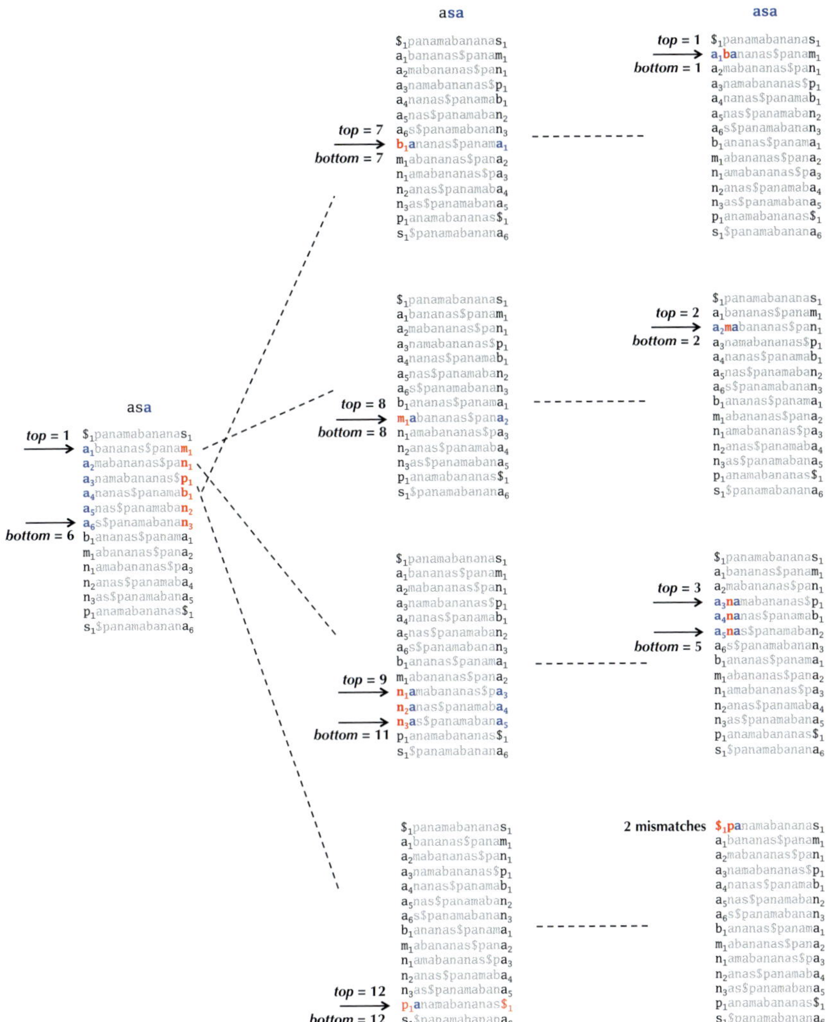

FIGURE 9.19 Using the Burrows-Wheeler transform to find approximate pattern matches of "asa" in "panamabananas$" with 1 mismatch. (Left) We identify six occurrences of "a". (Middle) Working backwards, we find four different inexact partial matches: "ba", "ma", "na" (three occurrences), and "pa". (Right) Additional mismatches are not allowed in the partial matches that we have found, but these partial matches can be extended to yield five inexact matches of "asa".

Charging Stations

Constructing a suffix tree

To construct a suffix tree, we will first modify the construction of the suffix trie as follows. Although each edge *edge* in a suffix trie is labeled by a single symbol SYMBOL(*edge*), it is unclear where this symbol came from in *Text*. We will therefore add another label for each edge (denoted POSITION(*edge*)) referring to the position of this symbol in *Text*. If an edge in the trie corresponds to more than one position in *Text*, then we will assign it its minimum starting position.

For example, consider the modified suffix trie for *Text* = "panamabananas$" shown in Figure 9.20. There are five edges labeled by "m" (colored purple), all of which are

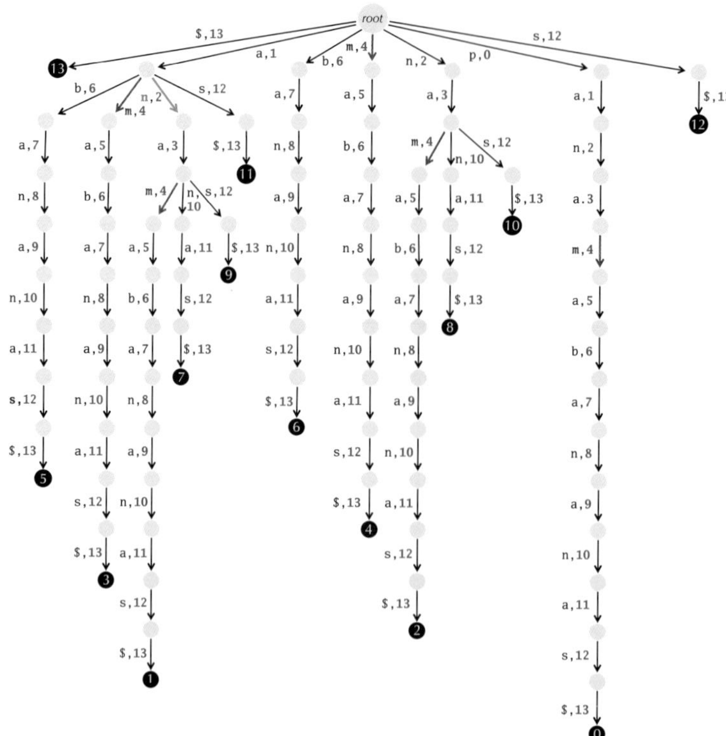

FIGURE 9.20 The modified suffix trie of *Text* = "panamabananas$". Each edge is assigned the minimum position to which the edge's symbol corresponds in *Text* (shown in blue).

labeled by position **4**, because this is the only occurrence of "m" in *Text*. On the other hand, the green edge corresponding to "**n**" is a different story. By following every path from this edge down to the leaves, we see that it corresponds to occurrences of "**n**" in the suffixes "a**n**amabananas$", "a**n**anas$", and "a**n**as$", at positions **2**, **8**, and **10**. As a result, we assign this edge the minimum of these positions.

The following pseudocode constructs the modified suffix trie of a string *Text* by traversing the suffixes of *Text* from longest to shortest. Given a suffix, it attempts to spell the suffix by moving downward in the tree, following edge labels as far as possible until it can go no further. At that point, it adds the rest of the suffix to the trie in the form of a path to a leaf, along with the position of each symbol in the suffix.

```
MODIFIEDSUFFIXTRIECONSTRUCTION(Text)
    Trie ← a graph consisting of a single node root
    for i ← 0 to |Text| − 1
        currentNode ← root
        for j ← i to |Text| − 1
            currentSymbol ← j-th symbol of Text
            if there is an outgoing edge from currentNode labeled by currentSymbol
                currentNode ← ending node of this edge
            else
                add a new node newNode to Trie
                add an edge newEdge connecting currentNode to newNode in Trie
                SYMBOL(newEdge) ← currentSymbol
                POSITION(newEdge) ← j
                currentNode ← newNode
        if currentNode is a leaf in Trie
            assign label i to this leaf
    return Trie
```

We can now transform a modified suffix trie into a suffix tree as follows. Note in Figure 9.6 (page 480) that each edge *edge* in SUFFIXTREE("panamabananas$") is labeled by a string of symbols, denoted STRING(*edge*). As we mentioned in the main text, storing all these strings is memory-intensive, and so we will instead label *edge* by two integers: the starting position of the first occurrence of STRING(*edge*) in *Text*, denoted POSITION(*edge*), and its length, denoted LENGTH(*edge*). For the modified suffix tree of *Text* = "panamabananas$" shown in Figure 9.21, these two integers are colored blue and red, respectively. For example, the edge labeled "mabananas$" in Figure 9.6 is labeled by POSITION(*edge*) = **4** and LENGTH(*edge*) = **10** in Figure 9.21. There are two edges labeled "na" in Figure 9.6, and both of them are labeled by POSITION(*edge*) = **2** and LENGTH(*edge*) = **2** in Figure 9.21.

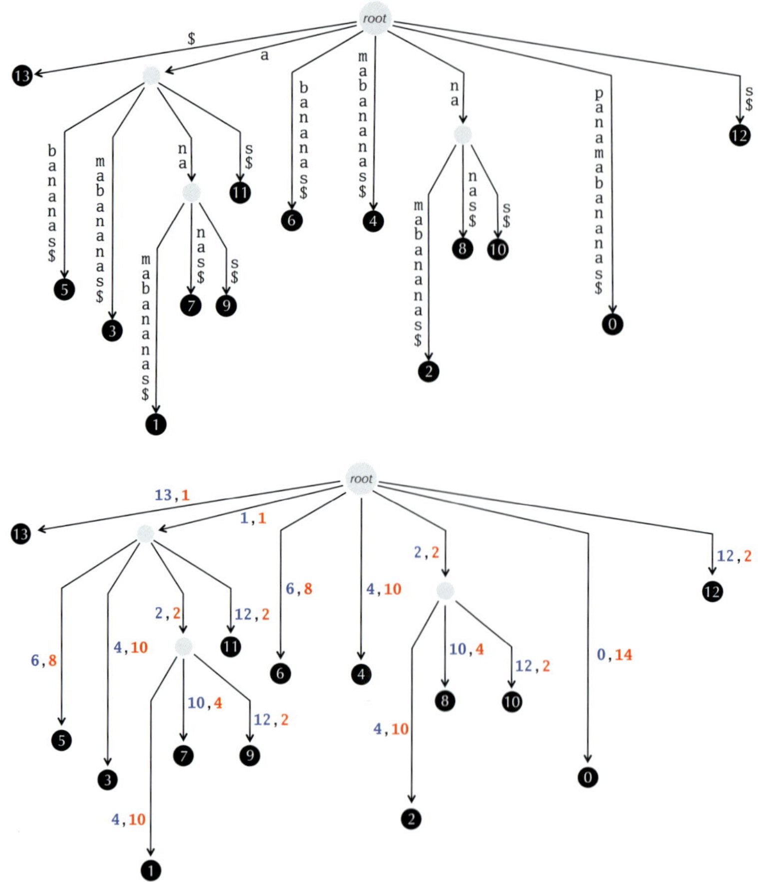

FIGURE 9.21 (Top) The suffix tree of *Text* = "panamabananas$. (Bottom) The modified suffix tree of *Text*. For each edge, the initial position of the substring to which it corresponds in *Text* is shown in blue, and the length of this substring is shown in red.

The following pseudocode constructs a suffix tree using the modified suffix trie constructed by **MODIFIEDSUFFIXTREECONSTRUCTION**. This algorithm will consolidate each non-branching path (i.e., a path whose intermediate nodes have indegree and outdegree equal to 1) of the modified suffix trie into a single edge.

```
SUFFIXTREECONSTRUCTION(Text)
    Trie ← MODIFIEDSUFFIXTRIECONSTRUCTION(Text)
    for each non-branching path Path in Trie
        substitute Path by a single edge e connecting the first and last nodes of Path
        POSITION(e) ← POSITION(first edge of Path)
        LENGTH(e) ← number of edges of Path
    return Trie
```

Solving the Longest Shared Substring Problem

A naive approach for finding a longest shared substring of strings $Text_1$ and $Text_2$ would construct one suffix tree for $Text_1$ and another for $Text_2$. Instead, we will add "#" to the end of $Text_1$, add "$" to the end of $Text_2$, and then construct the single suffix tree for the concatenation of $Text_1$ and $Text_2$ (Figure 9.22). We color a leaf in this suffix tree blue if it is labeled by the starting position of a suffix starting in $Text_1$; we color a leaf red if it is labeled by the starting position of a suffix starting in $Text_2$.

We also color the remaining nodes of the suffix tree blue, red, and purple according to the following rules:

- a node is colored blue or red if all leaves in its subtree (i.e., the subtree beneath it) are all blue or all red, respectively;
- a node is colored purple if its subtree contains both blue and red leaves.

We use COLOR(v) to denote the color of node v.

There are three purple nodes in Figure 9.22 (other than the root), and the strings spelled from the root to each of these nodes are "a", "ana", and "na". Note that these three substrings are shared by $Text_1$ = "panama" and $Text_2$ = "bananas". This is no accident.

Exercise Break: Prove that a path ending in a purple node in the suffix tree of $Text_1$ and $Text_2$ spells out a substring shared by $Text_1$ and $Text_2$.

Exercise Break: Prove that a path ending in a blue (respectively, red) node in the suffix tree of $Text_1$ and $Text_2$ spells out a substring that appears in $Text_1$ but not in $Text_2$ (respectively, $Text_2$ but not in $Text_1$).

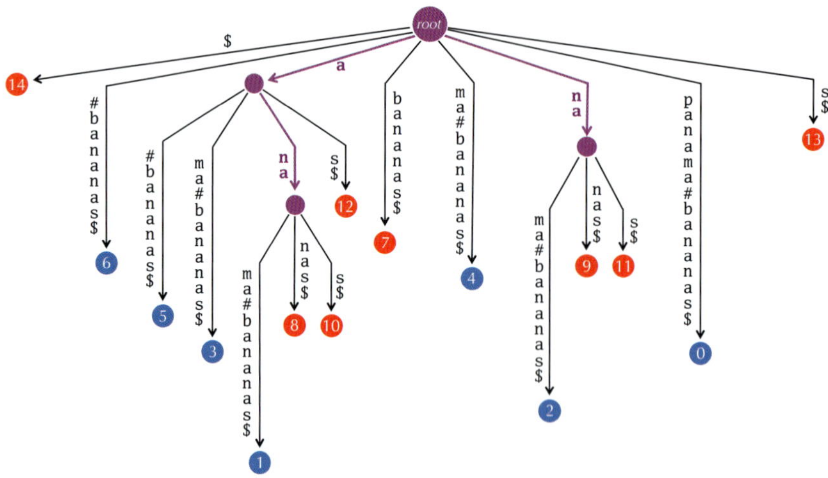

FIGURE 9.22 SUFFIXTREE("**panama#bananas$**"), constructed for $Text_1$ = "**panama**" and $Text_2$ = "**bananas**". Leaves corresponding to suffixes starting in "**panama#**" are colored blue; leaves corresponding to suffixes starting in "**bananas$**" are colored red. Every string of symbols spelled from the root to a purple node corresponds to a substring shared by $Text_1$ and $Text_2$.

The previous two exercises imply that in order to find the longest shared substring between $Text_1$ and $Text_2$, we need to examine all purple nodes as well as the strings spelled by paths leading to the purple nodes. A longest such string yields a solution to the Longest Shared Substring Problem.

TREECOLORING, which is illustrated in Figure 9.23, colors the nodes of a suffix tree from the leaves upward. This algorithm assumes that the leaves of the suffix tree have been labeled "`blue`" or "`red`" and all other nodes have been labeled "`gray`". A node in a tree is called **ripe** if it is gray but has no gray children.

9P

```
TREECOLORING(ColoredTree)
    while ColoredTree has ripe nodes
        for each ripe node v in ColoredTree
            if there exist differently colored children of v
                COLOR(v) ← "purple"
            else
                COLOR(v) ← color of all children of v
    return ColoredTree
```

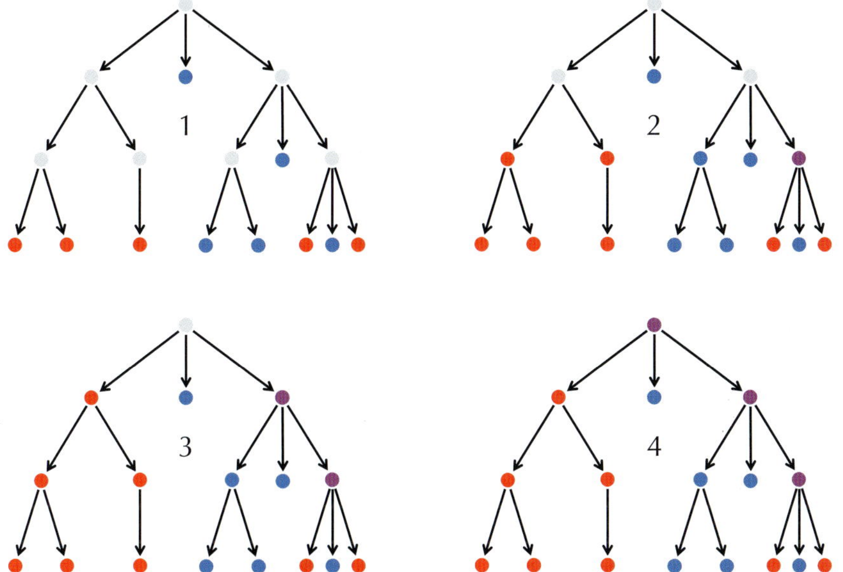

FIGURE 9.23 An illustration of the steps taken by **TREECOLORING** for a tree with an initial coloring of leaves as red or blue (top left).

Exercise Break: Modify the coloring algorithm above to find the Longest Shared Substring of more than two strings.

Partial suffix array construction

To construct the partial suffix array SUFFIXARRAY$_K$(*Text*), we first need to construct the full suffix array and then retain only the elements of this array that are divisible by K, along with their indices i. This is illustrated in Figure 9.24 for *Text* = "panamabananas$" and $K = 5$, where SUFFIXARRAY$_K$(*Text*) corresponds to the bold elements.

i	0	**1**	2	3	4	5	6	7	8	9	10	**11**	**12**	13
SUFFIXARRAY(*Text*)	13	**5**	3	1	7	9	11	6	4	2	8	**10**	**0**	12

FIGURE 9.24 Partial suffix array construction.

Detours

The reference human genome

When genomes are assembled from DNA taken from a number of donors, the reference genome represents a mosaic of donor genomes. The existing reference human genome was derived from thirteen volunteers in the United States; it continues to be improved by fixing errors and filling in the remaining gaps (there are currently over a hundred).

The reference genome is often used as a template on which new individual genomes can be rapidly assembled. Comparison of the reference genome and individual human genomes typically reveals about three million SNPs, and about 0.1% of an individual human genome cannot be matched to the reference genome at all.

In regions with high diversity, the reference genome may differ significantly from individual genomes. An example of a high-diversity region of the human genome is the **major histocompatibility complex**, a family of genes powering immune systems. Since these genes have an unusually large number of alternate forms, two individuals hardly ever have exactly the same set of genes from this complex.

Needless to say, a reference genome derived from just thirteen people in the United States is a very biased representative of the billions of people living worldwide. This bias limits the conclusions that can be drawn from the reference genome in the emerging field of personalized medicine. For this reason, researchers are now working on a new form of reference genome containing thousands of individual human genomes. This reference **pan-genome** can be represented as a giant graph, in which individual genomes correspond to paths.

Rearrangements, insertions, and deletions in human genomes

Until recently, biologists focused primarily on small-scale mutations in the human genome, assuming that rearrangements and indels are relatively rare. In 2005, Evan Eichler surprised biologists when he found hundreds of rearrangements and indels separating the genomes of two individuals. This finding was important because rearrangements and indels are often hallmarks of disease; for example, repeated insertions of the nucleotide triplet `CAG` increases the severity of Huntington's disease.

In 2013, Gerton Lunter revealed the true extent of indels in the human population by identifying over a *million* indels in a cohort of over a hundred individuals. Intriguingly, he found that over half of the indels occurred in just 4% of the genome; in other words, some regions of the human genome represent "indel hotspots". As the catalog of human rearrangements and indels grows, biologists are gaining the ability to identify frequently mutated genes as well as implicate rearrangements and indels when diagnosing complex disorders.

The Aho-Corasick algorithm

The **Aho-Corasick algorithm** for the Multiple Pattern Matching Problem was invented by Alfred Aho and Margaret Corasick in 1975. The runtime of their algorithm is $\mathcal{O}(|Patterns| + |Text| + m)$, where m is the number of output matches.

Imagine sliding the trie in Figure 9.25 against $Text$ = "bantenna". Similarly to TRIEMATCHING, the Aho-Corasick algorithm starts at the root and attempts to build a path spelling a prefix of "bantenna". This attempt fails after three nodes ("**bant**enna"). In TRIEMATCHING, we would begin again at the root and attempt to find a match starting at the second symbol of "bantenna".

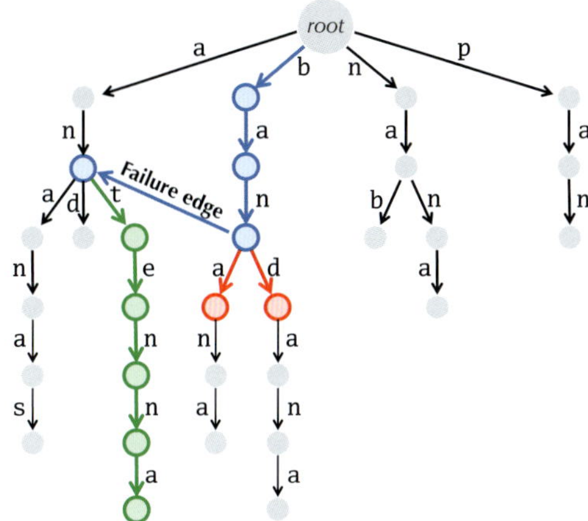

FIGURE 9.25 The trie from Figure 9.1 with an additional failure edge. The failure edge allows us to jump from the path beginning with "ban" to the path beginning with "an".

However, we have thrown away some important information: we *already matched* the first two symbols of "**an**tenna" when searching through "**ban**tenna". Thus, a more sensible strategy is to jump directly to the node "an" and then continue downward in the trie. Indeed, this strategy will eventually match the pattern "**antenna**". We can implement this "jumping ahead" strategy by augmenting the tree with a **failure edge** connecting node "ban" to node "an" (Figure 9.25). More generally, failure edges are formed by connecting node v to node w if w is the longest suffix of v that appears in the trie. After constructing all failure edges, the Aho-Corasick algorithm then follows failure edges during pattern matching whenever a mismatch is found in order to avoid going all the way back to the root, thus saving time.

Exercise Break: Construct all failure edges for the trie in Figure 9.25.

From suffix trees to suffix arrays

We can construct SUFFIXARRAY(*Text*) from SUFFIXTREE(*Text*) by applying a **preorder traversal** of SUFFIXTREE(*Text*) as long as the outgoing edges from every node of the suffix tree are arranged lexicographically.

Given a rooted tree, a node w is called a **child** of a node v if there is an edge connecting v to w in the tree. The preorder traversal of a tree involves visiting a node of the tree, starting at the root, and then recursively preorder traversing the subtrees rooted at each of its children from left to right (Figure 9.26), which is accomplished by the following pseudocode. By taking the order of leaves visited in a preorder traversal of the suffix tree, we obtain the suffix array (consult Figure 9.6 and Figure 9.7).

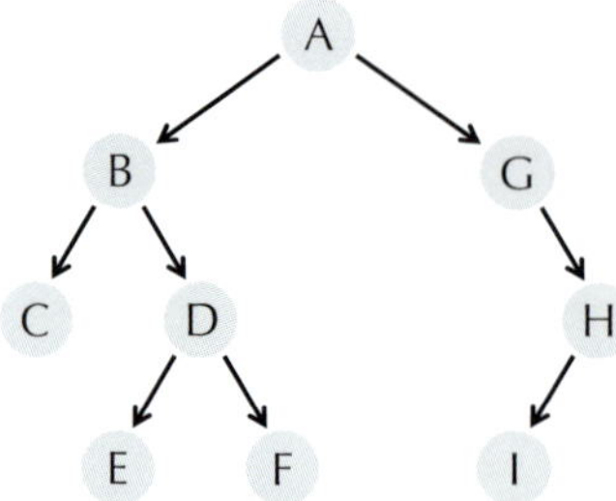

FIGURE 9.26 The preorder traversal of the above tree visits its nodes in increasing order of their labels.

```
PREORDER(Tree, Node)
    visit Node
    for each child Node' of Node from left to right
        PREORDER(Tree, Node')
```

Conversely, SUFFIXTREE(*Text*) can be constructed from SUFFIXARRAY(*Text*) in linear time by using the **longest common prefix (LCP) array** of *Text*, LCP(*Text*), which stores the length of the longest common prefix shared by consecutive lexicographically ordered suffixes of *Text*. For example, LCP("panamabananas$") is (0, 0, 1, 1, 3, 3, 1, 0, 0, 0, 2, 2, 0, 0), as shown in Figure 9.27, which also contains SUFFIXARRAY("panamabananas$").

LCP Array	Sorted Suffixes	Suffix Array
0	$	13
0	abananas$	5
1	amabananas$	3
1	anamabananas$	1
3	ananas$	7
3	anas$	9
1	as$	11
0	bananas$	6
0	mabananas$	4
0	namabananas$	2
2	nanas$	8
2	nas$	10
0	panamabananas$	0
0	s$	12

FIGURE 9.27 The LCP array of "panamabananas$" is formed by sorting the suffixes of "panamabananas$" lexicographically and then finding the length of the longest common prefix shared by consecutive suffixes according to the lexicographic order.

Suffix Tree Construction from Suffix Array Problem:
Construct a suffix tree from the suffix array and LCP array of a string.

Input: A string *Text*, its suffix array, and its LCP array.
Output: The suffix tree of *Text*.

From suffix arrays to suffix trees

Given the suffix array SUFFIXARRAY and LCP array LCP of a string *Text*, the suffix tree SUFFIXTREE(*Text*) can be constructed in linear time using the algorithm illustrated in Figure 9.28. After constructing a **partial suffix tree** for the i lexicographically smallest suffixes (denoted SUFFIXTREE$_i$(*Text*)), this algorithm iteratively inserts the $(i+1)$-th suffix into this tree to form SUFFIXTREE$_{i+1}$(*Text*).

We define the **descent** of a node v in a suffix tree, denoted DESCENT(v), as the length of the concatenation of all path labels from the root to this node. For example, the three leaves in Figure 9.28 (top left) have descents equal to 1, 9, and 11, whereas the purple internal node has descent 1. We assume that the descents of all nodes in the growing partial suffix tree have been precomputed.

We start with SUFFIXTREE$_0$(*Text*), which we define as the tree consisting only of the root. To insert the $(i+1)$-th suffix (corresponding to the element SUFFIXARRAY$(i+1)$ in the suffix array of *Text*) into SUFFIXTREE$_i$(*Text*), we need to know where the path representing this suffix "splits" from the already constructed partial suffix tree SUFFIXTREE$_i$(*Text*). For example, this split happens at the purple node while inserting "anamabananas$" into SUFFIXTREE$_3$(*Text*) (Figure 9.28 (top)) and at the purple edge labeled "namabananas$" while inserting "ananas$" into SUFFIXTREE$_4$(*Text*), thus breaking this edge into edges labeled "na" and "mabananas$" in SUFFIXTREE$_5$(*Text*) (Figure 9.28 (bottom)).

STOP

STOP and Think: How would you find the node or edge where the partial suffix tree splits while constructing the suffix tree from the suffix array and LCP array?

To find the node/edge where the partial suffix tree splits, we will walk up the rightmost path (i.e., the last added path in the partial suffix tree) in the partial suffix tree beginning at the previously inserted leaf, labeled SUFFIXARRAY(i), to the root. We stop when we encounter the first node v such that DESCENT$(v) \leq$ LCP$(i+1)$. Afterwards, we have to consider two cases depending on whether DESCENT$(v) =$ LCP$(i+1)$ (the split occurs at the node v) or DESCENT$(v) <$ LCP$(i+1)$ (the split occurs on the edge leading from v):

- DESCENT$(v) =$ LCP$(i+1)$: the concatenation of the labels on the path from the root to v equals the longest common prefix of suffixes corresponding to SUFFIXARRAY(i) and SUFFIXARRAY$(i+1)$. We insert SUFFIXARRAY$(i+1)$ as a new leaf x connected to v, and we label the edge (v, x) with the suffix of *Text*

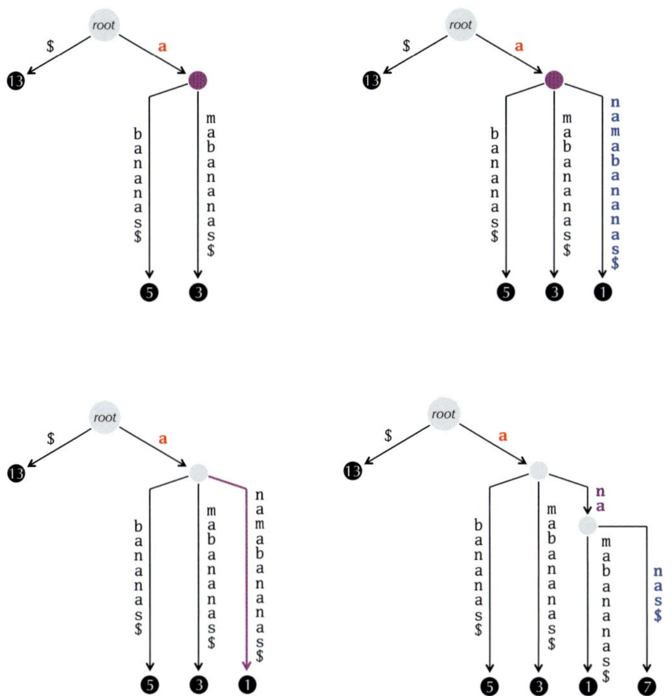

FIGURE 9.28 (Top left) SUFFIXTREE$_3$(*Text*) for *Text* = "panamabananas$". (Top right) The fourth lexicographically ordered suffix of *Text* is "**anamabananas$**", and we can thread only its first letter into this suffix tree. Our stopping point is at the purple node, so we create a new node branching off this node with edge labeled "**namabananas$**" to obtain SUFFIXTREE$_4$(*Text*). (Bottom left) Preparing to insert "ananas$" into SUFFIXTREE$_4$(*Text*). The fifth lexicographically ordered suffix of *Text* is "**ananas$**", the first three symbols of which we can thread into SUFFIXTREE$_4$(*Text*) until we reach a stopping point in the middle of the edge "**na**mabananas$". (Bottom right) To form SUFFIXTREE$_5$(*Text*), we create a new node in the middle of the purple edge and branch from this node into two edges: one labeled "mabananas$" in order to retain the suffix "anamabananas$" and another labeled "**nas$**" to spell the new suffix "**ananas$**" (symbols on the newly added edge are colored blue). In general, the number of red and purple symbols that we spell down in SUFFIXTREE$_i$(*Text*) to form SUFFIXTREE$_{i+1}$(*Text*) is given by the $(i+1)$-th entry in the LCP array from Figure 9.27. For example, the number of red or purple symbols that we spell down in SUFFIXTREE$_4$(*Text*) to form SUFFIXTREE$_5$(*Text*) is given by the fifth entry in the LCP array (which is equal to 3).

starting at position SUFFIXARRAY$(i+1)$ + LCP$(i+1)$. Thus, the edge label consists of the remaining symbols of the suffix corresponding to SUFFIXARRAY$(i+1)$ that are not already represented by the concatenation of the labels of the path connecting the root to v. This completes the construction of the partial suffix tree SUFFIXTREE$_{i+1}$(*Text*) (see Figure 9.28 (top) for an example).

- DESCENT(v) $<$ LCP$(i+1)$: the concatenation of the labels on the path from the root to v has fewer symbols than the longest common prefix of the suffixes corresponding to SUFFIXARRAY(i) and SUFFIXARRAY$(i+1)$. The question therefore arises of how to recover these missing symbols. We denote the rightmost edge leading from v in SUFFIXTREE$_i$(*Text*) as (v, w) and argue that the missing symbols represent a prefix of this edge's label. In this case, we split this edge and construct SUFFIXTREE$_{i+1}$(*Text*) as described below (see Figure 9.28 (bottom) for an example):

  1. Delete the edge (v, w) from SUFFIXTREE$_i$(*Text*).
  2. Add a new internal node y and a new edge (v, y) labeled by a substring of *Text* starting at position SUFFIXARRAY$(i+1)$ + DESCENT(v). The new label is formed by the final LCP$(i+1)$ − DESCENT(v) symbols of the longest common prefix of SUFFIXARRAY(i) and SUFFIXARRAY$(i+1)$. Thus, the concatenation of the labels on the path from the root to y is now the longest common prefix of SUFFIXARRAY(i) and SUFFIXARRAY$(i+1)$.
  3. Define DESCENT(y) as LCP$(i+1)$.
  4. Connect w to the newly created internal node y by an edge (y, w) that is labeled by a substring of *Text* starting at position SUFFIXARRAY(i) + LCP$(i+1)$ and ending at position SUFFIXARRAY(i) + DESCENT$(w) - 1$. The new label consists of the remaining symbols of the deleted edge (v, w) that were not used as the label of edge (v, y).
  5. Add SUFFIXARRAY$(i+1)$ as a new leaf x as well as an edge (y, x) that is labeled by a suffix of *Text* beginning at position SUFFIXARRAY$(i+1)$ + LCP$(i+1)$. The label of this edge consists of the remaining symbols of the suffix corresponding to SUFFIXARRAY$(i+1)$ that are not already represented by the concatenation of the labels on the path from the root to v.

Exercise Break: Prove that the running time of this algorithm is $\mathcal{O}(|Text|)$.

Binary search

The game show *The Price is Right* features a timed challenge called the "Clock Game" in which a contestant makes repeated guesses at the price of an item, with the host telling the contestant only whether the true price is higher or lower than the most recent guess.

An intelligent strategy for the Clock Game is to pick a sensible range of prices within which the item's price must fall, and then guess a price halfway between these two extremes. If this guess is incorrect, then the contestant has immediately eliminated half of the set of possible prices. The contestant then makes a guess in the middle of the remaining possible prices, eliminating half of them again. Iterating this strategy quickly yields the price of the item.

This strategy for the Clock Game motivates a binary search algorithm finding the position of an element *key* within a sorted array ARRAY. This algorithm, called **BINARYSEARCH**, is initialized by setting *minIndex* equal to 0 and *maxIndex* equal to the last index of ARRAY. It sets *midIndex* equal to $(minIndex + maxIndex)/2$ and then checks to see whether *key* is greater than or less than ARRAY$(midIndex)$. If *key* is larger than this value, then **BINARYSEARCH** iterates on the subarray of ARRAY from *minIndex* to $midIndex - 1$; otherwise, **BINARYSEARCH** iterates on the subarray of ARRAY from $midIndex + 1$ to *maxIndex*. Iteration eventually identifies the position of *key*.

For example, if $key = 9$ and ARRAY $= (1, 3, 7, 8, 9, 12, 15)$, then **BINARYSEARCH** would first set *minIndex* equal to 0, *maxIndex* equal to 6, and *midIndex* equal to 3. Because *key* is greater than ARRAY$(midIndex) = 8$, we examine the subarray whose elements are greater than ARRAY$(midIndex)$ by setting *minIndex* equal to 4, so that *midIndex* is recomputed as $(4 + 6)/2 = 5$. This time, *key* is smaller than ARRAY$(midIndex) = 12$, and so we examine the subarray whose elements are smaller than this value. This subarray consists of only a single element, which is *key*.

```
BINARYSEARCH(ARRAY, key, minIndex, maxIndex)
    while maxIndex ≥ minIndex
        midIndex ← ⌊(minIndex + maxIndex)/2⌋
        if ARRAY(midIndex) = key
            return midIndex
        else if ARRAY(midIndex) < key
            minIndex ← midIndex + 1
        else
            maxIndex ← midIndex − 1
    return "key not found"
```

Bibliography Notes

The Aho-Corasick algorithm was introduced by Aho and Corasick, 1975. Suffix trees were introduced by Weiner, 1973. Suffix arrays were introduced by Manber and Myers, 1990. The Burrows-Wheeler Transform was introduced by Burrows and Wheeler, 1994. An efficient implementation of the Burrows-Wheeler transform was described by Ferragina and Manzini, 2000. The genetic cause of Ohdo syndrome was elucidated by Clayton-Smith et al., 2011. Rearrangements and indels in the human genome were studied by Tuzun et al., 2005 and Montgomery et al., 2013. BLAST, the dominant database search tool in molecular biology, was developed by Altschul et al., 1990.

WHY HAVE BIOLOGISTS STILL NOT DEVELOPED AN HIV VACCINE?

Hidden Markov Models

Classifying the HIV Phenotype

How does HIV evade the human immune system?

In 1984, US Health and Human Services Secretary Margaret Heckler declared that an HIV vaccine would be available within two years, stating, "Yet another terrible disease is about to yield to patience, persistence and outright genius."

In 1997, Bill Clinton established a new research center at the National Institutes of Health with the goal of developing an HIV vaccine. In his words, "It is no longer a question of *whether* we can develop an AIDS vaccine, it is simply a question of *when*."

In 2005, Merck began clinical trials of an HIV vaccine but discontinued them two years later after learning that the vaccine actually *increased* the risk of HIV infection in some recipients.

Today, despite enormous investment and ongoing clinical trials, we are still far from an HIV vaccine, and 35 million people are living with the disease. Scientists have made great progress in developing a successful **antiretroviral therapy**, a drug cocktail that stabilizes an infected patient's symptoms. However, this therapy does not cure AIDS and cannot prevent the spread of HIV, and so it does not hold the promise of a true vaccine for containing the AIDS epidemic.

Classical vaccines against viruses are often made from the surface proteins of a virus. These vaccines stimulate the human immune system to recognize viral envelope proteins as foreign, destroy them, and keep a record of it, so that the immune system can identify and eradicate the virus in a later encounter.

However, HIV viral envelope proteins are extremely variable because the virus must mutate rapidly in order to survive (see **DETOUR: The Red Queen Effect**). The HIV PAGE 582 population in a *single* infected individual rapidly evolves to evade the human immune system (Figure 10.1), not to mention that HIV strains taken from *different* patients represent multiple highly diverged subtypes. Therefore, a successful HIV vaccine must be broad enough to account for this variability.

In an effort to counteract HIV's variability, we could create a single peptide that contains the least variable segments of the envelope proteins taken from all known HIV strains and use this peptide as the basis for a universal vaccine fighting all HIV strains. However, not only do HIV envelope proteins mutate fast, but they are also "masked" by **glycosylation**, a post-translational modification that often makes these proteins invisible to the human immune system (see **DETOUR: Glycosylation**). As a PAGE 582 result, all attempts at developing an HIV vaccine have thus far failed.

HIV has just nine genes, and in this chapter we will focus on the rapidly mutating *env* gene, which has a mutation rate of 1 to 2% per nucleotide per year. The protein

```
VKKLGEQFR-NKTIIFNQPSGGDLEIVMHSFNCGGEFFYCNTTQLFN----------NSTES------DTITL
VKKLGEQFR-NKTIIFNQPSGGDLEIVMHSFNCGGEFFYCNTTQLFN----------NSTDNG-----DTITL
VKKLGEQFR-NKTIIFNQPSGGDLEIVMHSFNCGGEFFYCNTTQLFD----------NSTESNN----DTITL
VDKLREQFGKNKTIIFNQPSGGDLEIVMHTFNCGGEFFYCNTTQLFNSTWNS---TGNGTESYNGQENGTITL
VDKLREQFGKNKTIIFNQPSGGDLEIVMHTFNCGGEFFYCNTTQLFNSTWNG---TNTT--GLDG--NDTITL
VDKLREQFGKNKTIIFNQSSGGDLEIVTHTFNCGGEFFYCNTTQLFNSNWTG---NSTE--GLHG--DDTITL
VKKLGEQFG-NKTIIFNQSSGGGLEIVMHSFNCGGEFFYCNTTQLFNN--TR-----NSTESNNGQGNDTTTL
VKKLREQFGKNKTIIFKQSSGGDLEIVTHTFNCAGEFFYCNTTQLFNSNWTE-----NSITGLDG--NDTITL
VGKLREQFGK-KTIIFNQPSGGDLEIVMHSFNCQGEFFYCNTTRLFNSTWDNSTWNSTGKDKENGN-NDTITL
```

FIGURE 10.1 A multiple alignment of a short region of gp120 proteins sampled from a single HIV-positive patient at nine different time points. Almost half of the columns (shown in darker text) are not conserved across all time points, illustrating how quickly HIV evolves, even within an individual host. Amino acids differing from the most common symbol in a column are shown in blue.

encoding the *env* gene then gets cut into **glycoprotein gp120** (approx. 480 amino acids) and **glycoprotein gp41** (approx. 345 amino acids). Together, gp120 and gp41 form the **envelope spike**, which mediates entry of the HIV virus into human cells.

Since HIV mutates so fast, different HIV isolates may have different phenotypes, thus requiring different drug cocktails. For example, HIV viruses can be divided into fast-replicating **syncytium-inducing (SI)** isolates and slow-replicating **non-syncytium-inducing (NSI)** isolates. During infection, viral proteins like gp120 that are used by HIV to enter the cell are transported to the cell surface, where they can cause the host cell membrane to fuse with neighboring cells. This causes dozens of human cells to fuse their cell membranes into a giant, nonfunctional **syncytium**, or abnormal multinucleate cell (Figure 10.2). This mechanism allows an SI virus to kill many human cells by infecting only one.

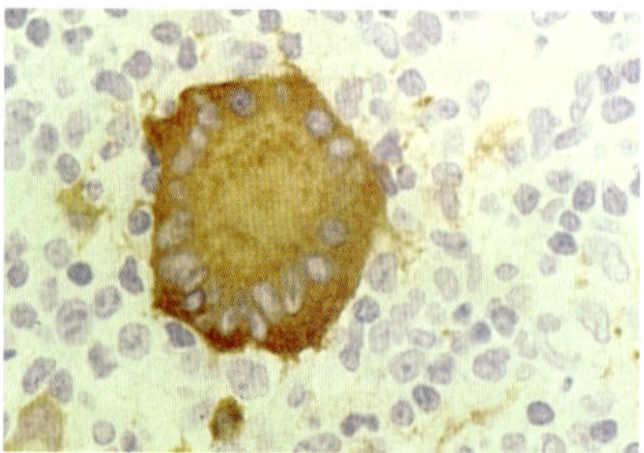

FIGURE 10.2 A syncytium with multiple nuclei in an HIV patient.

Because gp120 is important in classifying a virus as SI or NSI, biologists are interested in determining which amino acids in gp120 can be used for this classification.

In 1992, Jean-Jacques De Jong analyzed a multiple alignment of the **V3 loop** region in gp120 (Figure 10.3 (top)) and devised the **11/25 rule**, which asserts that an HIV strain is more likely to have an SI phenotype if the amino acid at either positions 11 or 25 of its V3 loop is arginine or lysine. It later was shown that many other positions influence the SI/NSI phenotype.

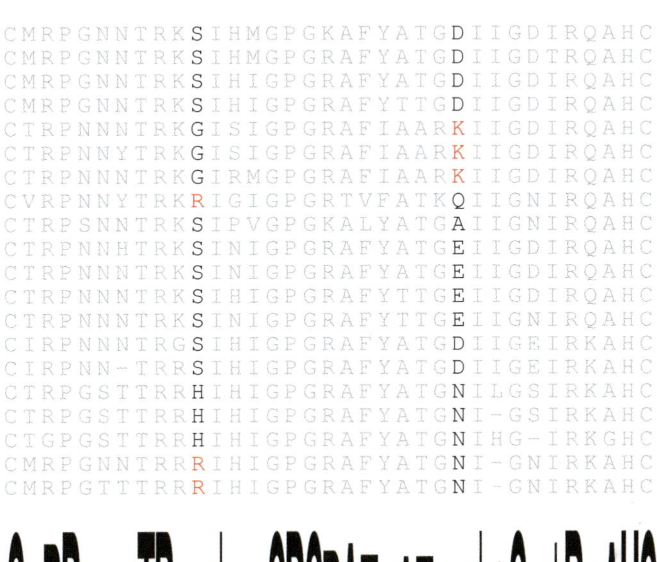

FIGURE 10.3 A multiple alignment of the V3 loop region taken from twenty HIV patients, and the motif logo of this alignment. The alignment's 11th and 25th columns are shown in darker text; occurrences of arginine (R) or lysine (K) in these columns are shown in red. The 11/25 rule will classify six of the patients as infected with an SI isolate. Although the V3 loop is an important and rather conserved segment of gp120, the level of conservation varies across different positions. For example, whereas the first and last positions are extremely conserved, positions 11 and 25 exhibit high variation.

Limitations of sequence alignment

Before biologists could even start to examine the question of predicting HIV phenotypes from gp120 sequences, they faced the problem of constructing accurate multiple alignments of these sequences. Indeed, a single misalignment, placing an incorrect amino acid at a position influencing the SI/NSI phenotype, could lead to a faulty classification of HIV phenotypes. And we already know from Chapter 5 that constructing a multiple alignment of highly diverged sequences is a difficult algorithmic problem.

Figure 10.3 (bottom) shows a motif logo from the V3 loop of gp120 and illustrates that some positions in gp120 are relatively conserved, whereas others are extremely variable. Furthermore, this motif logo does not account for insertions and deletions, which are prevalent in other regions of gp120 that are less conserved than the V3 loop. These insertions and deletions make analyzing gp120 even more complex.

Because the columns of the multiple alignment in Figure 10.3 have varying levels of conservation, we question the wisdom of using the *same* amino acid scoring matrix (as well as indel penalties) across different columns of an alignment. A better approach would use a *different* scoring approach across different columns. For example, an amino acid differing from R in position 3 of the alignment in Figure 10.3 should incur a larger penalty than an amino acid deferring from S in position 11.

In other words, the *problem formulation* of multiple sequence alignment introduced in Chapter 5 does not offer an adequate translation of the biological problem of HIV classification into an algorithmic problem. We must therefore devise a new problem formulation for sequence alignment that will lead to a statistically solid analysis of gp120 proteins. But first, we will ask you to join us in our time machine for one more trip.

Gambling with Yakuza

The Japanese crime syndicates called yakuza descend from groups of 18th Century traveling gamblers called bakuto. (In fact, "yakuza" is the name of a losing hand in a Japanese card game.) One of the most popular games that the bakuto would host in their makeshift casinos is called **Chō-Han**. In this game, which literally translates as "even-odd", the dealer rolls two dice, and players wager on whether the sum of the dice will be even or odd.

Although playing Chō-Han in a yakuza gambling house would undoubtedly make for a fun evening, we can play an equivalent — albeit less exciting — game called "Heads or Tails" by flipping a coin and wagering on the outcome. Assume that for some strange reason, more people wager on tails than on heads in this game. Then a crooked dealer might use a biased coin that is more likely to result in heads than tails. We will assume that this biased coin results in heads with probability 3/4.

STOP and Think: Say that you play Heads or Tails 100 times, and the coin produces heads 63 times. Is the dealer cheating? Was the coin fair or biased?

This question is not well-formulated, since either coin could have produced any sequence of flips. But can we determine which coin is *more likely* to have been used?

We write the probabilities of tails ("T") and heads ("H") for the fair coin (F) as

$$\Pr_F(\text{"H"}) = 1/2 \qquad \Pr_F(\text{"T"}) = 1/2$$

and the probabilities for the biased coin (B) as

$$\Pr_B(\text{"H"}) = 3/4 \qquad \Pr_B(\text{"T"}) = 1/4$$

Since coin flips are independent events, the probability that n flips of the fair coin will generate a given sequence $x = x_1x_2\ldots x_n$ with k occurrences of "H" is

$$\Pr(x|F) = \prod_{i=1}^{n} \Pr_F(x_i) = (1/2)^n .$$

On the other hand, the probability that the biased coin will generate the same sequence is

$$\Pr(x|B) = \prod_{i=1}^{n} \Pr_B(x_i) = (1/4)^{n-k} \cdot (3/4)^k = 3^k/4^n.$$

If $\Pr(x|F) > \Pr(x|B)$, then the dealer more likely used a fair coin, and if $\Pr(x|F) < \Pr(x|B)$, then the dealer more likely used a biased coin. The numbers $(1/2)^n$ and $3^k/4^n$ are so small for large n that in order to compare them, we will use their **log-odds ratio**,

$$\log_2\left(\frac{\Pr(x|F)}{\Pr(x|B)}\right) = \log_2\left(\frac{2^n}{3^k}\right) = n - k \cdot \log_2 3.$$

Exercise Break: Show that $\Pr(x|F)$ is larger than $\Pr(x|B)$ when the log-odds ratio is positive (i.e., when $k/n < 1/\log_2 3$) and smaller than $\Pr(x|B)$ when the log-odds ratio is negative (i.e., when $k/n > 1/\log_2 3$).

Returning to our example of witnessing $k = 63$ heads in $n = 100$ flips, the log-odds ratio is positive, since

$$k/n = 0.63 < 1/\log_2 3 \approx 0.6309.$$

It follows that $\Pr(x|F) > \Pr(x|B)$, and so the dealer most likely used a fair coin, even though 63 is closer to 75 than it is to 50.

Two Coins up the Dealer's Sleeve

In bakuto gambling houses, a Chō-Han dealer would remove his shirt during play, displaying a tattooed chest, in order to reduce any suspicions of dice tampering. (These tattoos would later become a yakuza tradition.) We will assume, however, that in Heads or Tails, the crooked dealer is wearing a shirt and keeps both coins up a sleeve, secretly changing them back and forth whenever he likes during the sequence of flips. Since he does not want to be caught switching coins, he does so only occasionally.

We will assume that the crooked dealer switches coins with probability 0.1 after each flip. Given a sequence of coin flips, we must determine when the dealer used the biased coin and when he used the fair coin.

Casino Problem:
Given a sequence of coin flips, determine when the crooked dealer used a fair coin and when he used a biased coin.

> **Input**: A sequence $x = x_1x_2 \ldots x_n$ of coin flips made by two possible coins (F and B).
> **Output**: A sequence $\pi = \pi_1\pi_2 \ldots \pi_n$, with each π_i being equal to either F or B and indicating that x_i is the result of flipping the fair or biased coin, respectively.

Unfortunately, this problem is poorly stated, since either coin can generate any outcome. Instead, we need to determine the *most likely* sequence of coins used by the dealer.

STOP and Think: Can you reformulate the Casino Problem so that it makes sense?

A well-defined computational problem for finding the most likely sequence of coins used by the dealer should somehow grade different sequences π as better answers than others. One approach to guessing the most likely coin the dealer used for each flip would be to slide a window (of length $t < n$) along the sequence of flips $x = x_1 \ldots x_n$ and then calculate the log-odds ratio under each window. If the log-odds ratio of the window falls below zero, then the dealer most likely used the biased coin inside this window; otherwise, the dealer most likely used the fair coin.

STOP and Think: Do you see any problems with this method?

There are two issues with the window-sliding approach. First, we have no apparent choice for the length of the window. Second, overlapping windows may classify the same outcome as caused by both the fair and biased coins. For example, if $x =$ "HHHHHTTHHHTTTTT", then the window $x_1 \ldots x_{10} =$ "HHHHHTTHHH" has a negative log-odds ratio, and the window $x = x_6 \ldots x_{15} =$ "TTHHHTTTTT" has a positive log-odds ratio. So which coin did the dealer use on the flips $x_6 \ldots x_{10} =$ "TTHHH"?

Finding CG-Islands

In the next section, we will improve our method of grading sequences of coin flips. The solution will lead us to a computational paradigm that has been successfully applied to a wide array of bioinformatics problems, including HIV comparison. For now, however, you may still not believe how coin flipping could possibly relate to sequence comparison. Thus, we will briefly describe a different biological problem that more clearly relates to our coin flipping analogy.

In the early 20th Century, Phoebus Levene discovered the four nucleotides making up DNA. At this time, little was known about DNA (Watson & Crick's double helix paper was still half a century away). As a result, Levene doubted that DNA could store genetic information using just a four-letter alphabet, and he hypothesized that DNA comprised nearly equal amounts of adenine, cytosine, guanine, and thymine.

A century later, we know that complementary nucleotides on *opposing* strands of DNA have equal frequencies because of base pairing — ignoring extremely rare base-pairing errors. However, it is not true that nucleotide frequencies are approximately equal on a *single* strand of DNA. For example, different species have widely varying **GC-content**, or the percentage of cytosine and guanine nucleotides in a genome. For example, the human genome's GC-content is approximately 42%.

After accounting for the human genome's skewed GC-content, we might expect that each of the dinucleotides CC, CG, GC, and GG would occur in the human genome with frequency $0.21 \cdot 0.21 = 4.41\%$. However, the frequency of CG in the human genome is only about 1%! This dinucleotide is so rare because of **methylation**, the most common DNA modification, which typically adds a methyl group (CH_3) to the cytosine nucleotide within a CG dinucleotide. The resulting methylated cytosine has the tendency to further deaminate into thymine (see DETOUR: DNA Methylation). As a PAGE 582
result of methylation, CG is the least frequent dinucleotide in many genomes.

Nevertheless, methylation is often suppressed around genes in areas called **CG-islands**, where CG appears relatively frequently (Figure 10.4). If you were to sequence a mammalian genome that you knew nothing about, perhaps one of the first things you might do in order to find genes in this genome is look for CG-islands.

STOP and Think: How would you identify CG-islands in a genome?

	A	C	G	T
A	0.053	0.079	0.127	0.036
C	0.037	0.058	0.058	0.041
G	0.035	0.075	0.081	0.026
T	0.024	0.105	0.115	0.050

	A	C	G	T
A	0.087	0.058	0.084	0.061
C	0.067	0.063	0.017	0.063
G	0.053	0.053	0.063	0.042
T	0.051	0.070	0.084	0.084

FIGURE 10.4 Dinucleotide frequencies for a collection of CG-islands (left) and non-CG-islands (right) in the human genome computed for a single strand of the X chromosome. Frequencies of CG are shown in red.

A naive approach to search for CG-islands in a genome would slide a window down the genome, declaring windows with higher frequencies of CG as potential CG-islands. The disadvantages of this method are analogous to those of using a sliding window to determine which coin the crooked dealer most likely used at any given point in time. We do not know how long the window should be, and overlapping windows may simultaneously classify the same genomic position as belonging to a CG-island and as not belonging to a CG-island.

Hidden Markov Models

From coin flipping to a Hidden Markov Model

Our goal is to develop a single concept that models both the crooked dealer and the search for CG-islands in a genome. To this end, we will think about the crooked dealer not as a human but as a primitive machine. We do not know how this machine is constructed, but we do know that it proceeds in a sequence of steps; in each step, it is in one of two hidden states, F and B, and it emits a symbol, "H" or "T".

After each step, the machine makes two decisions:

- Which hidden state will I move to next?
- Which symbol will I emit in that state?

The machine answers the first question by choosing randomly among the F and B states, with probability 0.9 of remaining in its current state and probability 0.1 of changing states. The machine answers the second question by choosing between the symbols "H" and "T" with probabilities that depend on the state it is in. In our coin flipping example, the probabilities for state F (0.5 and 0.5) differ from the probabilities for state B (0.75 and 0.25). Our goal is to infer the machine's most likely sequence of states by analyzing the sequence of symbols that it emits.

We have just transformed the dealer into an abstract machine called a **Hidden Markov Model (HMM)**. The only difference between our specialized "coin flipping machine" and the general concept of an HMM is that the latter can have an arbitrary number of states and may have arbitrary probability distributions governing which state to move into and which symbols to emit. In general, an HMM (Σ, *States*, *Transition*, *Emission*) is defined by a set of four objects:

- an alphabet Σ of emitted symbols;
- a set *States* of **hidden states**;
- a $|States| \times |States|$ matrix $Transition = (transition_{l,k})$ of **transition probabilities**, where $transition_{l,k}$ represents the probability of moving from state l to state k;
- a $|States| \times |\Sigma|$ matrix $Emission = (emission_k(b))$ of **emission probabilities**, where $emission_k(b)$ represents the probability of emitting symbol b from alphabet Σ when the HMM is in state k.

For each state l,

$$\sum_{\text{all states } k} transition_{l,k} = 1$$

and

$$\sum_{\text{all symbols } b \text{ from } \Sigma} emission_l(b) = 1\,.$$

Exercise Break: What are Σ, *States*, *Transition*, and *Emission* for the HMM modeling the crooked dealer?

The HMM diagram

As illustrated in Figure 10.5, an HMM can be visualized using an **HMM diagram**, a graph in which every state is represented by a solid node. Solid directed edges connect every pair of nodes, as well as every node to itself. Each such edge is labeled with the transition probability of moving from one state to the other (or remaining in the same state). In addition, the HMM diagram has dashed nodes representing each possible symbol from the alphabet Σ and dashed edges connecting each state to each dashed node. Each such edge is labeled by the probability that the HMM will emit this symbol while in the given state.

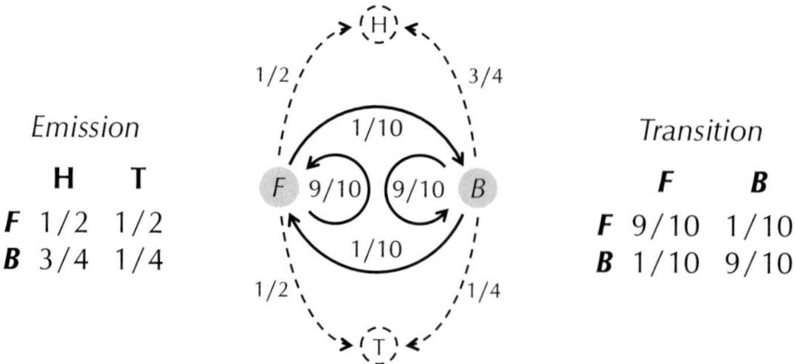

FIGURE 10.5 The transition and emission probability matrices for the crooked dealer HMM described by the HMM diagram shown in the center. This HMM has two states (gray nodes), *F* and *B*. In each state, the HMM can emit one of two symbols (dashed nodes), heads ("H") or tails ("T"), with the probabilities shown along dashed edges. Transition probabilities are shown on solid edges; the crooked dealer HMM transitions between states *F* and *B* with probability 1/10 and remains in the same state with probability 9/10.

A **hidden path** $\pi = \pi_1 \ldots \pi_n$ in an HMM is the sequence of states that the HMM passes through; such a path corresponds to a path of solid edges in the HMM diagram. Figure 10.6 presents an example in which the crooked dealer HMM produces a sequence of flips x = "THTHHHTHTTH" with hidden path $\pi = FFFBBBBBFFF$, i.e., the fair coin is used for the first three flips and last three flips, and the biased coin is used for the five intermediate flips.

Reformulating the Casino Problem

We can now rephrase the improperly formulated Casino Problem as finding the most likely hidden path π for a string x of symbols emitted by an HMM. To solve this problem,

i	1	2	3	4	5	6	7	8	9	10	11
x	T	H	T	H	H	H	T	H	T	T	H
π	F	F	F	B	B	B	B	B	F	F	F
$\Pr(\pi_i \rightarrow \pi_{i+1})$	$\frac{1}{2}$	$\frac{9}{10}$	$\frac{9}{10}$	$\frac{1}{10}$	$\frac{9}{10}$	$\frac{9}{10}$	$\frac{9}{10}$	$\frac{9}{10}$	$\frac{1}{10}$	$\frac{9}{10}$	$\frac{9}{10}$
$\Pr(x_i \mid \pi_i)$	$\frac{1}{2}$	$\frac{1}{2}$	$\frac{1}{2}$	$\frac{3}{4}$	$\frac{3}{4}$	$\frac{3}{4}$	$\frac{1}{4}$	$\frac{3}{4}$	$\frac{1}{2}$	$\frac{1}{2}$	$\frac{1}{2}$

FIGURE 10.6 A sequence x of emitted symbols along with a hidden path π for the crooked dealer HMM. $\Pr(\pi_i \rightarrow \pi_{i+1})$ denotes the probability $transition_{\pi_i, \pi_{i+1}}$ of transitioning from state π_i to π_{i+1}. $\Pr(\pi_0 \rightarrow \pi_1)$ is set equal to 1/2 to comply with the assumption that in the beginning, the dealer is equally likely to use the fair or biased coin. $\Pr(x_i|\pi_i)$ denotes the probability that the dealer produced symbol x_i from state π_i and is equal to $emission_{\pi_i}(x_i)$.

we will first consider the simpler problem of computing the probability $\Pr(x, \pi)$ that an HMM follows the hidden path $\pi = \pi_1 \ldots \pi_n$ and emits the string $x = x_1 \ldots x_n$. Note that

$$\sum_{\text{all strings of emitted symbols } x} \quad \sum_{\text{all hidden paths } \pi} \Pr(x, \pi) = 1.$$

STOP and Think: What is $\Pr(x, \pi)$ for the x and π in Figure 10.6?

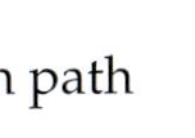

Each emitted string x has probability $\Pr(x)$, which is independent of the hidden path taken by the HMM:

$$\Pr(x) = \sum_{\text{all hidden paths } \pi} \Pr(x, \pi).$$

Each hidden path π has probability $\Pr(\pi)$, which is independent of the string that the HMM emits:

$$\Pr(\pi) = \sum_{\text{all strings of emitted symbols } x} \Pr(x, \pi).$$

The event "the HMM follows the hidden path π and emits x" can be thought of as a combination of two consecutive events:

- The HMM follows the path π. The probability of this event is $\Pr(\pi)$.
- The HMM emits x, given that the HMM follows the path π. We refer to the probability of this event as the **conditional probability** of x given π, denoted $\Pr(x|\pi)$.

Both of these events must occur for the HMM to follow path π and emit string x, which implies that

$$\Pr(x, \pi) = \Pr(x|\pi) \cdot \Pr(\pi).$$

PAGE 583 To learn more about this formula, see **DETOUR: Conditional Probability**.

To compute $\Pr(x, \pi)$, we will first compute $\Pr(\pi)$. As shown in Figure 10.6, we write $\Pr(\pi_i \rightarrow \pi_{i+1})$ to denote the transition probability of the HMM transitioning from state π_i to π_{i+1}. For simplicity, we assume that in the beginning, the dealer is equally likely to use the fair or biased coin, an assumption that is modeled by setting $\Pr(\pi_0 \rightarrow \pi_1) = 1/2$ in Figure 10.6, where π_0 is a "silent" **initial state** that does not emit any symbols. The probability of π is therefore equal to the product of its transition probabilities (green elements in Figure 10.6),

$$\Pr(\pi) = \prod_{i=1}^{n} \Pr(\pi_{i-1} \rightarrow \pi_i) = \prod_{i=1}^{n} transition_{\pi_{i-1}, \pi_i}.$$

Probability of a Hidden Path Problem:
Compute the probability of an HMM's hidden path.

Input: A hidden path π in an HMM (Σ, *States*, *Transition*, *Emission*).
Output: The probability of this path, $\Pr(\pi)$.

Note that we have already computed $\Pr(x|\pi)$ for the crooked dealer HMM when the dealer's hidden path consisted only of B or F, which we wrote as $\Pr(x|B)$ and $\Pr(x|F)$, respectively. To compute $\Pr(x|\pi)$ for a general HMM, we will write $\Pr(x_i|\pi_i)$ to denote the emission probability $emission_{\pi_i}(x_i)$ that symbol x_i was emitted given that the HMM was in state π_i (Figure 10.6). As a result, for a given path π, the HMM emits a string x with probability equal to the product of emission probabilities along that path,

$$\begin{aligned}\Pr(x|\pi) &= \prod_{i=1}^{n} \Pr(x_i|\pi_i) \\ &= \prod_{i=1}^{n} emission_{\pi_i}(x_i).\end{aligned}$$

Probability of an Outcome Given a Hidden Path Problem:
Compute the probability that an HMM will emit a string given its hidden path.

> **Input**: A string $x = x_1 \ldots x_n$ emitted by an HMM (Σ, *States*, *Transition*, *Emission*) and a hidden path $\pi = \pi_1 \ldots \pi_n$.
> **Output**: The conditional probability $\Pr(x|\pi)$ that x will be emitted given that the HMM follows the hidden path π.

Returning to our formula for $\Pr(x, \pi)$, the probability that an HMM follows path π and emits string x can be written as a product of emission and transition probabilities,

$$\begin{aligned}\Pr(x, \pi) &= \Pr(x|\pi) \cdot \Pr(\pi) \\ &= \prod_{i=1}^{n} \Pr(x_i|\pi_i) \cdot \Pr(\pi_{i-1} \rightarrow \pi_i) \\ &= \prod_{i=1}^{n} \textit{emission}_{\pi_i}(x_i) \cdot \textit{transition}_{\pi_{i-1}, \pi_i} .\end{aligned}$$

Exercise Break: Compute $\Pr(x, \pi)$ for the x and π in Figure 10.6. Can you find a better explanation for x = "THTHHHTHTTH" than $\pi = FFFBBBBBFFF$?

STOP and Think: Now that you have learned about HMMs, try designing an HMM that will model searching for CG-islands in a genome. What barriers do you encounter?

The Decoding Problem

The Viterbi graph

As we stated in the previous section, in both the crooked dealer and CG-island HMMs, we are looking for the most likely hidden path π for an HMM that emits a string x. In other words, we would like to maximize $\Pr(x, \pi)$ among all possible hidden paths π.

Decoding Problem:
Find an optimal hidden path in an HMM given a string of its emitted symbols.

> **Input**: A string $x = x_1 \ldots x_n$ emitted by an HMM (Σ, *States*, *Transition*, *Emission*).
> **Output**: A path π that maximizes the probability $\Pr(x, \pi)$ over all possible paths through this HMM.

In 1967, Andrew Viterbi used an HMM-inspired analog of a Manhattan-like grid to solve the Decoding Problem. For an HMM emitting a string of n symbols $x = x_1 \ldots x_n$, the nodes in the HMM's **Viterbi graph** are divided into $|States|$ rows and n columns (Figure 10.7 (middle)). That is, node (k, i) represents state k and the i-th emitted symbol. Each node is connected to all nodes in the column to its right; the edge connecting $(l, i-1)$ to (k, i) corresponds to transitioning from state l to state k (with probability $transition_{l,k}$) and then emitting symbol x_i (with probability $emission_k(x_i)$). As a result, every path connecting a node in the first column of the Viterbi graph to a node in the final column corresponds to a hidden path $\pi = \pi_1 \ldots \pi_n$.

We assign a weight of

$$\text{WEIGHT}_i(l, k) = transition_{\pi_{i-1},\pi_i} \cdot emission_{\pi_i}(x_i)$$

to the edge connecting $(l, i-1)$ to (k, i) in the Viterbi graph. Furthermore, we define the **product weight** of a path in the Viterbi graph as the product of its edge weights. For a path from the leftmost column to the rightmost column in the Viterbi graph corresponding to the hidden path π, this product weight is equal to the product of $n-1$ terms,

$$\prod_{i=2}^{n} transition_{\pi_{i-1},\pi_i} \cdot emission_{\pi_i}(x_i) = \prod_{i=2}^{n} \text{WEIGHT}_i(l, k).$$

STOP and Think: How does this expression differ from the formula for $\Pr(x, \pi)$ that we derived in the previous section?

The only difference between the above expression and the expression that we obtained for $\Pr(x, \pi)$,

$$\prod_{i=1}^{n} transition_{\pi_{i-1},\pi_i} \cdot emission_{\pi_i}(x_i),$$

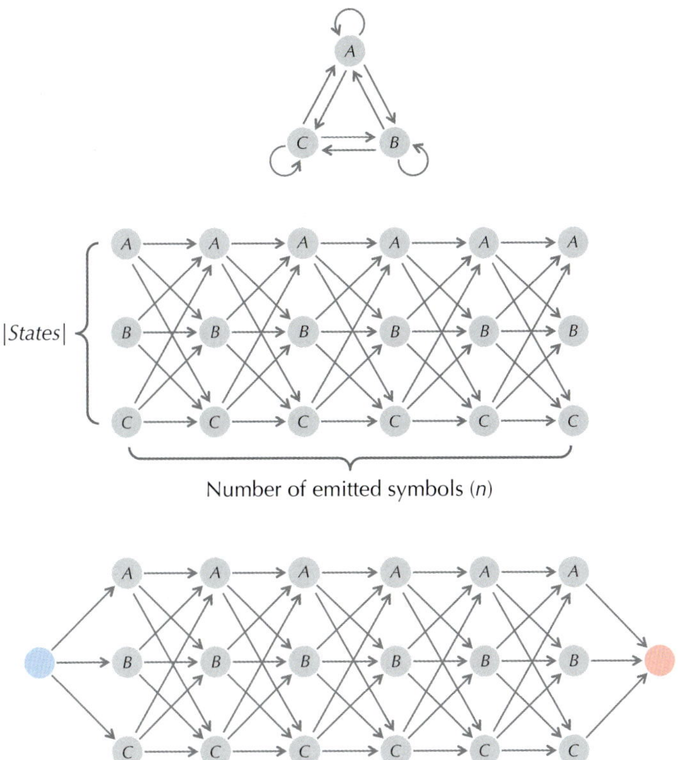

FIGURE 10.7 (Top) The diagram of an HMM with three states (emission/transition probabilities as well as nodes corresponding to emitted symbols are omitted). (Middle) Given a string of n symbols $x = x_1 \ldots x_n$ emitted by an HMM, Viterbi's Manhattan is a grid with $|States|$ rows and n columns in which each node is connected to every node in the column to its right. The weight of the edge connecting $(l, i-1)$ to (k, i) is $\text{WEIGHT}_i(l, k) = transition_{l,k} \cdot emission_k(x_i)$. Unlike the alignment graphs from Chapter 5, in which the set of valid directions was restricted to south, east, and southeast edges, every node in a column is connected by an edge to every node in the column to its right in the Viterbi graph. (Bottom) The Viterbi graph with additional source node (blue) and sink node (red). A path of largest product weight connecting the source to the sink corresponds to an optimal hidden path solving the Decoding Problem.

is the single factor $transition_{\pi_0,\pi_1} \cdot emission_{\pi_1}(x_1)$, which corresponds to transitioning from the initial state π_0 to π_1 and emitting the first symbol. To model the initial state, we will add a source node *source* to the Viterbi graph and then connect *source* to each node $(k, 1)$ in the first column with an edge of weight $\text{WEIGHT}_1(source, k) = transition_{\pi_0,k} \cdot emission_k(x_1)$. We will also assume that the HMM has another silent

terminal state that the HMM enters when it has finished emitting symbols. To model the terminal state, we add a sink node *sink* to the Viterbi graph and connect every node in the last column to *sink* with an edge of weight 1 (Figure 10.7 (bottom)).

Every hidden path π in the HMM now corresponds to a path from *source* to *sink* in the Viterbi graph with product weight $\Pr(x, \pi)$. Therefore, the Decoding Problem reduces to finding a path in the Viterbi graph of largest product weight over all paths connecting *source* to *sink*.

Exercise Break: Find the maximum product weight path in the Viterbi graph for the crooked dealer HMM when x= "HHTT".

The Viterbi algorithm

We will apply a dynamic programming algorithm to solve the Decoding Problem. First, define $s_{k,i}$ as the product weight of an optimal path (i.e., a path with maximum product weight) from *source* to the node (k, i). The **Viterbi algorithm** relies on the fact that the first $i-1$ edges of an optimal path from *source* to (k, i) must form an optimal path from *source* to $(l, i-1)$ for some (unknown) state l. This observation yields the following recurrence:

$$\begin{aligned} s_{k,i} &= \max_{\text{all states } l} \left\{ s_{l,i-1} \cdot (\text{weight of edge between nodes}(l, i-1) \text{ and } (k, i)) \right\} \\ &= \max_{\text{all states } l} \left\{ s_{l,i-1} \cdot \text{WEIGHT}_i(l, k) \right\} \\ &= \max_{\text{all states } l} \left\{ s_{l,i-1} \cdot \textit{transition}_{\pi_{i-1},\pi_i} \cdot \textit{emission}_{\pi_i}(x_i) \right\} \end{aligned}$$

Since *source* is connected to every node in the first column of the Viterbi graph,

$$\begin{aligned} s_{k,1} &= s_{source} \cdot (\text{weight of edge between } \textit{source} \text{ and } (k, 1)) \\ &= s_{source} \cdot \text{WEIGHT}_0(\textit{source}, k) \\ &= s_{source} \cdot \textit{transition}_{source,k} \cdot \textit{emission}_k(x_1) \end{aligned}$$

In order to initialize this recurrence, we set s_{source} equal to 1. We can now compute the maximum product weight over all paths from *source* to *sink* as

$$s_{sink} = \max_{\text{all states } l} s_{l,n} \,.$$

STOP and Think: How can we adapt our algorithm for finding a longest path in a DAG to find a path with maximum product weight?

How fast is the Viterbi algorithm?

We can interpret the Decoding Problem as yet another instance of the Longest Path in a DAG Problem from Chapter 5 because the path π maximizing the product weight $\prod_{i=1}^{n} \text{WEIGHT}_i(\pi_{i-1}, \pi_i)$ also maximizes the logarithm of this product, which is equal to $\sum_{i=1}^{n} \log(\text{WEIGHT}_i(\pi_{i-1}))$. Thus, we can substitute the weights of all edges in the Viterbi graph by their logarithms. Finding a longest path (i.e. a path maximizing the *sum* of edge weights) in the resulting graph will correspond to a path of maximum *product* weight in the original Viterbi graph. For this reason, the runtime of the Viterbi algorithm, which you are now ready to implement, is linear in the number of edges in the Viterbi graph. The following exercise shows that the number of these edges is $\mathcal{O}(|States|^2 \cdot n)$, where n is the number of emitted symbols.

Exercise Break: Show that the number of edges in the Viterbi graph of an HMM emitting a string of length n is $|States|^2 \cdot (n-1) + 2 \cdot |States|$.

Exercise Break: Apply your solution for the Decoding Problem to find CG-islands in the first million nucleotides from the human X chromosome. To help you design an HMM for this application, you may assume that transitions from CG-islands to non-CG-islands are rare, occurring with probability 0.001, and that transitions from non-CG-islands to CG-islands are even more rare, occurring with probability 0.0001. How many CG-islands do you find?

In practice, many HMMs have **forbidden transitions** between some states. For such transitions, we can safely remove the corresponding edges from the HMM diagram (Figure 10.8 (left)). This operation results in a sparser Viterbi graph (Figure 10.8 (right)), which reduces the runtime of the Viterbi algorithm, since the runtime of the algorithm for finding the longest path in a DAG is linear in the number of edges in the DAG.

Exercise Break: Let *Edges* denote the set of edges in the diagram of an HMM that may have some forbidden transitions. Prove that the number of edges in the Viterbi graph for this HMM is $|Edges| \cdot (n-1) + 2 \cdot |States|$.

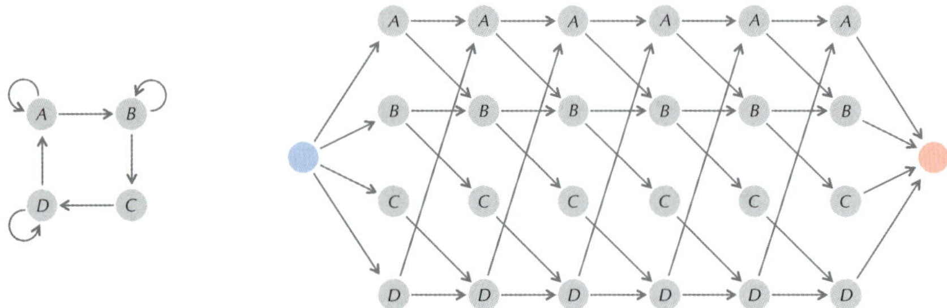

FIGURE 10.8 (Left) An HMM diagram for an HMM that has four states with some forbidden transitions, such as from *A* to *D* and from *C* to itself. Edges corresponding to forbidden transitions between states are not included in the HMM diagram. (Right) The Viterbi graph for this HMM emitting a string of length 6.

Finding the Most Likely Outcome of an HMM

Dynamic programming allows us to answer questions about HMMs extending beyond the most likely hidden path. For example, we have already computed the probability $\Pr(\pi)$ of a hidden path π. But what about computing $\Pr(x)$, the probability that the HMM will emit a string x?

Exercise Break: Which outcome is more likely in the crooked casino: "HHTT" or "HTHT"? How would you find the most likely sequence of four coin flips?

Outcome Likelihood Problem:
Find the probability that an HMM emits a given string.

> **Input**: A string $x = x_1 \ldots x_n$ emitted by an HMM (Σ, *States*, *Transition*, *Emission*).
> **Output**: The probability $\Pr(x)$ that the HMM emits x.

STOP and Think: To solve the Outcome Likelihood Problem, you can make a slight change to the Viterbi recurrence $s_{k,i} = \max_{\text{all states } l}\{s_{l,i-1} \cdot \text{WEIGHT}_i(l,k)\}$. What is the change?

We have already observed that $\Pr(x)$ is equal to the sum of $\Pr(x, \pi)$ over all hidden paths π. However, the number of paths through the Viterbi graph is exponential in the length of the emitted string x, and so we will use dynamic programming to develop a faster approach to compute $\Pr(x)$.

We denote the total product weight of all paths from *source* to node (k, i) in the Viterbi graph as $forward_{k,i}$; note that $forward_{sink}$ is equal to $\Pr(x)$. To compute $forward_{k,i}$, we will divide all paths connecting *source* to (k, i) into $|States|$ subsets, where each subset contains those paths that pass through node $(l, i-1)$ (with product weight $forward_{l,i-1}$) before reaching (k, i) for some l between 1 and $|States|$. Therefore, $forward_{k,i}$ is the sum of $|States|$ terms,

$$\begin{aligned} forward_{k,i} &= \sum_{\text{all states } l} forward_{l,i-1} \cdot (\text{weight of edge connecting } (l, i-1) \text{ and } (k, i)) \\ &= \sum_{\text{all states } l} forward_{l,i-1} \cdot \text{WEIGHT}_i(l, k)\,. \end{aligned}$$

Note that the only difference between this recurrence and the Viterbi recurrence,

$$s_{k,i} = \max_{\text{all states } l} \left\{ s_{l,i-1} \cdot \text{WEIGHT}_i(l, k) \right\},$$

is that the maximization in the Viterbi algorithm has changed into a summation symbol. We can now solve the Outcome Likelihood Problem by computing $forward_{sink}$, which is equal to

$$\sum_{\text{all states } k} forward_{k,n}\,.$$

10D

Now that we can compute $\Pr(x)$ for an emitted string x, a natural question is to find the most likely such string. In the crooked dealer example, this corresponds to finding the most likely sequence of flips over all possible sequences of fair and biased coins that the dealer could use.

Most Likely Outcome Problem:
Find a most likely string emitted by an HMM.

Input: An HMM $(\Sigma, States, Transition, Emission)$ and an integer n.
Output: A most likely string $x = x_1 \dots x_n$ emitted by this HMM, i.e., a string maximizing the probability $\Pr(x)$ that the HMM will emit x.

Exercise Break: Solve the Most Likely Outcome Problem (Hint: You may need to build a 3-dimensional version of Viterbi's Manhattan).

Profile HMMs for Sequence Alignment

How do HMMs relate to sequence alignment?

You may still be wondering what in the world HMMs have to do with our original problem of aligning sequences using a column-specific score. As we will see, HMMs offer an elegant solution to this problem.

Given a family of related proteins, we can check whether a newly sequenced protein belongs to this family by constructing pairwise alignments between the newly sequenced protein and each member of the family. If one of the resulting alignments scores above some stringent threshold, then we can assume that the new protein belongs to the family. However, this approach may fail to identify distantly related proteins, such as gp120 proteins taken from different HIV isolates, since these proteins may have scores falling below the threshold. If a sequence has weak similarities with many family members, then it most likely belongs to the family.

The problem, then, is to align a new protein to *all* members of the family at once. To do so, we must assume that we have already constructed a multiple alignment of a family of proteins. Fortunately, it will often be obvious that two proteins come from the same family (e.g., if the proteins are taken from closely related species). Accordingly, biologists often start by constructing an alignment of undeniably related proteins, which are typically easy to align even using the simple multiple alignment methods that we covered in Chapter 5.

Figure 10.9 (first panel) shows a 5×10 alignment *Alignment* representing a hypothetical family of proteins. Note that the sixth and seventh columns of this alignment contain many space symbols and likely do not represent meaningful characteristics of the family. Accordingly, biologists often ignore columns for which the fraction of space symbols is greater than or equal to a **column removal threshold** θ. Column removal results in a 5×8 **seed alignment** (Figure 10.9 (second panel)).

Given a seed alignment $Alignment^*$ representing a family of related proteins, our aim is to build an HMM that realistically models the propensities of symbols in $Alignment^*$ represented by the profile matrix PROFILE($Alignment^*$) (Figure 10.9 (third panel)). Rather than thinking about aligning the existing seed alignment to a given string *Text* (representing a new protein), we will instead think about computing the probability that the HMM emits *Text*. If the HMM is designed well, then the more similar *Text* is to the strings in $Alignment^*$, the more likely it will be emitted by the HMM.

We will first construct a simple HMM that treats the columns of $Alignment^*$ as k sequentially linked states called **match states** (Figure 10.9 (fourth panel)), denoted

	1	2	3	4	5			6	7	8
	A	C	D	E	F	A	C	A	D	F
	A	F	D	A	–	–	–	C	C	F
Alignment	A	–	–	E	F	D	–	F	D	C
	A	C	A	E	F	–	–	A	–	C
	A	D	D	E	F	A	A	A	D	F

	A	C	D	E	F	A	D	F
	A	F	D	A	–	C	C	F
*Alignment**	A	–	–	E	F	F	D	C
	A	C	A	E	F	A	–	C
	A	D	D	E	F	A	D	F

	A	1	0	1/4	1/5	0	3/5	0	0
	C	0	2/4	0	0	0	1/5	1/4	2/5
PROFILE(*Alignment**)	D	0	1/4	3/4	0	0	0	3/4	0
	E	0	0	0	4/5	0	0	0	0
	F	0	1/4	0	0	1	1/5	0	3/5

$M_1 \rightarrow M_2 \rightarrow M_3 \rightarrow M_4 \rightarrow M_5 \rightarrow M_6 \rightarrow M_7 \rightarrow M_8$

FIGURE 10.9 A 5×10 multiple alignment *Alignment* (first panel), its 5×8 seed alignment *Alignment** (second panel), the profile matrix PROFILE(*Alignment**) of the seed alignment (third panel), and the diagram of a simple HMM that models this profile (fourth panel). The seed alignment is obtained from the original alignment by ignoring poorly conserved columns (shaded gray); in this case, we ignore columns for which the fraction of space symbols is greater than or equal to the column removal threshold $\theta = 0.35$. To better illustrate the relationship between the alignment and its seed alignment, we have separated the first five columns in the seed alignment from its last three columns and numbered these columns above the original alignment. The match states MATCH(i) are abbreviated as M_i. The HMM only has one possible path; it is initially in state MATCH(1), the transition probability from state MATCH(i) to state MATCH($i+1$) is equal to 1 for all i, and all other transitions are forbidden. Emission probabilities are equal to frequencies in the profile, e.g., emission probabilities for M_2 are 0 for A, 2/4 for C, 1/4 for D, 0 for E, and 1/4 for F.

MATCH(1),...,MATCH(k). When the HMM enters state MATCH(i), it emits symbol x_i with probability equal to the frequency of this symbol in the i-th column of PROFILE(*Alignment**). The HMM then moves into state MATCH($i+1$) with transition probability equal to 1.

The **similarity score** between *Alignment** and *Text* is the probability Pr(*Text*) that the HMM for *Alignment** emits *Text*. This score is equal to the product of frequencies in PROFILE(*Alignment**) corresponding to each symbol of *Text*. For example, the probability that the HMM in Figure 10.9 emits ADDAFFDF is

$$1 \cdot \frac{1}{4} \cdot \frac{3}{4} \cdot \frac{1}{5} \cdot 1 \cdot \frac{1}{5} \cdot \frac{3}{4} \cdot \frac{3}{5} = 0.003375.$$

STOP

STOP and Think: What are the limitations of the HMM in Figure 10.9?

The HMM that we have proposed does score each column in Figure 10.9 differently, and to a degree, the more similar *Text* is to *Alignment**, the higher its similarity score. However, this HMM it is not in keeping with the spirit of HMMs because it has only one hidden path. Furthermore, it offers a simplistic view of multiple alignment because it does not account for insertions and deletions. Finally, it can only "align" *Text* against *Alignment** if the length of *Text* is exactly equal to the number of columns in *Alignment** (Figure 10.10). Yet we will use this limited HMM as the foundation of a more powerful HMM.

	A	C	D	E	F	A	D	F
	A	F	D	A	-	C	C	F
*Alignment**	A	-	-	E	F	F	D	C
	A	C	A	E	F	A	-	C
	A	D	D	E	F	A	D	F
Text	A	D	D	A	F	F	D	F
emission probability	1	1/4	3/4	1/5	1	1/5	3/4	3/5

FIGURE 10.10 Aligning *Text* = ADDAFFDF against the seed alignment *Alignment** represented as a simple HMM in Figure 10.9. This HMM is limited because we are not able to align a string of length other than 8. Indeed, there is no way to add space symbols to *Text* or to add symbols of *Text* "between" columns of *Alignment**.

Building a profile HMM

The improved HMM that we propose is called a **profile HMM**. Given a multiple alignment *Alignment* and a column removal threshold θ used to obtain a seed alignment *Alignment**, we will denote this profile HMM as HMM(*Alignment*, θ). Because the profile HMM will be constructed from the seed alignment, we will also informally refer to it as HMM(*Alignment**). Given a string *Text* to align against the existing seed alignment, our goal is to find an optimal hidden path in the profile HMM by solving the Decoding Problem for this HMM and the emitted string *Text*.

As with our first attempt at an HMM from Figure 10.9, the profile HMM will still traverse its states in an order consistent with traversing the columns of *Alignment** from left to right. However, to align strings *Text* of varying lengths, we will need more states in addition to the k match states.

First, we add $k+1$ **insertion states**, denoted INSERTION$(0), \ldots,$ INSERTION(k) (Figure 10.11). Entering INSERTION(i) allows the profile HMM to emit an additional symbol after visiting the i-th column of PROFILE(*Alignment**) and before entering the $(i+1)$-th column. Thus, we will connect MATCH(i) to INSERTION(i) and INSERTION(i) to MATCH$(i+1)$. Furthermore, to allow for multiple inserted symbols between columns of PROFILE(*Alignment**), we will connect INSERTION(i) to itself.

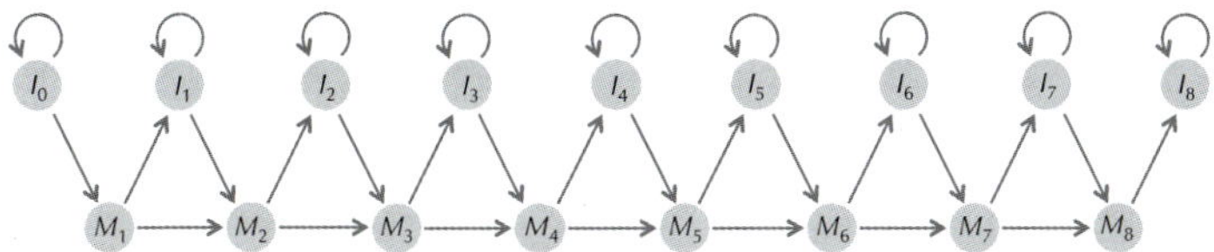

FIGURE 10.11 An HMM diagram for the seed alignment in Figure 10.9 with match and insertion states, abbreviated as M and I, respectively. The states I_0 and I_8 model insertions of symbols occurring before the beginning and end of *Alignment**, respectively.

STOP and Think: Can we use the HMM in Figure 10.11 to align a string *Text* of length less than 8?

After modeling insertions of new symbols in PROFILE(*Alignment**), we should also model "deletions" allowing the profile HMM to skip columns of PROFILE(*Alignment**). One way of modeling these deletions is to add edges connecting every state in the profile HMM to every state on its right (Figure 10.12).

STOP and Think: Revisit the Exercise Break on page 547 to recall that the running time of the Viterbi algorithm is proportional to the number of edges (with non-zero transition probabilities) in the HMM diagram. How many edges will the diagram in Figure 10.12 have? How can we reduce the number of edges in the HMM diagram?

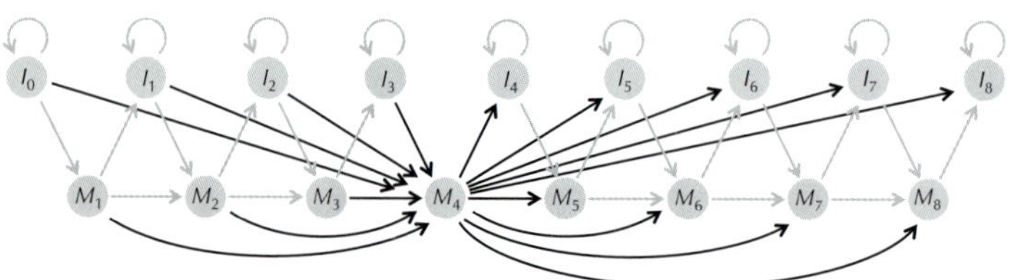

FIGURE 10.12 By adding edges connecting each state in the profile HMM from Figure 10.11 to every state on its right, we can skip columns of *Alignment* when comparing *Text* against this alignment. The above HMM diagram highlights all edges leading into and out of MATCH(4).

Instead of skipping states as in Figure 10.12, we can reduce the number of edges in the HMM diagram by introducing k silent **deletion states** DELETION$(1), \ldots,$ DELETION(k) (Figure 10.13). For example, instead of jumping from MATCH$(i-1)$ to MATCH$(i+1)$, we can make the transition MATCH$(i-1) \rightarrow$ DELETION$(i) \rightarrow$ MATCH$(i+1)$. Entering DELETION(i) allows the HMM to skip over a column of the alignment without emitting a symbol.

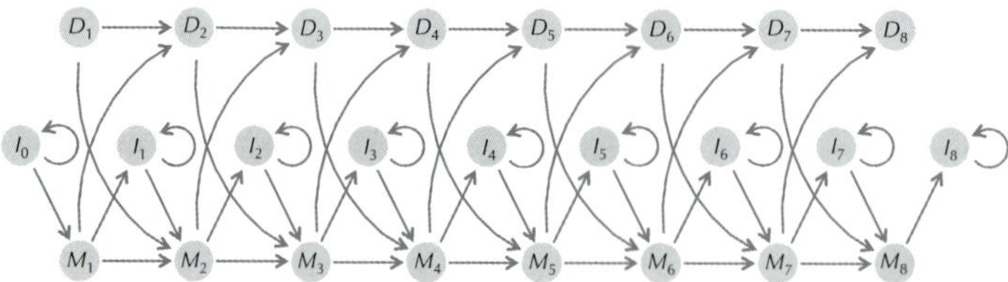

FIGURE 10.13 Adding silent deletion states (abbreviated as D_i) to the profile HMM diagram.

STOP and Think: Is the HMM in Figure 10.13 now adequate, or is there anything else that we have forgotten to add?

We can now transition back and forth between match states and insertion states, as well as back and forth between match states and deletion states, but we cannot transition between insertion states and deletion states. The profile HMM diagram should therefore include edges connecting INSERTION(i) to DELETION$(i+1)$ and connecting DELETION(i) to INSERTION(i) for each i. As a result, the profile HMM can move from any match/insertion state to any other match/insertion state on its right by sidetracking through intermediate deletion states. We obtain the complete profile HMM diagram shown in Figure 10.14 after connecting the initial state (S) to the first match/insertion/deletion states and connecting the final match/insertion/deletion states to the terminal state (E).

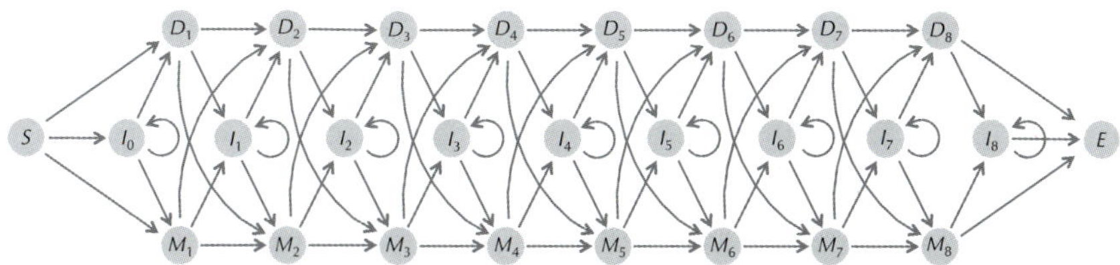

FIGURE 10.14 Adding transitions from insertion states to deletion states and vice-versa completes the profile HMM diagram for the profile matrix from Figure 10.9. Silent initial and terminal states are shown by S and E, respectively.

STOP and Think: Consider the following questions.

- How many edges does the HMM diagram in Figure 10.14 have? How does this compare to the HMM diagram in Figure 10.12?
- What does the Viterbi graph of the profile HMM in Figure 10.14 look like? How many nodes and edges does it have?

Transition and emission probabilities of a profile HMM

In Figure 10.15, we return to the multiple alignment *Alignment* from Figure 10.9 and represent each of the five colored rows of this alignment as a path in the diagram of HMM(*Alignment**). Symbols in the seed alignment *Alignment** (non-shaded columns) correspond to either a match state (non-space symbols) or a deletion state (space symbols). As for symbols not present in the seed alignment (shaded columns), space symbols are ignored, and non-space symbols are emitted from insertion states.

STOP and Think: How would you assign transition and emission probabilities for the profile HMM of the alignment in Figure 10.15?

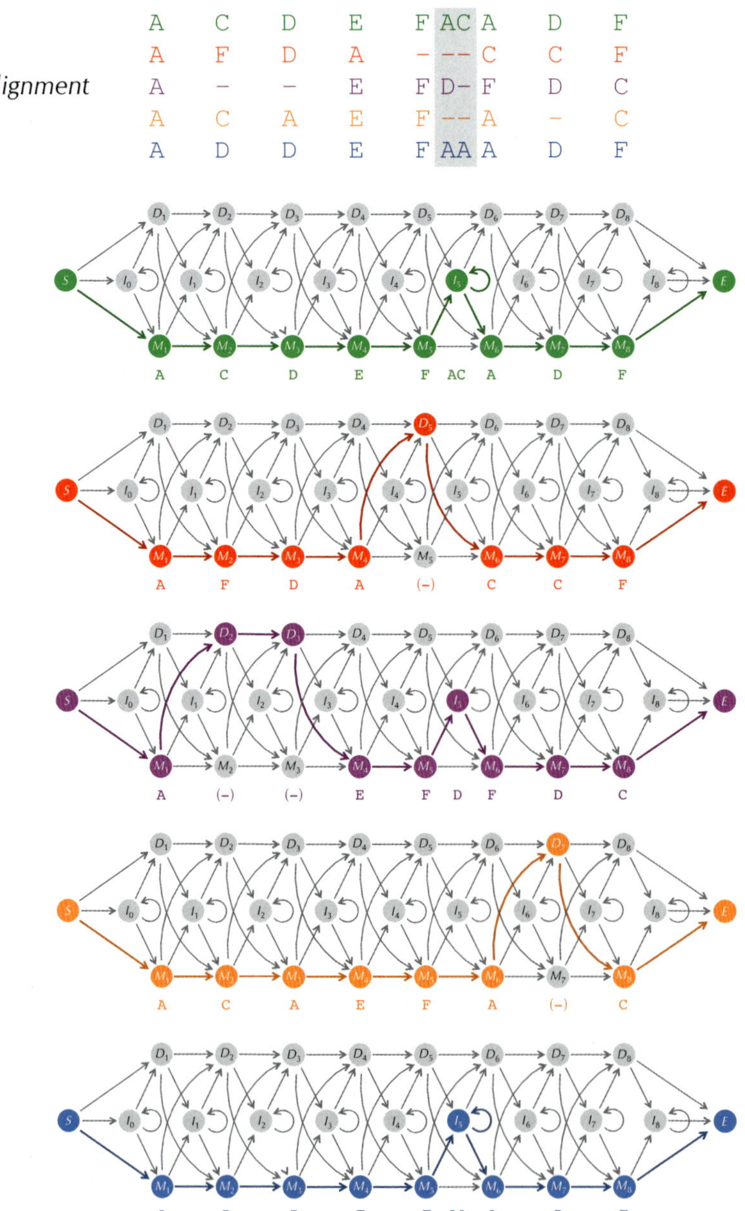

FIGURE 10.15 Five paths through the profile HMM corresponding to the five rows in the alignment from Figure 10.9. Space symbols below an HMM diagram correspond to deletion states and are shown in parentheses to indicate that they are not emitted by the HMM.

To assign the transition probability $transition_{l,k}$, we simply take the frequency of transitions from state l to state k made by these colored paths with respect to all paths that visited state l. For example, in Figure 10.15, four of the colored paths visit MATCH(5). Three of these paths then transition to INSERTION(5), and one transitions to MATCH(6). Thus, we set the following transition probabilities leaving MATCH(5):

$$\begin{aligned} transition_{\text{MATCH}(5),\,\text{INSERTION}(5)} &= 3/4 \\ transition_{\text{MATCH}(5),\,\text{MATCH}(6)} &= 1/4 \\ transition_{\text{MATCH}(5),\,\text{DELETION}(6)} &= 0 \end{aligned}$$

We can define the transition probabilities from the initial state analogously. For the multiple alignment in Figure 10.15, we enter MATCH(1) with probability 1; for a general profile HMM, the only other states we could enter from the initial state are INSERTION(0) and DELETION(1). The complete matrix of transition probabilities is shown in Figure 10.16.

STOP and Think: Due to the small number of strings in the alignment from Figure 10.15, many of the transition probabilities in the gray cells in Figure 10.16 are equal to zero. What are the possible negative consequences of these zeroes, and how would you address these consequences?

To assign the emission probability $emission_k(b)$, we divide the number of times that symbol b was emitted from state k by the total number of symbols emitted from state k. For example, in Figure 10.15, there are three occurrences of A, one occurrence of C, and one occurrence of D emitted from the state INSERTION(5). (Note: these symbols occur in shaded columns of *Alignment*, where the fraction of space symbols exceeds the column removal threshold θ.) Also, there are two occurrences of C, one occurrence of D, and one occurrence of F emitted from MATCH(2). We can therefore infer the following emission probabilities for these two states:

$$\begin{aligned} emission_{\text{INSERTION}(5)}(\texttt{A}) &= 3/5 & emission_{\text{MATCH}(2)}(\texttt{A}) &= 0 \\ emission_{\text{INSERTION}(5)}(\texttt{C}) &= 1/5 & emission_{\text{MATCH}(2)}(\texttt{C}) &= 2/4 \\ emission_{\text{INSERTION}(5)}(\texttt{D}) &= 1/5 & emission_{\text{MATCH}(2)}(\texttt{D}) &= 1/4 \\ emission_{\text{INSERTION}(5)}(\texttt{E}) &= 0 & emission_{\text{MATCH}(2)}(\texttt{E}) &= 0 \\ emission_{\text{INSERTION}(5)}(\texttt{F}) &= 0 & emission_{\text{MATCH}(2)}(\texttt{F}) &= 1/4 \end{aligned}$$

Exercise Break: Construct the 27×20 emission probability matrix for HMM(*Alignment*, 0.35) derived from *Alignment* in Figure 10.9.

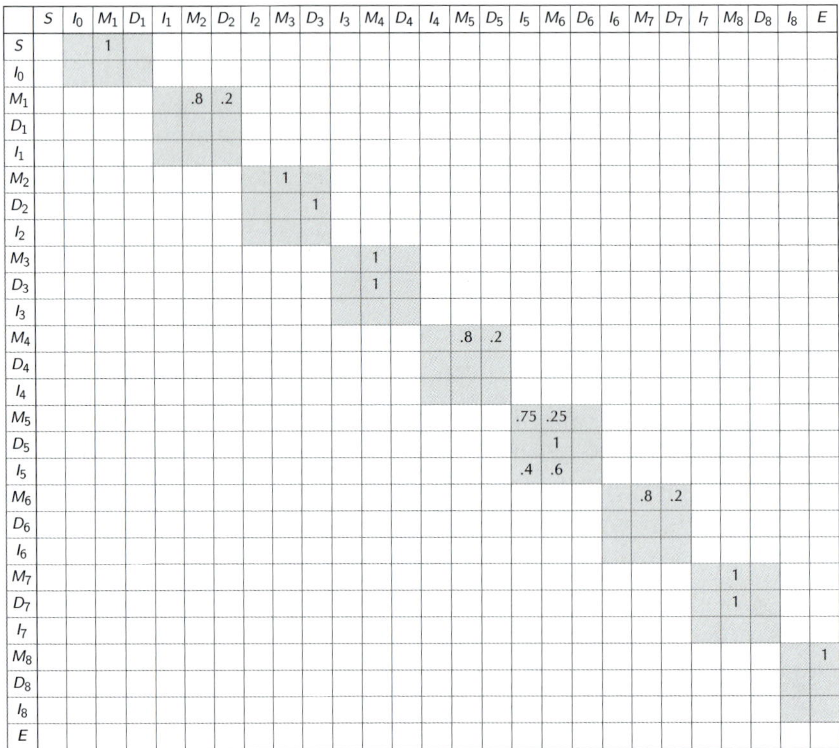

	S	I_0	M_1	D_1	I_1	M_2	D_2	I_2	M_3	D_3	I_3	M_4	D_4	I_4	M_5	D_5	I_5	M_6	D_6	I_6	M_7	D_7	I_7	M_8	D_8	I_8	E
S			1																								
I_0																											
M_1						.8	.2																				
D_1																											
I_1																											
M_2									1																		
D_2										1																	
I_2																											
M_3												1															
D_3												1															
I_3																											
M_4															.8	.2											
D_4																											
I_4																											
M_5																	.75	.25									
D_5																		1									
I_5																	.4	.6									
M_6																					.8	.2					
D_6																											
I_6																											
M_7																								1			
D_7																								1			
I_7																											
M_8																											1
D_8																											
I_8																											
E																											

FIGURE 10.16 The 27×27 matrix of transition probabilities for HMM(*Alignment*, 0.35), where *Alignment* is the multiple alignment from Figure 10.9 and 0.35 is the column removal threshold. Each row and column of this matrix corresponds to one of the 27 nodes in the HMM diagram in Figure 10.14. All values in empty cells are equal to zero. Cells shaded gray correspond to edges in the HMM diagram from Figure 10.14; cells shaded white correspond to forbidden transitions.

You are now ready to construct the profile HMM for an arbitrary multiple alignment.

Profile HMM Problem:
Construct a profile HMM from a multiple alignment.

Input: A multiple alignment *Alignment* and a threshold θ.
Output: HMM(*Alignment*, θ).

Exercise Break: Construct a profile HMM for the HIV sequences shown in Figure 10.1 with $\theta = 0.35$.

Classifying proteins with profile HMMs

Aligning a protein against a profile HMM

Given a protein family, represented by *Alignment*, we can now return to the problem of deciding whether a newly sequenced protein, represented by *Text*, belongs to the family. We first form HMM(*Alignment*, θ) for some parameter θ. As shown in Figure 10.17, a hidden path through HMM(*Alignment*, θ) corresponds to a sequence of match, insertion, and deletion states for aligning *Text* against *Alignment*.

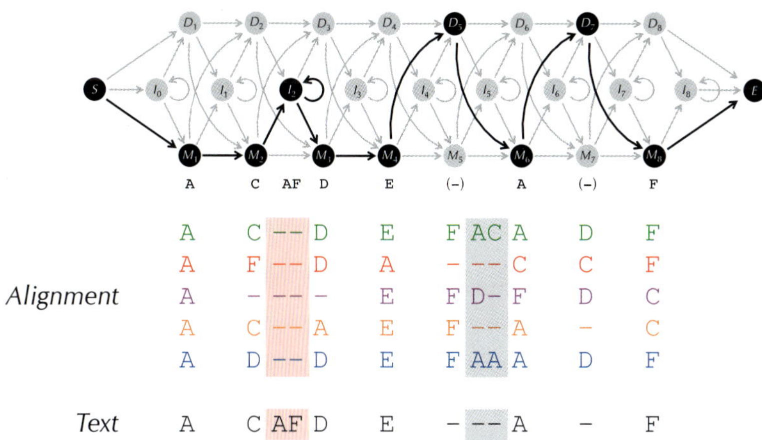

FIGURE 10.17 (Top) A path through HMM(*Alignment*, 0.35) for the multiple alignment from Figure 10.9 and the emitted string *Text* = ACAFDEAF. (Bottom) The emitted symbols correspond to aligning *Text* against *Alignment*. Specifically, the first two symbols are emitted from two match states and belong in the first two positions of the alignment. The next two symbols are emitted from an insertion state and belong in columns of their own (shown in pink). The space symbols in the seventh and eleventh columns above correspond to deletion states; these symbols are not emitted by the HMM. The space symbols in the gray columns do not correspond to any states and are passed over. The non-shaded columns form an augmented 6×8 seed alignment for comparison against newly sequenced proteins.

To find the "best" alignment of *Text* against *Alignment*, we simply need to apply the Viterbi algorithm to find an optimal hidden path in HMM(*Alignment*, θ). If the product weight of this optimal hidden path exceeds a predetermined threshold, then we may conclude that *Text* belongs to the protein family, in which case we augment the existing seed alignment with an additional row corresponding to *Text*. In this way, we can recruit more and more distant family members to a seed alignment, adding these new proteins to the growing multiple alignment, and thus making the resulting profile HMM more and more suitable for analyzing the protein family of interest.

STOP and Think: If the product weight for a new protein exceeds a threshold for more than one protein family, how would you classify this protein?

Profile HMMs have finally helped us achieve our original goal of scoring different columns of a multiple alignment differently based on the frequency of symbols in each column. For example, say that the seventh column of *Alignment** contains more occurrences of A than C, and the ninth column of *Alignment** contains more occurrences of C than A. A hidden path passing through MATCH(7) would be rewarded more for emitting A than C, whereas a hidden path passing through MATCH(9) would be rewarded more for emitting C than A.

The return of pseudocounts

The majority of transition probabilities in the gray cells of Figure 10.16 are equal to zero. (The same is true of emission probabilities.) These zeroes may cause problems; for example, the path in Figure 10.17 seems perfectly reasonable for *Text* = ACAFDEAF, and yet $\Pr(x, \pi)$ is equal to zero because the transition probability from MATCH(2) to INSERTION(2) for this profile HMM is zero.

As in Chapter 2, we will introduce pseudocounts by adding a small value σ to entries in the transition matrix that correspond to edges of the HMM diagram in Figure 10.14 (i.e., only the gray elements of Figure 10.16). Note that white cells in Figure 10.16, corresponding to forbidden transitions, are not affected by pseudocounts. The resulting matrix will then need to be normalized so that the elements in each row sum to 1.

Exercise Break: Compute the normalized matrix for the matrix in Figure 10.16 after adding the pseudocount $\sigma = 0.01$.

We will also add pseudocounts to the matrix of emission probabilities and normalize the resulting matrix. We refer to the profile HMM defined by the resulting normalized matrices of transition and emission probabilities as HMM($Alignment, \theta, \sigma$).

Profile HMM with Pseudocounts Problem:
Construct a profile HMM with pseudocounts from a multiple alignment.

Input: A multiple alignment *Alignment*, a threshold value θ, and a pseudocount value σ.
Output: HMM($Alignment, \theta, \sigma$).

STOP and Think: Since the HMM diagram in Figure 10.17 has 25 nodes — not including the start and end states — the Viterbi graph for the string emitted in this figure has 25 rows. How many columns does this Viterbi graph have?

We are now ready to align a string *Text* to a multiple alignment by constructing the Viterbi graph for this string (Figure 10.18) and solving the Decoding Problem to find the most likely hidden path.

STOP and Think: Find paths through the Viterbi graph corresponding to the bottom four hidden paths in Figure 10.15. What happens?

The troublesome silent states

If you reached this point without any questions about Figure 10.18, then we have successfully concealed from you that solving the Decoding Problem for HMMs with silent states is not as simple as it may appear: the graph in Figure 10.18 is not a Viterbi graph! To see why not, consider the path in Figure 10.19, which emits the same string as Figure 10.18 but passes through one fewer silent deletion state, thus reducing the number of columns by one. But the Viterbi graph is not allowed to change depending on the hidden path π, since we know nothing about the hidden path in advance! Instead, the number of columns in the Viterbi graph must equal the length of the *emitted string*, a condition that is violated in both Figure 10.18 and Figure 10.19.

STOP and Think: How can we modify the notion of the Viterbi graph for HMMs with silent states?

More generally, the Viterbi algorithm does not tolerate silent states other than the initial and terminal states. In other words, this algorithm assumes that node (k, i) in the

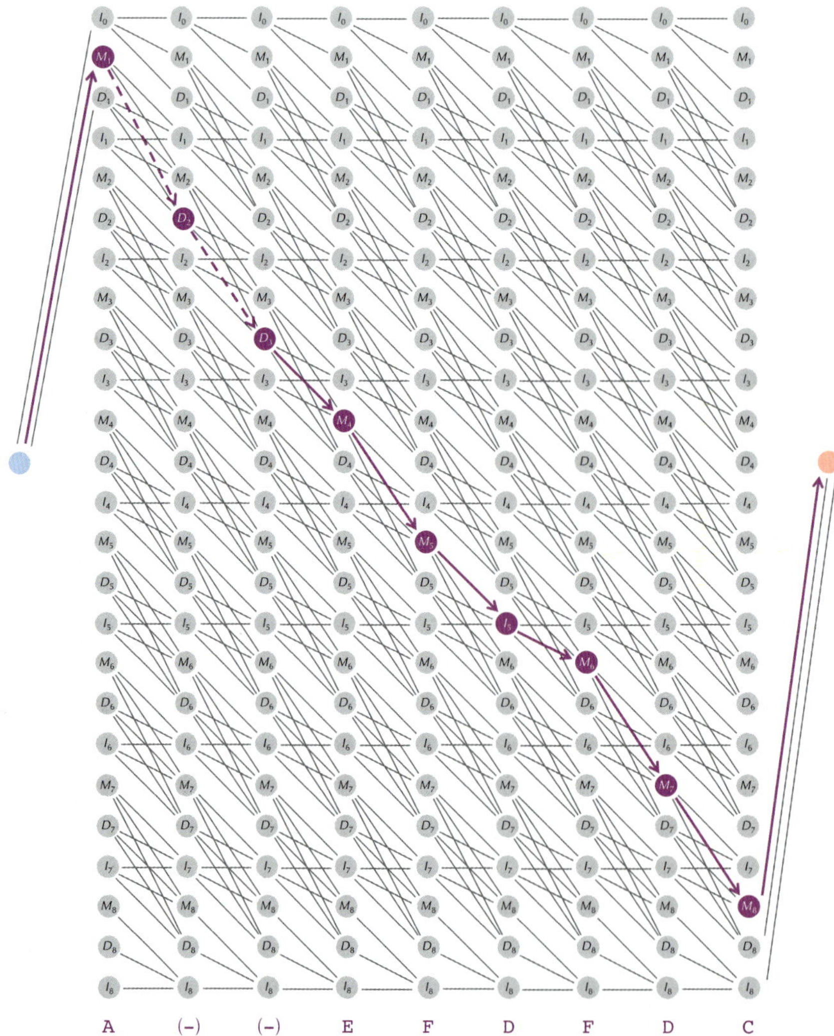

FIGURE 10.18 The Viterbi graph for HMM(*Alignment*, θ) and a path in this graph (shown in purple) corresponding to the hidden path for the emitted string **AEFDFDC** from Figure 10.15. Edges between columns correspond to allowed transitions in the HMM diagram from Figure 10.14 and have an implied rightward orientation. Edges entering nodes corresponding to deletion states are dashed. Emitted symbols are shown beneath each column.

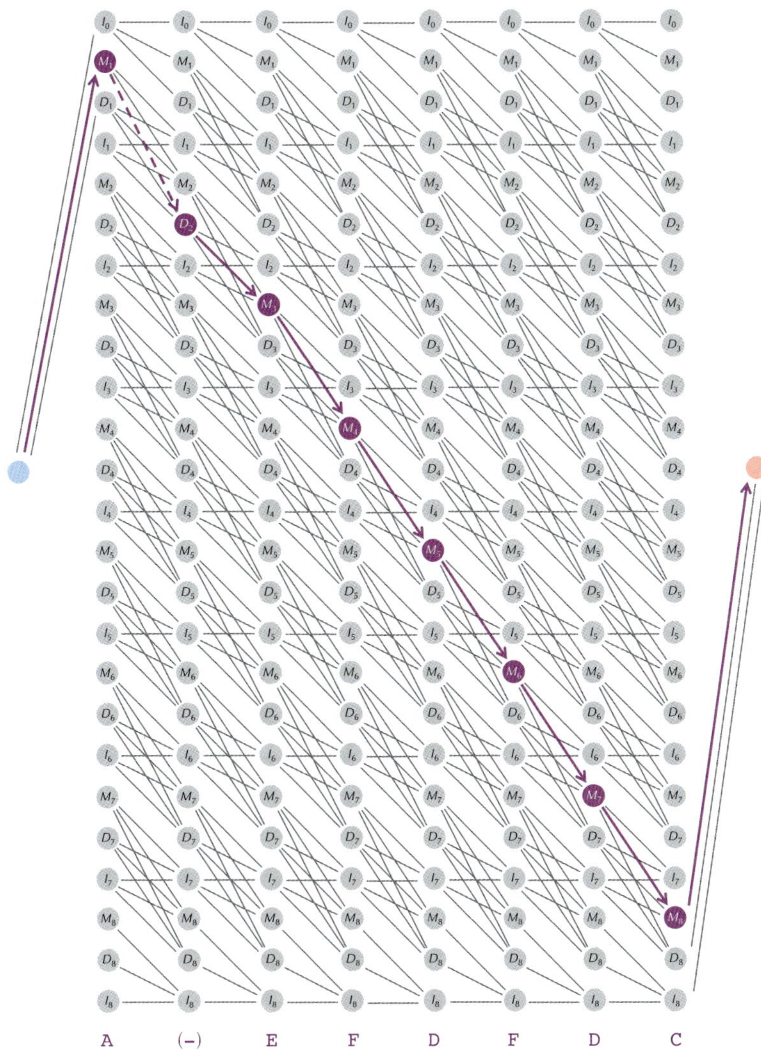

FIGURE 10.19 Another path through another "Viterbi graph" emitting the same string **AEFDFDC** as in Figure 10.18.

Viterbi graph describes the event "the HMM emitted symbol x_i when it was in state k". However, if k is a silent state, then the role of the node (k, i) in the Viterbi graph is poorly defined, as it is unclear how to define the weight of edges entering this node.

Fortunately, we can fix this issue in the case of profile HMMs by defining the Viterbi graph with $|States|$ rows and $|Text|$ columns (Figure 10.20). Every time the HMM moves into a deletion state, rather than crossing over to the next column of the Viterbi graph (as in Figure 10.18 and Figure 10.19), we will move *within* the same column. When the HMM moves into a match or insertion state, we will move to the next column. As a result, every column of the Viterbi graph corresponds to a single emitted symbol, even though a path can pass through more than one state in a given column.

Exercise Break: Show that the vertical edge connecting (i, l) to (i, k), where k is a deletion state, should be assigned weight equal to $transition_{l,k}$.

STOP and Think: Are there any remaining issues with the graph in Figure 10.20?

There is still a minor flaw with the graph in Figure 10.20. If the HMM moves from the initial state into DELETION(1), then the HMM will move through the first column without emitting a symbol. We will therefore transform the initial state into a column of silent states containing the initial state and all deletion states (Figure 10.21). This way, if the HMM enters DELETION(1) from the initial state, it can move downward through deletion states before transitioning to a match or insertion state in the first column.

You are now ready to use the profile HMM to align a sequence against a seed alignment. The only remaining snare is that when computing $s_{k,i}$ — or, equivalently, $\log(s_{k,i})$ — we must make sure that all incoming scores have been computed. We therefore suggest the top-down, column-by-column topological ordering for the profile HMM shown in Figure 10.22.

Sequence Alignment with Profile HMM Problem:
Align a new sequence to a family of sequences using a profile HMM.

Input: A multiple alignment *Alignment*, a threshold θ, a pseudocount value σ, and a string *Text*.
Output: An optimal hidden path emitting *Text* in HMM$(Alignment, \theta, \sigma)$.

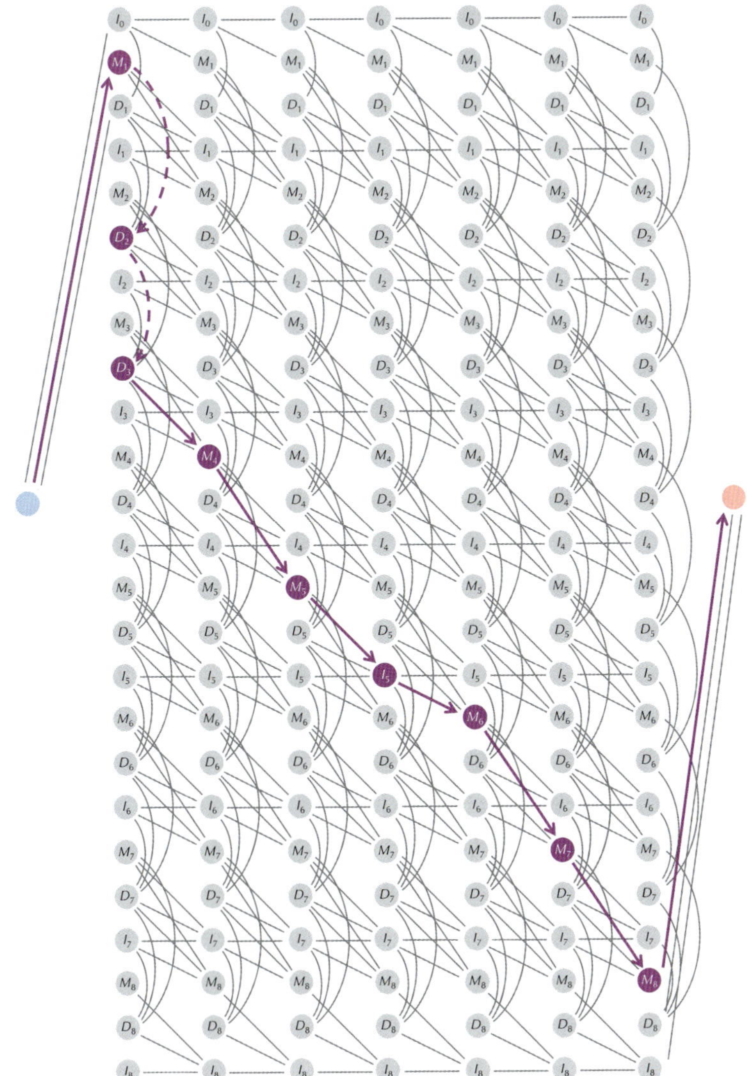

FIGURE 10.20 The Viterbi graph with |*States*| rows and |*Text*| columns for the profile HMM from Figure 10.14 emitting a string *Text* of length 7 so that edges entering deletion states are drawn downward within the same column instead of between columns as in Figure 10.18 and Figure 10.19. The purple path corresponds to the path through the HMM in Figure 10.15 emitting **`AEFDFDC`**.

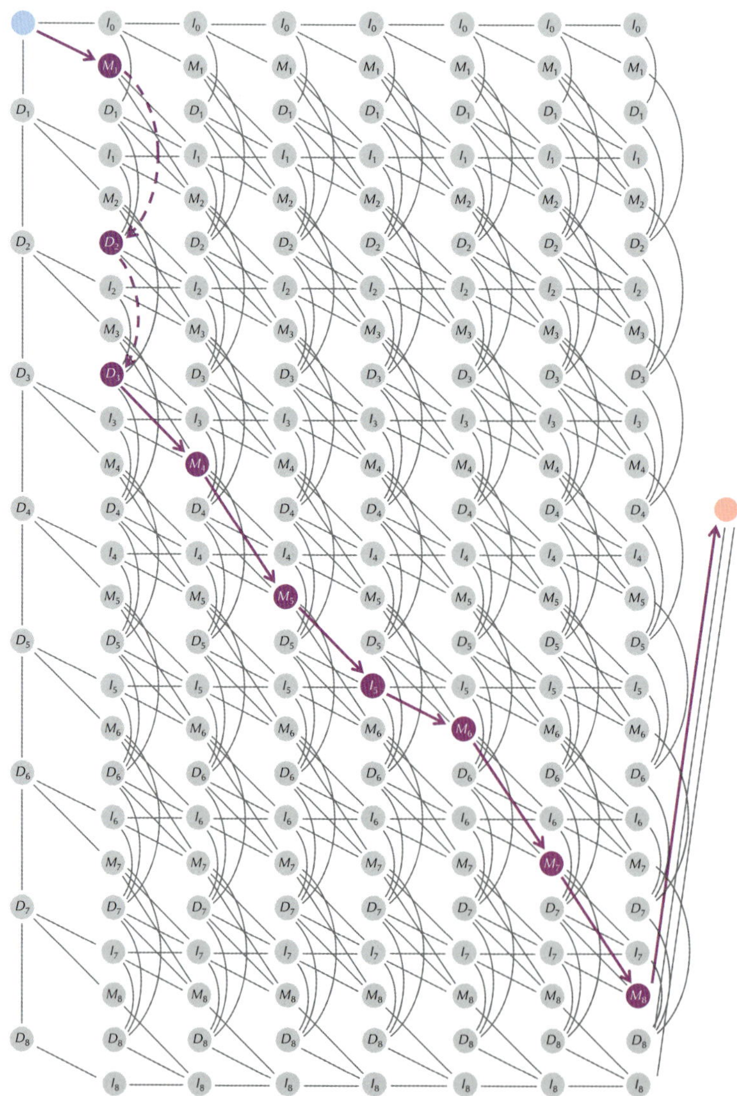

FIGURE 10.21 The final Viterbi graph for the profile HMM in Figure 10.14 emitting a string of length 7. Edges within the same column have a downward orientation; edges between columns have a rightward orientation. Once again, the purple path corresponds to the path through the HMM in Figure 10.15 emitting **AEFDFDC**.

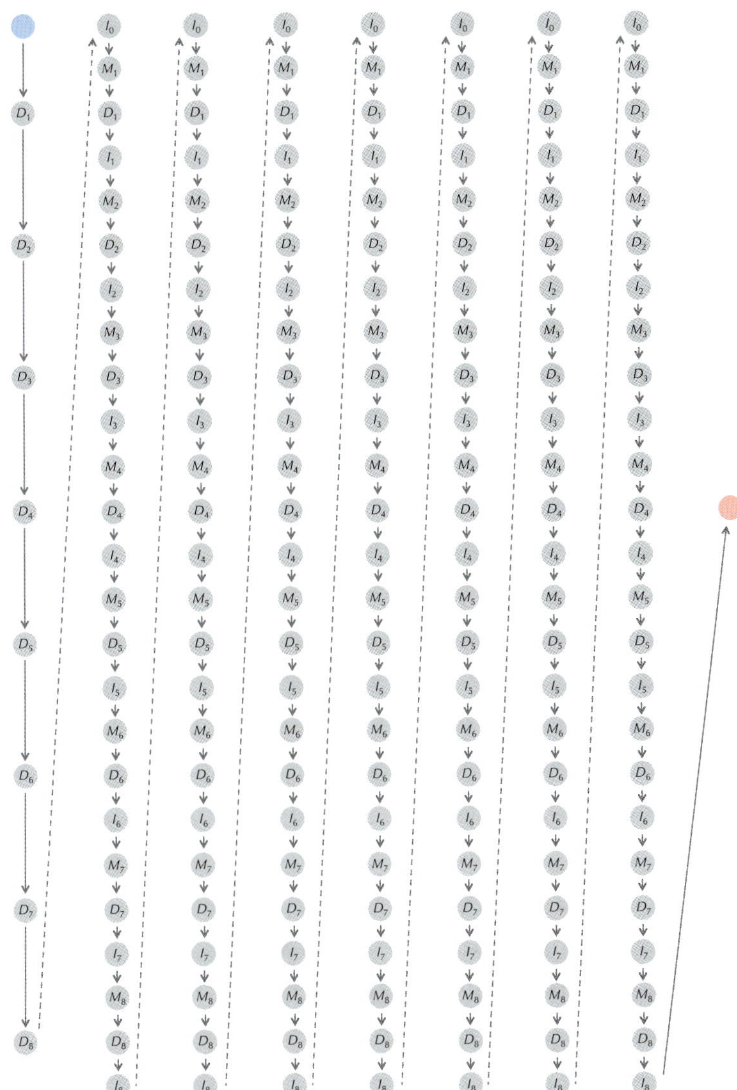

FIGURE 10.22 The topological order of the Viterbi graph from Figure 10.21, which proceeds top-down and column-by-column.

Exercise Break: Solve the Sequence Alignment with Profile HMM Problem for the profile HMM you constructed for the multiple alignment in Figure 10.1 along with a gp120 protein taken from chimpanzee Simian Immunodeficiency Virus (SIV).

STOP and Think: How would you build the Viterbi graph for an arbitrary HMM with silent states? In which situations will it be impossible to construct a Viterbi graph of an HMM with silent states?

Are profile HMMs really all that useful?

The Viterbi algorithm applies for any HMM, but we will describe how it works for profile HMMs in order to make an important point. Define $s_{\text{MATCH}(j),i}$ as the probability of the most likely hidden path for the prefix $x_1 \ldots x_i$ of x that ends at state MATCH(j), and define $s_{\text{INSERTION}(j),i}$ and $s_{\text{DELETION}(j),i}$ analogously. Because there are only three edges entering MATCH(j), the Viterbi recurrence states that

$$s_{\text{MATCH}(j),i} = \max \begin{cases} s_{\text{MATCH}(j-1),i-1} \cdot \text{WEIGHT}_i(\text{MATCH}(j-1), \text{MATCH}(j)) \\ s_{\text{INSERTION}(j-1),i-1} \cdot \text{WEIGHT}_i(\text{INSERTION}(j-1), \text{INSERTION}(j)) \\ s_{\text{DELETION}(j-1),i-1} \cdot \text{WEIGHT}_i(\text{DELETION}(j-1), \text{DELETION}(j)) \end{cases}$$

After taking the logarithm of both sides, the resulting recurrence is very similar to the standard recurrence relation for global pairwise alignment because it is the maximum of three sums:

$$\log\left(s_{\text{MATCH}(j),i}\right) =$$

$$\max \begin{cases} \log\left(s_{\text{MATCH}(j-1),i-1}\right) + \log\left(\text{WEIGHT}_i(\text{MATCH}(j-1), \text{MATCH}(j))\right) \\ \log\left(s_{\text{INSERTION}(j-1),i-1}\right) + \log\left(\text{WEIGHT}_i(\text{INSERTION}(j-1), \text{INSERTION}(j))\right) \\ \log\left(s_{\text{DELETION}(j-1),i-1}\right) + \log\left(\text{WEIGHT}_i(\text{DELETION}(j-1), \text{DELETION}(j))\right) \end{cases}$$

Figure 10.23 shows how a path in a Manhattan-like alignment graph corresponds to a path through the profile HMM. Diagonal edges, vertical edges, and horizontal edges in the Manhattan-like graph correspond to match states, insertion states, and deletion states, respectively.

Figure 10.23 may make it seem that we have wasted your time introducing HMMs, since it appears that a profile HMM is somehow equivalent to pairwise sequence alignment. However, keep in mind that the choice of edges in Figure 10.23 is based on varying transition and emission probabilities. By deriving individual scoring parameters for each column in the alignment matrix, profile HMMs allow us to capture subtle similarities that can fly under the radar of the simple scoring approaches from Chapter 5.

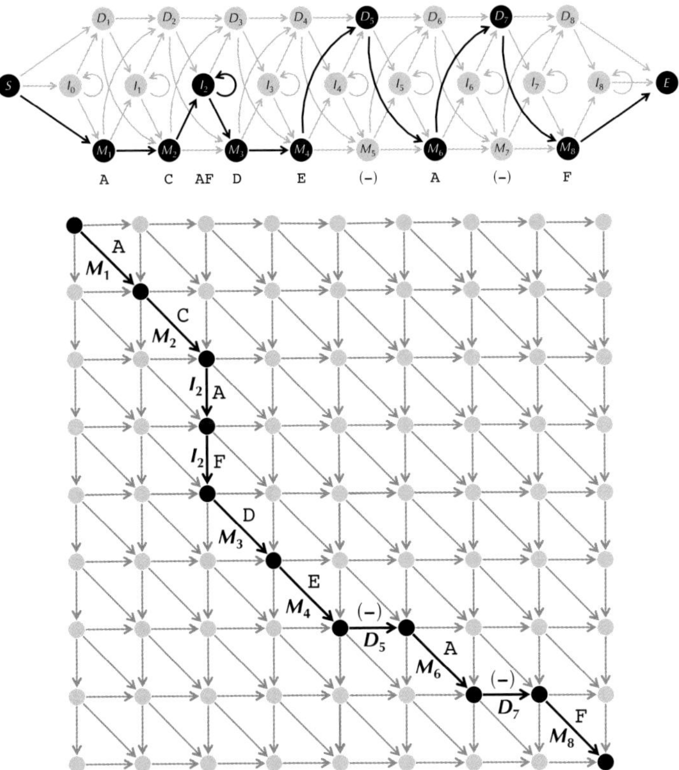

FIGURE 10.23 (Top) The hidden path through the profile HMM in Figure 10.17 (top) emitting `ACAFDEAF`. (Bottom) The path through a Manhattan-like graph corresponding to this hidden path.

Learning the Parameters of an HMM

Estimating HMM parameters when the hidden path is known

Thus far, our analysis has assumed that we know the parameters of an HMM, i.e., its transition and emission probabilities. We have described a naive — and not necessarily optimal — heuristic for choosing these parameters for a profile HMM, but it is not clear how to select parameters for an arbitrary HMM.

Indeed, the largest complication when modeling biological problems with an HMM is estimating the HMM's parameters from data. In terms of the crooked casino, imagine that you know that the dealer is using two coins to cheat, but you don't know the bias of the coins or the probability that the dealer switches coins at any given time. Can you infer these parameters just from the sequence of coin flips?

STOP and Think: Say that you observe the sequence of coin flips "HHTHHHTHHTTTH". What is your best guess for the biases of the two coins and for the probabilities of switching from one coin to the other? Would your best guess change if you knew that the hidden path were $\pi = FFFBBFFFFFBBB$?

We will collectively refer to the matrices *Transition* and *Emission* as *Parameters*. Our goal is to find *Parameters* and π when we are only given the emitted string x. We will work toward this goal by assuming that we are given x as well as either *Parameters* or π, and we must infer the remaining component. If x and *Parameters* are known, then we can find the most likely hidden path π using the Viterbi algorithm. However, we have not yet considered how to estimate *Parameters* if we know x and the hidden path π.

HMM Parameter Estimation Problem:
Find optimal parameters explaining the emitted string and the hidden path of an HMM.

> **Input**: A string $x = x_1 \ldots x_n$ emitted by an HMM with unknown transition and emission probabilities following a known hidden path $\pi = \pi_1 \ldots \pi_n$.
> **Output**: A transition matrix *Transition* and an emission matrix *Emission* that maximize $\Pr(x, \pi)$ over all possible transition and emission matrices.

If we know both x and π, then we can compute empirical estimates for the transition and emission probabilities using a method similar to one we used for estimating parameters for profile HMMs. If $T_{l,k}$ denotes the number of transitions from state l to state k in the hidden path π, then we can estimate the probability $transition_{l,k}$ by computing the ratio of $T_{l,k}$ to the total number of transitions leaving state l,

$$transition_{l,k} = \frac{T_{l,k}}{\sum_{\text{all states } j} T_{l,j}}\ .$$

Likewise, if $E_k(b)$ denotes the number of times symbol b is emitted when the hidden path π is in state k, then we can estimate the probability $emission_k(b)$ as the ratio of $E_k(b)$ to the total number of emitted symbols from state k,

$$emission_k(b) = \frac{E_k(b)}{\sum_{\text{all symbols } c \text{ in the alphabet}} E_k(c)} .$$

It turns out that the above two formulas for computing *Transition* and *Emission* result in parameters solving the HMM Parameter Estimation Problem.

Viterbi learning

If we know x and *Parameters*, then we can construct the most likely path π by applying the Viterbi algorithm to solve the Decoding Problem:

$$(x, ?, Parameters) \rightarrow \pi$$

On the other hand, if we know x and π, then reconstructing *Parameters* amounts to solving the HMM Parameter Estimation Problem:

$$(x, \pi, ?) \rightarrow Parameters$$

STOP and Think: What do the expressions $(x, \pi, ?) \rightarrow Parameters$ and $(x, ?, Parameters) \rightarrow \pi$ remind you of?

HMM Parameter Learning Problem:
Estimate the parameters of an HMM explaining an emitted string.

Input: A string $x = x_1 \ldots x_n$ emitted by an HMM with unknown transition and emission probabilities.
Output: A transition matrix *Transition* and an emission matrix *Emission* that maximize $\Pr(x, \pi)$ over all possible transition and emission matrices and over all hidden paths π.

Unfortunately, the HMM Parameter Learning Problem is intractable, and so we will instead develop a heuristic that is analogous to the Lloyd algorithm for k-means clustering from Chapter 8. In that algorithm, illustrated in Figure 8.12, we iterated two steps, "From Centers to Clusters",

$$(Data, ?, Centers) \rightarrow HiddenVector,$$

and "From Clusters to Centers",

$$(Data, HiddenVector, ?) \rightarrow Centers.$$

As for HMM parameter estimation, we begin with an initial random guess for *Parameters*. Then, we use the Viterbi algorithm to find the optimal hidden path π:

$$(x, ?, Parameters) \rightarrow \pi$$

Once we know π, we will question our original choice of *Parameters* and apply our solution to the HMM Parameter Estimation Problem to update *Parameters* based on x and π:

$$(x, \pi, ?) \rightarrow Parameters'$$

We then iterate over these two steps, hoping that the estimated parameters are getting closer and closer to the parameters solving the HMM Parameter Learning Problem:

$$\begin{aligned}(x, ?, Parameters) \rightarrow (x, \pi, Parameters) & \rightarrow (x, \pi, ?) \\ \rightarrow (x, \pi, Parameters') & \rightarrow (x, ?, Parameters') \\ \rightarrow (x, \pi', Parameters') & \rightarrow (x, \pi', ?) \\ \rightarrow (x, \pi', Parameters'') & \rightarrow \ldots\end{aligned}$$

This approach to learning the HMM's parameters is called **Viterbi learning**.

STOP and Think: Can $\Pr(x, \pi)$ decrease during Viterbi learning? When would you decide to stop the Viterbi learning algorithm?

Note that we have not specified how Viterbi learning should terminate. In practice, there are various stopping rules to control its running time. For example, the algorithm can be stopped if the number of iterations exceeds a predefined threshold or if $\Pr(x, \pi)$ changes very little from one iteration to another.

Also, because Viterbi learning is dependent on the initial guess for *Parameters*, it may become stuck in a local optimum. Like other heuristics, it is often run many times, retaining the best choice of *Parameters*.

Exercise Break: Apply Viterbi learning to learn parameters for an HMM modeling CG-islands as well as for the profile HMM for the gp120 HIV alignment in Figure 10.1.

Soft Decisions in Parameter Estimation

The Soft Decoding Problem

In Chapter 8, we introduced a "soft" clustering algorithm, based on the more general expectation maximization algorithm, that relaxed the Lloyd algorithm's rigid assignment of points to clusters. Analogously, by generating a single optimal hidden path, the Viterbi algorithm provides a rigid "yes" or "no" answer to the question of whether an HMM was in state k at time i. But how certain are we that this was the case?

Returning to the crooked casino analogy once more, say that the i-th coin flip is heads. If this flip occurs in the middle of ten consecutive heads, then you should be relatively confident that the biased coin was used. But what if, of the ten flips surrounding the i-th flip, six are heads and four are tails? In this case, you should be less certain that the biased coin was used.

In the case of an arbitrary HMM, we would like to compute the conditional probability $\Pr(\pi_i = k|x)$ that the HMM was in state k at time i given that it emitted string x.

Soft Decoding Problem:
Find the probability that an HMM was in a particular state at a particular moment given its emitted string.

Input: A string $x = x_1 \ldots x_n$ emitted by an HMM.
Output: The conditional probability $\Pr(\pi_i = k|x)$ that the HMM was in state k at step i given that it emitted x.

The unconditional probability that a hidden path will pass through state k at time i and emit x can be written as the sum

$$\Pr(\pi_i = k, x) = \sum_{\text{all paths } \pi \text{ with } \pi_i = k} \Pr(x, \pi)\,.$$

The conditional probability $\Pr(\pi_i = k|x)$ is equal to the proportion of paths that pass through state k at time i and emit x with respect to all paths emitting x:

$$\begin{aligned}\Pr(\pi_i = k|x) &= \frac{\Pr(\pi_i = k, x)}{\Pr(x)} \\ &= \frac{\sum_{\text{all paths } \pi \text{ with } \pi_i = k} \Pr(x, \pi)}{\sum_{\text{all paths } \pi} \Pr(x, \pi)}\,.\end{aligned}$$

STOP and Think: If the Viterbi algorithm for the crooked casino emits a path $\pi = \pi_1\pi_2 \ldots \pi_n$ with $\pi_i = B$, is the dealer more likely to have used a biased coin at step i? Is it possible that $\pi_i = B$ but that $\Pr(\pi_i = B|x)$ is smaller than $\Pr(\pi_i = F|x)$?

The forward-backward algorithm

We note that $\Pr(\pi_i = k, x)$ is equal to the sum of product weights $\Pr(\pi, x)$ of all paths π through the Viterbi graph for x that pass through the node (k, i). As shown in Figure 10.24 (top), we can break each such path into a blue subpath from *source* to (k, i), which we denote π_{blue}, and a (red) subpath from (k, i) to *sink*, which we denote π_{red}. Writing $\text{WEIGHT}(\pi_{\text{blue}})$ and $\text{WEIGHT}(\pi_{\text{red}})$ as the respective product weights of these subpaths yields the recurrence

$$\begin{aligned}\Pr(\pi_i = k, x) &= \sum_{\text{all paths } \pi \text{ with } \pi_i = k} \Pr(x, \pi) \\ &= \sum_{\text{all paths } \pi_{\text{blue}}} \sum_{\text{all paths } \pi_{\text{red}}} \text{WEIGHT}(\pi_{\text{blue}}) \cdot \text{WEIGHT}(\pi_{\text{red}}) \\ &= \sum_{\text{all paths } \pi_{\text{blue}}} \text{WEIGHT}(\pi_{\text{blue}}) \cdot \sum_{\text{all paths } \pi_{\text{red}}} \text{WEIGHT}(\pi_{\text{red}}) .\end{aligned}$$

We have already computed the sum of product weights of all blue subpaths; it is just $forward_{k,i}$, which we encountered when solving the Outcome Likelihood Problem. Now we would like to compute the sum of product weights of all red subpaths, which we denote as $backward_{k,i}$, so that the preceding equation becomes

$$\Pr(\pi_i = k, x) = forward_{k,i} \cdot backward_{k,i} .$$

The name of $backward_{k,i}$ derives from the fact that to compute this value, we can simply *reverse* the directions of all edges in the Viterbi graph (Figure 10.24 (bottom)) and apply the same dynamic programming algorithm used to compute $forward_{k,i}$. Since the reversed edge connecting $(l, i+1)$ to (k, i) has weight $\text{WEIGHT}_i(k, l) = transition_{k,l} \cdot emission_l(x_{i+1})$, we have that

$$backward_{k,i} = \sum_{\text{all states } l} backward_{l,i+1} \cdot \text{WEIGHT}_i(k, l).$$

STOP and Think: How should this recurrence be initialized?

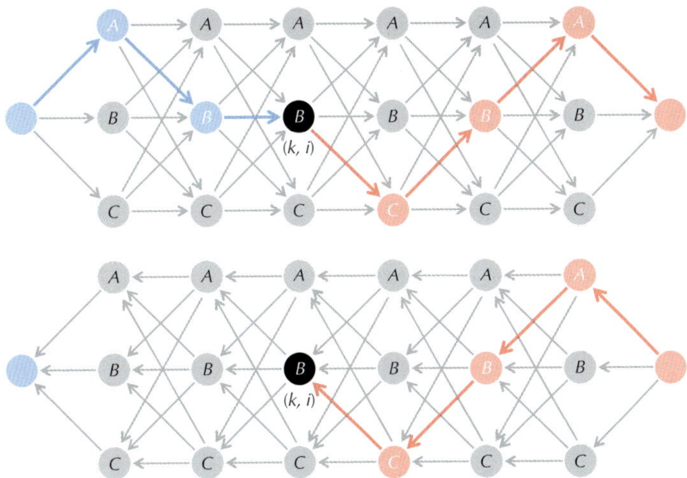

FIGURE 10.24 (Top) Each path from *source* to *sink* passing through the (black) node (k, i) in the Viterbi graph from Figure 10.7 (bottom) can be partitioned into two subpaths, one from *source* to (k, i) (shown in blue) and another from (k, i) to *sink* (shown in red). (Bottom) A "reversed Viterbi graph" in which all edges have been reversed, with a path from *sink* to (k, i) highlighted in red. The recurrence for $backward_{k,i}$ is based on computing $backward_{l,i+1}$ for every state l.

The resulting dynamic programming approach for computing $\Pr(\pi_i = k, x)$ is called the **forward-backward algorithm**. Combining the forward-backward algorithm with our solution to the Outcome Likelihood Problem for computing $\Pr(x)$ yields that

$$\Pr(\pi_i = k|x) = \frac{\Pr(\pi_i = k, x)}{\Pr(x)} = \frac{forward_{k,i} \cdot backward_{k,i}}{forward(sink)},$$

and so we are ready to solve the Soft Decoding Problem.

Exercise Break: Consider the following questions.

- For the crooked dealer HMM, compute $\Pr(\pi_i = k|x)$ for x = "THTHHHTHTTH" and each value of i. How does your answer change if x = "HHHHHHHHHHH"?
- Apply your solution for the Soft Decoding Problem to find CG-islands in the first million nucleotides from the human X chromosome. How does your answer differ from the solution given by the Viterbi algorithm?

We have just seen how to compute the conditional probability $\Pr(\pi_i = k|x)$ that the HMM passes through node (k, i) in the Viterbi graph given that the HMM emits x. But what about the conditional probability $\Pr(\pi_i = l, \pi_{i+1} = k|x)$ that the HMM passes through the edge connecting (l, i) to $(k, i + 1)$ given that the HMM emits x? As with the forward-backward algorithm, we can divide every path through the edge in question into a blue path from *source* to this edge and a red path from this edge to *sink* (Figure 10.25).

Exercise Break: Prove that $\Pr(\pi_i = l, \pi_{i+1} = k|x)$ is equal to $forward_{l,i} \cdot \text{WEIGHT}_i(l, k) \cdot backward_{k,i+1}/forward(sink)$.

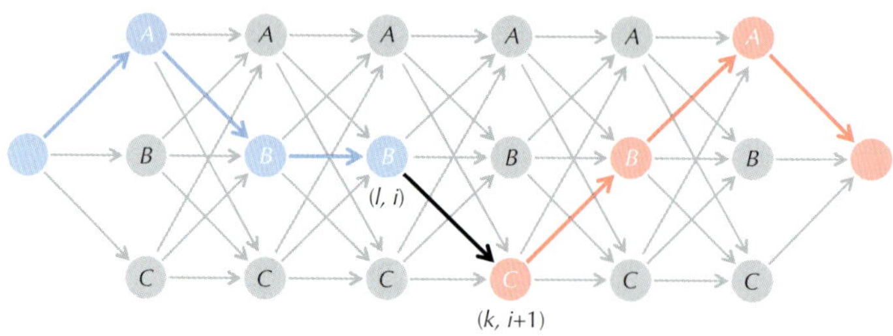

FIGURE 10.25 Each path from *source* to *sink* in the Viterbi graph passing through the (black) edge $(l, i) \rightarrow (k, i + 1)$ in the Viterbi graph can be partitioned into two subpaths, one from *source* to (l, i) (shown in blue) and another from $(k, i + 1)$ to *sink* (shown in red).

The probabilities $\Pr(\pi_i = k|x)$ can be put into a $|States| \times n$ responsibility matrix Π^*, where $\Pi^*_{k,i}$ corresponds to a *node* in the Viterbi graph and is equal to $\Pr(\pi_i = k|x)$. Figure 10.26 (top) shows the "responsibility" matrix Π^* for the crooked casino.

The probabilities $\Pr(\pi_i = l, \pi_{i+1} = k|x)$ can be put into another $|States| \times |States| \times (n - 1)$ responsibility matrix Π^{**}, where $\Pi^{**}_{l,k,i}$ corresponds to an *edge* in the Viterbi graph and is equal to $\Pr(\pi_i = l, \pi_{i+1} = k|x)$ (Figure 10.26 (bottom)). For brevity, we use Π to collectively refer to the matrices Π^* and Π^{**}.

Exercise Break: What is the complexity of an algorithm computing the matrices Π^* and Π^{**}?

	T	H	T	H	H	H	T	H	T	T	H
F	0.636	0.593	0.600	0.533	0.515	0.544	0.627	0.633	0.692	0.686	0.609
B	0.364	0.407	0.400	0.467	0.485	0.456	0.373	0.367	0.308	0.314	0.391

	1	2	3	4	5	6	7	8	9	10
FF	0.562	0.548	0.507	0.473	0.478	0.523	0.582	0.608	0.643	0.588
FB	0.074	0.045	0.093	0.059	0.037	0.022	0.045	0.025	0.049	0.098
BF	0.031	0.053	0.025	0.042	0.066	0.104	0.051	0.084	0.043	0.022
BB	0.333	0.354	0.374	0.426	0.418	0.351	0.322	0.282	0.265	0.293

FIGURE 10.26 (Top) The responsibility matrix Π^*, where x = "THTHHHTHTTH" and the emission/transition matrices *Parameters* are taken from the crooked dealer HMM in Figure 10.5. $\Pi^*_{k,i}$ is equal to $\Pr(\pi_i = k|x)$. (Bottom) The responsibility matrix Π^{**} , where $\Pi^{**}_{l,k,i} = \Pr(\pi_i = l, \pi_{i+1} = k|x)$ for the same emitted string and emission/transition matrices.

Baum-Welch Learning

The expectation maximization algorithm for parameter estimation, called **Baum-Welch learning**, alternates between two steps. In the E-step, it estimates the responsibility profile Π given the current parameters:

$$(x, ?, Parameters) \rightarrow \Pi$$

Then, in the M-step, it re-estimates the parameters from the responsibility profile:

$$(x, \Pi, ?) \rightarrow Parameters$$

We have already implemented the E-step of the expectation maximization algorithm, but the question remains how to design the M-step.

When we know the hidden path, the previously defined estimators for *Parameters*, reproduced below, define optimal choices for a given hidden path π:

$$transition_{l,k} = \frac{T_{l,k}}{\sum_{\text{all states } j} T_{l,j}} \qquad emission_k(b) = \frac{E_k(b)}{\sum_{\text{all symbols } c \text{ in the alphabet}} E_k(c)}.$$

Here, $T_{l,k}$ is the number of transitions from state l to state k in the hidden path π, and $E_k(b)$ is the number of times symbol b is emitted when the hidden path π is in state k.

Exercise Break: How would you redefine these estimators when the hidden path is unknown?

To see how to estimate $transition_{l,k}$ and $emission_k(b)$ when the hidden path is unknown, we will compute $T_{l,k}$ and $E_k(b)$ for a known path π in a slightly different way to make the transition from hard to soft choices more apparent. First, define the following binary variables:

$$T^i_{l,k} = \begin{cases} 1 \text{ if } \pi_i = l \text{ and } \pi_{i+1} = k \\ 0 \text{ otherwise} \end{cases} \qquad E^i_k(b) = \begin{cases} 1 \text{ if } \pi_i = k \text{ and } x_i = b \\ 0 \text{ otherwise} \end{cases}$$

With this notation, the formulas computing $T_{l,k}$ and $E_k(b)$ can be rewritten as

$$T_{l,k} = \sum_{i=1}^{n-1} T^i_{l,k} \qquad E_k(b) = \sum_{i=1}^{n} E^i_k(b)$$

When the hidden path is unknown, we will substitute the binary variables $T^i_{l,k}$ and $E^i_k(b)$ for new variables that are computed in terms of the conditional probabilities that a hidden path will pass through a given node or edge of the Viterbi graph:

$$\begin{aligned} T^i_{l,k} &= \Pr(\pi_i = l, \pi_{i+1} = k|x) \qquad & E^i_k(b) &= \Pr(\pi_i = k|x) \\ &= \Pi^{**}_{l,k,i} & &= \Pi^*_{k,i} \text{ if } x_i = b \text{ and } 0 \text{ otherwise} \end{aligned}$$

Armed with these probabilities computed in the previous section, we can compute new estimates for *Parameters* that often perform better in practice than the estimates provided by Viterbi learning:

$$T_{l,k} = \sum_{i=1}^{n-1} \Pi^{**}_{l,k,i} \qquad E_k(b) = \sum_{\text{all } i \text{ such that } x_i = b} \Pi^*_{k,i}$$

Substituting these values into the above formulas for $T_{l,k}$ and $E_k(b)$ yields updated transition and emission probabilities. For the example in Figure 10.26, we obtain

$$\begin{aligned} T_{F,F} &= 0.562 + 0.548 + \cdots + 0.588 &&= 5.512 \\ T_{F,B} &= 0.074 + 0.045 + \cdots + 0.098 &&= 0.547 \\ T_{B,F} &= 0.031 + 0.053 + \cdots + 0.022 &&= 0.521 \\ T_{B,B} &= 0.333 + 0.354 + \cdots + 0.293 &&= 3.418 \end{aligned}$$

$$\begin{aligned} E_F(H) &= 0.593 + 0.533 + 0.515 + 0.544 + 0.633 + 0.609 &&= 3.427 \\ E_F(T) &= 0.636 + 0.600 + 0.627 + 0.692 + 0.686 &&= 3.241 \\ E_B(H) &= 0.407 + 0.467 + 0.485 + 0.456 + 0.367 + 0.391 &&= 2.573 \\ E_B(T) &= 0.364 + 0.400 + 0.373 + 0.308 + 0.314 &&= 1.759 \end{aligned}$$

Using the formulas previously given for *transition* and *emission*, we obtain

$$\textit{transition}_{F,F} = \frac{5.512}{5.512 + 0.547} = 0.910 \qquad \textit{emission}_F(H) = \frac{3.427}{3.427 + 3.241} = 0.514$$
$$\textit{transition}_{F,B} = \frac{0.547}{5.512 + 0.547} = 0.090 \qquad \textit{emission}_F(T) = \frac{3.241}{3.427 + 3.241} = 0.486$$
$$\textit{transition}_{B,F} = \frac{0.521}{3.418 + 0.521} = 0.132 \qquad \textit{emission}_B(H) = \frac{2.573}{2.573 + 3.241} = 0.594$$
$$\textit{transition}_{B,B} = \frac{3.418}{3.418 + 0.521} = 0.868 \qquad \textit{emission}_B(T) = \frac{1.759}{2.573 + 1.759} = 0.406$$

Exercise Break: Use Baum-Welch learning to learn parameters for the HMM modeling CG-islands and for the HIV profile HMM. Compare these parameters with parameters derived by applying Viterbi learning.

The Many Faces of HMMs

Profile HMMs for multiple sequence alignment and HMMs finding CG-islands are just two examples of many applications of HMM in bioinformatics. Furthermore, applications of HMMs to HIV analysis are not limited to profile HMMs but also include analysis of HIV resistance against antiviral drug therapies.

Early in the chapter, we mentioned that patients infected with HIV are treated with a cocktail of several drugs. These drugs attempt to suppress replication of the virus, but HIV often mutates into drug-resistant strains that eventually dominate the virus population in a host, making a drug cocktail progressively ineffective. HIV viruses are often sequenced after drug therapy has failed in order to decide how to reformulate the drug cocktail. Thus, understanding HIV's pathways to drug resistance is important to design an effective cocktail.

Yet modeling HIV resistance pathways is a difficult task. Two mutations that are advantageous for the virus may interact synergistically, causing the double mutation to be fixed more often than we might predict from the frequencies of the individual substitutions. Mutations can also interact antagonistically, resulting in mutants that are less fit than we might predict.

In 2007, Niko Beerenwinkel and Mathias Drton introduced an HMM-based model for HIV evolution and developing drug resistance. However, their HMM is far too complex to explain here. We nevertheless mention it here in order to emphasize the power of HMMs. Even though they may seem like simple machines that flip coins and emit symbols, HMMs can be applied to tackle complex bioinformatics problems ranging from gene prediction to regulatory motif finding.

Epilogue: Nature is a Tinkerer and not an Inventor

The sequence of amino acids in a protein encodes its 3-D structure, which often defines the biological function of a protein. For example, a **zinc finger** is an element of the 3-D structure of **zinc finger proteins** (Figure 10.27). By arranging two cysteines and two histidines close to each other in a zinc finger protein's amino acid sequence, the protein is able to "grab" a zinc ion and fold tightly around it. Zinc fingers are so useful that they are found in thousands of human proteins. Furthermore, zinc finger proteins are used for more than binding zinc, as many of these proteins bind to other metals or even non-metals.

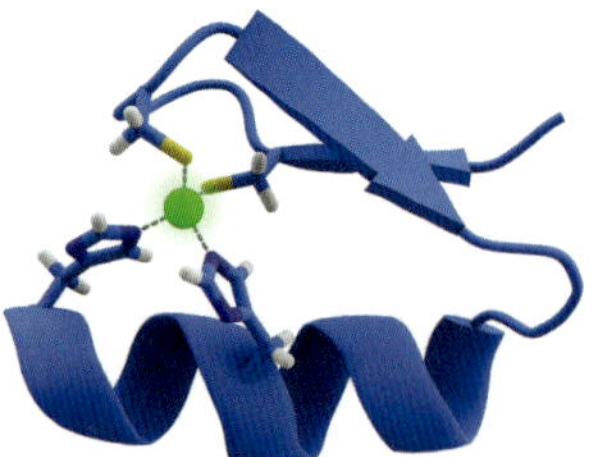

FIGURE 10.27 A zinc ion (shown in green) is held in place by two histidine residues and two cysteine residues within a zinc finger.

Over 100,000 experimentally determined protein structures are currently known, but many of them represent very similar structures or share segments with very similar structure. A **protein domain** is a conserved part of a protein that can function independently of the rest of the protein. Domains vary in length, but the average domain length is approximately 100 amino acids (zinc finger domains are only 20-30 amino acids long.) Many proteins consist of several domains, and the same domain may appear (with variations) in many proteins.

Nobel Laureate François Jacob famously said in 1977: "Nature is a tinkerer and not an inventor." In accordance with this principle, nature uses domains as building blocks, shuffling them into different arrangements to create **multi-domain proteins**. Most domains once existed as independent proteins; for example, many domains belonging to human multi-domain proteins can be found as single-domain proteins in bacteria.

Multi-domain proteins occur naturally when a genome rearrangement creates a new protein-coding sequence containing parts of the coding sequences from two different genes. Association of two domains into a single protein often provides an evolutionary advantage, such as when both domains are enzymes, in which case it may be beneficial for the cell to ensure a fixed one-to-one ratio of the enzymes' activities.

Since proteins are often built from multiple domains with different structures and functions, biologists commonly analyze individual domains instead of entire proteins in order to understand evolutionary relationships. Since sequence similarities between domains with similar structures can be extremely low, classifying domains into structural families can be difficult. The **Pfam database**, which contains over 10,000 HMM-derived multiple alignments of protein domain families, can be used to analyze new protein sequences.

Challenge Problem: Using the Pfam HMM for gp120 (constructed from a seed alignment of just 24 gp120 proteins), construct alignments of all known gp120 proteins and identify the "most diverged" gp120 sequence.

Detours

The Red Queen Effect

The **Red Queen Effect** is the hypothesis that evolution is necessary not only to equip organisms with an advantage in a *fixed* environment, but also to help them survive in response to *changing* environments. Its name derives from a statement that the Red Queen made to Alice in Lewis Carroll's *Through the Looking-Glass*:

> *Now, here, you see, it takes all the running you can do, to keep in the same place.*

The Red Queen Effect is often seen in predator-prey relationships. For example, an adaptation may help wolves run a little faster, and caribou must evolve in turn to survive. The result is that wolves and caribou appear to run at the same speed, with the slowest wolves starving and the slowest caribou being eaten.

Glycosylation

Cells have a dense coating of sugar chains, called **glycans**, on their surface. Glycans are often post-translational modifications of **glycoproteins**, which modulate interactions with other cells in a multicellular organism or between the cell and another organism (e.g., between human cells and a virus). For example, influenza infection begins with an interaction between the proteins on the virus's surface and glycans on the host cell's surface.

Glycans are constructed from a family of building blocks called **monosaccharides**. Each monosaccharide can be linked with other monosaccharides to form complex, tree-like structures (Figure 10.28).

DNA methylation

DNA methylation results in the addition of a methyl group (CH_3) to a cytosine or guanine nucleotide (Figure 10.29), which often alters the expression of nearby genes. Genes that acquire a high concentration of methylated residues in their upstream regions have suppressed expression. DNA methylation is vital to development, and both DNA hypermethylation and hypomethylation have been linked to various cancers.

DNA methylation is important in the process of **cell differentiation**, in which embryonic stem cells become specialized tissues. The change is often permanent, preventing a cell from reverting to a stem cell or converting to a different cell type. Methylation is inherited during cell division but is usually removed during zygote formation.

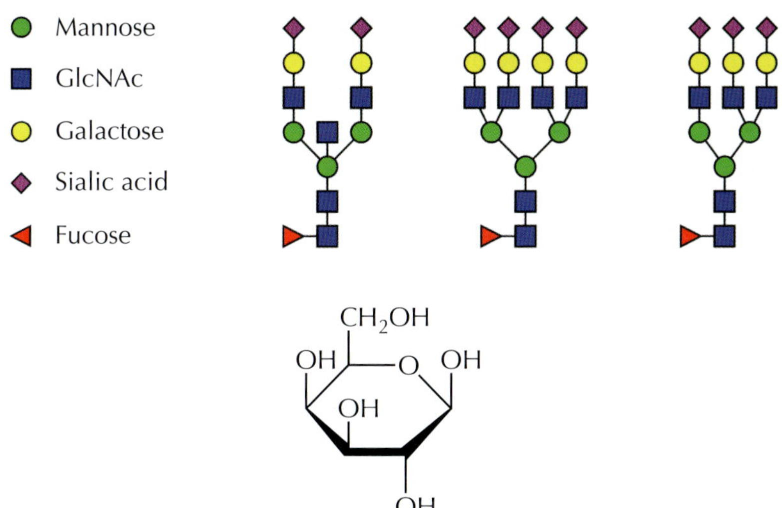

FIGURE 10.28 (Top) Five types of monosaccharides along with three examples of how these monosaccharides are assembled into glycans in humans. (Bottom) The chemical formula for the monosaccharide galactose.

FIGURE 10.29 DNA methylation of a cytosine nucleotide base, with the methyl group shown in blue.

Conditional probability

Let's return to the game of Chō-Han and analyze the sum s of two standard six-sided dice. Let A be the event that s is odd and B be the event that s is larger than 10. The probability of A is equal to $1/2$ because half of the 36 possible outcomes for rolling two dice produce an odd sum. The probability of B is equal to $3/36$ because there are three outcomes ($5+6$, $6+5$, and $6+6$) for which $s > 10$.

STOP and Think: If we tell you that s is larger than 10 (but you cannot see the dice), is s more likely to be odd or even?

The **conditional probability** of event A given event B, denoted $\Pr(A|B)$, is the probability that event A will occur given that event B has occurred. For the dice example, since B corresponds to two tosses with s odd ($6+5$ and $5+6$) and one toss with s even ($6+6$), $\Pr(A|B) = 2/3$. Note that $\Pr(A|B)$ is completely different from the probability $\Pr(A, B)$ that both events A and B will occur, which is equal to $2/36$ (A and B only occur together for the sums $6+5$ and $5+6$).

More generally, the conditional probability $\Pr(A|B)$ is often defined using the following formula:

$$\Pr(A|B) = \frac{\Pr(A,B)}{\Pr(B)}.$$

Exercise Break: To test your knowledge of conditional probability, consider the following puzzle, called the "Monty Hall Problem", which originally appeared in a letter to *American Statistician* in 1975 and has stumped many aspiring mathematicians over the years:

> *Suppose you're on a game show, and you're given the choice of three doors: Behind one door is a car; behind the others, goats. You pick a door, say No. 1, and the host, who knows what's behind the doors, opens another door, say No. 3, which has a goat. He then says to you, "Do you want to pick door No. 2?" Is it to your advantage to switch your choice?*

Bibliography Notes

The decoding algorithm was developed by Viterbi, 1967. The first algorithms for HMM parameter estimation were developed by Baum et al., 1970. Churchill, 1989, Krogh et al., 1994, and Baldi et al., 1994 pioneered the application of HMMs in computational biology. Bateman et al., 2002 described applications of profile HMM alignments for developing a database of protein domain families Pfam. De Jong et al., 1992 discovered the 11/25 rule. Beerenwinkel and Drton, 2007 constructed an HMM for analyzing HIV resistance.

WAS T. REX JUST A BIG CHICKEN?
Computational Proteomics

Paleontology Meets Computing

Growing up in Montana in the 1950s, Jack Horner was shy and introverted. He progressed so slowly in reading and mathematics that other kids called him stupid. However, his high school project on dinosaurs took top honors at the local science fair and was noticed by a University of Montana professor, who helped Jack gain admission to the university.

Yet Horner's grades did not improve in college; after failing five consecutive quarters, he dropped out. Years later, Horner would learn that he suffers from **dyslexia**, a developmental disorder that is often characterized by difficulty with reading comprehension and mathematics despite normal or above-average intelligence.

Fortunately for Horner, he eventually found his calling. After being drafted during the Vietnam War and later working as a truck driver, he accepted a job as a technician at Princeton's Natural History Museum, where he quickly established a reputation among his peers as a brilliant researcher. He would go on to become the world's most famous paleontologist, providing inspiration for one of the main characters of the bestselling novel *Jurassic Park* and advising Steven Spielberg for the film adaptation.

Horner was able to succeed despite dyslexia partly because paleontology has not traditionally required mathematical fluency. However, Horner's own student would show that even paleontology is not immune from computing. In 2000, Horner was exploring his favorite dinosaur graveyard in Montana and discovered a 68 million year-old *Tyrannosaurus rex* leg bone fossil. Three years later, he gave a small chunk of this fossil to his student, Mary Schweitzer, who dissolved it in a demineralizing bath to study its components but left it in for too long (remember Alexander Fleming?). When she returned, all that remained was a fibrous substance. Schweitzer then sent this material to a mass spectrometrist (John Asara) in the hope of detecting *T. rex* peptides that had miraculously survived inside the bone.

In 2007, after analyzing thousands of spectra, Asara and Schweitzer published a paper in *Science* announcing the discovery of *T. rex* peptides that closely matched chicken peptides. Their result provided the first molecular evidence for the controversial hypothesis that birds evolved from dinosaurs.

The fact that proteins could survive for millions of years was so amazing that it led to many grandiose claims. Paleontologist Hans Larsson suggested that dinosaurs would "enter the field of molecular biology and really slingshot paleontology into the modern world". *The Guardian* projected that "scientists may one day be able to emulate *Jurassic Park* by cloning a dinosaur." Horner himself even published a book called *How to Build a Dinosaur*, detailing his plan to recreate a dinosaur by genetically modifying the chicken genome.

Yet some scientists remained skeptical. Whereas previous dinosaur studies did not require much computation, Asara's *T. rex* analysis was powered by an algorithm relying on complicated statistics. In 2008, *Science* published rebuttals arguing that Asara and Schweitzer had failed to prove that some of their peptides are not simply statistical artifacts. But how can we know which side is correct? In this chapter, we will investigate the *T. rex* peptides by delving into some algorithms for analyzing spectra.

Which Proteins Are Present in This Sample?

Only four scientists have ever won two Nobel Prizes. One of them is Frederick Sanger, whose assembly of the first genome in 1977 we mentioned in Chapter 3. Yet Sanger had already won his first Nobel prize two decades earlier for determining the sequence of 52 amino acids making up insulin, the protein needed to absorb glucose in the blood. Similarly to how scientists sequence genomes, Sanger broke multiple molecules of insulin into short peptides, sequenced these peptides, and then assembled them into the amino acid sequence of insulin (Figure 11.1).

Although protein sequencing was very difficult in the 1950s, DNA sequencing was impossible. Today, it has become essentially trivial to generate millions of reads for DNA sequencing, but protein sequencing remains difficult. For this reason, most proteins are discovered by first sequencing a genome and then predicting all of the genes that PAGE 626 this genome encodes (see **DETOUR: Gene Prediction**). By translating the nucleotide sequence of each protein-coding gene into an amino acid sequence, biologists derive a putative **proteome** of a species, i.e., the set of all its proteins.

However, different cells in an organism express different proteins. For example, brain cells express proteins giving rise to neuropeptides, whereas other cells do not. An important problem in the study of proteins, or **proteomics**, is to identify which *specific* proteins are present in each biological tissue under different conditions, and how these proteins interact.

For example, suppose we are studying the chicken ribosome, a complex molecular machine consisting of many proteins. Knowing the chicken proteome does not tell us which specific proteins compose the ribosome complex. Instead, we can isolate the ribosome, break it apart, and identify which proteins it contains. In practice, merely confirming that a 10 amino acid-long peptide from a known chicken protein is present in a sample is usually sufficient to confirm this protein's presence in the sample. The

```
GIVEECCA
GIVEECCASV
GIVEECCASVC
GIVEECCASVCSL
GIVEECCASVCSLY
                      SLYELEDYC
                            ELEDY
                            ELEDYCD
                              LEDYCD
                                EDYCD
                                          FVDEHLCG
                                          FVDEHLCGSHL
                                                  HLCGSHL
                                                          SHLVEA
                                                                VEALY
                                                                        YLVCG
                                                                          LVCGERGF
                                                                           LVCGERGFF
                                                                                       GFFYTPK
                                                                                             YTPKA
```

GIVECCASVCSLYELEDYCDFVDEHLCGSHLVEALYLVCGERGFFFYTPKA

FIGURE 11.1 The peptide assembly that Frederick Sanger used to determine the amino acid sequence of insulin.

process of confirming that a peptide from a *known* proteome is present in a sample is called **peptide identification**. But how could we form a *T. rex* proteome?

Although peptide identification dominates modern proteomics studies, the proteomes of many species, including extinct species like *T. rex*, remain unknown. In this case, biologists rely on *de novo* **peptide sequencing**, or inferring the amino acid sequence of a peptide without relying on a proteome, and so this is where we will begin.

Decoding an Ideal Spectrum

You may be experiencing *déjà vu*, since we already discussed *cyclic* peptide sequencing in Chapter 4. So we will first remind you of the basics of mass spectrometry, with an emphasis on its application to *linear* peptide sequencing.

Given a large number of identical copies of a peptide in a sample — which will typically contain millions of cells — a mass spectrometer breaks each copy into two smaller fragments, where different copies of the same peptide may break differently. For example, one copy of `REDCA` may break into `RE` and `DCA`, and another may break into

RED and CA. The fragments RE and RED are called **prefixes** of REDCA, whereas DCA and CA are called **suffixes** of REDCA. Figure 11.2 shows the integer masses of amino acids.

G	A	S	P	V	T	C	I	L	N	D	K	Q	E	M	H	F	R	Y	W
57	71	87	97	99	101	103	113	113	114	115	128	128	129	131	137	147	156	163	186

FIGURE 11.2 The integer mass table of the 20 standard amino acids (reproduced from Chapter 4).

The algorithmic question that we first pose is analogous to the one that we asked about cyclic antibiotics but now applied to linear peptides: if we weigh each prefix and suffix of an unknown peptide, can we reconstruct the peptide? Given an amino acid string *Peptide*, its ideal spectrum, denoted IDEALSPECTRUM(*Peptide*), is the collection of integer masses of all its prefixes and suffixes (Figure 11.3 (top)). Note that an ideal spectrum may have repeated masses; for example, IDEALSPECTRUM(GPG) = {0, 57, 57, 154, 154, 211}. We say that an amino acid string *Peptide* **explains** a collection of integers *Spectrum* if IDEALSPECTRUM(*Peptide*) = *Spectrum*.

Fragment	""	R	RE	RED	REDC	REDCA	EDCA	DCA	CA	A
Mass	0	156	285	400	503	574	418	289	174	71

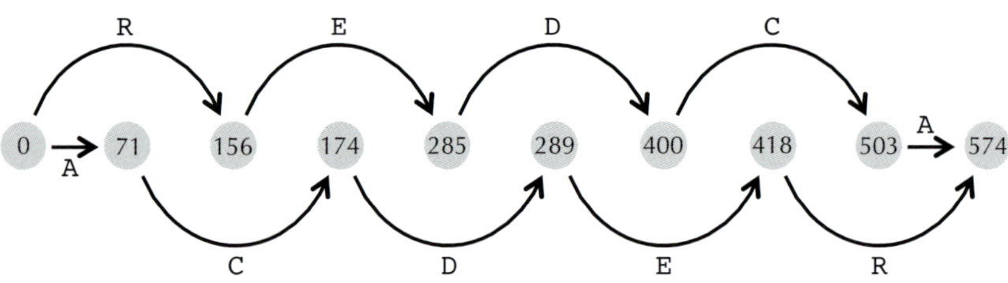

FIGURE 11.3 (Top) Masses of prefixes and suffixes of REDCA form IDEALSPECTRUM(REDCA) = {0, 71, 156, 174, 285, 289, 400, 418, 503, 574}. (Bottom) The DAG GRAPH(IDEALSPECTRUM(REDCA)) in which each mass in the spectrum is assigned to a node and two nodes are connected with a directed edge if the difference in their masses is the mass of an amino acid.

Decoding an Ideal Spectrum Problem:
Reconstruct a peptide from its ideal spectrum.

Input: A collection of integers *Spectrum*.
Output: An amino acid string *Peptide* that explains *Spectrum*.

We would like to separate masses in a spectrum into those derived from prefix and suffix peptides, but it is unclear how to do so. Instead, note that if two masses are "one amino acid mass apart", then it is likely that they correspond to two prefixes or two suffixes that differ by a single amino acid. For example, we would not know that masses 400 and 503 correspond to the prefixes `RED` and `REDC` (Figure 11.3 (top)). But we could hypothesize that because the difference between these masses is 103 (the mass of `C`), these masses correspond to prefixes or suffixes differing in a single occurrence of `C`.

This idea motivates a graph-based approach to solving the Decoding an Ideal Spectrum Problem. We represent the masses in a spectrum as a sequence *Spectrum* of integers $s_1, \ldots, s_m$ in increasing order, where s_1 is zero and s_m is the total mass of the (unknown) peptide. We define a labeled graph GRAPH(*Spectrum*) by forming a node for each element of *Spectrum*, then connecting nodes s_i and s_j by a directed edge labeled by an amino acid *a* if $s_j - s_i$ is equal to the mass of *a* (Figure 11.3 (bottom)). As we assumed when sequencing antibiotics, we do not distinguish between amino acids having the same integer masses (i.e., the pairs `K/Q` and `I/L`).

Exercise Break: Prove that for any choice of *Spectrum*, GRAPH(*Spectrum*) is a DAG.

Figure 11.3 (bottom) shows that GRAPH(*IdealSpectrum*(`REDCA`)) consists of two paths connecting *source* = 0 to *sink* = s_m. Concatenating the amino acids along these paths spells out `REDCA` and its reverse `ACDER`, both of which represent solutions of the Decoding an Ideal Spectrum Problem. The sequencing approach based on spelling a path from *source* to *sink* in GRAPH(*Spectrum*) is described by the following pseudocode.

```
DECODINGIDEALSPECTRUM(Spectrum)
    construct GRAPH(Spectrum)
    find a path Path from source to sink in GRAPH(Spectrum)
    return the amino acid string spelled by labels of Path
```

Exercise Break: Decode the ideal spectrum {0, 57, 114, 128, 215, 229, 316, 330, 387, 444}.

If you attempted this exercise, then there is a good chance that you found a path corresponding to a peptide with an incorrect spectrum (Figure 11.4). It is true that each peptide explaining *Spectrum* corresponds to a path from *source* to *sink* in GRAPH(*Spectrum*). However, not every path from *source* to *sink* in this graph corresponds to a peptide explaining *Spectrum*; consider, for example, the path spelling out GGDTN in Figure 11.4. For this reason, we must rewrite the above faulty pseudocode as follows.

```
DECODINGIDEALSPECTRUM(Spectrum)
    construct GRAPH(Spectrum)
    for each path Path from source to sink in GRAPH(Spectrum)
        Peptide ← the amino acid string spelled by the edge labels of Path
        if IDEALSPECTRUM(Peptide) = Spectrum
            return Peptide
```

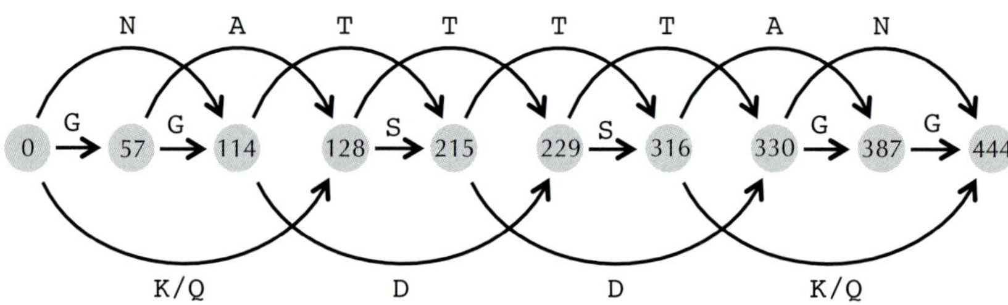

FIGURE 11.4 The DAG GRAPH(*Spectrum*) for *Spectrum* = {0, 57, 114, 128, 215, 229, 316, 330, 387, 444}. Only eight of the 32 paths from *source* to *sink* in this graph correspond to peptides explaining *Spectrum*.

Exercise Break: Which masses in the spectrum in Figure 11.4 does the peptide GGDTN fail to explain?

Although **DECODINGIDEALSPECTRUM** solves the Decoding an Ideal Spectrum Problem, exploring all paths in a DAG may be time-consuming, since the number of such paths may be exponential in the number of masses in the spectrum (see **DETOUR: Finding All Paths in a Graph**).

From Ideal to Real Spectra

We already know from our analysis of antibiotics in Chapter 4 that the realities of peptide sequencing are more harsh than reconstructing a peptide from an ideal spectrum. After breaking each copy of a peptide into two smaller fragments, a mass spectrometer ionizes them, resulting in electrically charged **fragment ions**. It measures each fragment ion's **mass/charge ratio** as well as its **intensity**, or the number of fragment ions detected at that mass/charge ratio (peptides may break frequently at some bonds and hardly ever at other bonds). As a result, a spectrum is represented as a collection of peaks in a chart, where a peak's x-coordinate represents its mass/charge ratio, and its height represents its intensity (Figure 11.5 (top)).

Modern mass spectrometers have limitations on the range of mass/charge ratios that they can detect, making it difficult to analyze entire proteins by mass spectrometry. As a result, proteins are usually analyzed by first breaking them into shorter peptides using enzymes called **proteases**. The most popular protease used in proteomics, and the one used in the *T. rex* study, is called **trypsin**. This protease typically breaks a protein after the amino acids R and K and results in peptides of average length 14.

Figure 11.5 shows a mass spectrum for one of the *T. rex* spectra (henceforth referred to as *DinosaurSpectrum*), along with its two putative interpretations, ATKIVDCFMTY and GLVGAPGLRGLPGK. Once we infer the peptide that generated a given spectrum, we can **annotate** the spectrum by establishing a correspondence between the masses of spectral peaks and the masses of the peptide's prefixes/suffixes. To comply with standard mass spectrometry terminology, a peak annotated as the prefix of length i is labeled b_i, and a peak annotated as the suffix of length i is labeled y_i.

STOP and Think: Which of the two interpretations of *DinosaurSpectrum* in Figure 11.5 do you think better explains *DinosaurSpectrum*?

STOP

Sequencing a peptide from its real spectrum is even more difficult than it may already seem. Mass spectra often have "noisy" peaks that contribute to false masses, which may have higher intensities than peaks corresponding to true prefixes and suffixes. Since some peptide bonds hardly ever break, the intensities at different mass/charge ratios may differ by orders of magnitude. As a result, a spectrum may not have peaks corresponding to true prefixes and suffixes. For example, in Figure 11.5 (bottom), there are no peaks annotated as b_5 or y_9 in *DinosaurSpectrum*. For this reason, although we ignored intensities when sequencing antibiotics, we will take them more seriously in this chapter.

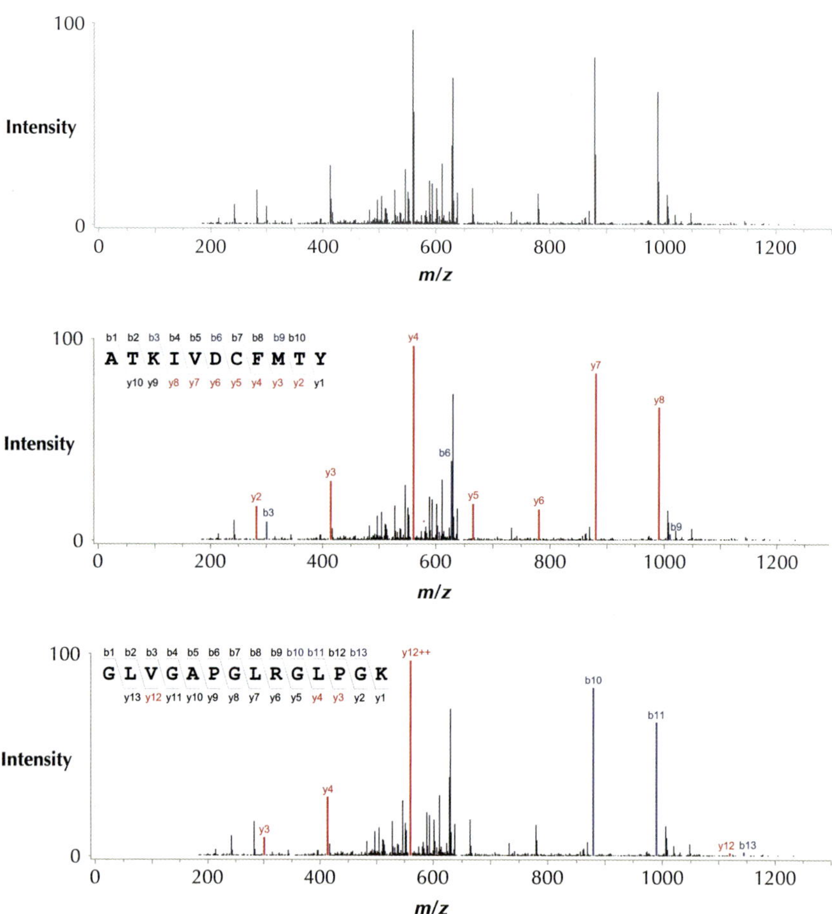

FIGURE 11.5 (Top) The spectrum *DinosaurSpectrum*. (Middle) The same spectrum annotated by ATKIVDCFMTY. (Bottom) The same spectrum annotated by GLVGAPGLRGLPGK. The peak corresponding to the prefix peptide of length i is annotated as b_i, and the peak corresponding to the suffix peptide of length i is annotated as y_i. For example, the peak annotated as b_{10} corresponds to GLVGAPGLRG, and the peak annotated as y_3 corresponds to PGK. Most annotated peaks have charge +1, but some (such as the one denoted y_{12}++) have charge +2. The charge of the fragment ion represented by a given peak in the spectrum is not known in advance but can often be inferred after the peptide that generated the spectrum has been identified. Only six peaks in *DinosaurSpectrum* are annotated by GLVGAPGLRGLPGK; peaks b_{10}, b_{11}, and b_{13} are annotated by prefix peptides, and peaks y_3, y_4, and y_{12} are annotated by suffix peptides.

Exercise Break: REDCA is one possible peptide that explains some but not all of the masses in the spectrum shown in Figure 11.6, which has false and missing masses. Can you find another peptide that explains even more masses in this spectrum?

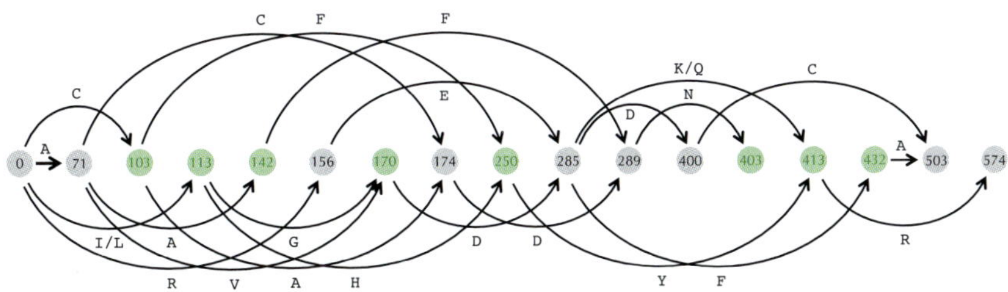

FIGURE 11.6 The DAG GRAPH(*Spectrum*) constructed from *Spectrum* = {0, 71, 103, 113, 142, 156, 170, 174, 250, 285, 289, 400, 403, 413, 432, 503, 574} with one missing mass (418) and eight false masses (shown in green) compared to the ideal spectrum of REDCA.

Exercise Break: List all peptides spelled by paths from *source* to *sink* for the DAG GRAPH(*Spectrum*) shown in Figure 11.6, and compute how many masses in *Spectrum* each of the identified peptides explains. Do any of these peptides explain *Spectrum* better than REDCA?

The issue of false and missing masses is only one of the many complications in mass spectrometry. When a mass spectrometer breaks a peptide, small parts of the resulting fragments may be lost, thus lowering their mass. For example, when breaking REDCA into RE and DCA, RE might lose a water molecule (H_2O) of mass 1 + 1 + 16 = 18, and DCA might lose an ammonia molecule (NH_3) of mass 1 + 1 + 1 + 14 = 17. The respective integer masses of the resulting fragments will be equal to MASS(RE) − 18 and MASS(DCA) − 17.

Because of the many practical complications of mass spectrometry, we will need to make some simplifying assumptions in order to move toward a computational problem modeling peptide sequencing. Instead of trying to account for the great variety of different fragmentation patterns when reconstructing a peptide, we will treat them as noise. We will also assume that all peaks have charge +1 and that the spectra are discretized (i.e., that all masses are integers).

Peptide Sequencing

Scoring peptides against spectra

Consider an imaginary world in which peptides are built from just two amino acids, X and Z, having respective masses 4 and 5. For example, given the peptide XZZXX, prefixes have masses 4, 9, 14, 18, and 22, while suffixes have masses 22, 18, 13, 8, 4.

Now consider the hypothetical spectrum

$$(0, 0, 0, 3, 8, 7, 2, 1, 100, 0, 1, 4, 3, 500, 2, 1, 3, 9, 1, 2, 2, 0)$$

arising from this peptide, where the i-th element of this vector corresponds to the intensity detected at mass i. The prefixes of XZZXX annotate peaks with intensities 3, 100, 500, 9, and 0, whereas the suffixes annotate peaks with intensities 3, 1, 3, 9, and 0. Our goal is to develop an approach for scoring peptides against spectra in the hope that we will be able to find a peptide that generated the spectrum by simply finding the peptide with maximum score against this spectrum.

STOP and Think: How would you score a peptide against a spectrum?

One scoring approach is the **intensity count**, or the sum of intensities of all peaks annotated by a peptide. For example, we would score peptide XZZXX against the spectrum above as the sum of intensities of all peaks annotated by XZZXX, i.e., $3 + 100 + 500 + 9 + 0 + 0 + 8 + 0 + 2 + 1$. However, the intensity count does not work well in practice because peak intensities vary widely. As a result, the tallest peaks in the spectrum (100 and 500 in our toy example) may dominate the score. Since the highest peaks may represent noise and since lower intensity peaks often represent correct prefix/suffix peptides, the intensity count is not a good scoring function in practice.

Another scoring approach, called the **shared peaks count**, simply counts the number of "tall" peaks annotated by a peptide, i.e., the annotated peaks with intensities exceeding a predefined threshold. Taking the intensity threshold 5 in our ongoing example, the shared peak count is 4, since the prefixes of XZZXX annotate the tall peaks with intensities 100, 500, and 9, and the suffixes annotate the tall peak with intensity 8. The peptide in Figure 11.5 (middle) has shared peak count equal to 10, whereas the peptide in Figure 11.5 (bottom) has shared peak count equal to 6.

Although the shared peak count works better than the intensity count in practice, it is still far from ideal; a better approach would account for intensities of peaks without letting the tallest peaks dominate the score. To achieve this goal, we will convert peptides and spectra into vectors, then define a scoring function that is the dot product of these vectors.

First, given an amino acid string $Peptide = a_1 \ldots a_n$ of length n, we will represent its prefix masses using a binary **peptide vector** $\overrightarrow{Peptide}$ with $\text{MASS}(Peptide)$ coordinates. This vector contains a 1 at each of the n **prefix coordinates**

$$\text{MASS}(a_1), \text{MASS}(a_1 a_2), \ldots, \text{MASS}(a_1 a_2 \ldots a_n),$$

and it contains a 0 in each of the remaining noise coordinates. The toy peptide XZZXX, whose prefix masses are 4, 9, 14, 18, and 22, corresponds to the peptide vector (0, 0, 0, 1, 0, 0, 0, 0, 1, 0, 0, 0, 0, 1, 0, 0, 0, 1, 0, 0, 0, 1) of length 22 (Figure 11.7).

Converting a Peptide into a Peptide Vector Problem:
Convert a peptide into a peptide vector.

> **Input**: An amino acid string *Peptide*.
> **Output**: The peptide vector $\overrightarrow{Peptide}$.

	1	2	3	4	5	6	7	8	9	10	11	12	13	14	15	16	17	18	19	20	21	22
peptide vector	0	0	0	**1**	0	0	0	0	**1**	0	0	0	0	**1**	0	0	0	**1**	0	0	0	**1**
spectral vector	0	0	0	**4**	-2	-3	-1	-7	**6**	**5**	3	2	1	**9**	3	-8	0	3	1	2	1	0

FIGURE 11.7 The peptide vector of XZZXX and a hypothetical spectral vector generated by this peptide (assuming that X and Z have masses 4 and 5, respectively). Prefix coordinates of the peptide vector are shown in boldface. Amplitudes exceeding the threshold 3 in the spectral vector are shown in color. These bold entries correspond to three prefix coordinates (blue) and one noise coordinate (red). Prefix coordinates correspond to ones in the peptide vector, whereas noise coordinates correspond to zeroes.

Since a peptide vector uniquely defines the peptide that it originated from, we will use the terms "peptide vector" and "peptide" interchangeably.

Converting a Peptide Vector into a Peptide Problem:
Convert a peptide vector into a peptide.

> **Input**: A binary vector P.
> **Output**: A peptide whose peptide vector is equal to P (if such a peptide exists).

Where are the suffix peptides?

You may be wondering why peptide vectors only model prefix peptides, since both prefix and suffix peptides contribute to spectral annotations. Does it not make more sense to define the peptide vector of XZZXX as (0, 0, 0, 1, 0, 0, 0, 1, 1, 0, 0, 0, 1, 1, 0, 0, 0, 1, 0, 0, 0, 1) in order to reflect both its prefix masses (4, 9, 14, 18, and 22) and suffix masses (4, 8, 13, 18, 22)?

Indeed, any peak of mass s in a spectrum may be interpreted as either a prefix mass or suffix mass of an unknown peptide *Peptide* that generated the spectrum. Moreover, its **twin peak**, with mass equal to $\text{MASS}(Peptide) - s$, may be interpreted as either a mass of a suffix or a mass of a prefix of the same peptide.

To deal with this uncertainty, mass spectrometrists convert a spectrum *Spectrum* into a **spectral vector** $\overrightarrow{Spectrum}$ that consolidates information about the intensities of each peak and its twin into a single value, called an **amplitude**, at the coordinate representing the mass of the hypothetical prefix peptide for these twins. Why? Because algorithms interpreting a consolidated spectrum become easier than algorithms attempting to account for both twins (see **DETOUR: The Anti-Symmetric Path Problem**). To make matters more complicated, amplitudes in the spectral vectors also account for intensities of ion fragments with various charges and intensities of ion fragments with water and ammonia molecule loss.

PAGE 628

Figure 11.8 shows a spectral vector constructed from *DinosaurSpectrum* and illustrates that amplitudes in a spectral vector may be negative. Negative amplitudes typically correspond to positions in spectra without peaks or with low intensity peaks. The correspondence between intensities in a spectrum and amplitudes in its spectral vector is complex, but in general, the amplitude at mass i reflects the *likelihood* that the (unknown) peptide that generated the spectrum has a prefix with mass i (see **DETOUR: Transforming Spectra into Spectral Vectors**).

PAGE 629

After a peptide *Peptide* has been transformed into a peptide vector $\overrightarrow{Peptide} = (p_1, \ldots, p_m)$ and a spectrum *Spectrum* has been transformed into a spectral vector

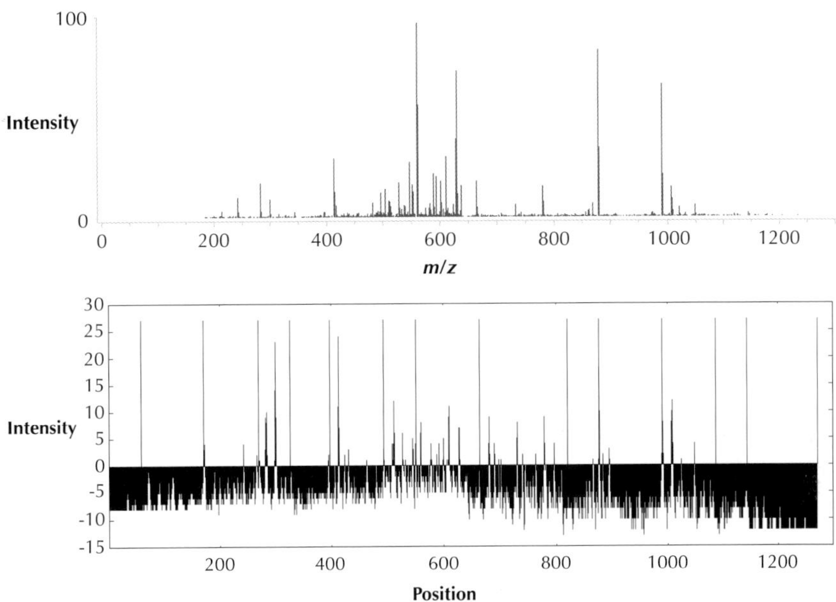

FIGURE 11.8 (Top) *DinosaurSpectrum* reproduced from Figure 11.5. (Bottom) The spectral vector of *DinosaurSpectrum*. The positions of ones in the peptide vector of GLVGAPGLRGLPGK (at coordinates 57, 170, 269, 326, 397, 494, 551, 664, 820, 877, 990, 1087, 1144, and 1272) are shown by red lines. The amplitudes of the spectral vector at these coordinates are equal to -8, +1, -4, - 6, -6, +3, +1, -4, -8, +18, +11, -10, -7, and 0, respectively. SCORE(GLVGAPGLRGLPGK, *DinosaurSpectrum*) is the dot product of the peptide vector and spectral vector, $-8+1-4-6-6+3+1-4-8+18+11-10-7+0=-19$. Because most amplitudes are negative, the fact that SCORE(GLVGAPGLRGLPGK, *DinosaurSpectrum*) is negative does not necessarily imply that the spectral interpretation above is incorrect.

$\overrightarrow{Spectrum} = (s_1, \ldots, s_m)$ of the same length, we define $\text{SCORE}(Peptide, Spectrum) = \text{SCORE}(\overrightarrow{Peptide}, \overrightarrow{Spectrum})$ as the dot product of $\overrightarrow{Peptide}$ and $\overrightarrow{Spectrum}$,

$$\text{SCORE}(Peptide, Spectrum) = p_1 \cdot s_1 + \ldots + p_m \cdot s_m .$$

Note that $\text{SCORE}(\overrightarrow{Peptide}, \overrightarrow{Spectrum})$ is simply an "amplitude count", the sum of amplitudes in $\overrightarrow{Spectrum}$ that are "annotated" by $\overrightarrow{Peptide}$. However, this score does not suffer from the limitations of the intensity count because we have transformed intensities in a spectrum into amplitudes in its spectral vector. As a result, high-intensity peaks in the spectrum contribute to the score but do not dominate it. The score of the peptide vector and spectral vector in Figure 11.7 is $4+6+9+3+0=22$; the score of the peptide vector and spectral vector in Figure 11.8 is -19.

STOP and Think: Can you find a peptide vector that scores higher than 22 against the spectral vector in Figure 11.7?

In the remainder of this chapter, we will work with spectral vectors instead of spectra. Given a spectral vector $\overrightarrow{Spectrum}$, our goal is to find a peptide *Peptide* maximizing $\text{SCORE}(\overrightarrow{Peptide}, \overrightarrow{Spectrum})$. Since the mass of a peptide and the parent mass of the spectrum that it generates should be the same, a peptide vector should have the same length as the spectral vector under consideration. We will therefore define the score between a peptide vector and a spectral vector of different length as $-\infty$.

Peptide Sequencing Problem:
Given a spectral vector, find a peptide with maximum score against this spectrum.

Input: A spectral vector $\overrightarrow{Spectrum}$.
Output: An amino acid string *Peptide* that maximizes $\text{SCORE}(\overrightarrow{Peptide}, \overrightarrow{Spectrum})$ among all possible amino acid strings.

Peptide sequencing algorithm

Given a spectral vector $\overrightarrow{Spectrum} = (s_1, \ldots, s_m)$, we will construct a DAG on $m+1$ nodes, indexed with the integers from 0 (*source*) to m (*sink*), and then connect node i to node j by a directed edge if $j-i$ is equal to the mass of an amino acid (Figure 11.9). We will further assign weight s_i to node i (for $1 \leq i \leq m$) and assign weight zero to node 0.

STOP and Think: How does this DAG compare to the DAG GRAPH(*Spectrum*) that we constructed to decode an ideal spectrum?

Any path connecting *source* to *sink* in this DAG corresponds to an amino acid string *Peptide*, and the total weight of nodes on this path is equal to $\text{SCORE}(\overrightarrow{Peptide}, \overrightarrow{Spectrum})$. We have therefore reduced the Peptide Sequencing Problem to the problem of finding a maximum-weight path from *source* to *sink* in a node-weighted DAG.

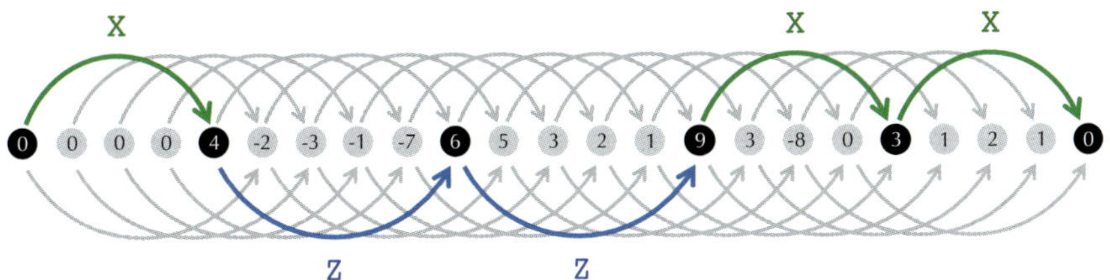

FIGURE 11.9 The node-weighted DAG for a spectral vector of length $m = 22$ and the amino acid alphabet {X, Z} with respective masses 4 and 5. The path from 0 to m representing peptide **XZZXX** (corresponding to the peptide vector (0, 0, 0, **1**, 0, 0, 0, 0, **1**, 0, 0, 0, 0, **1**, 0, 0, 0, **1**, 0, 0, 0, **1**) with score $0 + 4 + 6 + 9 + 3 + 0$ is highlighted. The labels of the DAG's nodes represent spectral vector amplitudes.

STOP and Think: In Chapter 5, we developed an algorithm for finding a path of maximum weight in an *edge-weighted* DAG. How can we modify this algorithm to find a path of maximum weight in a *node-weighted* DAG?

Exercise Break: Apply your algorithm for the Peptide Sequencing Problem to $\overrightarrow{DinosaurSpectrum}$.

By applying an algorithm solving the Peptide Sequencing Problem to $\overrightarrow{DinosaurSpectrum}$, we find the peptide ATKIVDCFMTY with score 96 (Figure 11.5 (middle)). However, Asara proposed a different peptide, GLVGAPGLRGLPGK with score -19, which we will call *DinosaurPeptide* (Figure 11.5 (bottom)). This peptide has much lower score than ATKIVDCFMTY; in fact, *billions* of peptides outscore *DinosaurPeptide*!

STOP and Think: Why do you think that Asara proposed *DinosaurPeptide* instead of the higher-scoring ATKIVDCFMTY?

Peptide Identification

The Peptide Identification Problem

If you followed our struggles to sequence antibiotic peptides, then you will agree that we should be wary of jumping to the conclusion that the highest-scoring peptide for *DinosaurSpectrum* must have generated this spectrum.

Despite many attempts, researchers have still not devised a scoring function that reliably assigns the highest score to the biologically correct peptide, i.e., the peptide that generated the spectrum. Fortunately, although the correct peptide often does not achieve the highest score among *all* peptides, it typically does score highest among all peptides *limited* to the species's proteome. As a result, we can transition from peptide sequencing to peptide identification by limiting our search to peptides present in the proteome, which we concatenate into a single amino acid string *Proteome*.

Exercise Break: How does the number of all peptides of length 10 (which we must explore in peptide sequencing) compare with the number of peptides of length 10 in the human proteome? (Note: there are approximately 20,000 protein-coding genes in the human protein, and the average length of a human protein is approximately 400 amino acids.)

Peptide Identification Problem:
Find a peptide from a proteome with maximum score against a spectrum.

Input: A spectral vector $\overrightarrow{Spectrum}$ and an amino acid string *Proteome*.
Output: An amino acid string *Peptide* that maximizes SCORE($\overrightarrow{Peptide}$, $\overrightarrow{Spectrum}$) among all substrings of *Proteome*.

STOP and Think: In practice, the input to the Peptide Identification Problem is a set of proteins rather than a single string *Proteome*. What are the potential pitfalls of concatenating all proteins as opposed to analyzing each protein separately?

Identifying peptides in the unknown T. rex proteome

You may be wondering why we have returned to peptide identification, since we do not know the *T. rex* proteome. It may therefore seem that we cannot apply an algorithm for the Peptide Identification Problem to the spectra obtained from the *T. rex* fossil.

STOP and Think: How could we form a protein database to search for *T. rex* peptides?

Approximately 90% of proteins making up animal bones are **collagen**. Dinosaur bones undoubtedly contained collagen, and there is little chance that other proteins could have survived for millions of years. Since the amino acid sequences of collagens are often conserved across different species, Asara reasoned that any proteins that had survived in the *T. rex* fossil would likely be similar to collagens from present-day species.

As a sanity check, Asara compared the *T. rex* spectra against the entire **UniProt** database, containing proteins from hundreds of species and totaling almost 200 million amino acids. He also included some mutated versions of collagens from present-day species in order to model possible differences between these collagens and collagens in *T. rex* (we will call the resulting protein database **UniProt+**). It turned out that most of the high-scoring peptides identified in this database were chicken collagens, which supported the hypothesis that birds evolved from dinosaurs.

In fact, *DinosaurPeptide* is only one mutation apart from a chicken collagen peptide. But how can we test whether this peptide is the correct interpretation of *DinosaurSpectrum*?

Searching for peptide-spectrum matches

Like peptide sequencing algorithms, peptide identification algorithms may return an erroneous peptide, particularly if the score of the highest-scoring peptide found in the proteome is much lower than the score of the highest-scoring peptide over all peptides. For this reason, biologists usually establish a score threshold and only pay attention to a solution of the Peptide Identification Problem if its score is at least equal to the threshold.

Given a set of spectral vectors *SpectralVectors*, an amino acid string *Proteome*, and a score threshold *threshold*, we will solve the Peptide Identification Problem for each vector $\overrightarrow{Spectrum}$ in *SpectralVectors* and identify a peptide *Peptide* having maximum score for this spectral vector over all peptides in *Proteome* (ties are broken arbitrarily). If $\text{SCORE}(\overrightarrow{Peptide}, \overrightarrow{Spectrum})$ is greater than or equal to *threshold*, then we conclude that *Peptide* is present in the sample and call the pair $(Peptide, \overrightarrow{Spectrum})$ a **peptide-spectrum match (PSM)**. The resulting collection of PSMs for *SpectralVectors* is denoted $\text{PSM}_{threshold}(Proteome, SpectralVectors)$.

PSM Search Problem:
Identify all peptide-spectrum matches scoring above a threshold for a set of spectra and a proteome.

> **Input:** A set of spectral vectors *SpectralVectors*, an amino acid string *Proteome*, and an integer *threshold*.
> **Output:** The set $\text{PSM}_{threshold}(Proteome, SpectralVectors)$.

The following pseudocode solves the PSM Search Problem using an algorithm that you just implemented to solve the Peptide identification Problem, which we call **PEPTIDEIDENTIFICATION**.

11G

```
PSMSEARCH(SpectralVectors, Proteome, threshold)
    PSMSet ← an empty set
    for each vector Spectrum→ in SpectralVectors
        Peptide ← PEPTIDEIDENTIFICATION(Spectrum→, Proteome)
        if SCORE(Peptide, Spectrum→) ≥ threshold
            add the PSM (Peptide, Spectrum→) to PSMSet
    return PSMSet
```

DinosaurPeptide turned out to be the highest scoring peptide for *DinosaurSpectrum* among all peptides in the UniProt+ database. But do the billions of peptides not occurring in this database that outscore *DinosaurPeptide* imply that the database that Asara formed is incomplete and that *DinosaurSpectrum* arose from a different peptide?

The reality is that the highest scoring peptide in a proteome is commonly outscored by billions of peptides not belonging to the proteome. However, this phenomenon does not imply that **PSMSEARCH** has identified the wrong peptide, because the total number of peptides with the same mass may be measured in the trillions or even quadrillions. In other words, the billions of peptides that outscore *DinosaurPeptide* represent a small fraction of all peptides having the same mass as this peptide. Thus, we need to complement **PSMSEARCH** with an evaluation of the statistical significance of its identified PSMs.

STOP and Think: Suppose that we search 1,000 spectra from a chicken sample against the chicken proteome, and identify 100 PSMs whose score surpasses a threshold. How would you estimate the percentage of erroneous PSMs among these 100 PSMs?

Peptide Identification and the Infinite Monkey Theorem

False discovery rate

To estimate the number of spurious PSMs in $\text{PSM}_{threshold}(Proteome, SpectralVectors)$, we will construct a **decoy proteome** *DecoyProteome*, a randomly generated amino acid string having the same length as *Proteome* (with the probability of generating any amino acid at each position equal to 1/20). We will then solve the PSM Search Problem for *DecoyProteome* instead of *Proteome* for the same score threshold.

We are not interested in the PSMs identified in the randomly generated decoy proteome, which are nothing more than biologically irrelevant artifacts. The point is that the *number* of these PSMs will provide a rough estimate for the number of erroneous PSMs identified in our biologically relevant search against the real proteome. We will therefore define the **false discovery rate (FDR)** of a PSM search as the ratio of the number of decoy PSMs to the number of PSMs identified with respect to the real proteome,

$$\frac{|\text{PSM}_{threshold}(DecoyProteome, SpectralVectors)|}{|\text{PSM}_{threshold}(Proteome, SpectralVectors)|}.$$

For example, if a search against *Proteome* results in 100 PSMs, and a search against *DecoyProteome* results in just five PSMs, then the FDR would be 5%, and we would conclude that approximately 95% of identified PSMs are likely valid. On the other hand, if searching against *DecoyProteome* returned close to 100 PSMs, then the FDR would be close to 1, and we would have a hard time making the argument that any peptides identified in our search against *Proteome* are biologically relevant.

Exercise Break: Estimate the FDR when searching all *T. rex* spectra against the UniProt+ database with *threshold* = 80.

STOP and Think: If the FDR turns out to be very high for a given value of *threshold*, can we still find reliable PSMs?

Even if the FDR is high, then we should not conclude that our spectral dataset is worthless or that we are searching against the wrong protein database. We may simply have selected the wrong score threshold for analyzing our data, since the FDR can vary widely depending on the choice of this threshold (Figure 11.10).

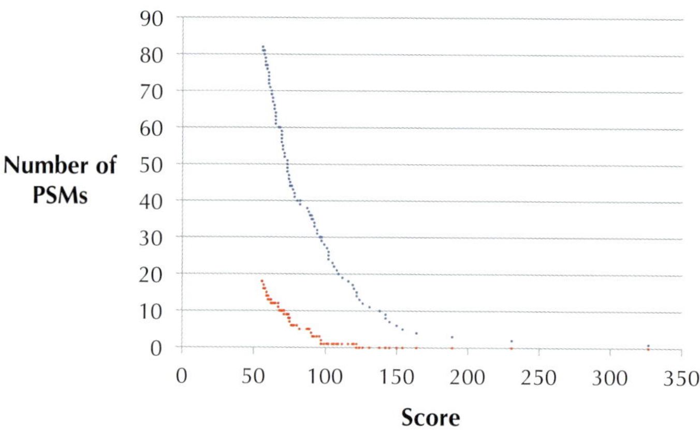

FIGURE 11.10 The number of PSMs identified by searching *DinosaurSpectrum* against the amino acid string derived from the UniProt+ database (blue) and a decoy proteome of the same length (red) depending on the score threshold.

For the *T. rex* spectral dataset, we find 27 PSMs in the amino acid string *Proteome* derived from UniProt+ and only one PSM in *DecoyProteome* with score at least equal to *threshold* = 100 (FDR = 3.7%). Unfortunately, we cannot automatically conclude that we have discovered two dozen dinosaur peptides because many of these PSMs correspond to common laboratory contaminants. Our goal is to figure out whether the remaining few, including *DinosaurPeptide*, indeed correspond to dinosaur peptides.

Specifically, FDR helps us analyze the *entire* set of identified PSMs for all *T. rex* spectra, but what about the statistical significance of an *individual* PSM? In particular, can we determine whether the PSM (*DinosaurPeptide*, $\overrightarrow{\textit{DinosaurSpectrum}}$), which we will call *DinosaurPSM*, is statistically significant? To answer this question, we first need to quantify what we mean by "statistically significant".

The monkey and the typewriter

Imagine that we have locked you in a room with a monkey and a typewriter. The monkey quickly tires of your company and starts banging away on the typewriter, creating strings of symbols (we will assume that the monkey particularly likes the space bar). As you begin to lose your sanity, you test each new string generated by the monkey to see if some of them are correctly spelled English words. (Figure 11.11). After all, according to the **infinite monkey theorem**, the monkey will eventually type *Hamlet* (see **DETOUR: The Infinite Monkey Theorem**).

PAGE 630

FIGURE 11.11 Searching for William Shakespeare.

You would be shocked if the monkey immediately typed "To be, or not to be". But would you be as surprised if, after cranking out a million strings, the monkey had typed a dozen English words?

Given a set of strings *Dictionary*, we define $\mathrm{E}(Dictionary, n)$ as the expected number of occurrences of strings from *Dictionary* in a randomly generated string of length n, where the probabilities of generating every letter at each position of the string are the same. Let *EnglishDictionary* denote the set of all English words. If it turns out that, after typing n symbols, the monkey types significantly more English words than $\mathrm{E}(EnglishDictionary, n)$, then we have every reason to believe that the monkey can spell! On the other hand, if the monkey types about the same number of words as $\mathrm{E}(EnglishDictionary, n)$, then the monkey is probably not Shakespeare reincarnate.

The Monkey and the Typewriter Problem:
Find the expected number of strings from a dictionary appearing in a randomly generated text.

Input: A set of strings *Dictionary* and an integer n.
Output: $\mathrm{E}(Dictionary, n)$.

STOP and Think: What do the monkey and the typewriter have to do with mass spectrometry?

Exercise Break: What is the expected number of times that the string `SHAKESPEARE` appears in a randomly generated English string (with no spaces) of length 200 million?

Statistical significance of a peptide-spectrum match

Now imagine that instead of a monkey typing words, we have an algorithm generating the set of all peptides scoring at least *threshold* against a spectral vector $\overrightarrow{Spectrum}$. We will henceforth call this set of high-scoring peptides a **spectral dictionary**, denoted

$$\text{DICTIONARY}_{threshold}(\overrightarrow{Spectrum}) .$$

For a PSM $(Peptide, \overrightarrow{Spectrum})$, we will use the term **PSM dictionary**, denoted $\text{DICTIONARY}(Peptide, \overrightarrow{Spectrum})$, to refer to the spectral dictionary

$$\text{DICTIONARY}_{\text{SCORE}(Peptide, \overrightarrow{Spectrum})}(\overrightarrow{Spectrum}) .$$

For *DinosaurPSM*, the PSM dictionary is $\text{DICTIONARY}_{-19}(\overrightarrow{DinosaurSpectrum})$.

Instead of checking which words generated by the monkey occur in an English dictionary, we will match peptides from a spectral dictionary against a proteome. If we find matches, then we must decide whether these matches represent biologically valid PSMs or just statistical artifacts. To make this decision, we consider

$$\text{E}(\text{DICTIONARY}_{threshold}(\overrightarrow{Spectrum}), n) ,$$

the expected number of peptides in a decoy proteome of length n that would occur in $\text{DICTIONARY}_{threshold}(\overrightarrow{Spectrum})$. If this number is larger than 1, then there is nothing surprising in finding a peptide that scores *threshold* against $\overrightarrow{Spectrum}$. We have therefore formulated our statistical significance test as a special case of the Monkey and the Typewriter Problem.

Expected Number of High-Scoring Peptides Problem:
Find the expected number of high-scoring peptides against a given spectrum in a decoy proteome.

> **Input**: A spectral vector $\overrightarrow{Spectrum}$ and integers *threshold* and n.
> **Output**: $\mathrm{E}(\text{DICTIONARY}_{threshold}(\overrightarrow{Spectrum}), n)$.

To solve this problem, we will begin with a spectral dictionary consisting of a single amino acid string *Peptide*, which we will attempt to match against a randomly generated string *DecoyProteome* of length n. Because *DecoyProteome* was randomly generated, the probability that *Peptide* matches the string beginning at a given position of *DecoyProteome* is $1/20^{|Peptide|}$. We call this expression the **probability** of *Peptide*. Therefore, the expected number of times that *Peptide* occurs in *DecoyProteome* is

$$\frac{n - |Peptide| + 1}{20^{|Peptide|}} \approx n \cdot \frac{1}{20^{|Peptide|}}.$$

Next, assume that a set of peptides *Dictionary* contains multiple amino acid strings of arbitrary lengths. Using the above approximation, the expected number of matches between strings in *Dictionary* and *DecoyProteome* can be approximated as

$$\mathrm{E}(Dictionary, n) \approx n \cdot \left(\sum_{\text{each peptide } Peptide \text{ in } Dictionary} \frac{1}{20^{|Peptide|}} \right).$$

We will refer to the sum inside the parentheses above as the **probability** of *Dictionary*, denoted $\Pr(Dictionary)$, so that the preceding approximation can be written as

$$\mathrm{E}(Dictionary, n) \approx n \cdot \Pr(Dictionary).$$

We have thus reduced the statistical analysis of a PSM $(Peptide, \overrightarrow{Spectrum})$ to the computation of $\Pr(\text{DICTIONARY}(Peptide, \overrightarrow{Spectrum}))$, the probability of the PSM dictionary.

STOP and Think: Can $\Pr(\text{DICTIONARY}(Peptide, \overrightarrow{Spectrum}))$ be greater than 1?

You may be wondering why we have used the probabilistic notation $\Pr(Dictionary)$. To learn why $\Pr(\text{DICTIONARY}(Peptide, \overrightarrow{Spectrum}))$ is indeed a probability (and thus cannot exceed 1), see **DETOUR: The Probabilistic Space of Peptides in a Spectral Dictionary**.

Probability of Spectral Dictionary Problem:
Find the probability of a spectral dictionary for a given spectrum and score threshold.

> **Input**: A spectral vector $\overrightarrow{Spectrum}$ and an integer *threshold*.
> **Output**: The probability of $\text{DICTIONARY}_{threshold}(\overrightarrow{Spectrum})$.

It seems that we are finally ready to test the statistical significance of *DinosaurPSM*. We simply need to first construct the PSM dictionary DICTIONARY(*DinosaurPSM*) and then compute $n \cdot \Pr(\text{DICTIONARY}(DinosaurPSM))$, where n is the length of the string formed by concatenating the UniProt+ database. If this value is small (say, 0.001), then we will be able to argue that *DinosaurPeptide* is a *T. rex* peptide rather than a statistical artifact.

Unfortunately, DICTIONARY(*DinosaurPSM*) contains over 200 *billion* peptides, and generating it would be extremely time-consuming. Can we somehow compute the probability of this dictionary without having to generate it?

Spectral Dictionaries

We will first compute the number of peptides in a spectral dictionary, since this simpler problem will provide insights on how to compute the probability of a spectral dictionary.

Size of Spectral Dictionary Problem:
Find the size of the spectral dictionary for a given spectral vector and score threshold.

> **Input**: A spectral vector $\overrightarrow{Spectrum}$ and an integer *threshold*.
> **Output**: The number of peptides in $\text{DICTIONARY}_{threshold}(\overrightarrow{Spectrum})$.

We will use dynamic programming to solve the Size of Spectral Dictionary Problem. Given a spectral vector $\overrightarrow{Spectrum} = (s_1, \ldots, s_m)$, we define its ***i*-prefix** (for i between 1 and m) as $\overrightarrow{Spectrum_i} = (s_1, \ldots, s_i)$ and introduce a variable $\text{SIZE}(i, t)$ as the number of peptides *Peptide* of mass i such that $\text{SCORE}(Peptide, \overrightarrow{Spectrum_i})$ is equal to t. For example, consider the spectral vector $\overrightarrow{Spectrum}$ = (4, -3, -2, 3, 3, -4, 5, -3, -1, -1, 3, 4, 1, 3) of length 14 and the toy amino acid alphabet consisting of amino acids X and Z with respective masses 4 and 5. There are only three peptides of mass 13 (XXZ, XZX, and ZXX); the first two peptides have score 1 against $\overrightarrow{Spectrum}_{13}$, and the third has score 3. Thus, $\text{SIZE}(13,1) = 2$, $\text{SIZE}(13,3) = 1$, and $\text{SIZE}(13,t) = 0$ for all values of t other than 1 and 3.

The key to establishing a recurrence relation for computing $\text{SIZE}(i,t)$ is to realize that the set of peptides contributing to $\text{SIZE}(i,t)$ can be split into 20 subsets depending on their final amino acid a. Each peptide ending in a specific amino acid a results in a shorter peptide with mass $i - |a|$ and score $t - s_i$ if we remove a from the peptide (here, $|a|$ denotes the mass of a). Thus, as illustrated in Figure 11.12,

$$\text{SIZE}(i,t) = \sum_{\text{all amino acids } a} \text{SIZE}(i - |a|, t - s_i) \,.$$

Since there is a single "empty" peptide of length zero, we initialize $\text{SIZE}(0,0) = 1$. We also define $\text{SIZE}(0,t) = 0$ for all possible scores t, and set $\text{SIZE}(i,t) = 0$ for negative values of i. Using the above recurrence, we can compute the size of a spectral dictionary of $\overrightarrow{Spectrum} = (s_1, \ldots, s_m)$ as

$$\left|\text{DICTIONARY}_{threshold}(\overrightarrow{Spectrum})\right| = \sum_{t \geq threshold} \text{SIZE}(m,t) \,.$$

STOP and Think: The above formula requires computing $\text{SIZE}(m,t)$ for all values of t greater than threshold. Given a spectral vector $\overrightarrow{Spectrum}$, can you find a value T such that $\text{SIZE}(m,t)$ is equal to zero when $t > T$?

Exercise Break: Compute $|\text{DICTIONARY}(DinosaurPSM)|$.

Note that the equation for the *probability* of a dictionary,

$$\Pr(Dictionary) = \sum_{\text{each peptide } Peptide \text{ in } Dictionary} \frac{1}{20^{|Peptide|}} \,,$$

is similar to an equation for the *size* of a dictionary,

$$|Dictionary| = \sum_{\text{each peptide } Peptide \text{ in } Dictionary} 1 \,.$$

This similarity suggests that we can derive a recurrence for the probability of a dictionary using arguments similar to those used to find the size of a dictionary.

Define $\Pr(i,t)$ as the sum of probabilities of all peptides with mass i for which $\text{SCORE}(\overrightarrow{Peptide}, \overrightarrow{Spectrum}_i)$ is equal to t. The set of peptides contributing to $\Pr(i,t)$ can

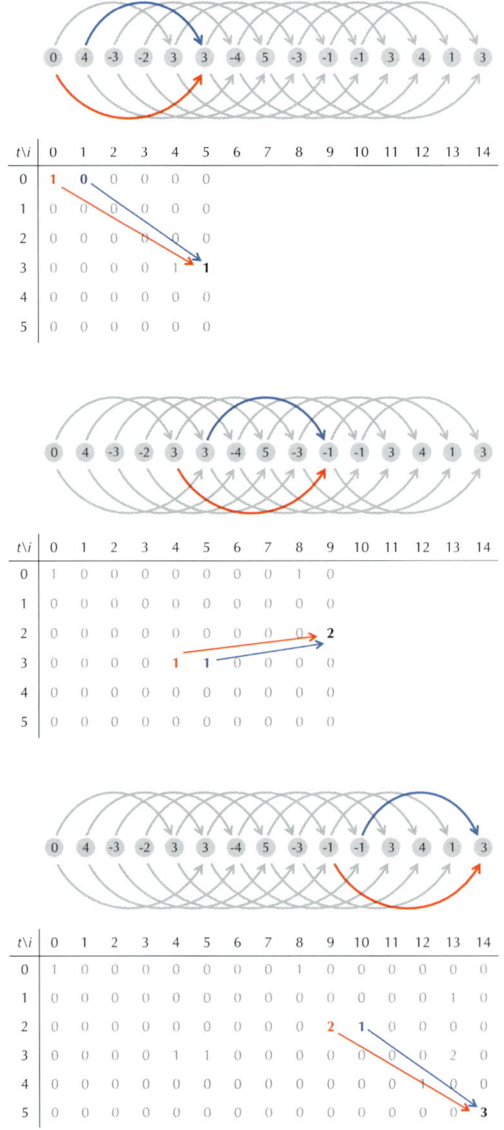

t\i	0	1	2	3	4	5	6	7	8	9	10	11	12	13	14
0	1	0	0	0	0	0									
1	0	0	0	0	0	0									
2	0	0	0	0	0	0									
3	0	0	0	0	1	**1**									
4	0	0	0	0	0	0									
5	0	0	0	0	0	0									

t\i	0	1	2	3	4	5	6	7	8	9	10	11	12	13	14
0	1	0	0	0	0	0	0	0	1	0					
1	0	0	0	0	0	0	0	0	0	0					
2	0	0	0	0	0	0	0	0	0	**2**					
3	0	0	0	0	1	1	0	0	0	0					
4	0	0	0	0	0	0	0	0	0	0					
5	0	0	0	0	0	0	0	0	0	0					

t\i	0	1	2	3	4	5	6	7	8	9	10	11	12	13	14
0	1	0	0	0	0	0	0	0	1	0	0	0	0	0	0
1	0	0	0	0	0	0	0	0	0	0	0	0	0	1	0
2	0	0	0	0	0	0	0	0	0	2	1	0	0	0	0
3	0	0	0	0	1	1	0	0	0	0	0	0	0	2	0
4	0	0	0	0	0	0	0	0	0	0	0	0	1	0	0
5	0	0	0	0	0	0	0	0	0	0	0	0	0	0	**3**

FIGURE 11.12 Computing SIZE(i, t) for the toy alphabet consisting of the amino acids X and Z with respective masses 4 and 5 and the spectral vector (4, -3, -2, 3, 3, -4, 5, -3, -1, -1, 3, 4, 1, 3). The boldfaced black entries in the dynamic programming matrices are computed by summing blue and red entries according to the formula SIZE(i, t) = SIZE$(i - 4, t - s_i)$ + SIZE$(i - 5, t - s_i)$. For example, in the matrix at the bottom, SIZE$(14, 5)$ = SIZE$(14 - 4, 5 - 3)$ + SIZE$(14 - 5, 5 - 3)$ = SIZE$(10, 2)$ + SIZE$(9, 2)$.

be split into 20 subsets depending on their final amino acid. Each peptide *Peptide* ending in a specific amino acid a results in a shorter peptide $Peptide_a$ if we remove a; $Peptide_a$ has mass $i - |a|$ and score $t - s_i$. Since the probability of *Peptide* is 20 times smaller than the probability of $Peptide_a$, the contribution of *Peptide* to $\Pr(i, t)$ is 20 times smaller than contribution of $Peptide_a$ to $\Pr(i - |a|, t - s_i)$. Therefore, $\Pr(i, t)$ can be computed as

$$\Pr(i,t) = \sum_{\text{all amino acids } a} \frac{1}{20} \cdot \Pr(i - |a|, t - s_i)\,,$$

which differs from the recurrence for computing $\text{SIZE}(i, t)$ only in the presence of the factor 1/20.

We can now compute the probability of a spectral dictionary as

$$\Pr(\text{DICTIONARY}_{threshold}(\overrightarrow{Spectrum})) = \sum_{t \geq threshold} \Pr(m, t)\,.$$

In particular, we find that DICTIONARY(*DinosaurPSM*) consists of 219,136,251,374 peptides and has probability 0.00018. We are therefore ready to test the statistical significance of *DinosaurPSM* found in searches against the UniProt+ database of length n = 194,613,142 (comprising 546,799 proteins).

Our aim is to compute $n \cdot \Pr(\text{DICTIONARY}(DinosaurPSM))$, as it approximates the number of peptides from DICTIONARY(*DinosaurPSM*) that we expect to find in a decoy proteome of length n. Since $\Pr(\text{DICTIONARY}(DinosaurPSM)) = 0.00018$, we have that

$$n \cdot \Pr(\text{DICTIONARY}(DinosaurPSM)) = 35{,}311\,.$$

We therefore expect to find tens of thousands of peptides scoring at least as high as *DinosaurPeptide* (against $\overrightarrow{DinosaurSpectrum}$) in a decoy database, and so there is nothing surprising about finding *DinosaurPSM* while searching the UniProt+ database! We therefore conclude that *DinosaurPeptide* is a statistical artifact rather than a real *T. rex* peptide. But what about the other *T. rex* peptides?

Exercise Break: Compute probabilities of spectral dictionaries for all other *T. rex* PSMs reported by Asara. Are these PSMs statistically significant?

T. rex Peptides: Contaminants or Treasure Trove of Ancient Proteins?

The hemoglobin riddle

Upon receiving criticism regarding the statistical foundations of his claims, Asara acknowledged some of the problems with his analysis, withdrew *DinosaurPeptide* as an explanation for *DinosaurSpectrum*, changed some of his previously proposed *T. rex* peptides, and released all 31,372 spectra from the *T. rex* fossil. Afterwards, other scientists re-analyzed all spectra and verified that although some of the originally reported *T. rex* PSMs are questionable, others are statistically solid (Figure 11.13).

However, Asara's release of *T. rex* spectra raised more questions than it answered. In these spectra, Matthew Fitzgibbon and Martin McIntosh identified an additional spectrum (Figure 11.14) that perfectly matched ostrich hemoglobin, thus adding another *T. rex* peptide to the seven collagen peptides in Figure 11.13. The hemoglobin PSM, which was missed by Asara, is an order of magnitude more statistically significant than any previously reported *T. rex* collagen peptide!

It would be shocking if the hemoglobin peptide indeed belonged to *T. rex* because hemoglobins are much less conserved than collagens. For example, human beta chain hemoglobin is 146 amino acids long and has 27, 38, and 45 amino acid differences with mouse, kangaroo, and, chicken respectively. Furthermore, intact hemoglobin peptides have never been found in much younger and widely available fossils, such as the bones of extinct cave bears. These fossils are so common in European caves that they were used as a source of phosphates to produce gunpowder during World War I.

ID	Peptide	Protein	Probability	$n \cdot$ Probability
P1	GL**V**GAPGLRGLPGK	Collagen α1t2	$1.8 \cdot 10^{-4}$	36,000
P2	GVVGLP$_{oh}$GQR	Collagen α1t1	$7.6 \cdot 10^{-8}$	16
P3	GVQGPP$_{oh}$GPQGPR	Collagen α1t1	$7.9 \cdot 10^{-11}$	$1.6 \cdot 10^{-2}$
P4	GATGAP$_{oh}$GIAGAP$_{oh}$GFPohGAR	Collagen α1t1	$3.2 \cdot 10^{-12}$	$6.4 \cdot 10^{-4}$
P5	GLPGESGAVGPAGPIGSR	Collagen α2t1	$9.9 \cdot 10^{-14}$	$2.0 \cdot 10^{-5}$
P6	GSAGPP$_{oh}$GATGFPohGAAGR	Collagen α1t1	$3.2 \cdot 10^{-14}$	$6.4 \cdot 10^{-6}$
P7	GAPGPQGPSGAP$_{oh}$GP**K**	Collagen α1t1	$7.0 \cdot 10^{-16}$	$1.4 \cdot 10^{-7}$
P8	VNVADCGAEALAR	Hemoglobin β	$7.8 \cdot 10^{-17}$	$1.6 \cdot 10^{-8}$

FIGURE 11.13 The seven candidate *T. rex* collagen peptides (P1 - P7) reported by Asara as well as a hemoglobin peptide (P8). The last column shows the probabilities of the PSM dictionaries formed by these peptides. Red symbols indicate mutated amino acids compared to peptides in the UniProt database. The amino acid P_{oh} stands for hydroxyproline, a modified form of proline that is common in collagens.

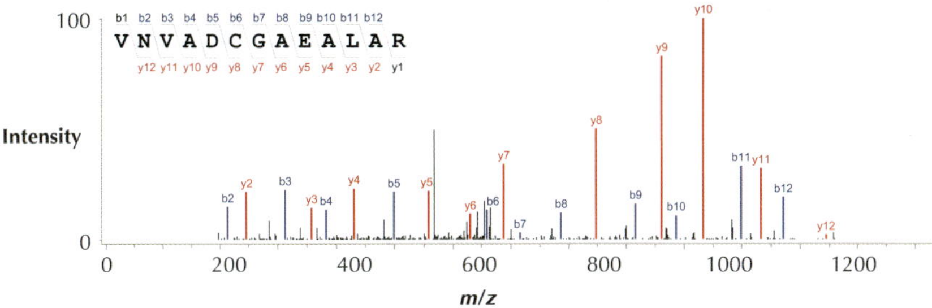

FIGURE 11.14 A high-quality *T. rex* spectrum matching the ostrich hemoglobin peptide `VNVADCGGAEAIAR`. Nearly all possible prefixes and suffixes are represented by high-intensity peaks; in fact, applying *de novo* sequencing to this spectrum results in the same peptide.

Because Asara had analyzed ostrich samples before analyzing the *T. rex* sample, Fitzgibbon and McIntosh argued that the hemoglobin peptide might indicate a contaminated sample in the form of a **carryover**, or the identification of leftover peptides hiding inside a mass spectrometer after a previous experiment. Contamination is a fact of life in every proteomics laboratory, as mass spectrometrists are never surprised when they identify human keratin in their samples: the air in any room typically contains millions of tiny human skin particles.

If the hemoglobin peptide is a carryover, then the entire *T. rex* sample has been contaminated, implying that all other *T. rex* peptides should be discarded. However, Asara maintained that there was no contamination in his experiment and that the ostrich hemoglobin must be a *T. rex* peptide, expanding the class of proteins that can survive for 68 million years beyond just collagens.

Yet if Horner's *T. rex* fossil is indeed a treasure trove of ancient proteins, and we believe that the hemoglobin peptide came from *T. rex*, then why should we limit our search to collagen peptides and their mutated variants? Why not search against all known proteins from all vertebrates? Of course, we should use criteria that are similar to ones that Asara used, such as allowing for up to one mutation. If we follow this criterion, then we should supplement Figure 11.13 with a surprisingly diverse set of peptides from ostrich, chicken, mouse, and human; some of these peptides are shown in Figure 11.15.

In light of these new peptides, Asara's claim about finding *molecular* evidence of the link between birds and dinosaurs becomes even weaker (for more on the debate surrounding the claim that birds evolved from dinosaurs, see DETOUR: Are Terrestrial Dinosaurs Really the Ancestors of Birds?). If we were to attempt to dismiss the

peptides in Figure 11.15 as statistical artifacts, then we might have to throw out the *T. rex* peptides in Figure 11.13 as well.

ID	Peptide	Protein	Probability	$n \cdot$ Probability
P9	EDCLSG**A**KPK	ATG7 (Chicken)	$3.2 \cdot 10^{-12}$	$6.4 \cdot 10^{-4}$
P10	ENAGEDPGLAR	DCD (Human)	$2.7 \cdot 10^{-12}$	$5.4 \cdot 10^{-4}$
P11	**E**GVDAGAAGDPER	TTL11 (Mouse)	$1.2 \cdot 10^{-12}$	$2.4 \cdot 10^{-4}$
P12	S**W**IHVALVTGGNK	CBR1 (Human)	$1.2 \cdot 10^{-12}$	$2.4 \cdot 10^{-4}$
P13	SSN**V**LSGSTLR	MAMD1 (Human)	$5.9 \cdot 10^{-13}$	$1.8 \cdot 10^{-4}$
P14	DEVTPA**Y**VVVAR	ASPM (Mouse)	$1.9 \cdot 10^{-13}$	$3.8 \cdot 10^{-5}$
P15	**R**NVADCGAEALAR	HBB (Ostrich)	$3.5 \cdot 10^{-15}$	$7.0 \cdot 10^{-7}$

FIGURE 11.15 Matching *T. rex* spectra against all vertebrate proteins in the UniProt database (allowing for up to 1 mutation) reveals a diverse set of peptides. Red symbols indicate mutated amino acids. Note the presence of another ostrich hemoglobin peptide (P15), which is slightly heavier (by 57 daltons) than the previously reported hemoglobin peptide in Figure 11.13 (P8). This change in mass may represent either a mutation of V into R (as shown above) or a modification of an amino acid.

The dinosaur DNA controversy

As the "*T. rex* peptides" paper continues to age, there is no end in sight to its controversy. Yet it was not the first paper to report the retrieval of genetic material from dinosaurs. In 1994, Scott Woodward announced that he had sequenced DNA from an 80 million year-old dinosaur bone. The most vehement critic of his finding was — believe it or not — Mary Schweitzer, who proved that Woodward had only sequenced contaminated human DNA.

The moral is that although we often present scientific discoveries as clear and incontrovertible, the reality is that some of the interesting avenues of modern science often fall short of this ideal. In a sense, the academic battleground is part of the appeal of becoming a scientist in the first place. But we also cannot help but wonder if we would have a conclusive answer to whether Horner's fossil really contained dinosaur peptides if it had originally been shared with dozens of independent researchers, who would have undoubtedly unearthed the shocking appearance of hemoglobin in the *T. rex* samples. Fittingly, in their criticism of Woodward's "dinosaur DNA" paper, Schweitzer wrote, "real advance in [paleontology] will come only when it is demonstrated that those studies can be replicated in independent laboratories."

Epilogue: From Unmodified to Modified Peptides

Post-translational modifications

The **PSMSEARCH** algorithm can only identify a peptide if it occurs in a proteome without mutations. Yet some of the peptides in Figure 11.13 are mutated.

STOP and Think: How could we generalize **PSMSEARCH** to find mutated peptides?

To find a highest-scoring peptide with up to k mutations matching a spectral vector, we could generate all mutated variants of all peptides in the proteome, concatenate them into an amino acid string *MutatedProteome*, and then run **PSMSEARCH** on *MutatedProteome*. Unfortunately, the number of mutated peptides will be so large that it may render **PSMSEARCH** impractical, even if we allow at most one mutation per peptide.

STOP and Think: How many mutated peptides with at most k mutations are there for a given peptide of length n?

In addition to searching for mutated peptides, we will also need to search for **post-translational modifications**, which alter amino acids after a protein has been translated from RNA. In fact, most proteins are modified after translation, and hundreds of types of modifications have been discovered. For example, the enzymatic activity of many proteins is regulated by the addition or removal of a phosphate group at a specific amino acid (Figure 11.16). This process, called **phosphorylation**, is reversible; **protein kinases** add phosphate groups, whereas **protein phosphatases** remove them.

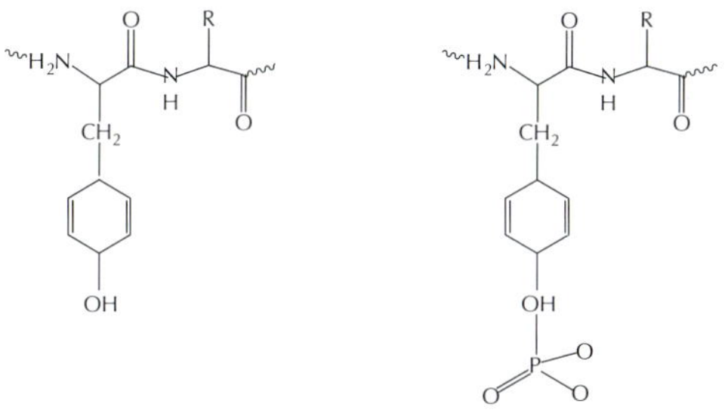

FIGURE 11.16 Tyrosine (left) and its post-translational modification into phosphorylated tyrosine (right).

In fact, you may have noticed in Figure 11.13 that most candidate *T. rex* peptides have a modification transforming proline (mass 97) into **hydroxyproline** (mass 113). Hydroxyproline is a major component of collagens that is important to collagen stability and that comprises roughly 4% of all amino acids in humans.

There are also important but rare post-translational modifications such as **diphthamide**. This modification of histidine only appears in a single protein (**protein synthesis elongation factor-2**), but it is universal across all eukaryotes! Researchers showed that diphthamide is the target for several toxins secreted by various pathogenic bacteria, which raises the question of why all eukaryotes would retain this modification if it makes them so vulnerable to pathogens — it must serve some important but still unknown function in normal physiology.

Searching for modifications as an alignment problem

A modification of mass δ applied to an amino acid results in adding δ to the mass of this amino acid. For example, $\delta = 80$ for phosphorylated amino acids (serine, threonine, and tyrosine), $\delta = 16$ for the modification of proline into hydroxyproline, and $\delta = -1$ for the modification of lysine into **allysin**. If δ is positive, then the resulting modified peptide has a peptide vector that differs from the original peptide vector $\overrightarrow{Peptide}$ by inserting a block of δ zeroes before the i-th occurrence of 1 in $\overrightarrow{Peptide}$. In the more rare case that δ is negative, the modified peptide corresponds to deleting a block of $|\delta|$ zeroes from $\overrightarrow{Peptide}$ (Figure 11.17).

```
     XZZXX  0 0 0 1 0 0 0 0 1 0 0 0 0 1 0 0 0 1 0 0 0 1
   XZ+3ZXX  0 0 0 1 0 0 0 0 0 0 0 1 0 0 0 0 1 0 0 0 1 0 0 0 1
XZ+3ZX-2X   0 0 0 1 0 0 0 0 0 0 0 1 0 0 0 0 1 0 1 0 0 0 1
```

FIGURE 11.17 Transforming the peptide XZZXX into the peptide $XZ^{+3}ZXX$ corresponds to inserting a block of three zeroes (shown in red) before the second occurrence of 1 in the peptide vector of XZZXX. Transforming the peptide $XZ^{+3}ZXX$ into the peptide $XZ^{+3}ZX^{-2}X$ corresponds to deleting a block of two zeroes (shown in green) before the fourth occurrence of 1 in the peptide vector of $XZ^{+3}ZXX$.

We will use the term **block indel** to refer to the addition or removal of a block of consecutive zeroes from a binary vector. Thus, applying k modifications to an amino acid string *Peptide* corresponds to applying k block indels to its peptide vector $\overrightarrow{Peptide}$. We define VARIANTS$_k$(*Peptide*) as the set of all modified variants of *Peptide* with up to k modifications.

Given a peptide *Peptide* and a spectral vector $\overrightarrow{Spectrum}$, our goal is to find a modified peptide from $\text{VARIANTS}_k(Peptide)$ with maximum score against $\overrightarrow{Spectrum}$.

Spectral Alignment Problem:
Given a peptide and a spectral vector, find a modified variant of this peptide that maximizes the peptide-spectrum score, among all variants of the peptide with up to k modifications.

> **Input**: An amino acid string *Peptide*, a spectral vector $\overrightarrow{Spectrum}$, and an integer k.
> **Output**: A peptide of maximum score against $\overrightarrow{Spectrum}$ among all peptides in $\text{VARIANTS}_k(Peptide)$.

A brute force approach to the Spectral Alignment Problem would score each peptide in $\text{VARIANTS}_k(Peptide)$ against the spectrum. We need to solve this problem more efficiently because our more ambitious goal is to solve the following problem, which will require multiple applications of an algorithm solving the Spectral Alignment Problem.

Modification Search Problem:
Given a spectrum and a proteome, find a peptide of maximum score against this spectrum among all modified peptides in the proteome with up to k modifications.

> **Input**: A spectral vector $\overrightarrow{Spectrum}$, an amino acid string *Proteome* and an integer k.
> **Output**: A peptide *Peptide* that maximizes $\text{SCORE}(\overrightarrow{Peptide}, \overrightarrow{Spectrum})$ among all modified variants of peptides from *Proteome* with up to k modifications.

Like modifications, mutations can also be viewed as block indels; for example, the mutation of V (integer mass 99) into R (integer mass 156) in peptide P9 from Figure 11.15 can be viewed as a block insertion with $\delta = 156 - 99 = 57$. We can therefore transform the Modification Search Problem into a "Mutation Search Problem" by simply substituting the word "modification" by "mutation" in the problem statement. In this new problem, a mutation of mass δ applied to an amino acid corresponds to the difference between the mass of this amino acid and another one. For example, mutations of valine (integer mass 99) correspond to modifications with the integer masses -42, -28, -12, -2, 2, 4, 14, 15, 16, 29, 30, 32, 38, 48, 57, 64, and 87.

Building a Manhattan grid for spectral alignment

The problem of finding a highest scoring peptide vector having up to k block indels recalls sequence alignment problems. This insight suggests that we should frame the Spectral Alignment Problem as an instance of the Longest Path in a DAG Problem.

STOP

STOP and Think: How would you build a DAG to solve this problem?

Consider an amino acid string $Peptide = a_1 \dots a_n$ of mass m and its modified variant $Peptide^{\text{mod}} = a'_1 \dots a'_n$ of mass $m + \Delta$. Construct the $(m+1) \times (m + \Delta + 1)$ Manhattan grid in which every node (i, j) is connected to every node (i', j') for $0 \leq i < i' \leq m$ and $0 \leq j < j' \leq m + \Delta$ (Figure 11.18). We call this graph SOUTHEAST$(m, m + \Delta)$ and call the nodes $(0, 0)$ and $(m, m + \Delta)$ *source* and *sink*, respectively.

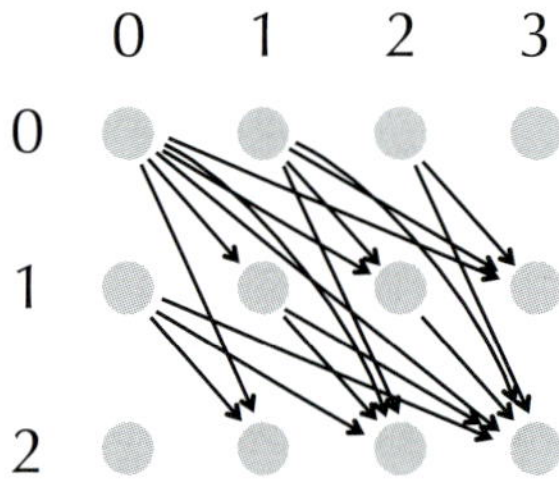

FIGURE 11.18 The graph SOUTHEAST$(2, 3)$. Each node in the graph is connected to every node lying to its south and east, with the exception of nodes in the same row and column.

Define the path PATH$(Peptide, Peptide^{\text{mod}})$ in SOUTHEAST$(m, m + \Delta)$ consisting of n edges:

$$\begin{aligned}(0,0) &\rightarrow (\text{MASS}(a_1), \text{MASS}(a'_1)) \\ &\rightarrow (\text{MASS}(a_1 a_2), \text{MASS}(a'_1 a'_2)) \\ &\rightarrow \dots \\ &\rightarrow (\text{MASS}(a_1 \dots a_n), \text{MASS}(a'_1 \dots a'_n)) \\ &= (m, m + \Delta).\end{aligned}$$

For example, consider the toy amino acid alphabet {X, Y, Z} with respective masses 2, 3, and 4. The blue path in Figure 11.19 (top) indicates PATH(XYYZX, **XY+2YZ-3X**):

$$(0,0) \rightarrow (2,2) \rightarrow (5,7) \rightarrow (8,10) \rightarrow (12,11) \rightarrow (14,13).$$

Except for the initial node (0, 0), every node (i, j) on this path indicates that the i-th element of $\overrightarrow{Peptide}$ = (0, 1, 0, 0, 1, 0, 0, 1, 0, 0, 0, 1, 0, 1) and the j-th element of $\overrightarrow{Peptide^{\text{mod}}}$ = (0, 1, 0, 0, 0, 0, 1, 0, 0, 1, 1, 0, 1) are both 1.

An edge connecting (i, j) to (i', j') in SOUTHEAST$(m, m+\Delta)$ is called **diagonal** if $i' - i = j' - j$ and **non-diagonal** otherwise. Note that if an amino acid of $Peptide^{\text{mod}}$ is unmodified, then the edge corresponding to this amino acid in PATH$(Peptide, Peptide^{\text{mod}})$ is diagonal. An amino acid a with modification δ in $Peptide^{\text{mod}}$ corresponds to a non-diagonal edge in this path connecting some node (i, j) with the node $(i + |a|, j + |a| + \delta)$.

We are now ready to solve the Spectral Alignment Problem for an amino acid string $Peptide = a_1 \ldots a_n$ of mass m and a spectral vector $\overrightarrow{Spectrum} = s_1 \ldots s_{m+\Delta}$. We know that the modified variant of *Peptide* solving this problem must have mass $m + \Delta$. We can thus represent all modified peptides of mass $m + \Delta$ in VARIANTS$_k(Peptide)$ as paths in SOUTHEAST$(m, m + \Delta)$ from *source* to *sink* with at most k non-diagonal edges (Figure 11.19 (top)). We refer to these modified peptides as VARIANTS$_k(Peptide, \overrightarrow{Spectrum})$.

STOP and Think: Does every path from *source* to *sink* in SOUTHEAST$(m, m + \Delta)$ correspond to a candidate modified peptide of mass $m + \Delta$?

STOP

Although every peptide in VARIANTS$_k(Peptide, \overrightarrow{Spectrum})$ corresponds to a path from *source* to *sink* in SOUTHEAST$(m, m + \Delta)$, many paths in this graph do not correspond to such modified peptides. Indeed, since *Peptide* is fixed, any path corresponding to a modified variant of *Peptide* will only pass through rows with indices

$$0, \text{MASS}(a_1), \text{MASS}(a_1 a_2), \ldots, \text{MASS}(a_1 \ldots a_n) = m\,,$$

shown as rows with darker nodes in Figure 11.19 (top).

Thus, nodes in other rows can be safely removed from SOUTHEAST$(m, m + \Delta)$, which results in the **PSM graph**, denoted PSMGRAPH$(Peptide, \overrightarrow{Spectrum})$, and shown in Figure 11.19 (bottom). Note that the $n + 1$ rows of nodes in the PSM graph have indices i equal to

$$0, \text{MASS}(a_1), \text{MASS}(a_1 a_2), \ldots, \text{MASS}(a_1 \ldots a_n)$$

rather than the indices $0, 1, \ldots, n$.

In the PSM graph, all edges entering into the row with index $i = \text{MASS}(a_1 \ldots a_t)$ originate at the row with index $\text{MASS}(a_1 \ldots a_{t-1})$. We therefore define DIFF(i) as the

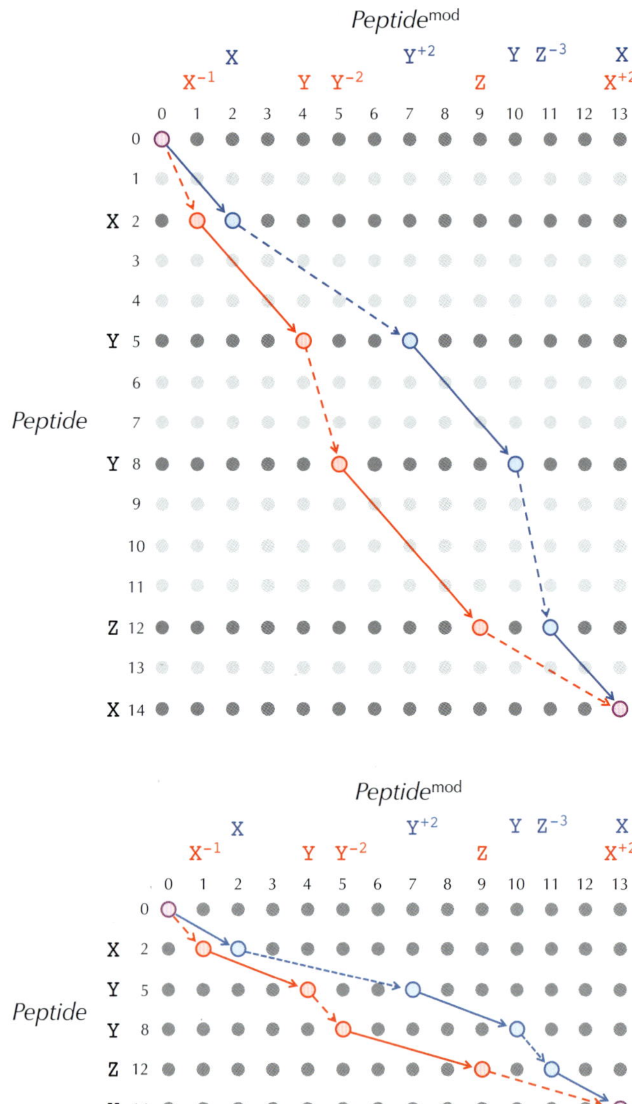

FIGURE 11.19 (Top) Two paths in SOUTHEAST(14, 13) formed by the modifications $\mathbf{XY^{+2}YZ^{-3}X}$ and $\mathbf{X^{-1}YY^{-2}ZX^{+2}}$ of XYYZX, where X, Y, and Z have respective masses 2, 3, and 4. Diagonal edges are solid, whereas non-diagonal edges are dashed. The indices of darker rows in the graph correspond to occurrences of 1 in the peptide vector of XYYZX. The darker nodes in this graph form the PSM graph. (Bottom) The PSM graph, obtained by removing the light nodes from the graph above.

mass of the amino acid a_t. For the peptide XYYZX, DIFF(2) = MASS(X) = 2, DIFF(5) = MASS(Y) = 3, DIFF(8) = MASS(Y) = 3, DIFF(12) = MASS(Z) = 4, and DIFF(14) = MASS(X) = 2.

STOP and Think: Can you assign weights to the nodes of PSMGRAPH($Peptide, \overrightarrow{Spectrum}$) so that the total weight of a path from *source* to *sink* corresponding to a modified peptide $Peptide^{\text{mod}}$ is equal to SCORE($Peptide^{\text{mod}}, \overrightarrow{Spectrum}$)?

Given a spectral vector $\overrightarrow{Spectrum} = (s_1, \ldots, s_{m+\Delta})$, we assign a weight of s_j to every node (i, j) in column j of the PSM graph. With this assignment of weights, the total weight of nodes in the path corresponding to $Peptide^{\text{mod}}$ is equal to SCORE($Peptide^{\text{mod}}, \overrightarrow{Spectrum}$). Solving the Spectral Alignment Problem is therefore equivalent to finding a path in the PSM graph with maximum total node weight among all paths with at most k non-diagonal edges.

Spectral alignment algorithm

We already know how to find a longest path in a node-weighted DAG. However, it is not clear how to find a longest path in a DAG under the additional constraint that this path has at most k non-diagonal edges.

As a workaround, we will convert the two-dimensional PSM graph from Figure 11.19 (bottom) into a three-dimensional **spectral alignment graph** consisting of $k + 1$ layers (Figure 11.20), where the node set in each layer coincides with the node set of the PSM graph. This graph will have nodes (i, j, t), where $0 \le i \le m$, $0 \le j \le m + \Delta$, and $0 \le t \le k$. Each such node inherits the weight of the node (i, j) in the PSM graph (i.e., the amplitude s_j from $\overrightarrow{Spectrum} = (s_1, \ldots, s_{m+\Delta})$).

As for edges, each of the $k + 1$ layers of the spectral alignment graph will inherit all diagonal edges from the PSM graph, i.e., each diagonal edge from (i, j) to $(i + x, j + x)$ in the PSM graph will correspond to the $k + 1$ edges (i, j, t) to $(i + x, j + x, t)$ for $0 \le t \le k$.

STOP and Think: The layers in the constructed graph are now disconnected. How should we connect them?

For each non-diagonal edge connecting (i, j) to (i', j') in the PSM graph, we will generate k edges connecting consecutive layers in the spectral alignment graph by connecting (i, j, t) to $(i', j', t + 1)$ for all t between 0 and $k - 1$. Every path in the spectral alignment graph from $(0, 0, 0)$ to $(m, m + \Delta, t)$ corresponds to a modified version of peptide with t

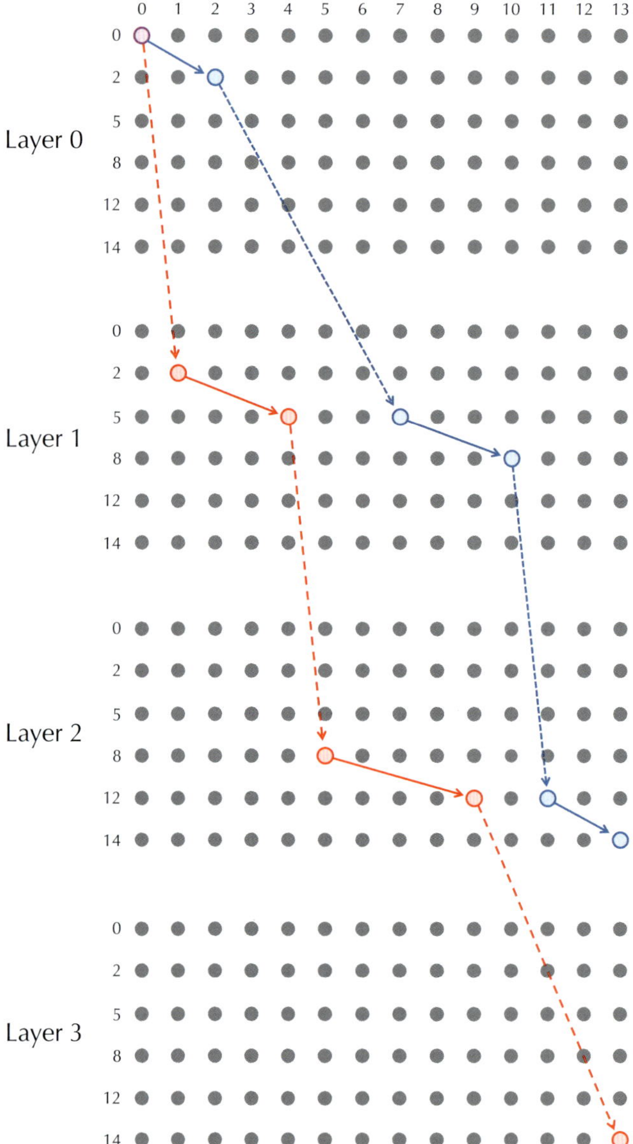

FIGURE 11.20 The PSM graph from Figure 11.19 (bottom) transformed into a spectral alignment graph with four layers (k = 3). The blue and red paths from the PSM graph in Figure 11.19 (bottom) correspond to the blue and red paths shown, which represent the respective modified variants $\mathtt{XY}^{+2}\mathtt{YZ}^{-3}\mathtt{X}$ and $\mathtt{X}^{-1}\mathtt{YY}^{-2}\mathtt{ZX}^{+2}$ of XYYZX. The blue path terminates in Layer 2 because it corresponds to a peptide with two modifications, whereas the red path terminates in Layer 3 because it corresponds to a peptide with three modifications.

modifications (Figure 11.20). The zero-th layer of this graph will store scores of peptides with no modifications, the first layer will store scores of peptides with one modification, and so on.

To solve the Spectral Alignment Problem, we will define $\text{SCORE}(i, j, t)$ as the maximum score of all paths connecting node $(0, 0, 0)$ to node (i, j, t) in the spectral alignment graph. Note that this score is equal to the weight s_j assigned to node (i, j, t) plus the maximum of the scores of all predecessors of node (i, j, t). One of these predecessors, $(i - \text{DIFF}(i), j - \text{DIFF}(i), t)$, is located in the same layer (t), and j other predecessors ($(i - \text{DIFF}(i), j', t-1)$ for $j' < j$) are located in the previous layer ($t - 1$). This reasoning results in the following recurrence for computing $\text{SCORE}(i, j, t)$,

$$\text{SCORE}(i, j, t) = s_j + \max_{j' < j} \begin{cases} \text{SCORE}(i - \text{DIFF}(i), j - \text{DIFF}(i), t) \\ \text{SCORE}(i - \text{DIFF}(i), j', t - 1) \end{cases} .$$

The maximum score of all peptides with at most k modifications is therefore the maximum value of $\text{SCORE}(m, m + \Delta, t)$ as t ranges from 0 to k. To initialize the recurrence, we assume that $\text{SCORE}(0, 0, 0) = 0$ and that $\text{SCORE}(0, 0, t) = -\infty$ for all $1 \leq t \leq k$.

Although the above recurrence computes the score of a modified peptide solving the Spectral Alignment Problem, we also need to reconstruct this peptide. In order to achieve this goal, we will need to implement a backtracking approach similar to the backtracking approach described in Chapter 5, which we leave to you as an exercise.

Exercise Break: What is the running time of the spectral alignment algorithm?

Challenge Problem: In addition to *T. rex*, Asara also analyzed a 200,000 year-old mastodon fossil. The extinction of the elephant-like mastodons 10,000 years ago was caused by a combination of climate change and hunting by humans armed with stone weapons. In contrast to dinosaurs, it is not surprising that scientists routinely identify peptides from recently extinct species like mastodons or cave bears.

Analyze the mastodon collagen peptides reported in Asara's 2007 paper and decide which of them form statistically significant PSMs. Can you identify other statistically significant mastodon peptides that this paper missed? Can you find non-collagen peptides (especially hemoglobin peptides) matching spectra from mastodons? Can you determine the number of different types of post-translational modifications of mastodon peptides by solving the Modification Search Problem?

Detours

Gene prediction

To predict split genes, researchers often attempt to recognize the locations of splicing signals at exon-intron junctions. For a simple example, the dinucleotides AG and GT on either side of an exon are highly conserved (Figure 11.21). To improve the accuracy of this approach (known as **statistical gene prediction**), researchers look for genomic features appearing frequently in exons and infrequently in introns.

Attempts to improve the accuracy of statistical gene prediction methods have led to **similarity-based gene prediction** approaches, which are based on the observation that a newly sequenced gene is often similar to a known gene in another species. For example, 99% of mouse genes have human analogs.

However, we cannot simply look for a similar sequence in the mouse genome based on known human genes, since the exon sequence and partition of a gene into exons in different species may be different. To address this complication, similarity-based approaches sometimes look for a set of putative exons in the mouse genome whose concatenation fits a known human protein.

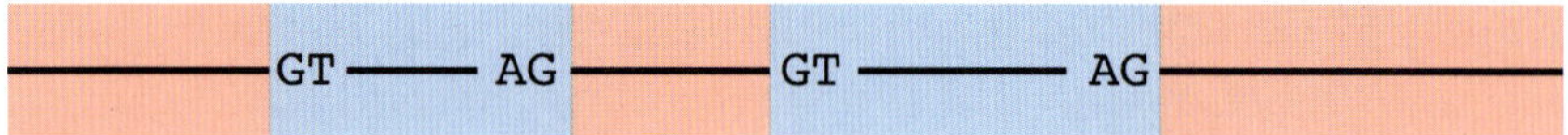

FIGURE 11.21 A split gene with exons (red) separated by introns (blue). Introns typically start with GT and end with AG.

Yet not all genes are split genes. In fact, bacteria do not have split genes at all, which simplifies bacterial gene prediction. Such genes begin with a start codon that codes for methionine (typically ATG but also sometimes GTG or TTG) and end with a stop codon (TAA, TAG, or TGA).

We can represent a genome of length n as a sequence of $n/3$ codons. The stop codons break this sequence into segments between every pair of consecutive stop codons. The suffixes of these segments that begin at the first start codon within a segment are called **open reading frames**, or **ORFs**. ORFs within a single genome may overlap because there are six possible reading frames.

Since there are three stop codons, each triplet of nucleotides in a randomly generated DNA string has probability $3/64$ of being a stop codon. The expected number of codons between two consecutive stop codons in a randomly generated nucleotide string (within a given reading frame) is therefore 64/3. This implies that we would expect to find a stop codon approximately every 64 nucleotides (within a given reading frame) in a randomly generated nucleotide string.

Yet the typical length of a bacterial gene is on the order of 1,000 nucleotides. A gene prediction algorithm might therefore select ORFs longer than some threshold length as candidate genes. Unfortunately, such an algorithm would also fail to detect short genes.

Many gene prediction algorithms also rely on subtle statistical differences between coding and non-coding regions, such as biases in **codon usage**, or the frequency of each codon. For example, there are six codons encoding leucine, but whereas `CUG` encodes 47% of all occurrences of leucine in *E. coli*, `CUA` encodes only 4%. Therefore, an ORF with many more occurrences of `CUG` than `CUA` is a candidate gene.

Bacterial gene prediction also takes advantage of several conserved motifs often found in genomic regions near start of RNA transcription. For example, the **Pribnow box** is a six-nucleotide sequence with consensus `TATAAT` that is an essential component for initiating transcription in bacteria.

Finding all paths in a graph

In the 19th Century, Charles Pierre Trémaux developed an algorithm for navigating mazes, described as follows. As you walk through a maze, drag a piece of chalk along the ground behind you. When you reach a junction, unmarked paths correspond to yet unexplored paths. So take an unexplored path for as long as you can, until you encounter a dead-end or a junction at which all the outgoing paths are marked. In this case, backtrack your steps until you encounter the exit or a junction with an unmarked path, or until you arrive at your starting point, in which case the maze has no solution.

Trémaux's maze algorithm is an example of **depth-first search (DFS)**, a technique for traversing the nodes of a graph. DFS starts at a given node and explores the graph as far as possible until it reaches a node that has no outgoing edges or from which we have already explored all edges. We then backtrack until we reach a node with unexplored edges. The preorder traversal that we encountered in **DETOUR: From Suffix Trees to Suffix Arrays** offers one example of a DFS applied to rooted trees. PAGE 522

The following recursive algorithm offers a DFS-inspired approach to finding all paths between a node v and a node *sink* in a DAG *Graph*.

```
ALLPATHS(Graph, v, sink)
    if v = sink
        Paths ← the set of paths consisting of the single-node path v
    else
        Paths ← an empty set of paths
        for all outgoing edges (v, w) from v
            PathsFromDescendant ← ALLPATHS(Graph, v, sink)
            add (v, w) as the first edge to each path in PathsFromDescendant
            add PathsFromDescendant to Paths
    return Paths
```

The Anti-Symmetric Path Problem

In the main text, we saw that not every path from *source* to *sink* in GRAPH(*Spectrum*) represents a solution of the Decoding an Ideal Spectrum Problem. This issue is caused by the fact that every mass in the spectrum may be interpreted as either a mass of a prefix or a mass of a suffix. Therefore, every node corresponding to mass s has a "twin" node (corresponding to MASS(*Peptide*) − s). Given an arbitrary node and its twin in GRAPH(*Spectrum*), a correct path from *source* to *sink* must pass through exactly one of these nodes in order to spell out a solution.

The Decoding an Ideal Spectrum Problem is a particular case of the following more general problem. Given a collection of **forbidden pairs** of nodes in a graph (when reconstructing peptides, forbidden pairs correspond to twins), a path in a graph is called **anti-symmetric** if it contains exactly one node from every forbidden pair.

Anti-Symmetric Path Problem:
Find an anti-symmetric path in a DAG.

> **Input**: A DAG with nodes *source* and *sink* and a set of forbidden pairs of nodes in this DAG.
> **Output**: An anti-symmetric path in this DAG from *source* to *sink*.

The Anti-Symmetric Path Problem is *NP*-Hard, but we should not give up hope of finding an efficient algorithm for the Decoding an Ideal Spectrum Problem because the latter is a specific instance of the former. In particular, forbidden pairs in peptide sequencing have the additional property that the sum of masses from each forbidden pair is equal to the mass of the entire peptide. In fact, there is a polynomial algorithm solving the Anti-Symmetric Path Problem for a DAG satisfying this additional property, but this algorithm is outside the scope of this book.

Transforming spectra into spectral vectors

Our goal is to develop a probabilistic model describing how a peptide vector generates an *integer-valued* spectrum and to use this model to transform a spectrum into a spectral vector. To address this problem, we will first introduce an abstract model that seemingly has nothing to do with peptide sequencing but rather describes a probabilistic process that transforms a peptide vector $P = (p_1, \ldots, p_m)$ into a *binary* vector $X = (x_1, \ldots, x_m)$ of the same length. We will see later how ideas developed for this model help us analyze real spectra.

We define the probability that P generates X as $\Pr(X|P) = \Pi_{i=1}^{m} \Pr(x_i|p_i)$, where $\Pr(x_i|p_i)$ is the probability that p_i in P generates x_i in X (Figure 11.22). For example, the probability that a 1 in P generates a 1 in X is written $\Pr(1|1)$ and is equal to some parameter ρ. The probability that a 0 in P generates a 1 in X is written $\Pr(1|0)$ and is equal to some parameter θ. The probability of a 1 in P generating a 0 in X is $\Pr(0|1) = 1 - \rho$, and the probability of a 0 in P generating a 0 in X is $\Pr(0|0) = 1 - \theta$.

		symbol in *P*	
		0	1
symbol in *X*	0	$1 - \vartheta$	$1 - \rho$
	1	ϑ	ρ

FIGURE 11.22 A matrix describing a probabilistic process that transforms a peptide vector *P* into a binary vector *X*.

For the toy amino acid alphabet containing just two amino acids with masses 2 and 3, Figure 11.23 illustrates the peptide vector $P = (0, 1, 0, 1, 0, 0, 1)$ generating the binary vector $X = (0, 0, 0, 1, 1, 0, 1)$ with probability

$$\Pr(X|P) = (1 - \theta) \cdot (1 - \rho) \cdot (1 - \theta) \cdot \rho \cdot \theta \cdot (1 - \theta) \cdot \rho.$$

Exercise Break: Which binary vector is more likely to be generated by the peptide vector P in Figure 11.23, $(0, 0, 0, 1, 1, 0, 1)$ or $(1, 0, 0, 0, 1, 0, 1)$?

We are interested in the following problem.

Most Likely Peptide Vector Problem:
Find a most likely peptide vector for a given binary vector.

> **Input**: A binary vector X and parameters ρ and θ such that $0 \leq \rho, \theta \leq 1$.
> **Output**: A peptide vector P that maximizes $\Pr(X \mid P)$ as defined by probabilities ρ and θ among all possible peptide vectors.

The solution of this problem is given in **DETOUR: Solving the Most Likely Peptide Vector Problem**.

peptide vector *P*	0	1	0	1	0	0	1
binary vector *X*	0	0	0	1	1	0	1
Pr(*X* | *P*)	$\Pr(0\|0)\ \cdot$	$\Pr(0\|1)\ \cdot$	$\Pr(0\|0)\ \cdot$	$\Pr(1\|1)\ \cdot$	$\Pr(1\|0)\ \cdot$	$\Pr(0\|0)\ \cdot$	$\Pr(1\|1)$
	$(1-\vartheta)\ \cdot$	$(1-\rho)\ \cdot$	$(1-\vartheta)\ \cdot$	$\rho\ \cdot$	$\vartheta\ \cdot$	$(1-\vartheta)\ \cdot$	ρ
Pr($X \| \overrightarrow{\mathbf{0}}$)	$\Pr(0\|0)\ \cdot$	$\Pr(0\|0)\ \cdot$	$\Pr(0\|0)\ \cdot$	$\Pr(1\|0)\ \cdot$	$\Pr(1\|0)\ \cdot$	$\Pr(0\|0)\ \cdot$	$\Pr(1\|0)$
	$(1-\vartheta)\ \cdot$	$(1-\vartheta)\ \cdot$	$(1-\vartheta)\ \cdot$	$\vartheta\ \cdot$	$\vartheta\ \cdot$	$(1-\vartheta)\ \cdot$	ϑ
LIKELIHOOD(*X* | *P*)	$1\ \cdot$	$\frac{\Pr(0\|1)}{\Pr(0\|0)}\ \cdot$	$1\ \cdot$	$\frac{\Pr(1\|1)}{\Pr(1\|0)}\ \cdot$	$1\ \cdot$	$1\ \cdot$	$\frac{\Pr(1\|1)}{\Pr(1\|0)}$
	$1\ \cdot$	$\frac{1-\rho}{1-\vartheta}\ \cdot$	$1\ \cdot$	$\frac{\rho}{\vartheta}\ \cdot$	$1\ \cdot$	$1\ \cdot$	$\frac{\rho}{\vartheta}$
$\log_2$(LIKELIHOOD(*X* | *P*))	0	$+\log_2 \frac{1-\rho}{1-\vartheta}\ +$	0	$+\log_2 \frac{\rho}{\vartheta}\ +$	0	$+\ 0$	$+\log_2 \frac{\rho}{\vartheta}$

FIGURE 11.23 A peptide vector $P = (0, 1, 0, 1, 0, 0, 1)$ generates a binary vector $X = (0, 0, 0, 1, 1, 0, 1)$ with probability $\Pr(X|P)$.

The infinite monkey theorem

In Jonathan Swift's *Gulliver's Travels*, a professor of the Grand Academy of Legado asks his students to generate random strings of letters by turning cranks on a machine. According to the professor, the Academy will eventually crank out brilliant works on all subjects.

Inspired by Swift's satirical treatment of certain academics, the **infinite monkey theorem** states that an immortal monkey typing an infinite sequence of symbols on a typewriter will one day reproduce *Hamlet*. In more technical terms, this theorem states that an infinite random string contains an arbitrary given text as a substring **almost surely**, or with probability equal to 1.

STOP and Think: In 2003, researchers placed a typewriter in a monkey enclosure, and found that the monkeys typed the letter "S" over and over. It is possible that an infinite random string generated by a monkey would contain only the letter "S". How, then, must this string contain *Hamlet* almost surely?

The probabilistic space of peptides in a spectral dictionary

In the main text, we defined the probability of *Peptide* as $1/20^{|Peptide|}$, and we defined the probability of a collection of peptides *Dictionary* as

$$\Pr(Dictionary) = \sum_{\text{each peptide } Peptide \text{ in } Dictionary} \frac{1}{20^{|Peptide|}} \cdot$$

But why have we used the probabilistic notation? After all, consider the following exercise, which indicates that it is possible for $\Pr(Dictionary)$ to be larger than 1.

Exercise Break: If *Dictionary* is the set of all peptides of length at most 10, what is $\Pr(Dictionary)$?

Yet recall that a peptide can match a spectrum if and only if its mass is equal to the mass of the spectrum. Thus, no peptide in a *spectral* dictionary can contain another peptide in the dictionary as a substring, i.e., a spectral dictionary forms a **substring-free set**.

Exercise Break: Prove that if *Dictionary* is a substring-free set, then $\Pr(Dictionary) \leq 1$.

However, you may still be wondering which event corresponds to $\Pr(Dictionary)$. More precisely, what is the "probabilistic space" of underlying outcomes from which *Dictionary* is formed? The probabilistic space that we propose contains all decoy proteomes of length n, where n is the length of the longest peptide in a spectral dictionary *Dictionary*. In this space, we will assume that each decoy proteome has the same probability. Thus, our probabilistic space consists of 20^n elements, each with probability $1/20^n$. Note the change: instead of considering the probabilistic space of all *peptides* in *Dictionary*, we have switched to considering all *decoy proteomes*.

Each string *Peptide* in *Dictionary* appears in exactly $20^{n-|Peptide|}$ decoy proteomes as their first peptide. The combined probabilities of all these decoy proteomes sum to

$$20^{n-|Peptide|} \cdot \frac{1}{20^n} = \frac{1}{20^{|Peptide|}},$$

which is $\Pr(Peptide)$. Since spectral dictionaries are substring-free, each decoy proteome has at most one peptide from the spectral dictionary starting at its first position. Thus, $\Pr(Dictionary)$ is simply the combined probability of all decoy proteomes that begin with one of the peptides in *Dictionary*.

Are terrestrial dinosaurs really the ancestors of birds?

Aside from the mysterious presence of hemoglobin peptides in Asara's *T. rex* spectra, scientists have recently expressed doubts about the hypothesis that birds evolved from terrestrial dinosaurs like *T. rex* and that flight was achieved from the biophysically improbable **ground-up model**. This hypothesis assumes that, in order to evolve into birds, dinosaurs first must have reduced their size while simultaneously developing feathers (arguably the most complex evolutionary invention for flight).

Most early dinosaur studies favored evidence for a small arboreal animal as the more logical interpretation for the bird ancestor. This conjecture assumes that before evolving a system for sustained flight, early birds used gravity-assisted aerodynamics such as parachuting and gliding (the latter is used by modern flying squirrels). Thus, the tiny *Scansoriopteryx* (Figure 11.24), whose fossils contain impressions of feathers and whose foot adaptations indicate an arboreal lifestyle, competes with *T. rex* for the honor of being the ancestor of birds.

FIGURE 11.24 An artistic recreation of *Scansoriopteryx*..

If you are interested in learning more about the evolutionary controversy surrounding the origin of birds, we suggest two papers, one on either side of the debate:

- "Jurassic archosaur is a non–dinosaurian bird" by Stephen Czerkas and Alan Feduccia.
- "Three crocodilian genomes reveal ancestral patterns of evolution among archosaurs" by Richard Green et al.

Solving the Most Likely Peptide Vector Problem

Define LIKELIHOOD$(X|P)$ as $\Pr(X|P)/\Pr(X|\vec{0})$, where $\vec{0}$ is an **all-zeroes vector** consisting of only zeroes. Figure 11.23 illustrates that a peptide vector $P = (0,1,0,1,0,0,1)$ generates a binary vector $X = (0,0,0,1,1,0,1)$ with

$$\begin{aligned}\text{LIKELIHOOD}(X|P) &= \frac{\Pr(0|1)}{\Pr(0|0)} \cdot \frac{\Pr(1|1)}{\Pr(1|0)} \cdot \frac{\Pr(1|1)}{\Pr(1|0)} \\ &= \frac{1-\rho}{1-\theta} \cdot \frac{\rho}{\theta} \cdot \frac{\rho}{\theta}.\end{aligned}$$

To avoid dealing with extremely small values that result from many multiplications in LIKELIHOOD$(X \mid P)$, we will instead use the **log-likelihood** $\log_2(\text{LIKELIHOOD}(X|P))$. Finding a peptide vector maximizing the log-likelihood is equivalent to finding a most probable peptide vector. Figure 11.23 illustrates that the peptide vector $(0,1,0,1,0,0,1)$ generates a binary vector $(0,0,0,1,1,0,1)$ with log-likelihood equal to

$$\log_2 \frac{1-\rho}{1-\theta} + \log_2 \frac{\rho}{\theta} + \log_2 \frac{\rho}{\theta}.$$

We will now transform a binary vector $X = (x_1, \ldots, x_m)$ into a spectral vector $S = (s_1, \ldots, s_m)$ by changing each occurrence of 0 into the amplitude $\log_2 [(1-\rho)/(1-\theta)]$ and each occurrence of 1 into the amplitude $\log_2(\rho/\theta)$. For example, the binary vector $(0,0,0,1,1,0,1)$ will be transformed into the spectral vector

$$\left(\log_2 \frac{1-\rho}{1-\theta}, \log_2 \frac{1-\rho}{1-\theta}, \log_2 \frac{1-\rho}{1-\theta}, \log_2 \frac{\rho}{\theta}, \log_2 \frac{\rho}{\theta}, \log_2 \frac{1-\rho}{1-\theta}, \log_2 \frac{\rho}{\theta}\right).$$

Note that $\log_2(\text{LIKELIHOOD}(X|P))$ is simply the dot product of the peptide vector $P = (p_1, \ldots, p_m)$ and the spectral vector $S = (s_1, \ldots, s_m)$,

$$P \cdot S = p_1 \cdot s_1 + \cdots + p_m \cdot s_m.$$

We denote $P \cdot S$ as SCORE(P, S) (Figure 11.25) and define the score between a peptide vector and a spectral vector of different length as $-\infty$. We have therefore transformed the Most Likely Peptide Vector Problem into the Peptide Sequencing Problem.

peptide vector *P*	0	1	0	1	0	0	1
binary vector *X*	0	0	0	1	1	0	1
spectral vector *S*	$\log_2 \frac{1-\rho}{1-\vartheta}$	$\log_2 \frac{1-\rho}{1-\vartheta}$	$\log_2 \frac{1-\rho}{1-\vartheta}$	$\log_2 \frac{\rho}{\vartheta}$	$\log_2 \frac{\rho}{\vartheta}$	$\log_2 \frac{1-\rho}{1-\vartheta}$	$\log_2 \frac{\rho}{\vartheta}$
SCORE(*P*, *S*)		$\log_2 \frac{1-\rho}{1-\vartheta}$	+	$\log_2 \frac{\rho}{\vartheta}$	+		$\log_2 \frac{\rho}{\vartheta}$

FIGURE 11.25 Scoring a peptide vector P against a spectral vector S as the dot product SCORE(P, S).

Our conversion of a binary vector X into a spectral vector S was based on a simple probabilistic model (describing how a peptide vector generates a binary vector) with just two parameters, ρ and θ. To see how spectra are transformed into spectral vectors in practice, see **DETOUR: Selecting Parameters for Transforming Spectra into Spectral Vectors**.

PAGE 634

Selecting parameters for transforming spectra into spectral vectors

If mass spectrometers generated binary spectra, then we could start by forming a large **training sample** of annotated spectra for which the peptides that generated these spectra are known. We could then estimate ρ (as the frequency of ones in binary spectra being generated by ones in peptide vectors) and θ (as the frequency of ones in binary spectra being generated by zeroes in peptide vectors) across all annotated spectra in the training sample. But since real mass spectrometers generate *integer-valued* rather than *binary* spectra, deriving parameters becomes more complex.

STOP

STOP and Think: Can you devise a probabilistic model that would convert the peptide vector (0, 1, 0, 1, 0, 0, 1) into the integer-valued vector (3, 4, 2, 6, 9, 4, 7)?

However, a similar probabilistic model will work if we define the probability of converting zeroes and ones in the peptide vector into various intensities in the real spectra (rather than into zeroes and ones as before). In fact, the conversion of real spectra into spectral vectors is based on a similar log-likelihood model that uses dozens of probabilistic parameters. Algorithms for converting real spectra into spectral vectors attempt to optimize these parameters so that amplitudes at prefix coordinates are maximized and amplitudes at noise coordinates are minimized.

To derive these parameters, we again need to build a large training sample of annotated spectra. We can consider all peaks with a certain intensity level in all spectra and compute which fraction of them are annotated by prefix or suffix peptides. For example, only 19% and 45% of the ten highest intensity peaks in Collision-Induced Dissociation spectra (which are similar to ones generated in Asara's laboratory) are explained by prefix and suffix peptides, respectively. The remaining high-intensity peaks are treated as noise. Given a spectrum generated by an unknown peptide vector $P = (p_1, \ldots, p_m)$, its spectral vector $S = (s_1, \ldots, s_m)$ is derived using these frequencies so that s_i is the log likelihood ratio $\log_2(\text{Pr}_1/\text{Pr}_0)$, where Pr_1 is an estimate of the probability that $p_i = 1$, and Pr_0 is an estimate of the probability that $p_i = 0$. A complete discussion of the details of the algorithm for generating spectral vectors is beyond the scope of this detour.

Figure 11.26 shows the set of prefix masses for *DinosaurPeptide* along with amplitudes of the spectral vector corresponding to these masses. Note from Figure 11.8 (bottom) that most amplitudes of the spectral vector are negative. The blue elements in Figure 11.26 correspond to positions that significantly exceed the average amplitude value.

	1	2	3	4	5	6	7	8	9	10	11	12	13	14
amino acid	G	L	V	G	A	P	G	L	R	G	L	P	G	K
mass	57	113	99	57	71	97	57	113	156	57	113	97	57	128
prefix mass	57	170	269	326	397	494	551	664	820	877	990	1087	1144	1272
amplitude	-8	+1	-4	-6	-6	+3	+1	-4	-8	+18	+11	-10	-7	0

FIGURE 11.26 Masses of amino acids in GLVGAPGLRGLPGK (second line), prefix masses for this peptide (third line), and the corresponding elements of the spectral vector for *DinosaurSpectrum*, shown in Figure 11.5 (top) (fourth line). Blue elements correspond to amplitudes that are significantly higher than the average, which is negative. Note that the second and third tallest peaks in the spectrum (labeled b_{10} and b_{11}) correspond to maximum amplitudes +18 and +11 in the spectral vector of *DinosaurSpectrum*. Also note that because there is no peak b_{12} (or y_2) in Figure 11.5 (bottom), $s_{1087} = -10$ is very small.

Bibliography Notes

T. rex peptides were reported by Asara et al., 2007 and faced criticism in Pevzner, Kim, and Ng, 2008 and Buckley et al., 2008. The "dinosaur DNA" paper by Woodward, Weyand, and Bunnell, 1994 was refuted by Hedges and Schweitzer, 1995. Czerkas and Feduccia, 2014 recently argued that birds did not evolve from dinosaurs, while Green et al., 2014 recently argued otherwise.

Chen et al., 2001 solved the Anti-Symmetric Path Problem in the case of graphs arising from mass spectra. Searching a protein database for the purpose of peptide identification in mass spectrometry was pioneered by Eng, McCormack, and Yates, 1994. The spectral alignment algorithm was introduced by Pevzner, Dančík, and Tang, 2000. The concept of spectral dictionary and the algorithm for evaluating statistical significance of PSMs were introduced by Kim, Gupta, and Pevzner, 2008 and Kim et al., 2009.

Appendix: Introduction to Pseudocode

What is Pseudocode?

An **algorithm** is a sequence of instructions to solve a well-formulated computational problem specified in terms of its **input** and **output**. An algorithm uses the input to generate the output. For example, the algorithm **PATTERNCOUNT** uses strings *Text* and *Pattern* as input to generate the number COUNT(*Text*, *Pattern*) as its output.

In order to solve a computational problem, you need to carry out the instructions specified by the algorithm. For example, if we want you to count how many times *Pattern* appears in *Text*, we could tell you to do the following:

1. Start from the initial position of *Text* and check whether *Pattern* appears in *Text* starting at this position.
2. If yes, draw a dot on a piece of paper.
3. Move to the next position of *Text* and check whether *Pattern* appears in *Text* starting at this position.
4. If yes, draw another dot on the same piece of paper.
5. Continue until you reach the end of *Text*.
6. Count the number of dots on the paper.

Since humans are slow, make mistakes, and hate repetitive work, we invented computers, which are fast, love repetitive work, and never make mistakes. However, while you may easily understand the above instructions for counting the number of occurrences of *Pattern* in *Text*, no computer in the world can execute them. The only reason you can understand them is because you have been trained for many years to understand human language. For example, we didn't specify that you should start from a *blank* piece of paper without any dots, but you assumed it. We didn't explain what it means to "reach the end of *Text*"; at what position of *Text* should we stop?

Because computers do not understand human language, algorithms must be rephrased in a **programming language** (such as Python, Java, C++, Perl, Ruby, Go, or dozens of others) in order to give the computer specific instructions. However, we don't want to describe algorithms in a specific language because it may not be your favorite

Our focus is on algorithmic ideas rather than on implementation details, which is why we will meet you halfway between human languages and programming languages by using **pseudocode**. By emphasizing ideas rather than implementation details, pseudocode is able to describe algorithms without being too formal, ignoring many of the tedious details that are required in a specific programming language. At the same time, pseudocode is more precise and less ambiguous than the instructions we gave above for counting a pattern in a text.

For example, consider the following pseudocode for an algorithm called **DISTANCE**, whose input is four numbers (x_1, y_1, x_2, y_2) and whose output is one number d. Can you guess what it does?

```
DISTANCE(x1, y1, x2, y2)
    d ← (x2 − x1)^2 + (y2 − y1)^2
    d ← √d
    return d
```

The first line of pseudocode specifies the name of an algorithm (**DISTANCE**), followed by a list of **arguments** that it requires as input (x_1, y_1, x_2, y_2). Subsequent lines contain the statements that describe the algorithm's actions, and the operation **return** reports the result of the algorithm.

We can invoke an algorithm by passing it the appropriate values for its arguments. For example, **DISTANCE**$(1,3,7,5)$ would return the distance between points $(1,3)$ and $(7,5)$ in two-dimensional space, by first computing

$$d \leftarrow (7-1)^2 + (5-3)^2 = 40$$

and then computing

$$d \leftarrow \sqrt{40}\ .$$

The pseudocode for **DISTANCE** uses the concept of a **variable**, which contains some value and can be assigned a new value at different points throughout the course of an algorithm. To assign a new value to a variable, we use the notation

$$a \leftarrow b\,,$$

which sets the variable a equal to the value stored in variable b. For example, in the pseudocode above, when we compute **DISTANCE**$(1,3,7,5)$, d is first assigned the value $(7-1)^2 + (5-3)^2 = 40$ and then is assigned the value $\sqrt{40}$.

Furthermore, we can use any name we like for variable names. For example, the following pseudocode is equivalent to the previous pseudocode for **DISTANCE**.

```
DISTANCE(x, y, z, w)
    abracadabra ← (z − x)^2 + (w − y)^2
    abracadabra ← √abracadabra
    return abracadabra
```

Whereas computer scientists are accustomed to pseudocode, we fear that some biologists reading this book might decide that pseudocode is too cryptic and therefore useless. Although modern biologists deal with algorithms on a daily basis, the language they use to describe an algorithm may be closer to a series of steps described in plain English. Accordingly, some bioinformatics books are written without pseudocode. Unfortunately, this language is insufficient to describe the complex algorithmic ideas behind various bioinformatics tools that biologists use every day.

To be able to explain complex algorithmic ideas, we will need to delve deeper into the details of pseudocode. As a result, you will be able not only to understand the algorithms in this book, but use pseudocode to design your own algorithms!

Nuts and Bolts of Pseudocode

We will now discuss some of the details of pseudocode that we use throughout this book. We will often avoid tedious details by specifying parts of an algorithm in English, using operations that are not listed below, or by omitting noncritical details.

*The **if** condition*

The algorithm **MINIMUM2**(a, b) shown below has two numbers (a and b) as its input and a single number as its output. What do you think that it does?

```
MINIMUM2(a, b)
    if a > b
        return b
    else
        return a
```

This algorithm, which computes the minimum of two numbers, uses the following construction:

```
if statement X is true
    execute instructions Y
else
    execute instructions Z
```

If statement X is true, then the algorithm executes instructions Y; otherwise, it executes instructions Z. For example, **MINIMUM2**$(1, 9)$ returns 1 because the condition "$1 > 9$" is false.

The following pseudocode computes the minimum of three numbers.

```
MINIMUM3(a, b, c)
    if a > b
        if b > c
            return c
        else
            return b
    else
        if a < c
            return c
        else
            return b
```

We may also use **else if**, which allows us to consider more than two different possibilities in the same **if** statement. For example, we can compute the minimum of three numbers as follows.

```
MINIMUM3(a, b, c)
    if a > c and b > c
        return c
    else if a > b and c > b
        return b
    else
        return a
```

Both of these algorithms are correct, but below is a more compact version that uses the **MINIMUM2** function that we already wrote as a **subroutine**, or a function that is called within another function. Programmers break their programs into subroutines in order to keep functions short and to improve readability.

```
MINIMUM3(a, b, c)
    if a > b
        return MINIMUM2(b, c)
    else
        return MINIMUM2(a, c)
```

STOP and Think: Try to **MINIMUM3** using just a single line of pseudocode.

Exercise Break: Write pseudocode for an algorithm **MINIMUM4**(a, b, c, d) that computes the minimum of four numbers.

Sometimes, we may omit the "**else**" statement (e.g., if we only need to test a condition once).

*The **for** loop*

Consider the following problem.

Summing Integers Problem:
Compute the sum of the first n positive integers.

> **Input**: A positive integer n.
> **Output**: The sum of the first n positive integers.

If n were a fixed number, then we could solve this problem using our existing framework. For example, the following program **SUM5** returns the sum of the first five integers (i.e., $1+2+3+4+5=15$).

```
SUM5()
    sum ← 0
    i ← 1
    sum ← sum + i
    i ← i + 1
    sum ← sum + i
    i ← i + 1
    sum ← sum + i
    i ← i + 1
    sum ← sum + i
    i ← i + 1
    sum ← sum + i
    return sum
```

We could then write **SUM6**, **SUM7**, and so on. However, we cannot endorse this programming style! After all, to solve the Summing Integers Problem for an *arbitrary* integer n, we will need an algorithm that takes n as its input. This is achieved by the following algorithm, which we call **GAUSS**.

```
GAUSS(n)
    sum ← 0
    for i ← 1 to n
        sum ← sum + i
    return sum
```

Gauss employs a **for** loop that has the following format:

```
for i ← a to b
    execute X
```

This **for** loop first sets i equal to a and executes instructions X. Afterwards, it increases i by 1 and executes X again. It repeats this process by increasing i by 1 until it becomes equal to b, when it makes a final execution of X. That is, i varies through the values $a, a+1, a+2, \ldots, b-1, b$ during execution of the loop.

*The **while** loop*

A different way of summing the first n integers, called **AnotherGauss**, is shown below.

```
AnotherGauss(n)
    sum ← 0
    i ← 1
    while i ≤ n
        sum ← sum + i
        i ← i + 1
    return sum
```

This algorithm uses a **while** loop having the following format:

```
while statement X is true
    execute Y
```

The **while** loop checks condition X; if X is true, then it executes instructions Y. This process is repeated until X is false. (Note: if X winds up always being true, then the **while** loop enters an **infinite loop**, which you should avoid at all costs, because your algorithm will never end.) In the case of **AnotherGauss**, the loop stops executing after n times through the loop, when i eventually becomes equal to $n+1$ and the numbers 1 through n have been added to *sum*.

Recursive algorithms

Below is yet another algorithm solving the Summing Integers Problem.

```
RECURSIVEGAUSS(n)
    if n > 0
        sum ← RECURSIVEGAUSS(n − 1) + n
    else
        sum ← 0
    return sum
```

You may be confused by the fact that RECURSIVEGAUSS(n) calls RECURSIVEGAUSS($n - 1$) as a subroutine. So imagine that you are computing the sum of the first 100 positive integers, but you are lazy and ask your friend to compute the sum of the first 99 positive integers for you. As soon as your friend has computed it, you will simply add 100 to the result, and you are done! Yet little do you know that your friend is just as lazy, and she asks her friend to compute the sum of the first 98 integers, to which she adds 99 and then passes to you. The story continues until a friend is assigned 1. Although every individual in this chain of friends is lazy, the group is nevertheless able to compute the sum.

RECURSIVEGAUSS presents an example of a **recursive algorithm**, which subcontracts a job by calling itself (on a smaller input).

Exercise Break: Can you write a function with just one line of pseudocode to solve the Summing Integers Problem?

You have probably been wondering why we have named all these summing algorithms "Gauss". In 1785, a primary school teacher asked his class to sum the integers from 1 to 100, assuming that this task would occupy them for the rest of the day. He was shocked when an 8 year old boy thought for a few seconds and wrote down the answer, 5,050. This boy was Karl Gauss, and he would go on to become one of the greatest mathematicians of all time. The following one-line algorithm implements his idea for solving the Summing Integers Problem. Why does it work?

```
GAUSS(n)
    return (n + 1) * n/2
```

Arrays

Finally, when writing pseudocode, we may also use an **array**, or an ordered sequence of variables. We often use a single letter to denote an array, e.g., $a = (a_1, a_2, \ldots, a_n)$.

Exercise Break: What does the **RABBITS** algorithm presented below do?

```
RABBITS(n)
    a₁ ← 1
    a₂ ← 1
    for i ← 3 to n
        aᵢ ← aᵢ₋₁ + aᵢ₋₂
    return a
```

RABBITS(n) computes an array holding the first n **Fibonacci numbers**, or the sequence $(1, 1, 2, 3, 5, 8, \ldots)$, where the first two elements are 1 and each subsequent element is formed by summing its preceding two elements. Why do you think that we called it **RABBITS**?

Glossary

0-based indexing a method of indexing symbols in a string or elements in an array, beginning at 0 rather than 1. 7

2-break a genome rearrangement that breaks a genome in two locations and joins the resulting fragments in a new way. 320

2-break distance the minimum number of 2-break operations transforming one genome into another. 320

2-break sorting the problem of finding a shortest sequence of 2-breaks transforming one sequence into another. 326

additive matrix a distance matrix for which there exists a tree fitting it. 361

adenylation domain (A-domain) a segment of an NRP synthetase that is responsible for adding a single amino acid to a non-ribosomal peptide. 225

adjacency consecutive elements $(p_i\ p_{i+1})$ in a signed permutation $P = (p_1 \ldots p_n)$ such that $p_{i+1} - p_i$ is equal to 1. 311

adjacency list a way of representing a graph by listing all nodes connected to each node. 129

adjacency matrix a way of representing a graph by using a matrix A such that $A_{i,j} = 1$ if an edge connects node i to node j, and $A_{i,j} = 0$ otherwise. 129

affine gap penalty a gap penalty that charges one penalty for the first symbol of the gap and another (smaller) penalty for each successive symbol added. 267

algorithm a sequence of instructions to solve a well-formulated computational problem specified in terms of its input and output. 640

alignment a two-row matrix comparing strings v and w such that the first row contains the symbols of v (in order), the second row contains the symbols of w (in order), and space symbols (called gap symbols and shown as dashes below) may be interspersed throughout both strings, as long as two space symbols are not aligned against each other; also, see "multiple alignment". 229

alignment graph a graph used to visualize all possible alignments of two strings, in which horizontal and vertical edges represent indels, and diagonal edges represent matches and mismatches. 237

alignment path a path from source to sink in an alignment graph. 237

Alu sequence an approximately 300 nucleotide-long sequence that occurs over a million times with minor variations in the human genome. 123

amino acid string a string formed by symbols representing 20 amino acids. 187

amplitude a value of a spectral vector. 600

antibiotic a circular antibacterial peptide. 185

antiretroviral therapy a drug cocktail that stabilizes an infected HIV patient's symptoms. 533

archaea unicellular organisms that make up one of the three domains of life, along with eukaryotes and bacteria. 36

backtracking pointer a reference to which node was used to compute the maximum score at the current node of an alignment graph. 253

Basic Local Alignment Search Tool (BLAST) a heuristic for finding all proteins in a database that are similar to a given protein. 512

Baum-Welch learning a method of parameter estimation based on expectation maximization. 579

Beltway Problem The problem of positioning a set of points to a circle given the pairwise distances between the points. 210

big-O notation a way of compactly describing the worst-case running time of an algorithm based on its input size. 53

binary search a search technique that finds a data point in a sorted array by iteratively dividing the data in half and determining the half in which the data point lies. 485

binary string a string formed from zeroes and ones. 54, 130

block indel the addition or removal of a block of consecutive zeroes from a binary vector. 620

breakpoint consecutive elements $(p_i \, p_{i+1})$ in a signed permutation $P = (p_1 \ldots p_n)$ that are "out of order" compared to the identity permutation, i.e., $p_{i+1} - p_i$ is not equal to 1; a breakpoint indicates that these two elements are "out of order" compared to the identity permutation. 311

breakpoint graph a graph used to compare the orderings of synteny blocks in two genomes. 321

Bridges of Konigsberg Problem the question of whether it was possible to cross every bridge of the Prussian city Königsberg exactly once and return to one's starting position. 139

bubble a structure within a de Bruijn graph in which two or more paths connect the same two nodes, arising from either sequencing errors or non-identical repeated sequences. 161

Burrows-Wheeler transform a transformation of a string by forming all possible cyclic rotations of the string, sorting them lexicographically, and taking the final column of the resulting matrix. 488

carryover the identification of leftover peptides hiding inside a mass spectrometer after a previous experiment. 617

cell differentiation the process by which embryonic stem cells become specialized tissues. 584

center a point serving as a central point of a cluster. 426

center of gravity a point whose i-th coordinate is the average of the i-th coordinates of all points in a cluster. 432

Central Dogma of Molecular Biology the flow of genetic information in the cell starting with DNA, which is then transcribed into RNA, which is then transcribed into protein. 186

CG-island a region of the genome where the dinucleotide CG appears relatively frequently. 539

character a feature that can be used to construct a phylogeny. 383

character table an $n \times m$ matrix representing a collection of n characters for m species. 383

character vector an array holding the values of m characters for a species. 383

Cho-Han a game in which the dealer rolls two dice, and players wager on whether the sum of the dice will be even or odd. 536

circadian clock an internal timekeeping mechanism that controls the daily schedules of living things. 67

clique a graph in which every pair of nodes are connected by an edge. 464

clique graph a graph whose connected components are all cliques. 464

cluster a collection of data points that are located close to each other. 422

codon a 3-mer of nucleotides that either codes for an amino acid or that signals the ribosome to stop translation of a protein. 187

codon usage the frequency of each codon in a genomic fragment. 629

complementary strand a strand of DNA synthesized from an existing strand that is "complementary" in that adenine binds with thymine and cytosine binds with guanine. 11

complexity the runtime of an algorithm, expressed with respect to the size of the input. 9

constructive proof a proof of a theorem that not only proves the desired result, but also provides us with a method for constructing the object we need. 146

contig a contiguous segment of the genome generated by a genome assembler. 160

coronavirus a family of viruses, including SARS, that received their name because the virus particle resembles the sun's corona. 353

coverage the coverage of a k-mer is the number of reads containing this k-mer. 159

cycle a path in a graph starting and ending at the same node. 175

cyclic rotation chopping off a suffix from the end of a string and adding it to the beginning of the string. 487

cystic fibrosis a genetic disease primarily affecting the lungs. 219

dalton a unit of molecular mass (abbreviated Da). 192, 219

de Bruijn graph a graph formed from a collection of k-mers for genome sequencing that assigns each k-mer to an edge connecting two nodes corresponding to the k-mer's prefix and suffix. 132

deamination the tendency for cytosine to mutate into thymine. 24

decoy proteome a randomly generated amino acid string used for estimating false discovery rates in proteomics. 607

degree the number of edges incident to a node. 175

depth-first search (DFS) a technique for traversing the nodes of a graph that starts at a given node and explores the graph as far as possible until it reaches a node that has no outgoing edges or from which we have already explored all edges. 629

diauxic shift a metabolic inversion in which yeast can use the ethanol that it produces from glucose as a food source in the presence of oxygen. 418

directed acyclic graph (DAG) a directed graph that does not contain cycles. 235

directed graph a graph whose edges have a direction. 127, 175

dispersive hypothesis the (untrue) hypothesis that a DNA molecule is broken into pieces before replication, so that each of the daughter molecules is a patchwork of old and new double-stranded DNA. 58

distance matrix a matrix that is symmetric, non-negative, and satisfies the triangle inequality. 356, 452

divide-and-conquer algorithm an algorithm consisting of two phases: a "divide" step splitting a problem instance into smaller instances, solving them; and a "conquer" step that stitches the smaller solutions into a solution to the original problem. 272

DNA array an early method of DNA sequencing in which all possible *k*-mers are synthesized on an array and fluorescence is used to detect where sequencing reads bind on the array. 171

DNA ligase an enzyme that sews together DNA, specifically, Okazaki fragments as part of DNA replication. 20

DNA polymerase an enzyme that facilitates the nucleotide-by-nucleotide copying of DNA. 4

DNA string a string of nucleotides. 4

DNA transposon a transposon that is duplicated by a transposase and then inserted into a new site in the genome. 173

Dollo's principle the hypothesis that when a species loses a complex organ, such as wings, the organ will not reappear in exactly the same form in the species's descendants. 385

domain of life the highest taxonomic division of life forms on Earth, separating bacteria, archaea, and eukaryotes. 36

dosage compensation the inactivation of one X chromosome in females to equalize gene expression between the sexes. 346

dynamic programming an algorithmic paradigm that is used to solve many problems involving a recurrence relation by building a matrix from smaller values to larger values. 238

dyslexia a developmental disorder that is often characterized by difficulty with reading comprehension and mathematics despite normal or above-average intelligence. 589

edit distance the minimum number of substitutions, insertions, and deletions of single symbols required to transform one string into another. 264

entropy a measure of the uncertainty of a probability distribution. 76

Euclidean distance the length of a line segment connecting two points $v = (v_1, \ldots, v_m)$ and $w = (w_1, \ldots, w_m)$ in m-dimensional space, given by $d(v, w) = \sqrt{\sum_{i=1}^{m}(v_i - w_i)^2}$. 426

Eulerian cycle a cycle that traverses each edge of a graph exactly once. 140

Eulerian graph a graph containing an Eulerian cycle. 140

Eulerian path a path in a graph that visits every edge exactly once. 134

evolutionary tree (phylogeny) a tree representing the evolutionary relationships among various species. 354

exon a contiguous segment of a protein-coding gene that is eventually translated into protein. 222

experimental spectrum the collection of all the fragment masses (or mass/charge ratios) generated by the mass spectrometer. 192

exponential algorithm an algorithm whose runtime on some datasets is exponential in the length of the input data. 141

false discovery rate (FDR) the fraction of false peptide-spectrum matches among all peptide-spectrum matches. 607

false mass a mass that is present in an experimental spectrum but absent from the theoretical spectrum of a peptide. 199

Fibonacci numbers the sequence $(1, 1, 2, 3, 5, 8, \ldots)$ for which the first two elements are 1 and each subsequent element is formed by summing its preceding two elements. 648

fission a genome rearrangement that breaks a single chromosome into two chromosomes. 316

forbidden transition a transition between states in an HMM that is not allowed. 549

forward half-strand a half-strand of a bacterial genome along which DNA polymerase must travel backward from *ter* to *ori* in order to move in the $3' \rightarrow 5'$ direction; also called a lagging half-strand. 19

forward-backward algorithm a dynamic programming algorithm for computing the probability that an HMM emitted a given string and was in a given state at a given time. 577

Fragile Breakage Model a model of genome evolution stating that every mammalian genome is a mosaic of long solid regions, which are rarely affected by rearrangements, as well as short fragile regions that serve as putative rearrangement hotspots and that account only for a small fraction of the genome. 329

fragment ion an electrically charged fragment of a peptide. 595

fusion a genome rearrangement that merges two chromosomes into a single chromosome. 316

gap a contiguous sequence of space symbols in a row of an alignment. 267

gap extension penalty a penalty assessed to each symbol after the first symbol in a gap. 267

gap opening penalty a penalty assessed to the first symbol of a gap. 267

gap symbol a space symbol ("-") used in an alignment to indicate an inserted/deleted symbol. 229

GC-content the percentage of cytosine and guanine nucleotides in a genome. 539

gene expression matrix an $n \times m$ matrix representing the expression of n genes at m different time points. 420

genetic code a rule converting each of the 64 possible RNA 3-mers into one of 20 different amino acids, or a stop codon. 187

genome assembly reconstructing a genome from short, overlapping DNA fragments called reads. 119

genome rearrangement a large-scale mutation that moves blocks of genes around in a genome. 34, 297

genome replication the duplication of a cell's genome. 3

genome sequencing determining the order of nucleotides in a genome. 117

genomic dot plot a way of comparing two genomes by plotting the positions of their shared k-mers in 2-D. 330

glycosylation a post-translational modification that often makes HIV envelope proteins invisible to the human immune system. 533

graph a network of nodes connected by edges. 127

graph theory the branch of mathematics considering the study of graphs. 141

Hamiltonian path a path in a graph that visits every node once. 130

Hamming distance the number of mismatched symbols between two strings of equal length. 28

heuristic a fast algorithm that trades accuracy for speed in order to find an approximate solution. 84

Hidden Markov Model (HMM) A set of four objects: an alphabet of emitted symbols, a set of hidden states, a matrix of transition probabilities between the states, and a matrix of emission probabilities representing the probability of emitting each symbol when the HMM is in each state. 541

hidden path an (unknown) sequence of states that an HMM passes through. 542

hierarchical clustering an algorithm that uses a distance matrix to organize a collection of data points into a tree, starting with the closest two points and moving upward. 453

HMM diagram a graph visualizing an HMM in which every state is represented by a node, and solid directed edges connect every pair of nodes, labeled with the transition probability of moving from one state to the other (or remaining in the same state). 542

horizontal gene transfer the exchange of genetic material between different species. 35

icosian game a game invented by Irish mathematician William Hamilton illustrating the Hamiltonian path problem. 176

ideal spectrum a spectrum of integer masses generated from a peptide that coincides with the peptide's theoretical spectrum. 194

identity permutation the signed permutation $(+1\ +2, \ldots, +n)$. 306

indegree the number of edges leading into a node in a directed graph. 142

indel penalty a score used for assessing a penalty to each gap symbol in an alignment. 256

infinite monkey theorem the statement that an immortal monkey typing an infinite sequence of symbols on a typewriter will one day reproduce *Hamlet*. 609, 632

initial state a silent state of an HMM that occurs at the beginning of the HMM's path and does not emit any symbols. 544

integer mass the approximate mass of a molecule, rounded to the closest integer (in Daltons). 192

intensity the number of fragment ions detected at a given mass/charge ratio. 208, 595

intensity count the sum of intensities of all peaks annotated by a peptide. 598

internal edge an edge in a tree that is not a limb (i.e., that connects two internal nodes). 365

internal node a node in a tree that has degree greater than 1 (i.e., is not a leaf). 357

intractable problem a problem for which no polynomial-time algorithm exists. 177

intron a contiguous segment of a gene that is not translated into protein. 222

k-mer composition the collection of all *k*-mer substrings of a string. 120

k-universal binary string a binary string containing every binary *k*-mer exactly once. 130

Las Vegas algorithm a randomized algorithm whose solutions are guaranteed to be correct. 92

leaf a node in a tree with degree 1. 357, 473

lexicographic order an ordering of strings according to how they would appear in a dictionary. 40, 120

limb an edge connecting a leaf to its parent. 357

Lloyd algorithm a clustering algorithm that chooses k arbitrary data points to serve as centers and then iteratively alternates between assigning data points to their nearest center and updating each center to be its cluster's center of gravity. 433

local alignment an alignment of substrings of two strings. 260

log-odds ratio a logarithm of the ratio of two probabilities. 537

longest common prefix (LCP) array an array storing the length of the longest common prefix shared by consecutive lexicographically ordered suffixes of a string. 525

longest common subsequence (LCS) a sequence of symbols appearing in the same order (although not necessarily consecutively) in two strings. 229

longest path a path of maximum weight connecting two fixed nodes in an edge-weighted graph. 233

loop an edge connecting a node to itself. 132

Manhattan Tourist Problem the problem of finding a legal path through a city that visits the most sights. 231

mass spectrometer an expensive molecular scale that shatters molecules into pieces and then weighs the resulting fragments. 192

mass/charge ratio the ratio of an ion's mass to its charge. 207, 595

messenger RNA (mRNA) an RNA molecule consisting of RNA transcribed from exons and in which the RNA transcribed from introns has been excised out. 223

methylation the most common DNA modification, which adds a methyl group (CH_3) to the cytosine nucleotide within a CG dinucleotide. 539

microarray a method for measuring the RNA expression of genes. 421

middle edge an edge in a longest alignment path that starts at a middle node. 278

middle node a node on a longest path in an alignment graph that belongs to a middle column in the alignment graph. 274

midpoint a point that is approximately equidistant from two centers in clustering. 439

mismatch penalty a score used for assessing a penalty to a mismatched symbol in an alignment. 256

missing mass a mass that is present in the theoretical spectrum of a peptide but absent from the experimental spectrum. 199

monoisotopic mass the sum of the masses of the atoms in a molecule, using the mass of the most abundant isotope for each element. 220

Monte Carlo algorithm a randomized algorithm that is not guaranteed to return an exact solution. 92

motif finding the problem of discovering a "hidden message" shared by a collection of strings. 68

motif logo a diagram for visualizing motif conservation that consists of a stack of letters at each position. 77

multi-domain protein a protein consisting of more than one independent domain. 582

multiple alignment a matrix with n rows representing an alignment of n strings for $n > 2$. 280

nearest neighbor a tree that is a single nearest neighbor interchange away from a given tree. 394

nearest neighbor interchange an operation on a tree resulting from removing an internal edge in the tree and rearranging the resulting four subtrees in a new way. 394

neighbor-joining algorithm a popular algorithm for evolutionary tree reconstruction. 377

neighbors leaf nodes in a tree sharing the same parent node. 362

NF-KB A transcription factor activating immunity genes in flies. 69

non-additive matrix a distance matrix for which there is no tree fitting it. 361

non-proteinogenic amino acid an amino acid not belonging to the 22 amino acids most commonly used to build proteins (the 20 standard amino acids, plus selenocysteine and pyrrolysine). 204

non-ribosomal code a code controlling the synthesis of nonribosomal peptides. 225

non-ribosomal peptide (NRP) a peptide that is synthesized by NRP synthetase. 191

NP-complete problem a computational problem belonging to a large class of equivalent problems for which no polynomial-time algorithm has been found, and which no one has demonstrated to be intractable. 177

Occam's razor the principle stating that when presented with some quandary, we should explain it using the simplest hypothesis that is consistent with what we already know. Also called "parsimony".. 304

Okazaki fragment a disjoint fragment of replicated DNA resulting from the asymmetry of replication when DNA polymerase replicates a forward strand one fragment at a time. 20

oncogene a gene in viruses that causes a cancer-like transformation of infected human cells. 227

open reading frame (ORF) a substring of a genome starting with a start codon, ending with a stop codon, and having no other stop codons in frame. 628

organelle a subunit within a cell performing some specific function. 164

outdegree the number of edges leading out of a node in a directed graph. 142

overlap alignment an alignment of a suffix of one string against a prefix of another. 266

overlap graph a graph on a set of reads whose nodes are the reads and in which edges connect overlapping reads. 128

overlapping words paradox the phenomenon that different strings have different probabilities of appearing in a random string. 38

parsimony score a scoring function for a tree produced by summing the weights of its edges. 387

partial suffix array a suffix array that only contains values that are multiples of some positive integer K. 508

path a sequence of edges in a graph, where each successive edge begins at the node where the previous edge ends. 175

peak a spike appearing in the output of a mass spectrometer, which indicates a particular mass/charge ratio. 208

penicillin the world's first discovered antibiotic. 185

peptide a short amino acid sequence. 185

peptide identification the process of confirming that a peptide from a known proteome is present in a sample. 591

peptide sequencing the process of determining the order of amino acids making up a peptide. 186, 591

peptide vector a binary vector possessing a 1 at the indices corresponding to each of a peptide's prefix masses. 599

peptide-spectrum match (PSM) a peptide (vector) and spectrum (vector) whose score exceed a predetermined threshold. 605

perfect coverage the idealistic assumption in genome assembly that every k-mer substring of the genome is generated as a read. 120

permutation a rearrangement of the elements 1 through n for some integer n; we use the term "permutation" to refer to a signed permutation. 304

Pfam database a database containing over 10,000 HMM-derived multiple alignments of protein domain families. 583

Philadelphia chromosome a chromosome formed by a translocation involving chromosomes 9 and 22 that is implicated in chronic myeloid leukemia. 315

phosphorylation a post-translational modification consisting of the addition of a phosphate group at a specific amino acid. 619

polynomial algorithm an algorithm whose runtime can be bounded by a polynomial in the length of the input data. 141

polytene chromosome a "superchromosome" resulting from multiple rounds of DNA replication without cell division. 346

post-translational modification a modification that alters the amino acids of a protein after it has been translated from RNA. 619

precursor mRNA (pre-mRNA) the RNA that is transcribed from a split gene, which consists of RNA transcribed from both introns and exons. 222

prefix a substring of a string that includes its first symbol. 126, 592

prefix reversal a reversal that flips a prefix of a permutation. 348

preorder traversal a way of traversing a tree by starting at the root, and at each node, visiting the node, then preorder traversing the subtrees rooted at each of its children from left to right. 524

primer a short segment of DNA that is complementary to a DNA polymerase that binds to the parent strand and jump starts the DNA polymerase. 17

probability distribution a collection of nonnegative numbers that sum to 1. 76

profile HMM a Hidden Markov Model used for sequence alignment. 554

programming language a language that provides a computer with specific instructions. 641

proof by contradiction a method of mathematical proof in which we assume the statement that we intend to disprove and then demonstrate how this assumption cannot be true. 328

protease an enzyme that breaks a peptide into shorter peptide fragments. 595

protein domain a conserved part of a protein that can function independently of the rest of the protein. 582

protein translation the process of converting RNA into a protein. 186

proteinogenic amino acid an amino acid belonging to the group of 22 amino acids most commonly used to build proteins (the 20 standard amino acids, plus selenocysteine and pyrrolysine). 203

proteome the set of all proteins for a given species. 590

proteomics the study of proteins. 590

pseudocode a universal method of describing algorithms that is more precise than human language but does not require us to get bogged down in the syntax of a specific programming language. 8, 641

pseudocount a small number used in randomized algorithms to ensure that low-probability events are sometimes chosen. 88

pseudogene a non-functional remnant of a gene that evolved from a previously functional gene. 212

PSM dictionary the spectral dictionary containing all peptides that outscore a given PSM with respect to the spectrum belonging to the PSM. 610

quorum sensing a communication method that allows bacteria to migrate to a better nutrient supply or defend against threats. 219

Random Breakage Model a model of genome evolution stating that the breakage points of rearrangements are selected randomly, implying that rearrangement hotspots in mammalian genomes do not exist. 301

read a DNA fragment that can be "read" by a sequencing machine. 119

read breaking the strategy of splitting reads into shorter k-mers in order to boost coverage. 159

read mapping a process by which sequencing reads taken from an individual are "mapped" against a reference species genome to identify mutations. 471

read-pair a pair of reads separated by a known distance in the genome; in practice, the distance is only approximately known. 151

reading frame a way of dividing a DNA string into triplets of nucleotides; each strand of DNA has three reading frames. 189

rearrangement hotspot a short region of a genome where multiple breakage points of rearrangements have occurred or been more likely to occur during the course of evolution. 300

recurrence relation an equation for a function $f(n)$ in terms of $f(m)$ for smaller values of m. 240

recursion tree a tree indicating the structure of recursive calls taken by a recursive algorithm. 292

recursive algorithm an algorithm that calls itself. 43, 647

Red Queen Effect the hypothesis that evolution is necessary not only to equip organisms with an advantage in a *fixed* environment, but also to help them survive in response to *changing* environments. 584

reference human genome a human genome that is used to map reads from individuals to determine how the individuals differ from the reference. 471

regulatory motif a short DNA interval to which a transcription factor binds. 68

regulatory protein a protein that controls the expression of (other) proteins in the cell. 67

replication fork a branching that occurs as two strands of DNA unzip and serve as templates for replication. 16

replication origin (ori) the region of a chromosome in which DNA replication begins. 4

replication terminus (ter) the point at which replication completes in a bacterial genome. 16

retrotransposon a transposon that is transcribed from DNA to RNA, which is then reverse transcribed into DNA and inserted at a new position. 173

reversal a genome rearrangement that inverts a segment of a chromosome, switching it to the opposite strand. 35, 299

reversal distance the minimum number of reversals required to transform one permutation into another. 305

reverse complement a DNA string formed from another string by reversing it and taking the complement of each nucleotide. 12

reverse half-strand a half-strand of a bacterial genome along which DNA polymerase can move from *ori* to *ter* in the $3' \rightarrow 5'$ direction; also called a lagging half strand. 19

reverse transcriptase an enzyme that "reverse" transcribes RNA back into DNA, running counter to the typical method of transcription from DNA to RNA indicated by the Central Dogma of Molecular Biology. 173, 461

ribonucleotide a building block of RNA, consisting of adenine, cytosine, guanine, and uracil. 186

ribosome a molecular machine that carries out protein translation. 191

RNA sequencing (RNA-seq) a process of determining gene expression by sequencing reads sampled from the RNA in a cell and using these reads to estimate the quantity of each RNA transcript in a sample. 461

RNA splicing the process of converting pre-mRNA into mRNA so that introns are excised out and the mRNA is ready for translation into protein. 223

RNA string a string formed from the four-letter alphabet {A, C, G, U}. 186

RNA transcription the process of transforming a strand of DNA into a strand of RNA. 186

RNA virus a virus possessing RNA instead of DNA. 354

rooted binary tree an unrooted binary tree that has a root (of degree 2) placed on one of its edges. 373

rooted tree a tree in which one node is designated as a special node, called the root, and the edges in the tree automatically inherit an implicit orientation away from the root. 357

scoring matrix a matrix used to represent the score for aligning any possible pair of symbols. 257

seed alignment an alignment resulting from a multiple alignment by eliminating all columns having a number of gap symbols above some threshold parameter. 552

semiconservative hypothesis the hypothesis, suggested by Watson and Crick and confirmed by the Meselson-Stahl experiment, that each parent strand of DNA acts as a template for the synthesis of a daughter strand. 58

Severe Acute Respiratory Syndrome (SARS) A viral respiratory illness that caused an international scare in 2003. 353

shared peaks count the number of peaks shared by two spectra, e.g., by an experimental and theoretical spectrum of a peptide. 598

signed permutation a permutation of the elements 1 through n for some integer n, where each element is assigned a positive or negative sign. 304

similarity-based gene prediction gene prediction approaches that are based on the observation that a newly sequenced gene is often similar to a known gene in another species. 628

simple tree a tree having no nodes of degree 2. 361

single nucleotide polymorphism (SNP) a single-nucleotide mutation that occurs in a significant percentage (typically 1%) of the human population. 471

sink node a designated ending node in a longest path problem. 233

skew diagram a diagram used to visualize the varying frequency of nucleotides across a genome. 26

sorting by reversals the computational problem of transforming a permutation into the identity permutation using the minimum number of reversals. 307

source node a designated starting node in a longest path problem. 233

spectral alignment graph a graph used to find the highest-scoring peptide over all peptides having up to k modifications with respect to a fixed peptide and spectrum. 625

spectral convolution the collection of all positive differences between pairs of masses in a spectrum. 205

spectral dictionary the collection of all peptides scoring above some threshold with respect to a spectrum. 610

spectral vector a vector that consolidates information about the intensity of each peak and its twin peak in a mass spectrum into a single value called an amplitude. 600

spectrum annotation establishing a correspondence between the masses of spectral peaks and the masses of the peptide's prefixes/suffixes. 595

spliceosome a molecular machine that facilitates RNA splicing. 223

split gene a gene formed by discontiguous intervals of DNA called exons (which are translated into protein) and introns (which are not). 222

squared error distortion the mean squared distance from each data point to its nearest center. 430

statistical gene prediction an approach in which researchers look for genomic features appearing frequently in coding regions and infrequently in non-coding regions of a genome. 628

stop codon a codon that does not translate into amino acids and serves to halt translation. 187

subpeptide a linear fragment of a peptide. 193

subroutine a function that is called within another function. 644

suffix a substring of a string that includes its final symbol. 126, 592

suffix array an array formed by sorting all suffixes of a string lexicographically and then compiling the list of their starting positions. 484

suffix tree a suffix trie with all edges on any non-branching path collapsed into a single edge. 481

suffix trie a trie formed by all suffixes of a string. 477

syncytium an abnormal multinucleate cell. 534

syncytium-inducing (SI) isolate a form of HIV in which the virus is able to kill many human cells at a time by causing dozens of human cells to fuse their cell membranes into a large, nonfunctional syncytium. 534

synteny block a segment of similar genes that occur contiguously in two different species. 298

targeted therapy a cancer treatment that attacks a specific molecule, often one occurring only in a specific type of cancer. 315

template strand a strand of DNA that is used as a "template" for DNA replication. 11

terminal state a silent state that an HMM enters when it has finished emitting symbols. 548

ternary string a string formed of the three-symbol alphabet $\{0, 1, 2\}$. 55

theoretical spectrum the collection of subpeptide masses of a peptide. 193

topological ordering a way of ordering the nodes $(a_1, \ldots, a_k)$ of a directed acyclic graph such that every edge (a_i, a_j) connects a node with a smaller index to a node with a larger index, i.e., $i < j$. 250

transcription factor a regulatory protein that controls the expression of other genes. 68

transcriptome the collection of all RNA transcripts present in a cell. 461

translocation a genome rearrangement in which two intervals of DNA on different chromosomes are excised and then reattached on the opposite chromosome. 315

transposon a DNA fragment that can change its position within the genome, often resulting in a duplication. 173

tree a connected graph without cycles. 357

trie a rooted tree representing a collection of patterns in which each edge is labeled by a letter of the alphabet, and paths from the root to nodes spell out the patterns in the collection. 473

trypsin a commonly used protease that typically breaks a protein after the amino acids `R` and `K`. 595

Turnpike Problem the problem of positioning a set of points on a line segment given the pairwise distances between the points. 211

twin peak a peak having mass equal to the mass of the peptide minus the mass of a given subpeptide. 600

Tyrocidine B1 an antibiotic produced by *Bacillus brevis*. 186

ultrametric tree a tree in which the distance from the root to any leaf is the same. 373

undirected graph a graph whose edges do not have directions. 127

UniProt a database containing proteins from various species. 605

UniProt+ our term for the augmentation of UniProt with some mutated versions of collagens from present-day species in order to model possible differences between these collagens and *T. rex* collagens. 605

unrooted binary tree a tree in which every node has degree equal to either 1 or 3. 373

unrooted tree a tree without a designated root node. 357

UPGMA a hierarchical clustering heuristic that introduces a molecular clock for constructing an ultrametric evolutionary tree. 373

viral vector a genetically modified virus. 4

Viterbi algorithm a dynamic programming algorithm for finding the most likely hidden path that an HMM has taken to produce a given emitted string. 548

Viterbi graph a graph defined for an HMM with respect to its output, where there is one node for each possible pair of state and emitted symbol, and edges connect every state in one column to every state in the next column, weighted by the product of the associated transition and emission probability. 546

Viterbi learning an approach for estimating the parameters of an HMM by starting with an initial guess for these parameters, and then alternating between using the Viterbi algorithm to find the optimal hidden path and recomputing the best choice of parameters for this hidden path. 574

whole genome duplication (WGD) a rare evolutionary event that duplicates an entire genome. 418

Bibliography

Aho, A. V. and M. J. Corasick (1975). "Efficient String Matching: An Aid to Bibliographic Search." *Communications of the ACM* Vol. 18: 333–340.

Alekseyev, M. A. and P. A. Pevzner (2009). "Breakpoint graphs and ancestral genome reconstructions." *Genome Research* Vol. 19: 943–957.

Alon, U., N. Barkai, D. A. Notterman, K. Gish, S. Ybarra, D. Mack, and A. J. Levine (1999). "Broad patterns of gene expression revealed by clustering analysis of tumor and normal colon tissues probed by oligonucleotide arrays." *Proceedings of the National Academy of Sciences* Vol. 96: 6745–6750.

Altschul, S. F., W. Gish, W. Miller, E. W. Myers, and D. J. Lipman (1990). "Basic local alignment search tool." *Journal of Molecular Biology* Vol. 215: 403–410.

Arthur, D. and S. Vassilvitskii (2007). "k-means++: The Advantages of Careful Seeding." In: *Proceedings of the Eighteenth Annual ACM-SIAM Symposium on Discrete Algorithms.* New Orleans, Louisiana: Society for Industrial and Applied Mathematics, 1027–1035.

Asara, J. M., M. H. Schweitzer, L. M. Freimark, M. Phillips, and L. C. Cantley (2007). "Protein Sequences from Mastodon and Tyrannosaurus rex Revealed by Mass Spectrometry." *Science* Vol. 316: 280–285.

Bafna, V. and P. A. Pevzner (1996). "Genome Rearrangements and Sorting by Reversals." *SIAM Journal on Computing* Vol. 25: 272–289.

Baldi, P., Y. Chauvin, T. Hunkapiller, and M. A. McClure (1994). "Hidden Markov models of biological primary sequence information." *Proceedings of the National Academy of Sciences* Vol. 91: 1059–1063.

Bateman, A., E. Birney, L. Cerruti, R. Durbin, L. Etwiller, S. R. Eddy, S. Griffiths-Jones, K. L. Howe, M. Marshall, and E. L. L. Sonnhammer (2002). "The Pfam Protein Families Database." *Nucleic Acids Research* Vol. 30: 276–280.

Baum, L. E., T. Petrie, G. Soules, and N. Weiss (1970). "A Maximization Technique Occurring in the Statistical Analysis of Probabilistic Functions of Markov Chains." *The Annals of Mathematical Statistics* Vol. 41: pp. 164–171.

Beerenwinkel, N. and M. Drton (2007). "A mutagenetic tree hidden Markov model for longitudinal clonal HIV sequence data." *Biostatistics* Vol. 8: 53–71.

Ben-Dor, A., R. Shamir, and Z. Yakhini (1999). "Clustering Gene Expression Patterns." *Journal of Computational Biology* Vol. 6: 281–297.

Bezdek, J. C. (1981). *Pattern Recognition with Fuzzy Objective Function Algorithms*. Kluwer Academic Publishers.

Buckley, M., A. Walker, S. Y. W. Ho, Y. Yang, C. Smith, P. Ashton, J. T. Oates, E. Cappellini, H. Koon, K. Penkman, B. Elsworth, D. Ashford, C. Solazzo, P. Andrews, J. Strahler, B. Shapiro, P. Ostrom, H. Gandhi, W. Miller, B. Raney, M. I. Zylber, M. T. P. Gilbert, R. V. Prigodich, M. Ryan, K. F. Rijsdijk, A. Janoo, and M. J. Collins (2008). "Comment on 'Protein Sequences from Mastodon and Tyrannosaurus rex Revealed by Mass Spectrometry'." *Science* Vol. 319: 33.

Burrows, D. and M. J. Wheeler (1994). "A block sorting lossless data compression algorithm." *Technical Report 124, Digital Equipment Corporation*.

Butler, J., I MacCallum, M. Kleber, I. A. Shlyakhter, M. Belmonte, E. S. Lander, C. Nusbaum, and D. B. Jaffe (2008). "ALLPATHS: *de novo* assembly of whole-genome shotgun microreads." *Genome Research* Vol. 18: 810–820.

Cann, R. L., M. Stoneking, and A. C. Wilson (1987). "Mitochondrial DNA and human evolution." *Nature* Vol. 325: 31–36.

Ceppellini, B. R., M. Siniscalco, and C. A. B. Smith (1955). "The Estimation of Gene Frequencies in a Random-Mating Population." *Annals of Human Genetics* Vol. 20: 97–115.

Chen, T., M.-Y. Kao, M. Tepel, J. Rush, and G. M. Church (2001). "A Dynamic Programming Approach to De Novo Peptide Sequencing via Tandem Mass Spectrometry." *Journal of Computational Biology* Vol. 8: 325–337.

Churchill, G. (1989). "Stochastic models for heterogeneous DNA sequences." *Bulletin of Mathematical Biology* Vol. 51: 79–94.

Clayton-Smith, J., J. O'Sullivan, S. Daly, S. Bhaskar, R. Day, B. Anderson, A. K. Voss, T. Thomas, L. G. Biesecker, P. Smith, A. Fryer, K. E. Chandler, B. Kerr, M. Tassabehji, S. A. Lynch, M. Krajewska-Walasek, S. McKee, J. Smith, E. Sweeney, S. Mansour, S. Mohammed, D. Donnai, and G. Black (2011). "Whole-exome-sequencing identifies mutations in histone acetyltransferase gene KAT6B in individuals with the Say-Barber-Biesecker variant of Ohdo syndrome." *The American Journal of Human Genetics* Vol. 89: 675–681.

Cohen, D. S. and M. Blum (1995). "On the Problem of Sorting Burnt Pancakes." *Discrete Applied Mathematics* Vol. 61: 105–120.

Conti, E., T. Stachelhaus, M. A. Marahiel, and P. Brick (1997). "Structural basis for the activation of phenylalanine in the non-ribosomal biosynthesis of gramicidin S." *The EMBO Journal* Vol. 16: 4174–4183.

Conti, E., N. P. Franks, and P. Brick (1996). "Crystal structure of firefly luciferase throws light on a superfamily of adenylate-forming enzymes." *Structure* Vol. 4: 287–298.

Cristianini, N. and M. W. Hahn (2007). *Introduction to Computational Genomics*. Cambridge University Press.

Cristianini, N. and M. W. Hahn (2006). *Introduction to Computational Genomics: A Case Studies Approach*. Cambridge University Press.

Czerkas, S. and A. Feduccia (2014). "Jurassic archosaur is a non-dinosaurian bird." *Journal of Ornithology* Vol. 155: 841–851.

De Jong, J., A. De Ronde, W. Keulen, M. Tersmette, and J. Goudsmit (1992). "Minimal Requirements for the Human Immunodeficiency Virus Type 1 V3 Domain To Support the Syncytium-Inducing Phenotype: Analysis by Single Amino Acid Substitution." *Journal of Virology* Vol. 66: 6777–6780.

de Bruijn, N. (1946). "A Combinatorial Problem." In: *Proceedings of the Section of Sciences, Koninklijke Akademie van Wetenschappen te Amsterdam*. Vol. 49: 758–764.

DeRisi, J. L., V. R. Iyer, and P. O. Brown (1997). "Exploring the metabolic and genetic control of gene expression on a genomic scale." *Science* Vol. 278: 680–686.

Do, C. B. and S. Batzoglou (2008). "What is the expectation maximization algorithm?" *Nature Biotechnology* Vol. 26: 897–899.

Doolittle, R. F., M. W. Hunkapiller, L. E. Hood, S. G. Devare, K. C. Robbins, S. A. Aaronson, and H. N. Antoniades (1983). "Simian sarcoma virus onc gene, v-sis, is derived from the gene (or genes) encoding a platelet-derived growth factor." *Science* Vol. 221: 275–277.

Drmanac, R., I. Labat, I. Brukner, and R. Crkvenjakov (1989). "Sequencing of megabase plus DNA by hybridization: Theory of the method." *Genomics* Vol. 4: 114–128.

Eisen, M. B., P. T. Spellman, P. O. Brown, and D. Botstein (1998). "Cluster analysis and display of genome-wide expression patterns." *Proceedings of the National Academy of Sciences* Vol. 95: 14863–14868.

Eng, J. K., A. L. McCormack, and J. R. Yates (1994). "An approach to correlate tandem mass spectral data of peptides with amino acid sequences in a protein database." *Journal of the American Society for Mass Spectrometry* Vol. 5: 976 –989.

Euler, L. (1758). "Solutio Problematis ad Geometriam Situs Pertinentis." *Novi Commentarii Academiae Scientarium Imperialis Petropolitanque*, 9–28.

Felsenstein, J. (2004). *Inferring Phylogenies*. Sinauer Associates.

Ferragina, P. and G. Manzini (2000). "Opportunistic Data Structures with Applications." In: *Proceedings of the 41st Annual Symposium on Foundations of Computer Science*. IEEE Computer Society, 390–398.

Gao, F. and C.-T. Zhang (2008). "Ori-Finder: A web-based system for finding *oriCs* in unannotated bacterial genomes." *BMC Bioinformatics* Vol. 9: 79.

Gao, F., E. Bailes, D. L. Robertson, Y. Chen, C. M. Rodenburg, S. F. Michael, L. B. Cummins, L. O. Arthur, M. Peeters, G. M. Shaw, P. M. Sharp, and B. H. Hahn (1999). "Origin of HIV-1 in the chimpanzee Pan troglodytes troglodytes." *Nature* Vol. 397: 436–441.

Gardner, M. (1974). "Mathematical Games." *Scientific American* Vol. 230: 120–125.

Gates, W. H. and C. H. Papadimitriou (1979). "Bounds for sorting by prefix reversal." *Discrete Mathematics* Vol. 27: 47–57.

Geman, S. and D. Geman (1984). "Stochastic Relaxation, Gibbs Distributions, and the Bayesian Restoration of Images." *IEEE Transactions on Pattern Analysis and Machine Intelligence* Vol. PAMI-6: 721–741.

Good, I. J. (1946). "Normal Recurring Decimals." *Journal of the London Mathematical Society* Vol. 21: 167–169.

Green, R. E., E. L. Braun, J. Armstrong, D. Earl, N. Nguyen, G. Hickey, M. W. Vandewege, J. A. St. John, S. Capella-Gutiérrez, T. A. Castoe, C. Kern, M. K. Fujita, J. C. Opazo, J. Jurka, K. K. Kojima, J. Caballero, R. M. Hubley, A. F. Smit, R. N. Platt, C. A. Lavoie, M. P. Ramakodi, J. W. Finger, A. Suh, S. R. Isberg, L. Miles, A. Y. Chong, W. Jaratlerdsiri, J. Gongora, C. Moran, A. Iriarte, J. McCormack, S. C. Burgess, S. V. Edwards, E. Lyons, C. Williams, M. Breen, J. T. Howard, C. R. Gresham, D. G. Peterson, J. Schmitz, D. D. Pollock, D. Haussler, E. W. Triplett, G. Zhang, N. Irie, E. D. Jarvis, C. A. Brochu, C. J. Schmidt, F. M. McCarthy, B. C. Faircloth, F. G. Hoffmann, T. C. Glenn, T. Gabaldón, B. Paten, and D. A. Ray (2014). "Three crocodilian genomes reveal ancestral patterns of evolution among archosaurs." *Science* Vol. 346:

Grigoriev, A. (1998). "Analyzing genomes with cumulative skew diagrams." *Nucleic Acids Research* Vol. 26: 2286–2290.

Grigoriev, A. (2011). "How do replication and transcription change genomes?" In: *Bioinformatics for Biologists*. Ed. by P. A. Pevzner and R. Shamir. Cambridge University Press, 111–125.

Guibas, L. and A. Odlyzko (1981). "String overlaps, pattern matching, and nontransitive games." *Journal of Combinatorial Theory, Series A* Vol. 30: 183 –208.

Hannenhalli, S. and P. A. Pevzner (1999). "Transforming Cabbage into Turnip: Polynomial Algorithm for Sorting Signed Permutations by Reversals." *Journal of the ACM* Vol. 46: 1–27.

Harmer, S. L., J. B. Hogenesch, M. Straume, H. S. Chang, B. Han, T. Zhu, X. Wang, J. A. Kreps, and S. A. Kay (2000). "Orchestrated transcription of key pathways in *Arabidopsis* by the circadian clock." *Science* Vol. 290: 2110–2113.

Hedges, S. and M. Schweitzer (1995). "Detecting dinosaur DNA." *Science* Vol. 268: 1191–1192.

Hertz, G. Z. and G. D. Stormo (1999). "Identifying DNA and protein patterns with statistically significant alignments of multiple sequences." *Bioinformatics* Vol. 15: 563–577.

Idury, R. M. and M. S. Waterman (1995). "A New Algorithm for DNA Sequence Assembly." *Journal of Computational Biology* Vol. 2: 291–306.

Ivics, Z., P. Hackett, R. Plasterk, and Z. Izsvák (1997). "Molecular reconstruction of Sleeping Beauty, a Tc1-like transposon from fish, and its transposition in human cells." *Cell* Vol. 91: 501–510.

Kellis, M., B. W. Birren, and E. S. Lander (2004). "Proof and evolutionary analysis of ancient genome duplication in the yeast Saccharomyces cerevisiae." *Nature* Vol. 428: 617–624.

Kim, S., N. Gupta, and P. A. Pevzner (2008). "Spectral Probabilities and Generating Functions of Tandem Mass Spectra: A Strike against Decoy Databases." *Journal of Proteome Research* Vol. 7: PMID: 18597511, 3354–3363.

Kim, S., N. Gupta, N. Bandeira, and P. A. Pevzner (2009). "Spectral dictionaries: Integrating de novo peptide sequencing with database search of tandem mass spectra." *Molecular & Cellular Proteomics* Vol. 8: 53–69.

Konopka, R. J. and S. Benzer (1971). "Clock mutants of Drosophila melanogaster." *Proceedings of the National Academy of Sciences of the United States of America* Vol. 68: 2112–2116.

Krogh, A., M. Brown, I. Saira Mian, K. Sjolander, and D. Haussler (1994). "Hidden Markov Models in Computational Biology: Applications to Protein Modeling." *Journal of Molecular Biology* Vol. 235: 1501 –1531.

Lawrence, C. E., S. F. Altschul, M. S. Boguski, J. S. Liu, A. F. Neuwald, and J. C. Wootton (1993). "Detecting subtle sequence signals: a Gibbs sampling strategy for multiple alignment." *Science* Vol. 262: 208–214.

Levenshtein, V. I. (1966). "Binary codes capable of correcting deletions, insertions, and reversals." *Soviet Physics Doklady* Vol. 10: 707–710.

Liachko, I., R. A. Youngblood, U. Keich, and M. J. Dunham (2013). "High-resolution mapping, characterization, and optimization of autonomously replicating sequences in yeast." *Genome Research* Vol. 23: 698–704.

Lloyd, S. (1982). "Least squares quantization in PCM." *IEEE Transactions on Information Theory* Vol. 28: 129–137.

Lobry, J. R. (1996). "Asymmetric substitution patterns in the two DNA strands of bacteria." *Molecular Biology and Evolution* Vol. 13: 660–665.

Lundgren, M., A. Andersson, L. Chen, P. Nilsson, and R. Bernander (2004). "Three replication origins in *Sulfolobus* species: Synchronous initiation of chromosome replication and asynchronous termination." *Proceedings of the National Academy of Sciences of the United States of America* Vol. 101: 7046–7051.

Lysov, Y., V. Florent'ev, A. Khorlin, K. Khrapko, V. Shik, and A. Mirzabekov (1988). "DNA sequencing by hybridization with oligonucleotides." *Doklady Academy Nauk USSR* Vol. 303: 1508–1511.

Ma, J., A. Ratan, B. J. Raney, B. B. Suh, W. Miller, and D. Haussler (2008). "The infinite sites model of genome evolution." *Proceedings of the National Academy of Sciences of the United States of America* Vol. 105: 14254–14261.

Manber, U. and G. Myers (1990). "Suffix Arrays: A New Method for On-line String Searches." In: *Proceedings of the First Annual ACM-SIAM Symposium on Discrete Algorithms*. Society for Industrial and Applied Mathematics, 319–327.

Martin, N., E. Ruedi, R. LeDuc, F.-J. Sun, and G. Caetano-Anolles (2007). "Gene-interleaving patterns of synteny in the Saccharomyces cerevisiae genome: Are they proof of an ancient genome duplication event?" *Biology Direct* Vol. 2: 23.

Maxam, A. M. and W. Gilbert (1977). "A new method for sequencing DNA." *Proceedings of the National Academy of Sciences of the United States of America* Vol. 74: 560–564.

Medvedev, P., S. K. Pham, M. Chaisson, G. Tesler, and P. A. Pevzner (2011). "Paired de Bruijn Graphs: A Novel Approach for Incorporating Mate Pair Information into Genome Assemblers." *Journal of Computational Biology* Vol. 18: 1625–1634.

Metzker, M. L., D. P. Mindell, X.-M. Liu, R. G. Ptak, R. A. Gibbs, and D. M. Hillis (2002). "Molecular evidence of HIV-1 transmission in a criminal case." *Proceedings of the National Academy of Sciences* Vol. 99: 14292–14297.

Montgomery, S. B., D. Goode, E. Kvikstad, C. A. Albers, Z. Zhang, X. J. Mu, G. Ananda, B. Howie, K. J. Karczewski, K. S. Smith, V. Anaya, R. Richardson, J. Davis, D. G. MacArthur, A. Sidow, L. Duret, M. Gerstein, K. Markova, J. Marchini, G. A. McVean, and G. Lunter (2013). "The origin, evolution and functional impact of short insertion-deletion variants identified in 179 human genomes." *Genome Research* Vol. 23: 749–761.

Nadeau, J. H. and B. A. Taylor (1984). "Lengths of chromosomal segments conserved since divergence of man and mouse." *Proceedings of the National Academy of Sciences of the United States of America* Vol. 81: 814–818.

Ng, J., N. Bandeira, W.-T. Liu, M. Ghassemian, T. L. Simmons, W. H. Gerwick, R. Linington, P. Dorrestein, and P. A. Pevzner (2009). "Dereplication and *de novo* sequencing of nonribosomal peptides." *Nature Methods* Vol. 6: 596–599.

O'Brien, S. J., W. G. Nash, D. E. Wildt, M. E. Bush, and R. E. Benveniste (1985). "A molecular solution to the riddle of the giant panda's phylogeny." *Nature* Vol. 317: 140–144.

Ohno, S. (1970). *Evolution by Gene Duplication*. Springer-Verlag.

Ohno, S. (1973). "Ancient linkage groups and frozen accidents." *Nature* Vol. 244: 259–262.

Park, H. D., K. M. Guinn, M. I. Harrell, R. Liao, M. I. Voskuil, M. Tompa, G. K. Schoolnik, and D. R. Sherman (2003). "Rv3133c/dosR is a transcription factor that mediates the hypoxic response of Mycobacterium tuberculosis." *Molecular Microbiology* Vol. 48: 833–843.

Pevzner, P. and G. Tesler (2003a). "Genome rearrangements in mammalian evolution: lessons from human and mouse genomes." *Genome Research* Vol. 13: 37–45.

Pevzner, P. and G. Tesler (2003b). "Human and mouse genomic sequences reveal extensive breakpoint reuse in mammalian evolution." *Proceedings of the National Academy of Sciences of the United States of America* Vol. 100: 7672–7677.

Pevzner, P. A. (1989). "1-Tuple DNA sequencing: computer analysis." *Journal of Biomolecular Structure and Dynamics* Vol. 7: 63–73.

Pevzner, P. A., H Tang, and M. S. Waterman (2001). "An Eulerian path approach to DNA fragment assembly." *Proceedings of the National Academy of Sciences of the United States of America* Vol. 98: 9748–53.

Pevzner, P. A., V. Dančík, and C. L. Tang (2000). "Mutation-Tolerant Protein Identification by Mass Spectrometry." *Journal of Computational Biology* Vol. 7: 777–787.

Pevzner, P. A., S. Kim, and J. Ng (2008). "Comment on 'Protein sequences from mastodon and Tyrannosaurus rex revealed by mass spectrometry'." *Science* Vol. 321: 1040.

Robinson, D. F. (1971). "Comparison of labeled trees with valency three." *Journal of Combinatorial Theory, Series B* Vol. 11: 105–119.

Rosenblatt, J. and P. Seymour (1982). "The Structure of Homometric Sets." *SIAM Journal on Algebraic Discrete Methods* Vol. 3: 343–350.

Saitou, N. and M. Nei (1987). "The neighbor-joining method: a new method for reconstructing phylogenetic trees." *Mol Biol Evol* Vol. 4: 406–425.

Sanger, F, S Nicklen, and A. Coulson (1977). "DNA sequencing with chain-terminating inhibitors." *Proceedings of The National Academy of Sciences of The United States Of America* Vol. 74: 5463–5467.

Sankoff, D. (1975). "Minimal Mutation Trees of Sequences." *SIAM Journal on Applied Mathematics* Vol. 28: 35–42.

Sedgewick, R. and P. Flajolet (2013). *An Introduction to the Analysis of Algorithms*. Addison-Wesley.

Sernova, N. V. and M. S. Gelfand (2008). "Identification of replication origins in prokaryotic genomes." *Briefings in Bioinformatics* Vol. 9: 376–391.

Smith, T. F. and M. S. Waterman (1981). "Identification of common molecular subsequences." *Journal of Molecular Biology* Vol. 147: 195–197.

Sokal, R. R. and C. D. Michener (1958). "A statistical method for evaluating systematic relationships." *University of Kansas Scientific Bulletin* Vol. 28: 1409–1438.

Solov'ev, A. (1966). "A combinatorial identity and its application to the problem about the first occurence of a rare event." *Theory of Probability and its Applications* Vol. 11: 276–282.

Southern, E. (1988). "Analysing Polynucleotide Sequences." Patent (United Kingdom).

Stachelhaus, T., H. D. Mootz, and M. A. Marahiel (1999). "The specificity-conferring code of adenylation domains in nonribosomal peptide synthetases." *Chemistry & Biology* Vol. 6: 493–505.

Studier, J. A. and K. J. Keppler (1988). "A note on the neighbor-joining algorithm of Saitou and Nei." *Mol Biol Evol* Vol. 5: 729–731.

Sturtevant, A. H. (1921). "A Case of Rearrangement of Genes in Drosophila." *Proceedings of the National Academy of Sciences of the United States of America* Vol. 7: 235–237.

Sturtevant, A. H. and T. Dobzhansky (1936). "Inversions in the Third Chromosome of Wild Races of Drosophila Pseudoobscura, and Their Use in the Study of the History of the Species." *Proceedings of the National Academy of Sciences of the United States of America* Vol. 22: 448–450.

Tang, Y. Q., J. Yuan, G. Osapay, K. Osapay, D. Tran, C. J. Miller, A. J. Ouellette, and M. E. Selsted (1999). "A cyclic antimicrobial peptide produced in primate leukocytes by the ligation of two truncated alpha-defensins." *Science* Vol. 286: 498–502.

Thomson, J. M., E. A. Gaucher, M. F. Burgan, D. W. De Kee, T. Li, J. P. Aris, and S. A. Benner (2005). "Resurrecting ancestral alcohol dehydrogenases from yeast." *Nature Genetics* Vol. 37: 630–635.

Tuzun, E., A. J. Sharp, J. A. Bailey, R. Kaul, V. A. Morrison, L. M. Pertz, E. Haugen, H. Hayden, D. Albertson, D. Pinkel, M. V. Olson, and E. E. Eichler (2005). "Fine-scale structural variation of the human genome." *Nature Genetics* Vol. 37: 727–732.

Venkataraman, N., A. L. Cole, P. Ruchala, A. J. Waring, R. I. Lehrer, O. Stuchlik, J. Pohl, and A. M. Cole (2009). "Reawakening retrocyclins: ancestral human defensins active against HIV-1." *PLoS Biology* Vol. 7: e95.

Viterbi, A. (1967). "Error bounds for convolutional codes and an asymptotically optimum decoding algorithm." *IEEE Transactions on Information Theory* Vol. 13: 260–269.

Wang, X., C. Lesterlin, R. Reyes-Lamothe, G. Ball, and D. J. Sherratt (2011). "Replication and segregation of an *Escherichia coli* chromosome with two replication origins." *Proceedings of the National Academy of Sciences* Vol. 108: E243–E250.

Weiner, P. (1973). "Linear Pattern Matching Algorithms." In: *Proceedings of the 14th Annual Symposium on Switching and Automata Theory*. IEEE Computer Society, 1–11.

Whiting, M. F., S. Bradler, and T. Maxwell (2003). "Loss and recovery of wings in stick insects." *Nature* Vol. 421: 264–267.

Wolfe, K. H. and D. C. Shields (1997). "Molecular evidence for an ancient duplication of the entire yeast genome." *Nature* Vol. 387: 708–713.

Woodward, N. Weyand, and M Bunnell (1994). "DNA sequence from Cretaceous period bone fragments." *Science* Vol. 266: 1229–1232.

Xia, X. (2012). "DNA replication and strand asymmetry in prokaryotic and mitochondrial genomes." *Current Genomics* Vol. 13: 16–27.

Yancopoulos, S., O. Attie, and R. Friedberg (2005). "Efficient sorting of genomic permutations by translocation, inversion and block interchange." *Bioinformatics* Vol. 21: 3340–3346.

Zaretskii, Z. A. (1965). "Constructing a tree on the basis of a set of distances between the hanging vertices." *Uspekhi Mat. Nauk* Vol. 20: 90–92.

Zerbino, D. R. and E. Birney (2008). "Velvet: Algorithms for *de novo* short read assembly using de Bruijn graphs." *Genome Research* Vol. 18: 821–829.

Zhao, H. and G. Bourque (2009). "Recovering genome rearrangements in the mammalian phylogeny." *Genome Research* Vol. 19: 934–942.

Zuckerkandl, E. and L. Pauling (1965). "Molecules as documents of evolutionary history." *Journal of Theoretical Biology* Vol. 8: 357–366.

Image Courtesies

Figure 1.28: AEvar Anfjörd Bjarmason
Figure 3.5: Dan Gilbert
Figure 3.45 (left): Royal Irish Academy
Figure 4.1: Open Clip Art
Figure 6.27: Wikimedia Commons user Skbkekas
Figure 6.1: Glenn Tesler
Figure 6.2: Glenn Tesler
Figure 6.3: Max Alekseyev
Figure 6.4: Max Alekseyev
Figure 6.5: Glenn Tesler
Figure 6.6: Glenn Tesler
Figure 6.22: Glenn Tesler
Figure 7.1 (right): Luc Viatour
Figure 7.14: Wikimedia Commons user Praveenp
Figure 7.15: Julie Langford (squirrel monkey), André Karwath (old world monkey), Frans de Waal (chimp), Kabir Bakie (gorilla)
Figure 7.21: Wikimedia Commons user Drägüs (winged stick insect), L. Shyamal (wingless stick insect), Katka Nemčoková (giant centipede)
Figure 7.30: Frans de Waal (chimp), Kabir Bakie (gorilla)
Figure 7.36: Wikimedia Commons user Praveenp
Figure 7.41: Alan D. Wilson (polar bear), Mark Dumont (spectacled bear), Peter Meenen (red panda), Wikimedia commons user Darkone (raccoon).
Figure 8.27: Guillaume Paumier (microarray)
Figure 10.27: Thomas Splettstoesser

Figure 10.28 (top): Wikimedia Commons user Dna 621
Figure 11.24: Matt Martyniuk